YOUTH TECHNICAL SESSIONS PROCEEDINGS

PROCEEDINGS OF THE VI YOUTH FORUM OF THE WORLD PETROLEUM COUNCIL - FUTURE LEADERS FORUM, SAINT PETERSBURG, RUSSIAN FEDERATION, 23-28 JUNE 2019

Youth technical sessions proceedings

Editor

Vladimir Litvinenko

Saint-Petersburg Mining University, Russian Federation

CRC Press
Taylor & Francis Group
Boca Raton London New York

CRC Press is an imprint of the
Taylor & Francis Group, an **informa** business
A BALKEMA BOOK

Published by:
CRCPress/Balkema
P.O. Box 447, 2300 AK Leiden, The Netherlands
e-mail: Pub.NL@taylorandfrancis.com
www.crcpress.com – www.taylorandfrancis.com

ISBN 13: 978-0-367-34668-3 (hbk)
ISBN 13: 978-0-429-32707-0 (ebk)
DOI: https://doi.org/10.1201/9780429327070

Visit the Taylor & Francis Web site at
http://www.taylorandfrancis.com

and the CRC Press Web site at
http://www.crcpress.com

Typeset by Integra Software Services Pvt. Ltd., Pondicherry, India

Library of Congress Cataloging-in-Publication Data

Youth Technical Sessions Proceedings – Litvinenko (ed)
© 2019 World Petroleum Council, London, UK, ISBN 978-0-367-34668-3

Table of contents

 ISBN 978-0-367-34668-3

Bridging the Generational Gap in the Oil & Gas Industry

We have once again reached that time in our calendar, the time to participate in the biggest Youth event of our industry - the 6th WPC Youth Forum. Almost two decades ago the WPC's management recognized that without the active involvement of young professionals as part of its various initiatives and projects, the WPC could not be a successful organisation. With this in mind, the first WPC Youth Forum was organised in Beijing China in 2004, encouraging young experts to present papers showcasing their achievements. A special committee then selected the best papers, recommending them to the Congress Programme Committee to be presented during the 2005 World Petroleum Congress in South Africa.

Since we held that first WPC Youth Forum in China in 2004, we have enjoyed the continuous involvement of young people in our organisation. Their participation in the Youth Committee helped us to better understand their views of the industry and our future, and also gave us an opportunity to build a dialogue with the next generation.

Connecting young professionals to each other and to industry leaders is a key objective for the WPC. Our global network provides them with an opportunity to build bridges across continents and between people.

That is why our mentoring program is considered so successful; during these "dialogues" the young professionals and the experienced industry leaders who are their mentors can exchange experiences on a broad range of versatile topics, and we hope many mentees and mentors will have the opportunity to meet during this Congress.

That first Youth Forum was followed by further editions organised in Paris, New Delhi and Calgary, all held with great success and creating some very good memories and initiatives. More recently the 5th event in this series, the 'Future Leaders Forum' was held in Rio de Janeiro in 2016 and once again we managed to gather a truly wonderful group of young professionals and students under the WPC banner.

Here we are now in St Petersburg and the petroleum industry's future faces many new challenges and opportunities, so the profound knowledge our Young Professionals gain during their studies and at the beginning of their professional life provides an important contribution, as the industry deals with our continually changing and highly competitive markets.

They are our successors and so we need to give them all the support we can to equip them with the right tools, knowledge and opportunities and thereby enable them to successfully lead us to the future.

Today young people have unprecedented access to information, providing them with real time information and an in-depth overview of industry options. This also means they face a lot more competition from their peers around the world compared to my time when I started my career many years ago.

We would like to acknowledge the Rector of the University V. Litvinenko and the Russian National Committee for all their efforts and also Minister Novak for his continued support. I think the content of this proceedings speak for themselves on the knowledge and expertise growing in our young people and we are certainly in good hands for the future. We wish them all the best of luck and success in their future professional life.

Tor Fjaeran, President, World Petroleum Council

Youth Technical Sessions Proceedings – Litvinenko (ed)
© 2019 World Petroleum Council, London, UK, ISBN 978-0-367-34668-3

Preface

Dear friends!

The collection offered to your attention is a unique publication that includes scientific articles by young oil and gas professionals from different countries and continents. They have decided to share their own research results and original conclusions with the participants of the 6th Future Leaders Forum of the World Petroleum Council.

The WPC Youth Forum is held every three years. This is an international platform, unprecedented in scale, level of representation and spectrum of issues discussed, for professional communication of talented and motivated young men and women from dozens of countries of the world. Such communication expands the horizons of interaction, gives impetus to fresh ideas and impels to the action.

This approach is consistent with the strategy and main lines of action of UNESCO and its Education programme. That is why the Future Leaders Forum is held with the support and under the auspices of UNESCO.

In order to maintain high demand and competitiveness in the energy markets, the modern oil and gas industry must find effective answers to new challenges arising from dynamically changing realities. Many of the challenges are global in nature. To solve them, it is necessary to combine intellectual potential and production resources. International cooperation is the most important success factor. Every year the complexity of production objectives increases, and the current students and young specialists, including delegates of the Future Leaders Forum, will have to deal with them very soon. The WPC Youth Forums are valuable, primarily because they help in finding and uncovering young talents, create conditions for the internationalization of industry-specific processes.

The industry today is in dire need of young qualified professionals. The main thing for each of you is to understand how important it is in the modern world not to stop in professional growth and learn throughout your life. Because the technologies that are currently used in the fuel and energy complex are often comparable to space technologies, and their development requires constant self-improvement from engineers.

Russia for the first time hosts the WPC Youth Forum. Russia's oldest educational center of engineers — St. Petersburg Mining University — rightly entrusted the role of a key organizer of the preparation and holding of the Forum. I truly believe that the 2019 Future Leaders Forum will leave a vivid trace in the memories of the participants and the history of the World Petroleum Council and will bring significant practical results.

The articles published in the collection have passed a competitive selection and reflect a wide range of scientific ideas and approaches, which characterizes images of a generation of future leaders about the industry. Many authors in this collection were published for the first time. This is a kind of their first step in big science and a successful career.

I wish our young authors to develop this success and make a significant contribution to the further development of the global oil and gas industry.

Rector **V. Litvinenko**

Youth Technical Sessions Proceedings – Litvinenko (ed)
© 2019 World Petroleum Council, London, UK, ISBN 978-0-367-34668-3

Organizers

World Petroleum Council

United Nations Educational, Scientific and Cultural Organization

Saint Petersburg Mining University

International Competence Centre for Mining Engineering Education under the auspices of UNESCO

Emerging Trends in Energy Industry

Youth Technical Sessions Proceedings – Litvinenko (ed)
© 2019 World Petroleum Council, London, UK, ISBN 978-0-367-34668-3

The blokchain system as an important digitization factor of world energy sector

V.M. Pshenichnyy
Gubkin Russian State University of Oil and Gas (National Research University), Moscow, Russia

ABSTRACT: To determine the main vectors of global energy development, from the point of view of digitalization, the digital system "Blockchain" was chosen, using as an example the main advantages and disadvantages of the digitalizing energy sector. The blockchain technology opens up the strengths of the smart contracts, the main business digitalization projects, and reveals the essence of increasing capitalization at the expense of digital systems.

1 DIGITALIZATION OF WORLD ENERGY BY INTRODUCING BLOCKCHAIN TECHNOLOGY

To date, one of the global trends that largely affects the world economy, society and other global factors of development is the transformation of the technological order – the so-called technological revolution 4.0. The displacement of the usual for all countries raw material economic foundations begins with the help of virtual-digital systems. Energy based on hydrocarbon resources has been a fundamental vector of economic development for a long time, but today it is on the threshold of transformation into a new system of generation and use of energy, which is called "Clean Energy", the main advantage of which is to improve the quality of life of people and countries as a whole.

Artificial intelligence, the Internet of things, smart homes, digitalization of energy, digital systems are the most important aspect of the energy business development strategy, where the prospects for the introduction of innovative technologies are closely related to improving the efficiency of business processes in energy companies, reducing costs, creating new intelligent systems.

One of these digital systems, which shows the rapid development in recent years, is the "Blockchain" system, a special technology that has created platforms for transactions between participants in different markets who have the same rights and perform their actions without intermediaries.

This system uses decentralized storage of information to reflect all transaction data, which together can provide a large number of commercial transactions in a variety of areas. «Blockchain» was created on the basis of Bitcoin system, which was offered by Satoshi Nakamoto in 2009. At the same time, the idea of blockchain appeared in the 1980s, when David Chowm presented his invention called DigiCash.

«Blockchain» is a distributed database consisting of a "chain of blocks", block storage devices are not connected to a common server, and the database allows you to control the reliability of transactions without the supervision of any financial regulators.

Transaction verification is carried out by the so-called miners – participants of the system, who confirm the authenticity of the committed actions and have access to information about any transaction ever carried out. Users act as a collective notary, which confirms the truth of the information in the database. Blockchain is a distributed and decentralized database formed by participants, in which it is impossible to falsify data due to the chronological recording and public confirmation by all participants of the transaction network. The main feature of the blockchain is the usage of algorithms for mathematical calculation, and the exclusion of "human" factor in the decision-making system "Blockchain" has an impact on reducing costs, increasing the level of flexibility of processes in any business sector, accelerates production processes and facilitates the life of society and people. For energy, the blockchain system is an innovation that market players seek to use. Many modern energy companies of the leading countries of America, Asia and the European Union, have and develop existing blockchain projects that are widespread in the energy sector and energy logistics, including this proven application in recent years, "smart contract", which attracts more and more attention from the major players in the energy and oil and gas markets.

A new era of "smart contracts" is coming, which will require an analysis of the prospects for the development of the blockchain system, consideration of project smart developments and study of their strategic potential.

The development of global energy is increasingly sensitive to new technological trends. The level of implementation of digital solutions in fuel and energy companies is becoming one of the determining factors of their competitiveness in the global arena. To date, all major Russian fuel and energy companies are actively involved in the technology race and are already demonstrating the first results of pilot projects. However, the current work is only the first local steps in the digital transformation of the Russian energy sector. The transition of the entire industry to digital rails requires a fundamentally new approach to the creation and implementation of intelligent solutions.

Table 1. Usage of digital technologies in the global energy sector in 2018. (*World Energy Council – World Energy Scenarios*, viewed 17.03.2019, https://www.worldenergy.org/wp-content/uploads/2016/10/World-Energy-Scenarios-2016_Full-Report.pdf)

Type of activity	Type of work/technology	Country-leader
Exploration	Aerospace survey of the Earth's surface, construction of virtual models of subsoil and production schemes (geological and mathematical modeling)	USA, Russia
Drilling of wells	With a significant departure from the vertical and high length (about 15 km), self-propelled, non-self-propelled, offshore drilling platforms	USA, France, Russia
Production of liquid and gaseous hydrocarbons	Hydraulic fracturing, production technologies in the Arctic, in the oceans on the shelf and considerable depths, underwater uninhabited production complexes	USA, Canada, Norway, France, Russia, China
Preparation for transportation and transportation of oil and gas	Cleaning, transfer to different phases, transshipment and transportation by land and sea	Russia, USA, Canada, Qatar, Australia, China
Oil and gas processing	Production of nano- and biopolymers, biotechnological, anti-corrosion, crystalline, fire-resistant materials, energy conversion materials, biosensors, pharmaceutical products	USA, Japan, Germany, UK, France, China, India, Brazil, Russia
Coal production and processing	Gasification, flexible robotic systems based on artificial intelligence, deserted production, unmanned transport (mines, sections, etc.), nano- and biotechnology of coal and waste processing	UK, Japan, USA, Germany, Czech Republic, Russia, South Africa

(*Continued*)

Table 1. (*Continued*)

Type of activity	Type of work/technology	Country-leader
Nuclear sector	Computer-aided design and product life cycle management, security systems, flexible robotic systems based on artificial intelligence, deserted production	Russia, USA, France, UK, Germany
Electroenergetics	Active-adaptive networks, transmission of electric power by direct current, transmission by alternating current on the basis of superconductors	USA, EU, Republic of Korea, China, India, Russia
Gas-hydrate production	On land, in the oceans	China, Japan, USA, Canada
Renewable energy	Solar, geothermal, wind, marine and land, biofuel technologies	USA, Brazil, Germany, Denmark, Norway, Spain, Iceland, Russia, China, Japan, Republic of Korea, Russia

At the moment, regulators and energy companies are faced with the challenge of meeting the growing demand as a condition of economic growth, social stability and profits and/or reproduction of electric power infrastructure. However, it is highly inappropriate for all actors to adopt a new investment "super-cycle" within the framework of the traditional electricity model due to its high cost, low efficiency and environmental friendliness, as well as long deployment time. That is why it is worth considering the leading key trends that will determine the shape of the energy of the near future, built primarily on the blockchain system:

- deep decentralization of energy production
- spread of energy saving technologies and practices
- spread of digital networks of intelligent control systems
- change of consumer behavior and the emergence of prosumers
- spread of new financial technologies

The main trend, according to experts, is the spread of new financial technologies, which in turn will entail others. It turns out a certain chain on which the energy of the leading world powers is already beginning to move.

In itself, the spread of digital technologies in the field of Finance is the emergence of new technologies in the financial sector (Blockchain, Smart Contract, Decentralized Autonomous Organizations), which creates opportunities for large – scale attraction of private investment in energy, monetization of consumer services, the formation of various practices of energy exchange. Due to the accumulation of "critical mass" of changes, it can be argued that from 2012-2014 the world — so far, mainly the vanguard of foreign countries — has finally moved to the implementation of the scenario of high-tech electricity. This process is natural, because, as an infrastructure industry, it reflects the transition of the world energy to more intensive scientific, technological and innovative development, as well as to the expansion of the world development space-by increasing the number of countries with rapidly growing industries. However, we should not assume that the new high — tech electric power industry is a solution only for the rich or the most actively growing economies with their powerful industrial sector and high level of welfare. Rapid urbanization, industrial development, the elimination of "energy poverty" and other challenges in developing economies can be addressed only through high-energy technologies. These economies simply will not be able to build a traditional energy system. This is a very complex and expensive system (for comparison, the cost of the US power system is more than $ 1 trillion). It requires complex competencies and a large number of skilled personnel for management and maintenance. In contrast, in full accordance with the General laws of the high-tech sector, new solutions in the power industry are becoming more accessible.

To date, there are about two hundred applications in the world-projects aimed at the energy sector, built on the blockchain system. Investments in the blockchain system in the energy sector increased by more than $ 150 million. That speaks about high interest of the market of investments and venture funds. Particular attention should be devoted to the usage of energy technologies.

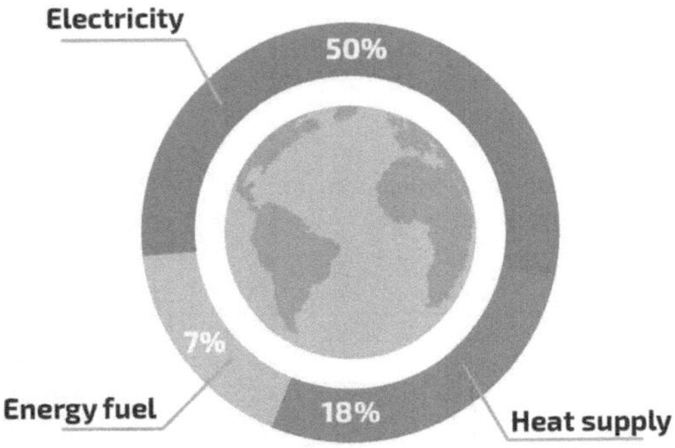

Figure 1. Use of blockchain technology by types of energy.

The figure shows that most of the projects on the blockchain platform are aimed at the electric power industry, the transformation of which is of interest to the leading world powers most of all. If we consider the basic assumptions that stimulate the development of blockchain technology in the financial sector, can be applied in the electricity sector. which can create the Foundation for further decentralization of the world's energy systems:

- Decentralized storage of transaction data increases the level of protection and provides a higher degree of independence from a single body that performs centralized administration;
- Blockchain technology has a wide range of applications. Chain blocks can facilitate payments made using the cryptocurrency, the digitization of contracts, management of digital content, the verification of transactions, execution of trading operations or can be applied in many other areas. It is expected that the next important step in the development of technology will be the introduction of "smart contracts»;
- New decentralized business models no longer require third-party intermediaries;
- Whether the technology will be successful will depend not only on the technical capabilities of the system, but also on the applicable regulatory framework, the scalability of the technology and its adaptability, and the return on investment.

At the moment, the development of all applications based on the blockchain technology in the electric power industry is still at the stage of concept formation or prototype testing. This applies both to the technology underlying the application data and to the possible use by consumers. Nevertheless, the technical potential of applications based on the blockchain technology is quite obvious today: in particular, a decentralized power supply system, as well as the conduct and reflection of transactions, is a real working structure, so the potential of blockchain technology in the electricity sector is promising. An example of this is projects that are already in use in Europe and North America.

Table 2. Leading blockchain projects in the field of energy. (*EIA – International Energy Outlook 2018, viewed 27.02.2019*, https://www.eia.gov/pressroom/presentations/capuano_07242018.pdf)

№	Project title	Description
1	Brooklyn Microgrid (TransactiveGrid) (USA)	The purpose of the project is to test how the blockchain technology can be used to carry out operations for the sale of electricity from solar panels directly between neighbors. The technology used in the project is based on the Ethereum blockchain
2	Vattenfall Company: Powerpeers (USA)	The strategic objective of the project is based on the idea of electricity exchange within the network of equal participants (P2P). Powerpeers customers can offer their own electricity to other participants and exchange electricity.
3	RWE & Slock.it: BlockCharge (Germany)	The essence of the project is to study the question of how a system based on blockchain technology with an integrated function "smart contracts" can be used to recharge electric vehicles.
4	Oneup Company: POWR (Holland)	Prototype of a decentralized energy transaction and supply system and tested it using data on ten households electricity consumption.
5	LO3 Energy: Exergy (USA)	Project on heating of houses with the help of heat energy generated by data centers.
6	GridSingularity Company (Austria)	The platform is based on blockchain, which is designed to organize the interaction of electricity producers, power grid operators, regulators and consumers.
7	Stock exchange Nasdaq (США)	A new authentication service that offers the possibility of obtaining certificates for electricity produced from solar energy through the blockchain system. The new service works by connecting solar panels to an IoT device (IoT – "Internet of Things"), which measures the power produced and supplied to the power grid.

All the above projects would not have come into the world without a strategic plan of transformation of legislation in the field of energy. For example, the German energy act contains provisions for energy supply contracts that are designed to balance the protection of consumers' rights and the interests of electricity suppliers. Some of the legal principles relating to this area are based on General civil law and, most importantly, on the provisions of the German Civil code and on the law enforcement practice that has developed on these issues. When implementing projects using blockchain technology, it is necessary to take into account this regulatory framework and a number of other laws and established requirements. Legislation in the field of consumer protection and data protection is complex, and it must be taken into account in the implementation of each blockchain project.

With the exception of specially noted cases, from the point of view of legal and regulatory aspects, considering the blockchain technology in the context of its main application, that is, the creation of a "decentralized system of energy transactions and energy supply".

In accordance with the applicable regulatory standards of Europe in relation to separation activities operation of the electricity transmission networks and activities in the production and (or) sale of electricity, electricity companies are obliged to separate their activities on the electricity transmission network (regulated business) of the activities on the supply of consumers (competitive activities). Consumers have the right, at their discretion, to choose an electricity supplier (or gas supplier) in a liberalized electricity market. In order to enable consumers to move seamlessly from one supplier to another, the concept of so-called balancing groups has been introduced. This approach has made it possible to assign each customer to the supplier without any difficulty, which is the principle of decentralization.

All the above-mentioned projects and legislative solutions in the field of electric power industry make it possible to create a paradigm of energy transition, for which the starting point is the blockchain technology. This paradigm should be focused on the large-scale use of distributed renewable energy, the involvement of private investment and the formation of decentralized markets, the intellectualization of infrastructure and the transition of consumers to active, pro-consumer models of behavior.

Particular attention should be paid to the opportunities that open up to the electric power industry after the introduction of blockchain technology:

1. Monitoring the operation of power grids with the help of "smart contracts". Smart contracts will signal to the system when transactions need to be initiated. The system will operate in accordance with pre – established rules, the purpose of which is to ensure that the control of all electricity flows intended for transmission and storage is carried out automatically in such a way as to balance supply and demand. For example, each time the volume of the produced electricity exceeds the existing needs, you can use "smart contracts" to ensure that data is surplus electricity is automatically sent to the vault. Conversely, electricity from storage can be used when the amount of electricity produced is insufficient. Thus, with the help of blockchain technology, it is possible to directly control the flow of electricity and storage that accumulate electricity. In addition, "smart contracts" can be used to manage the work on balancing supply and demand and to monitor the activities of virtual power plants, which is especially important for the Russian power industry.
2. Energy transaction and energy saving system. Decentralized system of energy transactions and power supply, which is a leading trend in the current conditions of transformation of the world power industry. The supply of electricity produced at small distributed power facilities to end users will be carried out through micro-networks created by the consumers themselves. The volumes of produced and consumed electricity will be measured with the help of "smart meters", and sale of electricity and payments in the cryptocurrency will be monitored through "smart contracts" and filled using the blockchain.
3. Billing customers for the energy they consume when heating rooms or heating water. The energy sector is large and other relevant applications outside the electricity sector can be considered to improve it, for example, when it comes to billing customers for energy they consume when heating premises or heating water, that is, activities currently carried out primarily by specialized meter reading service providers.
4. Creation of the Internet of energy. Thanks to blockchain technology, leading countries are considering the possibility of creating a system that will become the basis of the paradigm of the energy transition to an alternative energy system, which will radically solve the problems of the industry as a whole. The world's leading powers cannot ignore the fact of changing the technological model in the electric power industry and the associated challenges and shifts in the development of the industry. Otherwise, States will face a number of risks. One of them is the risk of getting more expensive and less flexible electricity compared to other industrialized countries, losing a number of new markets (for example, the energy storage market), significantly limiting the development of a new (digital) industry.

Based on the risks that can slow down the introduction of blockchain technology in the world electric power industry, it is necessary to develop a road map with the right strategic goals on the example of the experience of the leading world powers, which will allow in the shortest possible time to transform the former energy base into "Internet of energy".

Already in 2015-2017, several major energy companies announced the transition to new business models focused on "new energy": preferential investments in renewable energy, distributed generation, distribution networks, expansion of services for customers, radical innovations. An example of the implementation of such a business model is the new corporate strategy of E.ON (2015), in which the company has divided its business into two components: one division operates in the traditional energy markets (Uniper), the second-with new technological solutions and energy markets (E.ON).

After reviewing the experience of E.ON, a road map was created, which includes key strategic goals for developing countries in the field of introducing blockchain technology into the electric power industry, followed by the creation of the "Internet of energy".

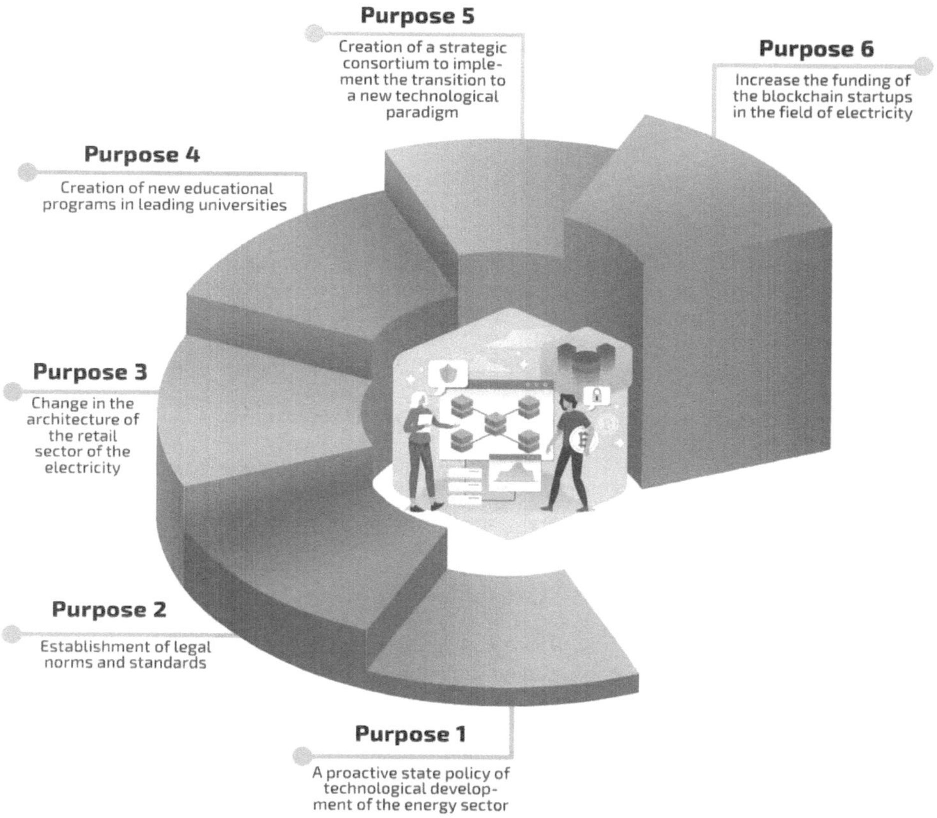

Figure 2. Strategic goals for rapid implementation of blockchain technology in the power industry

Achieving these goals will allow countries not only to quickly introduce blockchain technology into the electric power industry, but also to implement an energy transition to a new paradigm in the shortest possible time.

First, there will be the creation of a more flexible architecture of energy systems through distributed generation, the development of "smart" networks and the energy storage market, as well as through the emergence of active ("smart") consumers.

Secondly, the formation of a new business model of the electric power industry will appear instead of the traditional value chain "generation-trading — transfer — sale" model of the "Internet of energy" and the provision of services in a new interactive environment.

Third, there will be a change in government regulation — a gradual transition from supporting old sources of electricity generation to supporting the consumer, integrating local solutions; from supplying energy to "connecting capacities" within the "Internet of energy", which will create a new "flexible market".

2 CONCLUSION

The main conclusions of the work - Blockchain affects cost reduction, increasing the flexibility of processes in any branch of business, speeds up production processes and facilitates the life of society and people. For the power system, the blockchain system is an innovation that all market players seek to use. Many modern energy companies from leading countries in America, Asia and the European Union have and are developing existing blockchain projects that are widespread in the electricity and energy logistics sector, including the smart contract application that has proven itself in recent years, which is attracting more and more and more attention from the major players in the energy and oil and gas markets. A new era of "smart contracts" is coming, which will require an analysis of the prospects for the development of the blockchain system, the consideration of project-based smart developments and the study of their strategic potential.

REFERENCES

Balashov A., Dergachev V., Kolomiychenko M. parfentyev, I. RBC (article: Bitcoin and the blockchain). [Electronic resource.] Mode of access: https://www.rbc.ru/accessed: 26.02.2019

Burkalev D. D. points of economic and innovative growth: a model of organization of effective functioning of the region blockchain, 2016. P. 18-19.

EIA – International Energy Outlook 2018, viewed 27.02.2019, https://www.eia.gov/pressroom/presentations/capuano_07242018.pdf

Everything you wanted to know about the next generation of public key crypto/Nick Sullivan // Ars Technica – 25.10.2013

Genkin A., Mikheev A. Blockchain: How it works and what awaits us tomorrow // Alpina Publishing house, 2018. P. 20-22.

Kolesov V. A. The use of blockchain technology in the educational process for the protection of intellectual property // in the collection: distance learning technologies Materials of the II all-Russian scientific and practical Internet conference, 2017. P. 343-347.

Kuprijanovskij V.P., Sinjagov S.A., Klimov A.A., Petrov A.V., Namiot D.E. Cifrovye cepi postavok i tehnologii na baze blokchejn v sovmestnoj jekonomike [Digital supply chains and block-based technologies in a joint economy] // International Journal of Open Information Technologies, 2017. P. 80-95.

Soliev R.Ju. Vazhnejshie tendencii informacionno-kommunikacionnyh tehnologij v razvitii sovremennoj global'noj jekonomiki [The most important tendencies of information and communication technologies in the development of the modern global economy] // Jekonomika i upravlenie: nauchno-prakticheskij zhurnal [Economics and management: scientific and practical journal], 2017. P. 28-40

Zemlyanukhina E. M., Karpenko A. N. Blockchain as a mechanism of organization management // St. Petersburg state University of Economics, 2016. P. 35-38.

World Energy Council – World Energy Scenarios, viewed 17.03.2019, https://www.worldenergy.org/wp-content/uploads/2016/10/World-Energy-Scenarios-2016_Full-Report.pdf

Youth Technical Sessions Proceedings – Litvinenko (ed)
© 2019 World Petroleum Council, London, UK, ISBN 978-0-367-34668-3

Digitalization: Opportunity or challenge for energy industry?

S. Denisova
Moscow State Institute of International Relations (MGIMO University), Moscow, Russian Federation

ABSTRACT: Digitalization appears to be both, a challenge and an opportunity for the energy industry. Energy companies still hesitate to introduce digital technologies, although the benefits such as better safety performance, higher operational performance, exploration that is more efficient and further cost reduction are clear. The oil and gas industry strategies are challenged by the renewables branch that is becoming a more reliable energy source, having developed a range of competitive digital business models, an example of which are virtual power plants and distributed generation. What is more, given decarbonization policy alternative energy sources gain a competitive advantage over oil and gas, which further weakens market positions of the traditional energy branches.

1 INTRODUCTION

Nowadays digitalization is doubtlessly considered to be the key to business success, securing steady profit growth, cost reduction and process efficiency. The widespread digital transformation has facilitated emergence of new digital business models that could be classified into four different types, namely omni-channel, ecosystem driver, supplier and module producer (Weill & Woerner 2015; Weill & Woerner 2018). According to this classification, energy companies should develop omni-channel models concentrating on a creation of various communication channels with a client and integrated value chains. This approach would enable energy companies to safeguard their positions on the market against emerging startups.

Having been forced by falling energy prices to reduce the operating expenses, today, with the beginning recovery of the industry and growing profits, key players are able to streamline their business processes by increasing their investments into digital technologies and introducing new digital business models. However, usually companies hesitate to seize the chance, finding it hard to embrace the advantages of digitalization, and, thus, are exposed to pressure from new emerging actors on the energy market.

Our research will focus on defining in which regard digitalization as a trend disrupts the current business models of energy companies and on finding a solution to the question how new technologies can be implemented in current business strategies and how they could contribute to the transformation of the energy industry in general.

During the research, we will test the hypothesis that the impact of digitalization on each market player and the perception of digitalization as a threat or an opportunity differ across various segments of the energy industry and depend primarily on the nature of each segment, rigidness or flexibility of business models and incentives of each company to adapt.

2 LITERATURE REVIEW

Given the fact that digitalization has been the key innovation driver during the last years and gave rise to such technologies as block chain, internet of things, smart home/city, cloud technologies, big data and such new industries as fintech, its application in the energy sphere has been a subject of research both in a more academic, scientific literature as well as in more applied publications of advisory nature. As far as the energy industry in concerned, most materials focus either on general issues and conceptual frameworks of digitalization or new digital architecture (Filippov 2018; Teksler 2018; Holkin, Chausov 2018; Chase & Berzina 2018; Dutton & Fischer & Gaventa 2017). The rest is dedicated to the specific area related research such as oil and gas (Sergeeva & Larionova & Zaharova 2018; Dmitrievsky & Eremin 2018; Cavajal & Maucec & Cullick 2018; Cann & Goydan 2019), coal (Plakitkin & Plakitkina 2018; Sidorenko & Shaydullina & Kirakosyan 2018), renewables (Khokhov & Melnikov & Veselov & Kholkin & Datzko 2018) and electrical power generation and distribution (Filimonov & Chichirova & Filimonova 2018; Hovalova & Zholnerchik 2018; Knyaginina & Holkin 2017).

A small number of scientific papers deconstructs digital business models and financial implications of digitalization for the industry (Khoskovskaya & Sergeeva & Mukhametschin 2018). More practical approach to digitalization and changing business models can be found in reports of various consulting companies (Lawrence & Vrins 2018; PwC and GE Digital Alliance 2016; Unrealized potential of digital 2016; Geschäftsmodelle 2020 2016).

As it is shown above, digital transformation in each and every segment of the energy industry such as oil and gas, renewable energy and power generation has been a subject of scientific interest in many regards, with the typical for each segment features being thoroughly studied. Nevertheless, neither a substantial in-depth cross-segment study, nor a thorough analysis of digital business models of scientific nature has been conducted so far, that should otherwise be attached greater importance to, due to the fact that the energy complex is a closely integrated one and energy security results from the degree to which the key components are synchronized.

3 METHODOLOGY

To approach the objectives of the research we have elaborated a two-dimension matrix structure that will align digitalization impact and emerging business models with cross-segment analysis of basic energy areas, namely oil and gas, coal (or fossil fuels), renewable energy and electricity generation and distribution.

To address the problems we have used such methods as case study to get a deeper understanding of processes in each energy segment, comparison to compare them and find what influences the way the segment reacts to digitalization, and elements of SWOT analysis to explain the difference in perceiving digitalization as a threat or an opportunity.

To gain the necessary data, traditional methods of document analysis were used. The study is based on data and information acquired from the statistics, outlooks of analytical agencies and think tanks materials, specific papers of consulting companies such as McKinsey and Company, Boston Consulting Group, Ernst and Young, PricewaterhouseCoopers, Accenture, strategy papers and press releases of energy companies.

4 ANALYSIS

The key challenges the industry is facing nowadays amount to two key points. Firstly, players from other spheres and startups enter the market offering new services like Smart Home and forcing traditional market players out. Secondly, the clients take on a more active role, becoming prosumers, producers of their own energy sources, which for the energy companies leads to falling consumption and makes the whole energy system vulnerable. If the energy companies want to preserve their market positions, they need to adjust. There are several scenarios, according to which digital transformation may take place in the energy sector:

1) digitalization of client contacts by introducing new communication channels such as websites, social networks, apps;

2) digitalization of working processes by saving and managing energy, redistribution of energy in smart networks, marketing and data collection, customization;
3) digital partnerships with other companies;
4) measurement technologies for monitoring (big data etc.);
5) innovation facilitation by a new corporate culture and developing digital competencies;
6) new services (smart city, autonomous driving).

The choice of the strategy depends on particular challenges the company is confronted with, its capacities and the segment it belongs to. The table below presents the results of the conducted comparison of digitization impact on various energy segments.

Table 1. Analysis of digitalization impact on various segments of the energy industry

	Oil & Gas	Coal mining	Renewables	Electricity generation and distribution
Why digitalization is a disruptor for current business models?	New cheaper technologies for extraction such as shale gas, LNG, shrinking market share and increasing efficiency based competition.	Booming renewable energy industry and emergence of prosumers result in decreasing demand, hence, low prices. Industry is in decline.	Increasing share of small generation and prosumers leads to a deregulated market that is prone to blackouts and unstable power supply.	Small generation and emerging apps and technologies help one manage heating online or sell energy, become active participants of the energy market.
Why digitalization is an opportunity?	Digital technologies promote higher safety standards, prevent emergencies and increase efficiency. Examples: pipelines leak detection, predictive analytics for services and repair, drone delivery.	Digital technologies enable the segment to reduce production cost, operating risks. Examples: interconnected sensors, advanced analytics, quantum computing, market intelligence, teleremote equipment.	Digital technologies offer a solution to the key problems of balancing the supply and growing decentralization. Examples: virtual power plants, apps for selling excessive electricity.	Digital technologies secure new business fields for companies such as measurement services, repair and maintenance. Examples: e-mobility, demand management, microgrids.
Why digitalization can be perceived as a risk or a challenge?	Cybersecurity, secure positions in domestic market, difficulty with customization of products caused by less contact with a client.	There is no clear understanding of aims and objectives. The industry is characterized through high rigidness. Setting up a digital department will not help in addressing the problem.	Digitalization is the only solution, but the system and its output are complex. Although any emergency is unlikely to happen, its scope is going to be tremendous.	Fear of increasing electricity and heating prices that will significantly reduce the competitiveness of the industry.
Possible adjustment of business models	Oil & gas industry gains profit from repetitive processes, so the key is automatization and flexible workforces coupled with outsourcing.	Partnerships with university and data science centers, other companies outsourcing of extraction and processing. Integral changes along the	The companies that generate energy using renewable sources of energy shall emphasize internet of things and cloud technologies in their	Business models shall focus more on communication with the customer. Firstly, customer segmentation and demand management

(Continued)

Table 1. (*Continued*)

	Oil & Gas	Coal mining	Renewables	Electricity generation and distribution
	Technologies: smart well technology, online distant management, seismic modeling.	value chain aimed at fused models for better understanding of the resource base, equipment and material flow, failures prediction, mechanization, performance control.	business models, organizing virtual power plants and providing customers with monitoring services to ensure the stability of the energy ecosystem.	stimulates the clients to stay in the existing network by giving them additional benefits. Secondly, for those who have already become the market players as prosumers, energy firms can offer services or monitoring assistance.
Willingness of the actors to adapt	Insufficiently motivated	Insufficiently motivated	Highly motivated	Highly motivated
Main scenario	Digital processes, measurement, partnerships.	Digital processes, partnership and competencies, measurement.	Digital client communication, measurement.	Measurement and services.
Place in a new integrated energy ecosystem	Production and less distribution (B2B, B2C)	Production (B2B)	Production and distribution (B2B, B2C)	Distribution (B2C, B2B)

The comparison provides evidence to assume that the motivation of the industries to adapt is intertwined with their perception of digitalization. Initially digitalization being the disrupting innovation has been seen as a challenge to the industry. Nevertheless, only electricity generation sector has been directly affected by digitalization as its spread into the segment of renewable energy sources has dramatically undermined the business models of power generating and distributing companies causing them to lose profit. For the renewable sector digital transformation served as a solution to the basic problems such as blackouts and fluctuating supply, while the electricity segment had to adjust and use the opportunities of digitalization to remain competitive and maintain its financial performance. So, adoption of digital technologies was of vital importance for both segments.

The two remaining sectors belong more to the B2B sector and the main features they compete on are operating expenses. As reduction of costs does not necessarily imply substantial changes in business models, these companies approached digital transformation differently, emphasizing the risks of cybersecurity to the strategic infrastructure and continuing to treat digitalization as a threat. Therefore, digitalization acquires specific meanings, digital transformation is designed to find new cost-effective solutions rather than to increase sales by creating new ways to communicate with a client. Nevertheless, it is worth mentioning that a vertically integrated company, which is a classic organization model for the companies in extraction segment, has to adjust its processes on each stage to be a successful competitor in the digital environment.

The differentiating effects of digitalization across energy segments can be also explained by the nature of each segment. When a company-producer is concerned, or a sphere that is dominated rather by production processes than by communication with consumers, there are fewer companies that disrupt the economic landscape with their innovative, digital solutions.

As the emerging startups cannot operate independently, they offer their technological know-how to the established oil and gas companies. In the segment of renewable and electricity generation startups are able to reach out to the clients with their products and become an equal player on the market. Consequently, companies in the production dominated areas are less motivated to elaborate digital business models, being able to outsource parts of their business or acquire the startups. Logically their business models do not differ much in terms of communication with a client but leave the scope for horizontal links between companies and scientific community to exchange know-how and establish technological innovation partnerships to exchange expertise. In contrast, companies that operate in the renewables of electricity generation fields have to use the omni-channel digital business model, replacing their products or services with the new ones to secure new channels of communication with a client that are challenged by startups.

So, the companies in production-dominated segment are likely to address digitalization as a threat, applying its advanced technologies only during exploration and production stages, leaving customization and services beyond their attention, while companies in more customer-oriented segments tend to introduce digital business models, seeking for new ways to drive value and secure customer relations or to enlarge their product and service portfolio.

Given the decarbonization trends that have become drivers of technological and digital transformation, we may claim that the renewables segment is expected to compete with the production-dominated segment as well in the future. So, the production-oriented companies shall be recommended to revisit their digital strategies and find digital ways of marketing and selling their products. Otherwise they are likely to lose their market share to the renewables.

5 DIGITALIZATION IN RUSSIA

The key challenge for energy digitalization in Russia is excessive power generation capacities that prevent companies as well as end customers from introducing digital innovations as there is no finance based motivation to do that. Moreover, large distances and missing infrastructure cause additional difficulties, forcing the state to take on the leading role in the energy transformation.

According to the expert opinion expressed during the Russian energy week 2018, there are several issues that hinder digitalization process in Russia. Firstly, due to the unwillingness of the Russian companies to share technologies it is difficult to create a coherent digital infrastructure that will encompass all existing approaches in each segment of the energy sector. Secondly, despite import substitution policy the industry is still significantly dependent on imported technological solutions, partially because of lacking horizontal links between science and businesses. Thirdly, missing technological basis does not allow to structure the accessible data in a sensible way appropriate for the use of big data methods. Finally, neither do the stakeholders feel the need to adapt to the changing energy landscape, nor do they have the incentives to act in such a way.

A both effective and efficient response will involve introduction of signal digital standards that will provide the laying ground for the unique interconnected digital environment, increasing data transparence to make it possible for the third parties to process these data and develop new innovative approaches and services for the producers as well as end-customers.

Nevertheless, single top-down initiatives and specific strategies of companies may facilitate positive changes. The key state program, Energynet being part of the National Technological Initiative implies creation of complex high-tech smart energy systems, with "flexible" grids, distributed generation and prosumers being their pillars. So, the so called Internet of Energy suggests the emergence of new actors at the energy market that will be supported by flexible, self-organized grid systems and services, as well as development of energy storage technologies. The pilot projects in Kaliningrad Oblast have proved their economic viability.

6 CONCLUSION

Digitalization appears to be both, a challenge and an opportunity for the energy industry. Its perception depends on the segment of energy industry a specific company belongs to, resulting motivation and ways to implement digital components into their business models

Digitalization disrupts existing business models by developing new more cost-effective approach either by introducing new production technologies or by facilitating new ways for communication with a client.

The oil and gas industry strategies are challenged by the renewables branch that is becoming a more reliable energy source, having developed a range of competitive digital business models, an example of which are virtual power plants and distributed generation. What is more, given decarbonization policy alternative energy sources gain a competitive advantage over oil and gas, which further strengthens weakens market positions of the traditional energy branches.

In order to secure its position oil and gas has to offer its tailored digital strategies to emphasize not only digitalization of exploration, drilling and production processes, supply chain by means of cloud technologies, IoT, block chain, automation, big data etc. but also service-oriented approach that would complement their main business field.

REFERENCES

Cann G., Goydan R. 2019. Bits, Bytes, and Barrels: The Digital Transformation of Oil and Gas. MADCann Press. P. 290.

Carvajal G., Maucec M., Cullick S. 2017. Intelligent Digital Oil and Gas Fields: Concepts, Collaboration, and Right-time Decisions. Gulf Professional Publishing. P. 374.

Chase P., Berzina K. 2018. Transatlantic policy challenges of the digital energy nexus, viewed 19.03.2019, http://www.jstor.org/stable/resrep18948

Dellermann D., Fliaster A., Kolloch V. 2017. Innovation risk in digital business models: the German energy sector". *Journal of Business Strategy*, Vol. 38 Issue: 5, pp.35-43.

Digital will shape the future of oil and gas. 2018. Ernst & Young, viewed 19.03.2019, https://www.ey.com/Publication/vwLUAssets/ey-digital-will-shape-the-future-of-oil-and-gas/$FILE/ey-digital-will-shape-the-future-of-oil-and-gas.pdf

Dmitrievsky A.N., Eremin N.A. 2018. Big Geodata in the digital oil and gas ecosystem [Bolshie geodannye v ysifrovoy neftegazovoy ekosisteme]. *Energy policy*, vol. 2, pp. 31-39.

Dutton J. 2017. Infrastructure for a changing energy system: the next generation of policies for the European Union, viewed 19.03.2019, www.jstor.org/stable/resrep17732

Energy Cloud 4.0: Capturing Value through Disruptive Energy Platforms. 2018, viewed 19.03.2019, https://www.navigant.com/-/media/www/site/insights/energy/2018/energy-cloud-4-capturing-business-value.pdf

Filimonov A.G., Chichirova N.D., Chichirov A.A., Filimonova A.A. 2018. Implementation of digital economy elements in electric power industry [Vnedrenie elementov tsifrovoi ekonomiki v elekrtoenergetike]. *Reliability and energy safety*, vol. 11, no. 2, pp. 94-102.

Filippov, S. 2018. New Technological Revolution and Energy Requirements [Novaya technologicheskaya revolutsia i energeticheskie trebovania]. *Foresight and STI Governance*, vol. 12, no 4, pp. 20–33.

Holkin D., Chausov I. 2018. Digital transition in Russian power engineering: in search of meaning [Tsifromvoy perehod v energetike Rossii: v poiskah smysla]. *Energy policy*, vol. 5. pp. 7-16.

Holkin D.V., Knyaginina V.N. 2017. Digital Transition in the Electric Power Industry of Russia [Tsofrovoy perehod v elektroenergetike Rossii], viewed 19.03.2019, https://csr.ru/wp-content/uploads/2017/09/Doklad_energetika-Web.pdf

Hovalova T.V., Zholnerchik S.S. 2018. The effects of the introduction of smart grids [Effektii vnedrenia intellektualnykh elektroenergeticheskih setei]. *Strategic solution and risk management*, vol. 2, no. 105, pp. 92-101.

Improving oil and gas efficiency through digital. 2016. PwC, viewed 19.03.2019, https://www.strategyand.pwc.com/media/file/Improving-oil-and-gas-efficiency-through-digital.pdf

Key ideas of the Russian energy week 2018 [Klyuchevye momenty 2018], viewed 19.03.2019, https://rusenergyweek.com/upload/download/%D0%9A%D0%BB%D1%8E%D1%87%D0%B5%D0%B2%D1%8B%D0%B5%20%D0%BC%D0%BE%D0%BC%D0%B5%D0%BD%D1%82%D1%8B.pdf

Khodokovkaya Y.V., Sergeeva O.Y., Mukhametshin E.I. 2018. Business strategies of organization of Russia's competitive oil and gas market [Bisnes-strategii organizatsii konkurentosposobnogo nefteservisnogo rynka Rossii]. *Eurasian Law Journal*, vol.2, pp. 351.
Not your father's oil and gas business. Reshaping the future with upstream digitization. 2016. PwC, viewed 19.03.2019, https://www.strategyand.pwc.com/media/file/Not-your-fathers-oil-and-gas-business.pdf
Plakitkin Yu.A., Plakitkina L.S. 2018. Digitization of the Russian coal sector economy - from Industry 4.0 to Society 5.0 [Tsifrovizatsiaya ekonomiki ugolnoy promyshlennosti Rossii – ot "industrii 4.0" do "obshestva 5.0"]. *Mining*, vol. 4, pp. 22.
PwC and GE Digital Alliance. 2016. PwC, viewed 19.03.2019, https://www.pwc.com/us/en/about-us/alliances/ge-digital/assets/pwc-ge-digitial-upstream-oilgas.pdf
Tauber M., Bender J.P. 2018. New Technology Frontier in Mining. BCG, viewed 19.03.2019, https://www.bcg.com/publications/2018/new-technology-frontier-mining.aspx
Roadmap of national technological initiative "Energynet" [Doroznaya karta "Enerdzhinet" natsionalnoy technologicheskoy initsiativy", viewed 19.03.2019, https://minenergo.gov.ru/node/8916
Sergeeva O.A., Larionova O.A., Zaharova O.L. 2018. Digital transformation as an innovational approach of enhancing management efficiency at Russian oil and gas industry [Tsiofrovaya transformatsiya - innovatsionnyy put povysheniya effektivnosti upravlenia v NGK Rossii]. *Problems of economics and management of oil and gas complex*, vol. 9, pp. 33-37.
Sidorenko E.L. Shaydullina V.K., Kirakosyan S.A. 2018. Tokenization of coal industry: economic and criminological risks [Tokenizatsia ugolnoi promyshlennosti: ekonomicheskii i kriminilogischeskii riski]. *Coal*, vol. 12, no. 1113, pp. 54-59.
Teksler A. 2018. Power industry digitalization: from process automation to the digital transformation of the industry [Tsifrovizatsia energetiki: ot avtomatizatsii protsesov k tsifrovoy transformtsii otrasli]. *Energy policy*, vol. 5, pp. 3-6.
Transactive Energy Models. Business and regulatory models working group. 2016, viewed 19.03.2019, https://energy.wisc.edu/sites/default/files/2017-07/Transactive-Energy-Models.pdf
Unrealized potential of digital. 2016. PwC, viewed 19.03.2019, https://www.pwc.com/us/en/industries/energy/library/unrealized-potential-of-digital-the-imperative-of-oil-and-gas-digitization.html
Weill P. Woerner S.L. 2015. Thriving in an Increasingly Digital Ecosystem. MIT Sloan, viewed 19.03.2019, https://sloanreview.mit.edu/article/thriving-in-an-increasingly-digital-ecosystem/
Weill P. Woerner S.L. 2018. What's your digital business model? Harvard Business Review Press. P. 256.

Youth Technical Sessions Proceedings – Litvinenko (ed)
© 2019 World Petroleum Council, London, UK, ISBN 978-0-367-34668-3

Applying the principle of a sharing economy in large oil and gas companies

L. Khazipova & F. Khazipov
PJSC TATNEFT, Almetyevsk, Russian Federation

ABSTRACT: Under the conditions of limited oil production, the task of reducing the cost of oil production without key performance indicators degradation is becoming increasingly urgent. The model of oil and gas production divisions that are independent centers no longer meets modern requirements for efficiency. Improving the efficiency of existing resources is possible through the use of the principle of the sharing economy. In large oil and gas companies, the principle of the sharing economy will make it possible to save dozens of millions of dollars a year. In this case, two different approaches can be used:

1. Creation of responsibility centers with a clear delineation of services through the introduction of specialties.
2. Equipment rental instead of acquiring, by analogy with car sharing.

1 INTRODUCTION

From the time of the creation of barter, people shared with each other available benefits - they gave the horse for a while, borrowed a boat, i.e. shared things that they themselves did not use all the time. With the development of technology, society has returned to this practice, but now the exchange of things and services is not limited to a narrow circle of communication of a particular person, but has grown to the scale of the whole world. This phenomenon was called sharing economy – collaborative consumption - and was included in the list of ideas compiled by Time magazine that will change the world in the near future.

The concept of Collaborative consumption was proposed by economists Rachel Botsman and Roo Rogers in the book "What's Yours: The Rise of Collaborative Consumption". The idea is that it is often more profitable and more convenient for the consumer to pay for temporary access to the product than to own it (Botsman 2010).

The undoubted advantage of the new model is a much higher efficiency in the use of resources. It can be explained with a simple example: in the USA there are about 80 million hand drills, each of which is used for an average about 3% (Botsman 2010). If we come up with a scheme in which each tool will work more, a hand drill in every house will not be necessary. The idea is interesting to both the owner of the resource and the consumer, since it allows a multiple to increase the efficiency of using the available resources with little or no additional cost.

Thus, the business model of a collaborative consumption economy is not a manifestation of socialism in our day, as it may seem. Service providing companies and users are interested in making a profit and do not necessarily do it for free (Hayes 2017). "The economics of sharing is something different than mutual aid, given the fact that its members sell their services for money," emphasizes Arun Sundararajan, professor at New York University (Sundararajan 2016).

Despite the wide range of existing papers on different aspects of use sharing consumption (Miller 2018, Jackson 2018, LaBrecque 2014, Roh 2016), there are no papers discussing prospects of implementation of such principle in large oil and gas companies.

The advantages of the model of the sharing economy are obvious, but transition to it of large oil companies with a multi-billion dollar capitalization cannot happen quickly. This is due to the fact that large oil and gas companies tend to organize their structure not according to the traditional functional basis (marketing, finance, supply, production, shipments), but by activities, each of which is grouped around a certain type of product or service and includes in it required functional services. This trend highlighted the concept of a business segment as a part of an organization that works for an external consumer or is a relatively independent segment, according to which assets, the results of operations of the main and other activities for financial reporting can be separated from the organization as a whole. Such segments are often called business units or oil and gas divisions (NGDUs) in large oil companies.

2 DISADVANTAGES OF CURRENT STRUCTURE

Each NGDU, more recently, was an independent production and commercial division, combining almost all the functions and activities necessary for the production and sale (delivery) of oil. Each of these units is responsible for the results of its activities, i.e. in fact, they are independent responsibility centers.

In this case, the role of the corporate center is assumed by the executive office of the company. The role of the executive office lies primarily in the fact that it implements the function of a strategic manager and, accordingly, determines the list of functions, organizational structure of management, regulations, instructions and other methodological provisions for the effective management of business units. Development tasks can be easily set through the KPI system, either as physical indicators as an increase or stabilization of oil production while reducing costs and/or capital expenditures.

Despite the obvious advantages, the presence of a high level of independence of structural units has several significant drawbacks. Among the most critical are:

1. Inefficient use of resources. The independence of structural units leads to the need to possess all the necessary competencies and resources to ensure the realization of production tasks. Many types of competencies and resources have irregular demand, which leads to their inefficient use and, consequently, to significant losses.
2. Duplication of functions. The growth of production efficiency is impossible without solving of the strategic and business challenges facing the company. Due to the independence of structural units, these tasks are solved by them independently and unequally effectively. The absence of a single center of responsibility does not allow to distribute the tasks by divisions and completely eliminate duplication.
3. Availability of individual KPIs for the structural unit. Since each NGDU is faced with the task of continuous improvement of production indicators, this leads to the need to maximize the use of resources, regardless of the effectiveness of their use compare to other structural units. The solution to this problem is the most difficult, but the increase in the efficiency of resource use, at the same time, will be the greatest.

3 SOLUTION

To eliminate disadvantages listed above, it is necessary to single out responsibility centers not by binding them to the place of cost and profit origination, but by binding to the competencies, necessary to solve problems most effectively. Thus, the differentiation of the company into responsibility centers will bring the greatest effect in the divisional organizational structure, in which the divisions of the company have broad autonomy in making operational decisions, and the headquarters exercises only general control of the strategy and financial flows. The differentiation of the company into divisions can be carried out according to the types of products and services, groups of customers, geographic regions.

Let's explore the simplest example of applying the principle of a sharing economy, described in detail in the business model for companies operating in the segment of transport sharing and implemented in PJSC TATNEFT (Cohen 2014). Due to the extensive geography of the company's facilities, two years ago motor transport with a driver was widely used. All transport was assigned to workshops, departments of the NGDU and the executive office. The owner of the car and the employer of the driver was a motor company, payment for services was made on the day of use, regardless of the actual load of the driver.

If in the shops of the NGDU due to the presence of a shortage of vehicles, the load reaches an average of 85-90%, then the load in the departments of the NGDU and the executive office was often below 50% due to the peculiarities of the operating activities. A simple reduction in the number of motor vehicles did not allow to solve the problem due to the impossibility of ensuring a uniform load.

The solution to the problem was the rejection of the assigned vehicle in favor of the principle of "Taxi", when a car with a driver is ordered and paid only when there is a need for it. To achieve this goal, the following tasks were solved:

1. Refusal of assigned vehicles in the departments, except for cases when the percentage of involvement for production needs is high due to the specifics of the activity.
2. Creating a center of responsibility for the provision of vehicles at the request.
3. Establishment of a system for the effective use of motor transport as a resource. This task was accomplished in a similar manner to the Taxi principle used by people daily. Benefits of using shared vehicle are well known and showed in Figure 1.

To do this, the existing motor company, which owns a fleet of various vehicles from a Lux class sedan to an off-road vehicle for driving to distant oil production facilities, as well as an employer for drivers, was taken as a basis. At the second stage dispatch service - Call Center was created, which accepts and processes orders, as well as resolves an issues. To reduce dispatcher utilization and improve the ease of use of taxis, in the third stage, software was developed that runs under the Android and IOS operating systems, allowing employees to order a corporate taxi from a mobile device. The efficiency of use was ensured by assignment a limit on the use of motor vehicles assigned to the department of the NGDU.

As a result, the efficiency of the use of vehicles in the departments of the NGDU was increased drastically, which ensured a reduction in costs by more than 100 mln. rub. per year.

The use of the principle of sharing economy may not be limited to material resources, on the contrary, today there is a shortage of competent personnel, including their ineffective use.

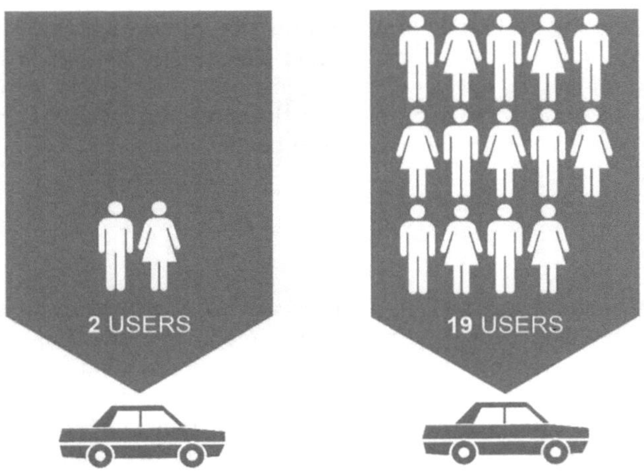

Figure 1. Advantages of use shared vehicle

The principle of using the company's strengths - competencies, is described in detail in the article «The Core Competence of the Corporation» (Prahalad 1990). This approach can be used to create responsibility centers in an existing company through restructuring.

Despite the seeming simplicity of the decision, the correct choice of the organizational and legal form of each business unit, its conformity to the goals and objectives of the restructuring is important and sometimes crucial for successful restructuring with the creation of a responsibility centers. The organizational and legal form of a business unit determines the management system in the business unit itself and in their complex, a level of responsibility for the results of activities, the prospects of existence and development, and the economic efficiency of allocating the center of responsibility.

Among the main stages of the restructuring in the form of creating responsibility centers are:

1. The existence of economic feasibility and evaluation of the possibility of restructuring.
2. Training of top and middle management of the basic principles and rules of the restructuring of the company and team building.
3. Development of the schedule of restructuring, including.
   a. determining the list of business units to be created;
   b. determining the order and timing of their creation;
   c. the choice of management structure;
   d. definition of the list and deadlines for development of regulatory documents (constituent contracts, regulations, instructions, etc.);
   e. holding meetings of the board, board of directors, general meeting (in case of reorganization);
   f. organization of planning, accounting, control, analysis and evaluation systems of business units;
   g. selection of leading cadres of business units, etc;
   h. registration and amendment of constituent documents, if necessary;
   i. support of the restructuring;
   j. evaluating of restructuring results and adjusting the process if necessary.

Evaluation of the economic results is one of the most difficult task, however, it must be necessarily made due to the significant costs of the restructuring project. To analyze and compare the effectiveness of projects, we propose to use the indicators outlined in the "Methodological guidelines for evaluating the effectiveness of investment projects" approved by the Ministry of Economy and the Ministry of Finance of the Russian Federation (Kossov 2000).

The organizational structure of the company in the process of its restructuring must be fixed (described) by an approved statement that determines:

1. the structure of the main activities;
2. the structure of the main management functions (marketing, organization of production, finance, economics, accounting, personnel management, information support, etc.);
3. type of management structure (linear, linear-functional, divisional, matrix, combined);
4. the structure of the organizational units and the functions assigned to them in the implementation of activities with the definition of the order of work for their implementation (rights, duties, time, incentives and penalties);
5. the procedure for monitoring the compliance of the organizational structure with the
6. activities and personification of responsibility for key processes of the unit and control them by KPI.

In accordance with the specified principles of formalizing the organizational structure of management, the functions of each centers and business units should be clearly defined, which are fixed in the developed statement.

PJSC TATNEFT has taken its own way of creating responsibility centers. At the first stage, the NGDU functions were identified that are directly related to the key function of the structural unit - development and production, and also the functions that serve direct customers in the face of the D & P block (development and production) were identified. It should be noted that the allocation of service functions was made not by creating new structural units in the company, but by creating responsibility centers in the structure of COB - a business service center. At the first stages, such areas as labor remuneration, personnel support, and book-keeping were identified.

The second stage included the centralization of production functions serving the D & P unit. These include: well repairs, surveying support, logistics, workplace infrastructure management, etc. The creation of responsibility centers allows at least to increase the efficiency of labor by introducing specialization. Meanwhile, the centralization of the production functions of the second stage of restructuring allowed to achieve more efficient use of resources throughout the company.

Let's take a closer look at the benefits of creating a single management company for the well workover that were previously in the structure of each NGDU.

The main task of the workshop of well workover is to bring the well equipment into working condition by replacing or repairing in well conditions. The effectiveness of the work is assessed through additional oil production from the well in comparison with the conditions before the start of work. The presence of specific conditions in each of the NGDU leads to a different specific effectiveness of the activities carried out. At the same time, the differences can reach two or more times, which naturally leads to the question: "Is there any sense to do well workovers in all NGDUs in approximately equal amounts if the effect is not the same?".

Only the creation of a center of responsibility in the form of a center of well workover does not allow to solve this problem. To begin with, a special module was developed into the ARMITS corporate information system, which allowed analyzing all the proposed types of geological and engineering measures in a single window for the company and ranking them by performance regardless of the references (geographic location, NGDU, well type, etc.). At the second stage, the conditions for the maintenance crews were given on the basis of current needs for the NGDU (similar to the Taxi principle). However, at the same time, payment for services for attracting additional crews is performed by the NGDU on the basis of the given cost limit.

4 CONCLUSION

The process of structural changes is always negative and it is important to make a qualitative analysis of risks and work to prevent them. For example, in order to reduce the difficulties during the transitional phase, the out stuffing of the personnel support service from the NGDU was carried out initially without moving the place of work of the personnel. At the same time, work was actively done on delineating areas of responsibility. To discuss issues of concern, a special group was created in the corporate social network.

Due to the fact that, despite the servicing functions, the HR service actively interacts with the staff of the NGDU, the "Front Office" was created, which made it possible to organize effective work directly in the structural unit.

Additionally, it is necessary to note that a lot of work done to reduce social tension. All work on centralization was carried out without staff reduction. All NGDU employees, whose positions were optimized in the new structure, were offered vacancies after restructuring.

Thus, despite the peculiarities of the existing organizational structure of large oil and gas companies, the application of the principle of a sharing economy can and should bring a profit to the company. In different conditions the method of application of the presented approach can vary considerably. To do this, all features of the company, including social, should be taken into account in order to prevent the loss of the company's most valuable resource - employees. It is very important to maintain the effectiveness of all processes before and after centralization. To do this KPIs should be assigned for each of the process, the responsibility matrix for KPIs have to be calculated in the new format of the responsibility center and the task is set to stabilize their effectiveness as the main indicator of the effectiveness of the transformations.

REFERENCES

Botsman, R. 2010. The Sharing Economy Lacks A Shared Definition, viewed 15.02.2019, https://www.fastcompany.com

Botsman, R. 2010. TEDxSydney Conference, viewed 15.03.2019, https://www.ted.com/talks/

Cohen, B.& Kietzmann, J. 2014. Ride On! Mobility Business Models for the Sharing Economy. *Organization & Environment* 27(3): 279-296.

Enterprise restructuring based on creation of responsibility centers [restrukturizatsiya predpriyatiya na osnove sozdaniya tsentrov otvetstvennosti], viewed 23.10.2018, http://econbooks.ru/books/part/16734

Hayes, A. 2017. The Economic Fundamentals of the Sharing Economy, viewed 15.03.2019, https://www.investopedia.com/

Jackson, R. 2018. Why Your Business Should Use The Sharing Economy, viewed 12.01.2019, https://channels.theinnovationenterprise.com/articles/why-your-business-should-use-the-sharing-economy

Joint consumption as a new economic model [Sovmestnoye potrebleniye kak novaya ekonomicheskaya model'], viewed 05.02.2019, https://kp.vedomosti.ru/article/2017/02/15/677751-shema-novogo-vremeni

Kossov, V., Livshits, V. & Shahnazarov, A. 2000. Guidelines for evaluating the effectiveness of investment projects. Ministry of Finance of Russian Federation. *Economy*: 11.

LaBrecque, S. 2014. Eight of the best sharing economy companies, viewed 12.01.2019, https://www.theguardian.com/sustainable-business/eight-best-sharing-economy-companies

Miller, D. 2016. What Is the Sharing Economy (and How Is it Changing Industries), viewed 23.01.2019, https://www.thebalancesmb.com/the-sharing-economy-and-how-it-changes-industries-4172234

Prahalad, C. & Hamel, G. 1990. The Core Competence of The Corporation. *Harvard Business Review* May-June: 4-9.

Roh, T. 2016. The sharing economy: Business cases of social enterprises using collaborative networks, viewed 18.01.2019, https://cyberleninka.org/article/n/669082.pdf

Sundararajan, A. 2016. The Sharing Economy: the End of Employment and the Rise of Crowd-Based Capitalism. U.S.: MIT Press.

What the shared economy is: the history of the term [Chto takoye delit'ekonomiku: istoriya termina], viewed 10.02.2019, https://rb.ru/story/share-it/

Youth Technical Sessions Proceedings – Litvinenko (ed)
 ISBN 978-0-367-34668-3

Renewable energy technologies application for the oil and gas industry in the Arctic Zone of the Russian Federation

T. Gabderakhmanova & M. Morgunova
Joint Institute for High Temperatures of the Russian Academy of Sciences, Moscow, Russia

ABSTRACT: The study investigates possibilities for renewable energy technologies application at the facilities of the oil and gas industry in the Arctic Zone of the Russian Federation (AZRF). Due to the relatively high potential for renewable energy sources in the Arctic region, it is possible to implement corresponding technologies to provide a more sustainable energy supply. The paper estimates the energy consumption of some of the installations of the oil and gas industry in the AZRF and diesel fuel replacement potential. Some of the prospective power generation system configurations based on alternative energy sources have been proposed. It has been shown that the solutions, which combine traditional and renewable energy technologies, are able to meet the essential reliability requirements while facilitating higher sustainability of oil and gas industry.

1 INTRODUCTION

Oil and gas industry is highly present in the Arctic region. One of the reasons is a huge potential of undiscovered technically recoverable oil and natural gas resources onshore and offshore Arctic (Gautier et al. 2009). The main part of proven reserves is considered to be in Yamalo-Nenetsky and Nenetsky autonomous districts (about 5 and 1 billion tons accordingly). More than 90% of the offshore Arctic and 53% of the onshore still have to be explored.

The industry is not only producing oil and natural gas, but it is also the main energy consumer. The power generation installed capacities at the remote fields can range from a few to more than 100 MW (Surgutneftegaz 2017, Gazprom 2017). Besides, more energy needs to be delivered to remote areas with low transport accessibility and harsh climate conditions, where the exploration is taking place. According to some estimations, around 1.7 million tons of oil equivalent (toe) is annually used in the regions of the Arctic and the Far East for the power generation.

Nowadays, the primary source of energy in the remote Arctic areas is diesel fuel. Diesel generators are actively used at the oil and natural gas facilities alone or in combination with gas turbines. Even though there is a considerable amount of large-scale gas turbine stations in some of the oil and natural gas producing regions in Russia (for example, in the Yamalo-Nenets Autonomous Okrug), which utilize associated gas from local oil fields, the consumption of diesel fuel is growing.

Along with a number of advantages, as high energy density, ease of use, etc., diesel fuel has serious disadvantages, such as environmental pollution. The fuel is also expensive due to the higher transportation costs, which is especially typical for the remote areas in the Arctic. The transport component in the specific cost of electricity produced using diesel fuel reaches up to 70%.

At the same time, some areas of the AZRF are characterized by a significant potential of renewable energy sources, such as wind and solar (Andreenko et.al. 2016), while the average cost of electricity produced from renewable energy sources has been decreasing significantly during recent years. For example, the average levelized cost of electricity from onshore wind in 2017 was about 50 $/MWh, which is 2.5 times less than in 2008 (Lazard 2008&2017).

Thus, it is of great importance to investigate the possibilities to optimize energy consumption of the oil and natural gas facilities in the Arctic region, as well as investigate the ways of efficient integration of renewable energy technologies into the energy infrastructure of the industry.

2 RENEWABLE ENERGY SOURCES POTENTIAL IN THE AZRF

Renewable energy generation technologies are efficiently used to power oil fields around the world (Choi et al. 2017) and have a high potential in the Arctic (Table 1), even though there are some natural and technical constraints (for further discussion on technical issues Solovyev et al. 2019).

2.1 *Solar energy*

Rather high solar radiation is typical for the north of Eastern Siberia and the Far East. Some territories of the AZRF have a high gross and technical potential of solar energy. The leaders in terms of the annual average global solar radiation on a horizontal surface, as well as the gross and technical potential, are the Republic of Sakha (Yakutia), the Krasnoyarsk Territory, the Chukotka Autonomous Okrug and the Yamalo-Nenets Autonomous Okrug. The diurnal sum of the global solar radiation to a tilted surface of the southern orientation (the slope angle is equal to the latitude of the area) from April to September is 4-4.8 kWh/m^2. Those numbers are promising and correspond with the solar radiation of more southern regions (Morgunova & Solovyev 2016).

Nevertheless, solar generation cannot be considered as a secure and stable resource to cover the all-year-round energy demand due to the long polar nights, especially as oil and natural gas industry has high consumption and high reliability requirements. However, in some cases, it can be effective even beyond the Arctic Circle in the combination with other energy sources (Gazpromneft 2017). There are also quite a few examples of powering small-scale seasonal consumers in the Russian Arctic using solar generation technologies (GIS 2018).

2.2 *Wind energy*

There are considerably higher possibilities for the use of wind energy technologies in the AZRF, especially in the coastal areas of the Arctic Ocean. The coastline of the Nenets Autonomous Okrug is characterized by relatively high average annual wind speed of more than 5 m/s, where wind speed reaches 8–10 m/s during almost half of the year. Besides, these

Table 1. Potential use of solar and wind energy in the AZRF

Region (Russian Federation)	Gross/Technical/Economic Potential (power generation) Solar, mm kWh	Wind, mm kWh
In total AZRF, including:	6200/2621/149	3012/7539/37585
Komi Republic	369/202/11	76/187/950
The Republic of Sakha (Yakutia), north	3183/1319/9	773/1929/9688
Krasnoyarsk Territory, north	728/303/67	177/440/2198
Arkhangelsk region	341/141/10	75/187/938
Murmansk region	116/49/17	124/309/1554
Nenets Autonomous Okrug	143/59/1	120/293/1498
Chukotka Autonomous Okrug	655/272/1	296/773/3664
Yamalo-Nenets Autonomous Okrug	666/277/33	1371/3419/17096

(adapted from Morgunova & Solovyev 2016)

figures are quite stable, as the yearly changes are relatively small and a coefficient of variation is in the 5-8% range (Andreenko et.al. 2016).

The efficient use of land-based wind turbines of the required power level in the Arctic conditions, precisely in the permafrost and harsh climate conditions, involves a number of risks. The modern megawatt-class wind generators are fairly large and heavy equipment that requires in most cases costly delivery and installation. Even though there are wind turbines specially designed for the Arctic (adapted for the crane installation) (Gsenger & Denisov 2017), there are limitations according to weight and size characteristics of the gearbox and other elements of the turbines. Besides, the construction of the facility on permafrost requires special approaches to the thermal insulation of the foundation. In particular, concrete, wood or other material with low thermal conductivity should serve as an intermediate element between the soil and metal structures of the tower, which in turn imposes restrictions on the weight characteristics of the wind turbine (for instance, Berdin et al. 2017).

The most promising option for the AZRF is the sea-based wind generators, which is also very much applicable to supplying electricity for the offshore field exploration in the AZRF (Steen & Weaver 2017). The advantages are the following: 1) possibility to deliver bulky weight by the sea; 2) a larger power capacity of an offshore wind power unit compared to a land-based; 3) higher wind speeds above the water surface. Offshore fields development expertise of oil and natural gas companies is considered to be one of the strongest arguments in finding the synergy in developing offshore wind. One of the recent examples is the floating wind farm Hywind constructed by Equinor (former Statoil, Statoil 2017).

In order to supply large-scale industrial consumers, several wind generation units of 5–7 MW each could work in conjunction with a gas turbine power plant. The autonomous power supply using wind turbines is considered to be technically feasible but requires some technical adjustments, such as a source of the reference voltage and frequency of the local network. The voltage and frequency are usually set by an internal combustion engine (for example, gas turbines, diesel generators) or by an electrical energy storage inverter. However, there are also some economic constraints, where such power plants require larger consumption volumes.

Nevertheless, there is a number of small-scale examples of combined energy generation solutions, which are successfully operating, particularly in the AZRF (GIS 2018).

2.3 *Fuel cells*

The introduction of fuel cells with solid oxide (SOFC) or melt-carbonate (RCTE) electrolyte is also one of the possible solutions to reduce diesel consumption. Their efficiency is higher than of heat engines In addition, RCTE systems are capable of capturing CO_2, which reduces the anthropogenic load on the environment.

The traditional niche for Peltier elements or SOFC fuel cells is cathodic protection stations for pipelines with a peak power of 0.1–3 kW. The stations have monotonous load curve, which positively affects fuel cells resource. At the same time, their widespread use is limited due to the necessity to allocate them in low-pressure pipelines of about 0.5-1 ATM.

3 ENERGY CONSUMPTION IN THE OIL AND NATURAL GAS INDUSTRY

There are many large and small types of energy consumers in the oil and natural gas industry (Table 2). All the electricity consumers can roughly be divided into the following groups: with an energy capacity of 1) 0.1–1 kW; 2) 1–10 kW; and 3) more than 10 kW. Systematization of consumers with a capacity of more than 10 kW is challenging since equipment in most cases has a wide range of installed capacity depending on the size of facilities.

The largest consumers are pump and compressor stations, where the most energy-intensive ones are pump installations with high-voltage electric motors for oil processing and transportation, as well as for the reservoir pressure maintenance. Port facilities are among the largest and most complex facilities requiring a reliable electricity supply. The power consumption of

Table 2. Oil and natural gas energy consumers by types and the installed capacity

Type of installation	Power range
Autonomous lighting, signalling devices	0.1 - 1 kW
Ground navigation and communication facilities	0.1 - 1 kW
Leak detection systems	0.1 - 1 kW
Telemechanics and communication systems	0.1 - 1 kW
Large navigation and communication systems	1-10 kW
Pipeline cathodic protection stations	1-10 kW
Industrial and repair stations	0.1-3 MW
Associated gas compressor stations	Up to 3 MW
Water intake and water lift	0.1 – 5 MW
Drilling rigs	1 – 5 MW
Processing plants	0.1 – 7 MW
Oil well pumps	Up to 10 MW
Oil treatment and pumping facilities	2 – 10 MW
Pump and compressor stations	0.1 – 20 MW
Electrical heating pipeline installations	5-50 MW
Pump and compressor stations (transportation)	5 – 50 MW

(adjusted by authors, Kudryashov & Kudryashova 2016)

electrical equipment of modern large sea ports (including loading systems, aspiration and ventilation systems, customs control facilities, water supply pump stations, alarm systems, navigation systems, etc.) is tens of MW. As the port infrastructure is under rapid development in the AZRF, its sustainable power supply is one of the critical questions, also for the oil and natural gas industry (Morgunova & Solovyev 2017).

3.1 *Reliability requirements*

All the facilities in the oil and gas industry involved in the production and transportation processes (for collecting and processing, maintaining reservoir pressure, transportation, etc.), have high requirements on the reliability of energy supply. The disruption of the production processes is not allowed and bears significant costs and risks. Thus, one of the necessary conditions of the energy supply for the oil and natural gas facilities is the use of mutually redundant power supplies, providing round-the-clock load coverage at any time of the year, as well as a stable supply of energy during short-term load increases. This implies that one or two power generation units must be in hot standby, as well as several units should be in operation.

The current practice for the energy supply of the energy-intensive consumers in the oil and natural gas fields is to use gas turbines and, more rarely, piston engines operating on several types of fuel (associated gas, natural gas and diesel fuel) (Gazpromneft 2017, Rosneft 2017). These solutions allow utilize associated gas and reduce the infrastructure costs for gas pipelines, compressor stations and transmission lines. Besides, it is possible to perform combined heat and power generation by these units. Diesel generators are often used in emergency operation mode or as a backup (Rosneft 2017). However, for the small-scale consumers with an installed capacity of up to several kW, the use of diesel generators as a backup is not always efficient if there is only a partial load. Electric power batteries are typically used to power tele- and communications systems, leak detection systems and other low power consumers.

3.2 *Features of operating modes and load curves*

The operation mode of certain categories of equipment determines the choice of the energy supply system. Typical consumer electrical load curves in oil and natural gas industry are quite dense and are characterized by the following indicators: for the daily schedule - the density coefficient of 0.88 and the coefficient of irregularity of 0.80; for the annual schedule - the density coefficient of 0.80 and the coefficient of irregularity of 0.62. This means that the basic load of the equipment in the industry is high in comparison to the installed capacity.

Besides, the installations have quite a different load curve character. The loads vary from a periodical (pump jacks) and regular (oil production) to irregular (drilling rigs). A load of drilling equipment is of a highly variable nature both because of the specifics of the drilling process and due to the specific features of the development site. The starting currents are up to 5 times higher than the nominal ones and there are significant power fluctuations (an example of the drilling rig load curve is presented in Kudryashov & Kudryashova (2016)). The power consumption of a pump or a compressor station is much higher but does not have significant fluctuations due to stable load. The load curves of cathodic protection systems, leak detection, telemetry and communication systems are monotonous. With regard to the operation modes of the oil and natural gas facilities, the power stations must be highly maneuverable.

4 THE POSSIBLE APPLICATION OF THE RENEWABLE ENERGY TECHNOLOGIES FOR THE OIL AND GAS INDUSTRY IN THE AZRF

In general, taking into account that high-voltage equipment in the industry, high-power wind generators can complement to the energy supply. With the existing renewable energy resource potential in the AZRF, it seems to be possible to supply that equipment by alternative energy sources.

However, the less critical are the reliability requirements and base load, the more are the opportunities to replace or supplement electricity from renewable energy sources. So, for instance, telemetry stations can be supplied from the compact power plants based on photovoltaic modules (Popel et al. 2015), electrical energy storage devices based on lithium-ion batteries, as well as a guaranteed power source (hydrogen-air fuel cells).

The reduction of diesel fuel consumption can be reached when implementing renewable energy installations to supply loads from more than 10 kW to several MW. In the Arctic conditions, traditional diesel generators can be replaced by combined wind-solar-diesel installations. In the case of pump stations in The Republic of Sakha (Yakutia), along with lower consumption of diesel fuel, the cost of electricity generation using such a combined installation may be significantly decreased.

To summarize, the most promising option to apply renewable energy technologies in the oil and natural gas industry in the AZRF is a combined renewable energy (particularly, wind) based power plant supplemented by fossil fuel based generation. This combination is capable to ensure proper level of energy supply reliability. Such technical solutions are not only the guaranteed way to optimize energy supply of the installations, but also suggest lower electricity production costs.

5 CONCLUSION

Renewable energy technologies can successfully be introduced in the oil and gas industry, even though there are some natural and technical constraints. They can efficiently be applied in the remote areas as the AZRF.

Combining traditional and renewable energy resources for oil and natural gas industry equipment power supply is the forefront solution, which also helps to reduce dependence on diesel fuel and meet all the reliability and operation regime requirements.

One of the most suitable solution for medium and large-scale consumers is a combined wind-diesel or wind-gas power plant. Smaller installations can be supplied from the compact power plants based on photovoltaic modules and power batteries. Those solutions are able to provide a stable and reliable energy supply, lower the diesel fuel consumption and decrease electricity generation costs.

The study was funded by RFBR research project No. 19-08-00945\19

REFERENCES

Andreenko, T.I., Gabderahmanova, T.S., Danilova, O.V., Ermolenko, G.V., Ermolenko, B.V. et al. 2015. Atlas of renewable energy resources in Russia (In Russian: Atlas resursov vozobnovljaemoj jenergii na territorii Rossii). Publishing house RHTU imeni D.I. Mendeleeva, 2015, 160 p. ISBN: 978-5-7237-1353-6

Berdin, V.Kh., Kokorin, A.O., Yulkin, G.M., Yulkin, M.A. 2017. Renewable energy sources in isolated settlements of the Russian Arctic (In Russian: Vozobnovljaemye istochniki jenergii v izolirovannyh naselennyh punktah Rossijskoj Arktiki), WWF. 80 p. ISBN 978-5-906599-35-3

Choi, Y., Lee, C., Song, J. 2017. Review of Renewable Energy Technologies Utilized in the Oil and Gas Industry. International journal of renewable energy research Vol.7(2):592-598.

Gautier, D., Bird, K., Charpentier, R., Grantz, A., et al. 2009. Assessment of Undiscovered Oil and Gas in the Arctic. Science (New York, N.Y.) 324 (5931):1175–1179.

Gazpromneft 2017. Annual report. Source: www.gazprom-neft.ru (Accessed 19 March 2019).

Gazpromneft 2017. Sustainable development report. Source: www.gazprom-neft.ru (Accessed 16 March 2019).

GIS 2018. Renewable Energy Sources of Russia (In Russian: Vozobnovlyaemye Istochniki Energii Rossii). Source: http://gisre.ru/ (accessed 13 March 2019).

Gsenger, S., Denisov, R. 2017. Prospects for the wind energy market in Russia (In Russian: Perspektivy vetroenergeticheskogorynka v Rossii). Friedrich Ebert Stiftung and World Wind Energy Association. 29 p.

Kudryashov, R.A., Kudryashova, O.M. 2016. Basics of calculating the power consumption of oil and gas fields (In Russian: Osnovy rascheta elektropotrebleniya neftyanykh i gazovykh promyslov). Study guide, Tyumen: TIU. 65 p. ISBN 978-5-9961-1374-3

Lazard 2008, 2017. Levelized Cost of Energy analysis. Source: https://www.lazard.com (Accessed 16 March 2019).

Morgunova, M.O., Solovyev, D.A. 2017. Challenges to overcome: energy supply for remote consumers in the Russian Arctic. Journal of Physics: Conference Series (JPCS) 891 (conference 1): 012157.

Morgunova, M.O., Solovyev, D.A. 2016. Energy supply of the Russian arctic: hydrocarbon or renewable energy? (In Russian: Jenergosnabzhenie Rossijskoj arktiki: uglevodorody ili VIJe?) Energeticheskaja politika 5:44-51.

Popel, O.S., Kiseleva, S.V., Morgunova, M.O., Gabderahmanova, T.S., et al. 2015. The use of renewable energy sources for power supply to consumers in the Arctic zone of the Russian Federation (In Russian: Ispol'zovanie vozobnovljaemyh istochnikov jenergii dlja jenergosnabzhenija potrebitelej v Arkticheskoj zone Rossijskoj Federacii). Arctic ecology and economy 1:65-69.

Rosneft 2017. Annual report. Source: www.rosneft.ru (accessed 19 March 2019)

Solovyev, D.A., Morgunova, M.O., Solovyev, A.A. 2019. Creating a Synergy: Renewable Energy to Develop the Russian Arctic. In V. Kharchenko, & P. Vasant (Eds.), Renewable Energy and Power Supply Challenges for Rural Regions (pp. 136-153). Hershey, PA: IGI Global. doi:10.4018/978-1-5225-9179-5.ch006

Statoil 2017. World's first floating wind farm has started production. Source: www.equinor.com (Accessed 18 March 2019) .

Steen, M., Weaver, T. 2017. Incumbents' diversification and cross-sectorial energy industry dynamic. Research Policy, 46:1071-1086.

Surgutneftegas 2017. Annual report. Source: www.surgutneftegas.ru (accessed 19 March 2019).

Energy Markets in Transition

Youth Technical Sessions Proceedings – Litvinenko (ed)
© 2019 World Petroleum Council, London, UK, ISBN 978-0-367-34668-3

Gas engine fuel market in the Russian Federation: consumption issues

L.F. Timirkhanova
L.F. Timirkhanova, Gubkin Russian State University, Moscow, Russian Federation

ABSTRACT: The paper discusses problems of development of a gas engine fuel segment in the Russian Federation, possible ways to resolve them, an attempt to perform a regression between number of gas engine fuel consumers in the world and other qualitative and quantitative factors. The topic of behavioral economics is considered as an explanation of stagnation of development, since economic transition barriers for consumers were not identified during this study.

1 INTRODUCTION

Despite huge variety of types of fuel, oil, natural gas and coal remain the main sources of energy. Types of fuel from oil possess wider circulation for vehicles owing to convenience of transportation. However the subject of use of natural gas as fuel for transport is often touched in a number of scientific works and strategic plans of public authorities and has the considerable prospects of development. At the same time there are barriers of active development of gas motor fuel (NGV) in the Russian market of production of transport fuel.

Undoubtedly, natural gas is the most multiple-purpose and available fuel capable to replace oil products. Let's list arguments pro gas motor fuel:

1. Gas more environmentally friendly fuel, than gasoline and diesel.
2. Russia - the unconditional leader in volumes of reserves
3. The economy needs cheaper type of fuel which will stimulate the market, having lowered the prices for transportation of goods and services, and will fuel growth of consumer ability
4. Legislature of the state uses leverage, by subsidizing and stimulation of use of the vehicles consuming NGV

But, despite these indisputable advantages of NGV, there is no large-scale transition to gas engine modes of transport.

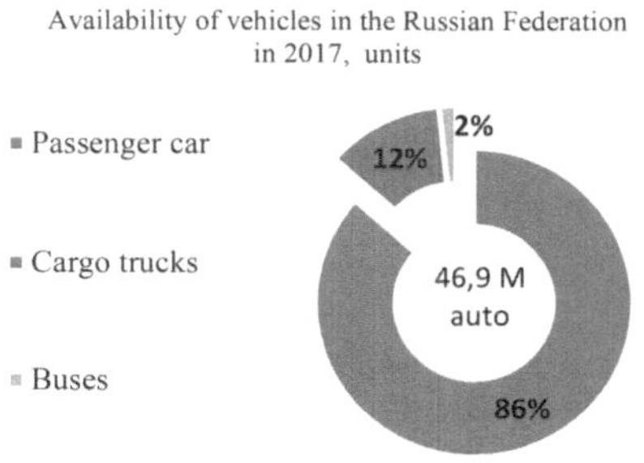

Figure 1. Availability of vehicles in the Russian Federationin 2017, units.

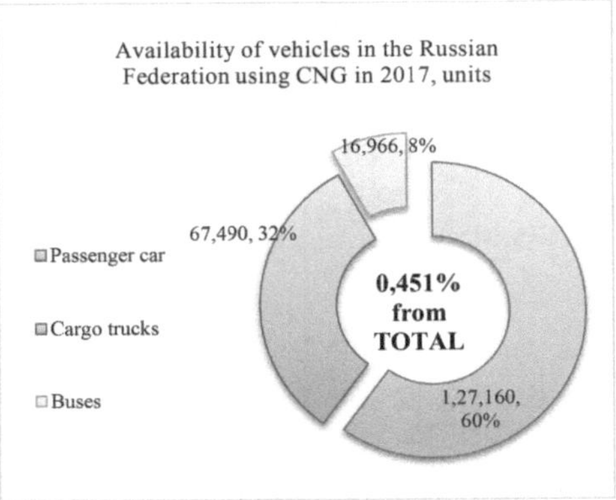

Figure 2. Availability of vehicles in the Russian Federationusing CNG in 2017, units

2 ANALYSIS

2.1 *Advantages of different types of gas motor fuel*

Liquefied Oil Gas (LOG)

This gas represents the mixture of C3H8 propane and C4H10 butane taken their associated oil gases from condensate fractions of natural gas, from gases of processes of stabilization of oil and condensate, from the petrofactory gases received from oil refining installations. At propane critical temperature of 96.8 °C, at butane 152.0 °C that allows to liquefy easily these gases and to store in liquid state with rather low pressure up to 1.6 MPas. It also means that the vessel for storage of the CIS will be rather easy, and can be stored liquefied gas as much as long on condition of full hermetic encapsulation of vessel. At the same time the cylinder for the CIS is pressure vessel, and it cannot give any form as, for example, to gasoline tank. This circumstance creates problems with seating of gas cylinder by car. Taking into account costs of the equipment the gas equipment, its weight and smaller cruising range at one gas station transfer of the car to the CIS remains favorable thanks to the price. Cars and low-tonnage trucks were and remain the engine of advance of the CIS in masses.

Compressed Natural Gases (CNG)

Critical temperature of methane is-82.3 °C, and its liquefaction is very expensive therefore methane as gas motor fuel is used generally in the compressed (compressed) look, at the same time gas is reduced in volume by 200–250 times. Bring the gas pipeline to automobile gas-filling compressor station (CNG filling station) and compress gas on the place. The supply of the main gas to fuel station is necessary that complicates expansion of network of gas stations a little. High pressure of CNG demands very strong, thick-walled, heavy cylinders. But there is more to come. On CNG it is possible to pass by 3.5 times smaller distance, than to the CIS at the equal volumes of gas cylinders.

Liquefied Natural Gas (LNG)

At liquefaction methane decreases in volume by 600 times – in it the main advantage of liquefaction the Same volume contains LNG three times bigger, than CNG.

Liquefaction takes place at temperature of-161.5 °C. Process energy-intensive also demands the cryogenic equipment. The liquefied methane is stored at temperature in the heat-insulated vessel from-160 to -196 °C. Very qualitative heat insulation is necessary.

The main advantages of gas motor fuel in its price, the price and once again the price. So far these advantages outweigh numerous and versatile shortcomings. In comparison with gasoline and diesel fuel the cost of CNG is 60-70% lower. Average average annual retail price for motor fuel in Europe has developed at the level: 1.27 euros for liter of diesel fuel; 1.4 euros for gasoline liter; 0.74 euros for cubic meter of natural gas. It is possible to count easily that in Russia monthly economy at run of 2500 km/month makes from 2000 to 4000 rub or 24000–48000 rub a year.

Conversion of vehicles to gas is favorable due to reducing costs of automobile fuel as to the commercial and state enterprises at which costs for the maintenance of vehicle fleets, and for citizens who can save on fuel decrease.

Table 1. Fuel prices the Russian Federation in 2016-2018

Type of fuel	Prrices, rub/per unit			Average	Growth rate (2018/2017), %
	2016	2017	2018		
AI-92	35,41	37,95	39,9	37,75	5%
AI-95	38,36	41,01	42,9	40,75	5%
Disel	36,27	40,24	47,00	41,64	17%
CNG (metan)	12,05	13,12	15,76	13,64	20%
LOF (propan)	18,33	17,7	22,14	19,39	25%

Reduction of harmful blowouts in the atmosphere

The fulfilled gases of internal combustion engines on the cars using gas motor fuel are several times less harmful, than exhausts of the engines using oil fuel. CNG at the moment are the most environmentally friendly types of motor fuel.

When using natural gas instead of oil fuel blowout of toxic substances to the environment decreases approximately by 8 - 10 times on carbon oxide, on nitrogen oxides – twice, on hydrocarbons 387 by 3 times. The formation of soot peculiar to diesel engines, is practically absent. Already now, rather small park of cars of Russia on CNG allows to reduce almost by 800 thousand tons blowouts of harmful substances to the environment

Use of gas motor fuel increases the term of operation of vehicles and reduces the price of their maintenance operation. Unlike petrol and diesel fuel, it has the best anti-detonation characteristics (octane number of gas motor fuel reaches 100-105 units), provides ease of preparation of uniform fuel-air mixture. The gas mixture burns down completely therefore the deposit on buckets, valves and spark plugs is not formed that reduces loads of the engine. Upon transition with liquid to gas motor fuel increase period of operation of the engine to overhaul and oil replacement term, is required replacement of spark plugs less often, the resource of the engine increases by 1.5-2 times. Naturally, there are also some features of use of CNG the reducing some operational characteristics of cars.

Safety. Gas is lighter than air, and in case of escape it right there disappears that considerably reduces risk of ignition. Besides, temperature of spontaneous ignition and the lower explosion limit at natural gas is much higher, than, for example, at motor petrol which in case of escape spreads under the car, forming in ground layer explosive mixture with air. Such mixture can easily ignite even from static electricity.

Available re-equipment of the engine. The service centers for installation of the gas-balloon equipment develop actively and modernized. The payback of the gas-balloon equipment installed in the vehicle is from 3 to 12 months. Besides, the car becomes two-fuel, i.e. keeps possibility of work, both on gas, and on gasoline.

Having listed all advantages of use of gas motor fuel, it is logical to expect growth of number of the cars using gas engine fuel, but practice shows that gas-balloon engines are not popular from the point of view of usual consumers, and business and city transport.(Kotov, Urazmetova 2013).

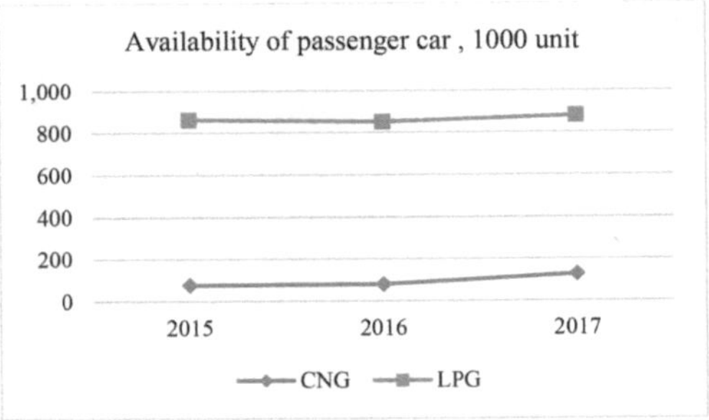

Figure 3. Availability of passenger cars in the Russian Federation using CNG in 2015-2017, units

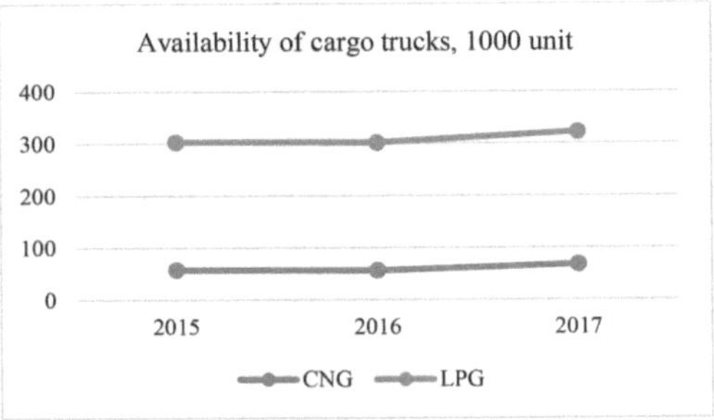

Figure 4. Availability of cargo trucks in the Russian Federation using CNG in 2015-2017, units

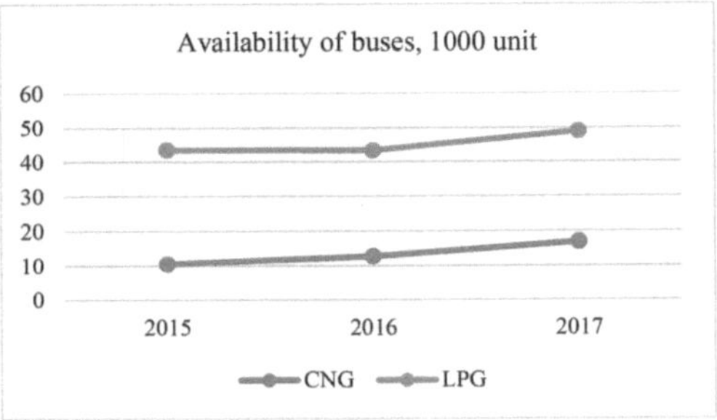

Figure 5. Availability of in the Russian Federation using CNG in 2015-2017, units

3 METHODOLOGY

We carried out multiple regression analysis for the purpose of identification of the quantitative parameters influencing number of the cars using gas motor fuel. As international experience and the number of cars significantly exceeds the Russian realities, the first analysis has been made from these various countries. Dependent variable quantity of gas motor fuel vehicles

When performing the following parameters have been considered:

- Fuel stations, unit
- GDP per capita, US $
- Gas production, billion cubic meters
- Population, people
- The cost of 1 liter of gas, $ USA
- The cost of 1l LPG, $ USA

Only «Fuel stations" and "Gas production" turned out to be significant.

The analysis in Russia was decided to be carried out on the basis of the number of cars distributed across the subjects of the federation.

Results were different from global analysis. The indicator "remoteness from gas stations" or "the number of the CNG filling station and gas station on the considered territorial subject of the Russian Federation" was the indicator having the greatest influence at this investigation phase. This indicator is significant also in the world analysis of the market of gas motor fuel. Also it has been revealed that the subsidized regions, that is those regions where exist and support of the federal center in use of the gas engine equipment in the form of subsidies was given, exist positive dependence of number of cars on NGV from the volume of the provided subsidies.

4 WORLD EXPERIENCE

It is obvious that without development of gears of the state stimulation, these reference points will not be reached. Unlike the countries where gasification of motor transport is state policy and additional expenses are compensated to buyers of gas cars by the state or stimulated in a different way, in Russia such practice is absent that is the essential factor containing growth of the market of NGV in Russia. International experience of stimulation of use of alternative fuel is presented in table 7 (Kotov, Urazmetova, 2013).

Table 7. International experience of stimulation of use of gas motor fuel

Country	Stimulation measures
Japan, Pakistan, Denmark, Italy	The ban on operation of the diesel cars operating in borders of settlements and/or nature protection areas
China, Italy, Great Britain, Chile	Entry into nature protection areas is allowed gas-balloon cars
Croatia	Discharging of gas-balloon cars of obligatory annual ecological testing
USA	Tax benefit of 50% for the new car using only alternative fuel

5 LONG-TERM FORECASTING

The long-term forecast of social and economic development prepared by the Ministry of Economic Development of the Russian Federation till 2030 gives the chance to estimate change of size of the given economic effect for the specified period. Sharp reduction of economic effect of transition to gas motor fuel since 2020 is connected, first of all, with alignment of the world prices and prices of domestic market of natural gas. (Gryaznov, 2015)

For complex assessment of economic effect for consumers we will consider order - to rank and structure of capital and current expenditure which arises at consumers in connection with transition to gas motor fuel.

Capital expenditure of consumers is formed due to the need feather - the equipment and retrofitting of cars by the gas-balloon systems allowing to use gas motor fuel as fuel for the internal combustion engine. Besides, number of domestic and foreign manufacturers already start - whether serial release of cars with the present gas-balloon system.

The additional current expenses connected with operation of cars on gas motor fuel are more frequent control visual inspections, scheduled works and engineering certifications of gas-balloon system. The specified expenses upon transition to gas motor fuel will average according to about 1.5 thousand rubles a year for cars and about 5.0 thousand rubles a year for passenger and cargo transport and also the special equipment.

As a result the payback period of capital expenditure from transition to gas motor fuel taking into account the size of capital and current expenditure makes for consumers from 1 to 2 years. At the same time optimum equivalent circuits are "motor petrol – LPG" and "diesel fuel – CNG".

Besides, calculations show that the equal given economic effect for LNG and CNG excludes expediency of use of LNG in connection with big capital expenditure therefore for ensuring uniform development of the specified types gas motor fuel it is necessary to support ratio of the given economic effects of LNG: CNG as 2:1. (Gryaznov, 2015)

According to Article 14 of the Federal law of 23.11.2009 No 261-FZ "About energy saving and about increase in power efficiency and about introduction of amendments to separate acts of the Russian Federation" the Government of the Russian Federation it has been established that regional and municipal programs in the field of energy saving and increase in power efficiency have to contain list of target indicators in the field of energy saving and increase in power efficiency. At the same time special attention is paid just public to transport as regulation of tariffs for transportation services belongs to competention of the territorial subject of the Russian Federation, the municipal unit.

6 DISCUSSION

Importance of development of the market of gas motor fuel is noted at the state level. Perspective reference points of development of this market are defined in the Energy strategy of Russia. It is specified that optimization of account part of fuel and energy balance provides evolutionary improvement of structure of demand for energy carriers, including:

1. Speedup of motorization of economy and the population with increase in consumption of motor fuels rates by 1.2 times higher, than rates of the general energy consumption at wider use of substitutes of oil products.
2. Significant increase in use as motor fuel of the liquefied and compressed natural gas (in equivalent up to 10-12 million tons in 2020).

In general the regulations existing in Russia, stimulate wide circulation of gas motor fuel insufficiently. Realization of the integrated approach considering the interests of all parties including suppliers of gas, owners of the CNG filling station, consumers of motor fuel, first of all the population is necessary. It is necessary to create the conditions allowing to increase the park of motor transport working at alternative types of motor fuel, to develop network of fuel stations, the accompanying infrastructure.

On the other hand the "methane" direction in Russia still has not gained the due development corresponding to potential opportunities of methane as environmentally friendly motor fuel. Such development is caused by insufficient technological level, small cost efficiency of the proposed solutions and, respectively, low competitiveness of gas cars in comparison with the cars using liquid oil fuels.

For the successful solution of problems on use of gas motor fuel coordination of efforts of all interested parties is necessary: gas workers, transport workers, producers of the gas-balloon equipment and automotive equipment, legislative and executive branches of the power of territorial subjects of the federation.

7 CONCLUSION

The further analysis is planned to continue assembled statistical data for search of new significant indicators, for the purpose of identification of acceptance of necessary measures for increase in number of cars on the gas-balloon equipment and developments of strategy of implementation of this program at the level of the country.

Intermediate conclusions of work are:

1) the program of promoting of NGV among the population is necessary;
2) the analysis and introduction of NGV in the new directions, except the motor transport is necessary (agricultural, public transport);
3) the state has to develop methods of stimulation of the market of gas motor fuel and consumers for transition to use of NGV.

REFERENCES

An article about communal, construction, road construction equipment[Stat'ya o kommunal'noy, stroitel'noy, dorozhno-stroitel'noy tekhnike]http://os1.ru/article/4345-vidy-gazomotornogo-topliva-ih-preimushchestva-i-nedostatki

Gordiyenko D.M., Malkin V.L., Ilyichev V.V., Smirnov E.V. 2015, *Features of fire hazard of stations with availability of compressed natural gas. Fire safety. №1.S* 60-62.

Gryaznov M.B.,2015 *Application of gas-motor fuel in the russian federation: problems and prospects. Bulletin of the Financial University, № 4 S 21-31*

Kirillov N.G. 2006. *Natural gas as a motor fuel [Prirodnyi gaz kak motornoe toplivo] Neft'gazpromyshlennost'. No 2. S. 40-65.*

Kotov D.V., Urazmetova L.R., 2014 *Current state, prospects*

and problems of development of the gas motor fuel market in Russia. Oil and gas business: online scientific magazine. №1 S 377-396

Natural gas as a motor fuel [Prirodnyi gaz kak motornoe toplivo] http://www.pro-gas.ru/fuel/news_bibl/10.htm.

OAO «Gazprom gazenergoset'».http://www.gazpromlpg.ru/index.php?Id=209

OAO «Gazprom Informatorii». What is compressed (compressed) natural gas[Chto takoe komprimirovannyi (szhatyi) prirodnyi gaz] http://www.gazprominfo.ru/articles/compressed-natural-gas/

Results of the Creon Energy Conference "SPG and CNG 2012" [Itogi Konferencii creonenergy «SPG i CNG 2012»]. http://neftegaz.ru/forum/showthread.php?Tid=5635.

Russia's energy strategy for the period of 2030[Energeticheskaya strategiya Rossii na period 2030 goda] http://www.atominfo.ru/files/strateg/strateg.htm

Russian Federation Federal State Statistic Services http://gks.ru

Sbornova E.R., Kolmakov V.S. 2012, *Fuel of the future [Toplivo budushego]. Gazovyi biznes №8.S.* 47-51.

Stadnik M. E., 2013. Negative impact of components of the transport system on state of environment. *Scientific dialogue.* № 12 (24) S. 38-47

Youth Technical Sessions Proceedings – Litvinenko (ed)
© 2019 World Petroleum Council, London, UK, ISBN 978-0-367-34668-3

Development of numerical methods for assessing the effects of tax reform in the Russian oil industry

E.M. Miller
The Russian Presidential Academy of National Economy and Public Administration, Moscow, Russia

ABSTRACT: Some positive results have been achieved since a tax reform was kicked off in the Russian oil sector late in 2014. The 2014–2017 period saw an increase in the average depth of oil refining in Russia. Despite ongoing disputes, now is a good time to complete the so-called "tax maneuver" and to create effective incentives for Russian oil refineries to enhance their operative efficiency. The key risk here lies in specific parameters of the reform, particularly in the amount, period and distribution mechanism of oil refineries subsidization. The mechanism does not work correctly, because it does not take into account the change in the structure of the market. The paper proposes a method that makes it possible to more effectively, in terms of public expenditures, determine the amount of subsidies needed by refineries to maintain their level of production and stimulus for modernization during the tax maneuver period.

1 INTRODUCTION

Over the past few years, a number of changes in the tax system have been carried out in the Russian oil sector. Its main goal was to increase investment in oil production, increase the utilization of refining capacity, and increase refining depth. However, today the configuration of the tax system involves the actual subsidization of the economy by undervalued oil prices (relative to world). Only part of the subsidy reaches the final consumer, while the other part remains in the oil industry (about 0.9% of GDP), which covers its inefficiency (Idrisov & Sinelnikov-Murylev 2012). The mechanism does not work, because it does not take into account the change in the structure of the market.

On 3.08.2018, final tax maneuver in the oil industry were adopted to complete the work begun:

- the Federal Act №301 introduces a correction factor that gradually reduces the rate of export customs duty on crude oil to zero in the period from 2019 to 2024;
- the formula for calculating the mineral extraction tax (MET) rate has changed since the beginning of 2019. A term is added to it, which increases the mineral extraction tax rate on the difference in export duties in 2018 and in the current tax period (Federal Act №305);
- an excise tax is imposed on petroleum raw materials owned by the taxpayer on the right of ownership and sent to them for processing. The Federal Act №305 provides for a tax deduction by a factor of 2 (reverse excise). Only refineries producing light oil products can get it;
- excise rates on petroleum products are adjusted upwards;
- additional income tax is introduced for a part of oil fields (including new ones) instead of MET (Federal Act №305).

Vertical-integrated oil companies (VIOCs) occupy a high percentage of the Russian oil market. VIOCs have production capacities at more than one stage of the value chain; therefore, their interaction with each other and other independent market participants is quite specific, for example, due to the existing restrictions related to the priority filling of their own capacities (refineries, storage facilities and gas stations).

It is interesting to compare two variant of calculation the final tax maneuver in terms of the cost of subsidizing the industry by the state. The first one does not taking into account the modernization programs for oil refinery and cost distribution of the VIOCs over the entire production chain, the second one do it (Kaukin & Miller 2017). It is assumed that the second option is more efficient in terms of the ratio of the benefits and costs of the state (Zhemkova et al 2019). Consequently, the main aim of the paper is modeling of the Russian oil market, taking into account the economy of individual refineries and changes in the tax regulation of the market.

2 MODEL

Oil product supply chain is large, complex and composed of several sub-problems. There are many works considering oil and oil products markets separately, but it leads to a non-optimal solution (Sahebi et al 2017). This study uses a new multi-period, multi-product and multi-transportation integrated oil product supply chain model of Amin-Naseri & Morandi Nasab (2016). The main feature of this model is to design an integrated supply chain model that considers both installation and capacity expansion of infrastructure and production facilities. For example in similar works (Neiro & Pinto 2004, Ghatee & Hashemi 2009, Yang et al 2010, Tong et al 2011) did not take into account it.

Due to data limitations, the calculations were carried out using a simplified model configuration. The equations did not include storage steps and considered only one type of transport for each stream: for the flow from the field to the refinery - pipeline, from the refinery to the consumers - railway. The model considered two types of petroleum products (light and dark), and it was also assumed that each company optimizes profits regardless of decisions made by other VIOCs and independent companies, since due to high and difficult to overcome entry barriers, exit costs, investment costs, this share for many years is a constant.

The objective function was the profit function (see equation 1), which is equal to the proceeds from oil exports, oil products exports, and the sale of oil products on the domestic market (see equation 2), minus the costs of oil production, oil refining and transportation (see equation 3). The costs of production and processing are individual for each oil and gas complex and refinery, respectively.

From the above we note that profit:

$$P = RF - CF \tag{1}$$

where RF = revenue function; CF = costs function, or we can write in detail:

$$RF = \sum_{p\in P}\sum_{m\in M}\sum_{t\in T} gm_{pm}^{t} PP_{p}^{t} + \sum_{p\in P}\sum_{t\in T} Ep_{p}^{t} EPP_{p}^{t} + \sum_{t\in T} Eco^{t} EOP^{t} \tag{2}$$

where gm_{pm}^{t} = volumes from the sale of petroleum products p to consumer m (domestic consumer) at time t, PP_{p}^{t} = prices of petroleum products p at time t, Ep_{p}^{t} = volumes from the sale of petroleum products p to export at time t, EPP_{p}^{t} = export prices of petroleum products p at time t, Eco^{t} = volumes from the sale of oil to export at time t, EOP^{t} = export prices of oil at time t.

$$\begin{aligned} CF = &\sum_{k\in K}\sum_{t\in T} qo_{k}^{t} ec_{k}^{t} + \sum_{n\in N}\sum_{t\in T} qp_{n}^{t} rc_{n}^{t} + \sum_{t\in T}\sum_{k\in K}\sum_{n\in N} ttube_{n}^{t} \\ &+ \sum_{t\in T}\sum_{n\in N}\sum_{r\in R} trail_{n}^{t} \end{aligned} \tag{3}$$

where CF consist of ec_{k}^{t} = development costs on oil field k at time t (qo_{k}^{t} = volumes of oil production,); rc_{n}^{t} = processing costs on oil refinery plant n at time t (qp_{n}^{t} = volumes of oil refinery), $ttube_{n}^{t}$ = oil transportation costs from oil field k to on oil refinery plant n at time t; $trail_{n}^{t}$ = petroleum products transportation costs from oil refinery plant n to region r at time t.

3 ANALYSIS RESULT

Open data sources were used for calculations. The volume of reserves and production of crude oil, unit costs for exploration and production, volumes of primary refining and production of petroleum products, unit costs for oil refining, unit costs for transporting oil to refineries, export volumes oil and oil products, oil prices (internal and external) were taken from annual reports of vertical-integrated companies. Transport matrixes between oil and gas basins and oil refineries, as well as between oil refineries and the region were compiled using information from analytical materials of consulting companies. Current and forecasted rates for the mineral extraction tax, export duties and excises from the Tax Code and the Tariff Act. The cost of transportation by rail was calculated using the freight base of Russian Railways.

As an example, let us consider the calculations for the company Lukoil, all of its mining and production facilities are presented in Figure 1. The company has facilities in three oil and gas basins (Timan-Pechora, West Siberia, Ural and Volga) and four refineries (Ukhta, Nizhny Novgorod, Perm and Volgograd). The ability to transport crude oil from the basin to the refinery is represented by blue solid lines. The company supplies domestically to the following regions: Astrakhan Oblast, Volgograd Oblast, Vologda Oblast, Vladimir Oblast, Kirov Oblast, Sverdlovsk Oblast, the Republic of Chuvashia, the Republic of Mari-El, and the Republic of Bashkortostan. Blue dotted arrows represent oil exports, and red dotted arrows represent oil products.

Solving the linear optimization problem for the Lukoil, the objective function of which was described above, gave optimal volumes of crude oil production at each oil and gas basin, exports and domestic consumption of oil, oil products production at each refinery, exports and domestic consumption of oil products by type. The calculations were carried out taking into account changes in the depth of refinery processing and customs and tax parameters in 2019-2024.

Next, it was necessary to calculate the indicator for which the comparison of calculation options was carried out. In the paper, it was the gross value added of the refinery, which was calculated as the cost of oil produced minus the cost of incoming oil. All refineries were divided into three groups according to the volume of available capacity for oil refining: large, medium and small. For all, the GVA was calculated for the maximum possible period of time, in this case it was 2013-2019. For each group, a minimum of average GVA was taken. These values were compared with the forecast values of the GVA calculated for the Lukoil refinery without taking into account the modernization programs and the structure of the production of the VIOCs and considering them.

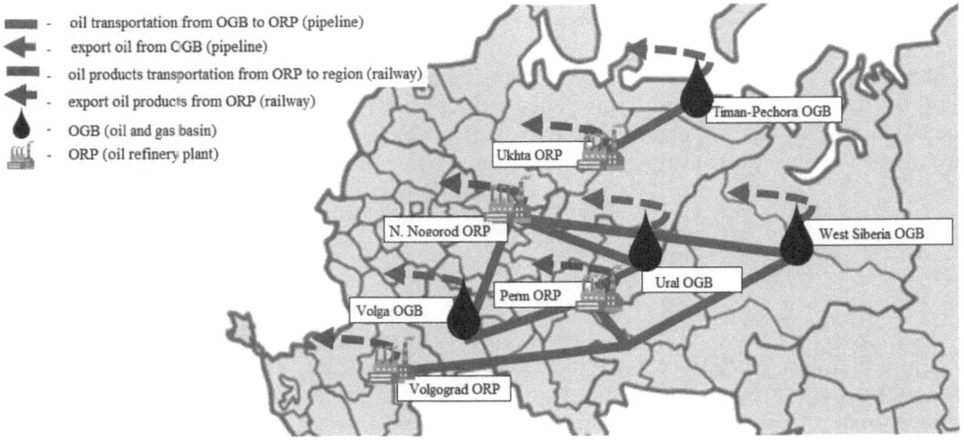

Figure 1. Structure of Lukoil's oil production supply chain

Figure 2 shows two options for calculating the required subsidies to refineries so that have GVA higher than the minimum value for the group during the period of the final tax maneuver (upper left chart and lower chart). The upper right chart shows that taking into account the structure of VIOCs gives a significant increase in efficiency in the calculation of subsidies.

Table 1 presents the amounts of subsidies calculated taking into account the modernization of facilities and the redistribution of costs, as well as its share of subsidies that do not take into account these parameters (calls it S1 in table 1).

Calculations demonstrate the possibility of the model were carried out according to all refineries and oil complexes of the Lukoil. The optimization problem solution have shown that taking into account the modernization of production capacity and profits (costs) generated by the company at past (subsequent) production stages leads to more efficient, in terms of the state budget, industry subsidies. It can be reduced subsidy by about 15% for the Lukoil (market share is 20%).

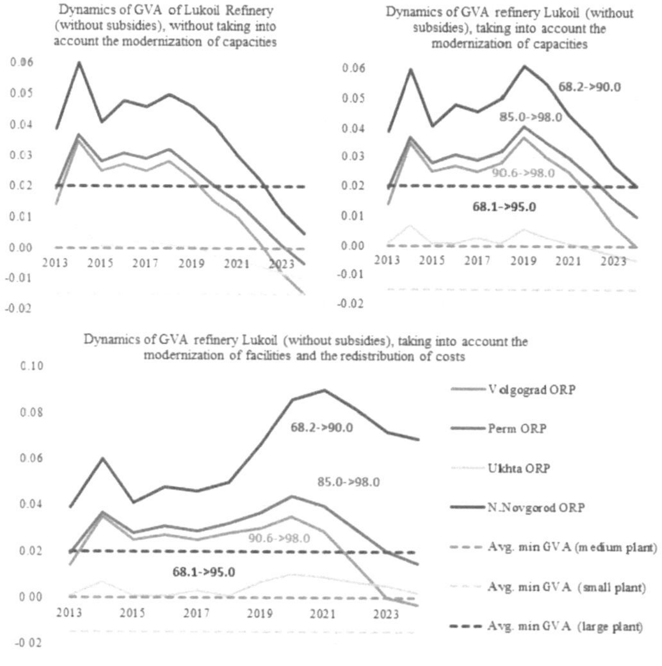

Figure 2. The forecast of GVA ORPs without subsidies

Table 1. Subsidies of oil refinery plant in 2019-2024

Oil Refinery Plant	Subsidies, trillion rubles 2019	2020	2021	2022	2023	2024
N. Novgorod ORP	0.01	0.02	0.02	0.01	0.00	0.03
Ukhta ORP	-0.00	-0.01	-0.01	-0.01	-0.01	-0.01
Volgograd ORP	0.00	0.00	0.00	-0.01	-0.01	-0.01
Perm ORP	0.00	0.00	0.00	0.00	-0.01	-0.01
The required subsidy	0.00	0.01	0.01	0.02	0.03	0.03
Share of S1 subsidy	0.00	86.98	83.33	80.73	80.37	82.83

4 CONCLUSION

The results of the optimization problem calculations for the Lukoil company's value chain showed that taking into account the modernization of production facilities and profits (costs) generated by the company at past (subsequent) production stages leads to more efficient budgets for the refining industry from the point of view of the state budget. The calculated subsidy, which does not take into account the change in the production basket and the optimization of the entire production chain, can be reduced by about 15% compared with calculations involving these improvements. A proven algorithm for calculating the required subsidies for Lukoil's companies can be applied to the entire industry with more detailed data from other companies, which will make it possible to more effectively adapt the parameters of the tax system to changing market structure without creating conditions for subsidizing inefficiency.

To realize this goal, the project "Development of a spatial and technological atlas of the energy system of the Russian Federation" has now been launched in RANEPA. The goal is to obtain data on all objects, including various types of power plants, developed sources of fuel and energy minerals, powerful resources for their processing, and taking into account data on the technologies used.

ACKNOWLEDGEMENTS

The work was performed in accordance with the state task of the RANEPA for 2018 in the research area "Energy, infrastructure industries, ecology, safety and information technologies", No. AAAA-A18-118013190243-1.

REFERENCES

Bobylev Yu. N. & Turuntseva M. Yu. 2010 Taxation of the mineral and raw materials sector of the economy. *Scientific works 140R of Inst. politicians them. E.T. Gaidar.*

Ghatee, M. & Hashemi, S. M. 2009 Optimal network design and storage management in petroleum distribution network under uncertainty. *Engineering Applications of Artificial Intelligence* 22(4-5): 796-807.

Idrisov G. I. & Sinelnikov-Murylev S. G. 2012 Modernization or conservation: the role of export duties on oil and petroleum products. *Economic Policy* 3: 5-19.

Federal Act No. 301-FZ "Concerning Amendments to Part II of the Tax Code of the Russian Federation" dated 3 August *2018* [O vnesenii izmineniy v chast' 2 Nalogovogo kodeksa Rossijskoj Federacii], viewed 18.03.2019, http://publication.pravo.gov.ru/Document/View/0001201808040008

Federal Act No. 305-FZ "Concerning Amendments to Article 3.1 of the Federal Customs Tariff Act" dated 3 August *2018* [O vnesenii izmineniy v stat'u 3.1 zakona Rossijskoj Federacii "O tamozhennom tarife"], viewed 18.03.2019, http://publication.pravo.gov.ru/Document/View/0001201808030078

Forecast of the long-term socio-economic development of the Russian Federation for the period up to 2030 [Prognoz dolgosrochnogo sozial'no-ekonomicheskogo razvitiya Rossijskoj Federacii na period do 2030 goda], viewed 18.03.2019, http://economy.gov.ru/minec/activity/sections/macro/prognoz/doc 20130325_06

Kaukin A. S. & Miller E. M. 2017 Tax maneuver in the oil industry. *Economic development of Russia* 10 (24): 60-64.

Moradi Nasab N. & Amin-Naseri M. R. 2016 Designing an integrated model for a multi-period, multi-echelon and multi-product petroleum supply chain. *Energy* 114: 708-733.

Reporting companies PAO "LUKOIL", viewed 18.03.2019, http://www.lukoil.ru/InvestorAndShareholderCenter/RegulatoryDisclosure/AnnualReport

Neiro, S. M. & Pinto, J. M. 2004 A general modeling framework for the operational planning of petroleum supply chains. *Computers & Chemical Engineering* 28(6-7): 871-896.

Sahebi H., & Nickel S., & Ashayeri J. 2017 Strategic and tactical mathematical programming models within the crude oil supply chain context—A review. *Computers & Chemical Engineering* 68: 56-77.

Tong K. & Feng Y. & Rong G. 2011 Planning under demand and yield uncertainties in an oilsupply chain. *Ind Eng Chem Res* 2(51): 814-834.

Yang J. & Gu H. & Rong G. 2010 Supply chain optimization for refinery with consider-ations of operation mode changeover and yield fluctuations. *Ind Eng Chem Res* 1(49): 276-287.

Zhemkova A. M. & Idrisov G. I. & Kaukin A. S. & Miller E. M. 2019 Freezing prices for oil products - stopping the tax maneuver? *Economic development of Russia* 1(26): 31-35.

Youth Technical Sessions Proceedings – Litvinenko (ed)
© 2019 World Petroleum Council, London, UK, ISBN 978-0-367-34668-3

Analysis of cluster initiatives in the energy sector of the EU

I. Meshkov
IMEMO RAN, Moscow, Russian Federation

ABSTRACT: European Union (EU) countries strive towards increasing their competitiveness and securing energy supplies by exploiting renewable sources of energy and reducing their impact on the environment. Innovation and regional clusters are one of the lenses through which energy transition in Europe and SMEs activity in the sector may be analysed. The article focuses on overviewing the 84 existing cluster organisations (COs) in the EU that specialise on the energy sector. Research has shown that such clusters exist in most EU countries and are growing in number, serving as platforms for communication and knowledge sharing between different stakeholders and focusing mostly on energy efficiency and renewable energy technologies. Such cluster organisations play an important role in energy transition projects in their regions, whilst the potential of other cluster initiatives still needs to be untapped.

1 INTRODUCTION

The topic of energy transition has become very popular in recent years, especially in countries dependent on the import of energy sources, as well as the EU. There are different approaches to defining energy transition, including the analytical assessment of historical shifts in energy systems (Smil 2005, 2010), the multi-level perspectives approach (Verbong and Geels 2007) and the application of a geographical framework (Bridge et al. 2012). While the first and the second approaches may provide a global overview of current trends in the energy sector, the geographical framework focuses on the spatial patterns of economic activity. The studies and common sense suggest that a single energy transition does not exist across Europe, as its nature is determined by local and national circumstances (Bouzarovski and Herrero, 2017).

Switching energy systems from fossil fuels to renewable energy sources (RES) requires financial, technical and social innovation (Lowitzsch 2019). It also requires new types of partnerships and engagement between businesses, consumers, governments and research institutions. One can analyse how this engagement is happening by looking at innovational ways regional clusters are developing.

We use classic definitions for clusters and cluster initiatives. Clusters are geographic concentrations of interconnected companies and institutions in a particular field (Porter, 1998). Cluster initiatives are organised efforts to increase growth and competitiveness of clusters within a region, involving cluster firms, government and/or the research community (Cluster Initiative Greenbook, 2003). Clusters become relevant for innovation when a critical mass of companies is achieved in a location or a sector and the companies go on to support each other through direct and indirect sharing of knowledge, skills, technologies, infrastructure etc. Cluster initiatives are increasingly managed by specialised institutions, known as cluster organisations, which take various forms, ranging from non-profit associations and public agencies to companies (EC Communication, 2008). Cluster initiatives and cluster organisations are institutions which focus on increasing interaction between actors in order to overcome network and collaboration failures.

The work not only presents an analysis of existing cluster organisations which coordinate the work of companies from the energy sector in EU regions, but also gives insight on how cluster initiatives may contribute to achieving the goal of energy transition in EU countries.

2 LITERATURE REVIEW

Despite a wide range of existing papers on cluster initiatives and organisations (Cluster Initiative Greenbook and Cluster Initiative Greenbook 2.0) and specific regional clusters in the energy sector (Foley et al 2015; Virkkala et al. 2014; Elola et al. 2012; Volkov 2012) there is little research on the overview of cluster initiatives in the energy sector as well as the relation between cluster initiatives and the process of energy transition (McCauley, Stephens, 2012; De Laurentis et al., 2014).

One of the main roles of cluster initiatives is to be an intermediary between local companies (Laur et al., 2012). As efficient cluster initiatives mobilise public efforts to improve the business environment and capture spillovers (Porter 2008) they can be compared with so-called "grassroots initiatives", also players in the renewable energy sector (Magnani and Osti, 2016; Kooij et al., 2017). The role of consumers' ownership in renewable energies is also considered important (Lowitzsch et al. 2019).

An overview of historical energy shifts can be found in Kander et al. (2014) and an overview of the current situation in the EU energy sector is provided in statistical materials such as "EU Energy in figures. Statistical pocketbook 2018" as well as in IRENA and REN21 reports. A broader review of the energy transition in European and developing countries has also been provided in various literature (Leach 1992; Gales et al. 2007; Droege, 2011) as well as how to better tackle the challenges behind energy transition (Pellerin-Carlin, 2017; Verbong and Loorbach, 2012). The different policy paths in energy transition that can be followed on the local level have also been analysed but mostly based on several case-studies (Emilianoff, 2013). A different view on the energy sector may also be explained by a shift in scientific paradigms away from single-sector thinking to a coherent energy system understanding on how to benefit from the integration of all sectors and infrastructure, as well as by the introduction of the so called Smart Energy Systems (Lund et al., 2017). The current paper builds on the analysis of cluster organisations in the EU, touching upon the broader topic of Smart Energy Systems development.

3 METHODOLOGY

One of the ways to analyse cluster initiatives is to look into the data of cluster organisations, which are institutional representations of such initiatives.

The task was approached by classifying the cluster organisations (CO) with companies engaged in the energy sector (Electric Power Generation and Transmission in specific) in the EU. The databases of the European clusters collaboration platform (ECCP) and European Cluster Observatory (ECO) were used in the study. It includes information on the year of creation, region, number of companies (SMEs and large ones), areas of specialisation, connection with local research institutions, cluster's partners, CO's strategy and other data.

A comparative analysis is also done between the existence of a CO in a region and local efforts to foster energy transition. The information on these efforts has been gathered taking into consideration public statements of regional authorities or existing regional strategic documents.

4 ANALYSIS

4.1 *Results of the cluster organisations overview*

Analysis of cluster development on the local/regional level provides a better understanding of the economic forces behind the energy transition idea.

Analysis based on the data of ECCP and ECO has shown that there are about 84 COs focusing on the energy sector of the EU. Such COs exist in 21 EU countries, with the biggest number in Spain (16), Germany (8) and France (7).

Taking into account the total number of EU regions (276 regions under NUTS 2 classification), the number of COs with energy focus may be considered small, but there is positive dynamic in growth of such initiatives in recent years. Most cluster organisations were founded in 2008-2017, with several peaks in 2008 (10 organizations) and 2011 (11 organisations).

About 80% of the clusters focus on energy efficiency and renewables energy technologies, whilst there are only 3 clusters that focus on fossil energy sources. Several clusters focus on automotive and maritime industries, with a small portion of companies engaged in energy activities.

More than the half of the clusters have a share of SMEs beyond 60%, with the total number exceeding 7 thousand companies. There are several transnational players of European origin which are active in supporting clusters – EDF, General Electric, Siemens Gamesa, Iberdrola, Schneider Electric, ABB and others. These companies tend to mostly participate in clusters located in the same place of origin – Iberdrola is present in 6 out of 16 Spanish COs, EDF is present in 4 out of 7 French COs and ABB is present in 2 out of 4 COs in Sweden.

Almost 60% of cluster organisations have local universities, research centres and R&D laboratories among stakeholders. They serve not only as a source of innovation and expert knowledge, but also as providers of specialised skills.

An interesting point is that 81% of COs have a defined development strategy with their focus, in most cases, on local companies' competitiveness growth, communication/collaboration between the members of the cluster (which comprise local authorities, research centres, big investors) and shared learning activities. Cluster managers provide additional opportunities to network with other local actors which is an important step to knowledge spillovers and partnerships. Lack of common strategy is one of the problems of "grassroot" initiatives in the community energy sector, which faces pressure from bigger institutions and players (Smith et al., 2013). Cluster initiatives in this case may provide necessary context for joint action.

A broader smart energy concept may encompass interaction between different cluster organisations. 67% of analysed COs have at least one other partner - cluster organisation, whilst many of them have more than 5 partner-organisations, including clusters initiatives in other sectors (from agriculture to automotive industry). Interconnections between clusters is a positive sign of possible knowledge spillovers not only within clusters, but on a broader transnational scale.

4.2 *Relationship of energy clusters with the local energy transition initiatives*

While 36 out of 84 regions where analysed cluster initiatives are located have directly or indirectly announced energy transition strategies or targets, only a few have a strong commitment to achieve 100% energy production from RES in the coming 20-30 years (Catalonia, Tirol, Lüneburg, Munster, Oslo, Andalucia Smart City, Grenoble, Northern German Energy Transition project). Barcelona ("capital of Catalonia") and Oslo were ranked amongst the top 5 global smart cities, according to Juniper Research in 2015 (Juniper, 2015). Cluster organisations in these regions have been active for the past 5 years and have a defined strategy, a large share of SMEs and include several companies-leaders in the industry.

Several regions (North eastern Scotland, Western Macedonia, Principado de Asturias, Extremadura and others) have embraced a "Just transition" idea, which implies a fair transition for workers and communities currently dependent on jobs in high carbon sectors. Germany is a vivid example of a policy shift from coal as the last coal mine of the country was closed in 2018 (while open-pit mining is continuing). After making this milestone decision in 2007, the State of North Rhine-Westphalia has focused on the development of emerging business clusters (Taylor, 2015). Since 2010 a project of efficient urban redevelopment, known as InnovationCity Ruhr – Modellstadt Bottrop, has been in place in Bottrop. Nevertheless, amongst the 41 regions in EU where coal is mined, only 8 have one or several COs.

The amount of regions suggests that country-specific conditions and policy framework require member state policies to play a leading role in fostering an energy transition (Creutzig et al., 2014). Leadership on the regional and city level has resulted in the adoption of the EU

Covenant of Mayors for Climate & Energy in 2008, which up to now has united more than 7700 local and regional authorities from 57 countries voluntarily committed to implementing EU climate and energy objectives. The sharing of "good practices" and visions on fulfilling common climate goals have resulted in better bottom-up interaction between actors and faster target achievement. Nonetheless, many cities still fail to see smart city programs as part of their long-term, comprehensive development plan and, consequently, they do not engage in methodological strategic planning (Angelidou, 2016).

EU-level programmes such as the European Innovation Partnership on Smart Cities and Communities (EIP-SCC) may be good examples of supranational support of local activities, although the direct impact of this support should be additionally studied.

5 DISCUSSION

The contribution of this work is in the overview of the ECCP data on cluster organizations, which are responsible for managing cluster initiatives in the EU energy sector. Regional level initiatives are important to consider when looking into paths of energy transition in EU countries. In this regard, composition of clusters, the way COs manage them, the role of biggest companies as well as other factors represent the available "soft" infrastructure, which, in the context of energy transition, may be considered even more important than the "hard" infrastructure.

As several examples show, cluster organisations may play significant role in bringing together different stakeholders involved in the regional development of a particular sector. Ambition of authorities on the EU, national, regional and city level to follow common targets and transform energy systems by creating a so called Energy Union (Energy Union package, 2015) should be a continuation of already existing bottom-up activities.

Further research on the relation between the existence of a regional cluster, cluster organisations specialising on the energy sector and energy mix of the region may deepen understanding of clusters' influence on regional energy transition. Efficiency of the cluster organisations may also be tested by considering the number of concluded projects, companies helped, networking events organised and local impact

6 CONCLUSION

Regional clusters with activity focusing on the energy sector should be taken into account whilst analysing energy transition in the EU. Being live models of innovation systems, clusters unite SMEs, research institutions, local authorities and investors, providing the necessary framework for them to focus on specific goals, important for a particular community. Overview of the existing cluster organisations in the energy sector has shown that they exist in most EU countries, serving as platforms for communication between different stakeholders and knowledge sharing, whilst focusing on energy efficiency and renewable energy technologies. As a further step, cluster initiatives have the potential to turn into regional or city level multi-stakeholder energy transition projects, vital for reshaping national energy systems.

The analysis of cluster initiatives in the energy sector has shown, that the existence of cluster organisations in a region doesn't necessarily result in a local joint action plan for energy transition. It is for this reason that the potential of clusters initiatives as platforms for communication and interaction is not fully exploited.

Most COs have a strategy focused on the support of local actors as well as a wide partnership network of COs that specialise on energy and other industries. The experience of several clusters proves that they can be adequate platforms for implementing "smart city" concept or supporting renewable energy projects in communities, which are considered as a middle way between individualism and supra-individual agencies (governments/corporations) (Aiken, 2012). Taking abovementioned factors into consideration, cluster initiatives may become "hotspots" on the EU energy transition map.

REFERENCES

Aiken, G. 2012. Community transitions to low carbon futures in the transition towns network (TTN). *Geography Compass*, 6(2),89-99.

Angelidou M. 2016. Four European Smart City Strategies. *International Journal of Social Science Studies*, vol. 4, No. 4.

Bouzarovski S., Herrero S.T. 2017 The energy divide: Integrating energy transitions, regional inequalities and poverty trends in the European Union. *European Urban and Regional Studies*, Vol. 24(1), pp. 69-86.

Bridge G., Bouzarovski S., Bradshaw M., Eyre N. 2013. Geographies of energy transition: Space, place and low-carbon economy. *Energy Policy*, 53, pp. 331-340.

Creutzig, F., Goldschmidt, J. C., Lehmann, P., Schmid, E., von Blücher, F., Breyer, C., ... & Susca, T. 2014. Catching two European birds with one renewable stone: Mitigating climate change and Euro-zone crisis by an energy transition. *Renewable and Sustainable Energy Reviews*, 38, 1015-1028.

De Laurentis, C., Morgan, K., & Eames, M. 2014. The role of regions in low carbon transitions dynamics: the 'Greening'of Puglia Region. In 9th International Conference on Regional Innovation Policies.

Droege, P. 2011. Urban energy transition: From fossil fuels to renewable power. *Elsevier*.

EC Communication. 2015. Energy Union Package. A Framework Strategy for a Resilient Energy Union with a Forward-Looking Climate Change Policy.

Emelianoff, C. 2014. Local energy transition and multilevel climate governance: the contrasted experiences of two pioneer cities (Hanover, Germany, and Växjö, Sweden). *Urban Studies*, 51(7),1378-1393.

European Commission. 2018. EU energy in figures.

Gales, B., Kander, A., Malanima, P., & Rubio, M. 2007. North versus South: Energy transition and energy intensity in Europe over 200 years. *European Review of Economic History*, 11(2),219-253.

Hansen T. and Coenen L. 2013. The Geography of Sustainability Transitions: A literature review, pp. 1-24. *Centre for Innovation, Research and Competence in the Learning* Economy, *Lund University*, Lund.

Hargreaves T., Hielscher S., Seyfang G., Smith A. 2013. Grassroots innovations in community energy: The role of intermediaries in niche development. *Global Environmental Change*, Volume 23, Issue 5, pp. 868-880.

Juniper Research (2015) Smart cities // press release available at: https://www.juniperresearch.com/press/press-releases/barcelona-named-global-smart-city-2015

Kander, A., Malanima, P., & Warde, P. 2014. Power to the people: energy in Europe over the last five centuries (Vol. 46). *Princeton University Press*.

Laur I., Klofsten M., Bienkowska D. 2012. Catching Regional Development Dreams: A Study of Cluster Initiatives as Intermediaries. *European Planning Studies*, 20:11, 1909-192.

Lindqvist, G., Ketels, C., & Sölvell, Ö. 2013. The cluster initiative greenbook 2.0. *Ivory Tower Publishers*.

Lowitzsch J. (ed.) 2019. Energy Transition: Financing Consumer Co-Ownership in Renewables. *Palgrave macmillan*.

Lund, H., Østergaard, P. A., Connolly, D., & Mathiesen, B. V. 2017. Smart energy and smart energy systems. *Energy*, 137, 556-565.

Magnani, N., & Osti, G. 2016. Does civil society matter? Challenges and strategies of grassroots initiatives in Italy's energy transition. *Energy Research & Social Science*, 13, 148-157.

McCauley S. M., Stephens J.C. 2012. Green energy clusters and socio-technical transitions: analysis of a sustainable energy cluster for regional economic development in Central Massachusetts, USA. *Sustainability Science*, Vol. 7, Issue 2, pp 213-225.

Porter M., Clusters and the New Economics of Competitiveness. 1998. *Harvard Business Review*, November - December issue

Sölvell, Ö., Lindqvist, G., & Ketels, C. 2003. The cluster initiative greenbook. *Stockholm: Ivory Tower*.

Smil, V., 2005. Energy at the Crossroads: Global Perspectives and Uncertainties, *MIT Press*.

Smil, V., 2010. Energy Transitions: Histories, Requirements, Prospects. *Praeger Publishers*.

Taylor R.P. 2015. A review of industrial restructuring in the Ruhr Valley and Relevan points for China. *Institute for Industral Productivity*, case study.

Verbong, G., Geels, F. 2007. The ongoing energy transition: lessons from a socio-technical, multi-level analysis of the Dutch electricity system (1960–2004). *Energy Policy* 35 (2), 1025–1037.

Verbong, G., & Loorbach, D. (Eds.). 2012. Governing the energy transition: reality, illusion or necessity? Routledge.

Volkov A.N. 2012. Cluster Policies in Energy Efficiency Management in Regional Innovative Strategy of Sustainable Development. *European researcher*, Vol. 32, №10-2.

Youth Technical Sessions Proceedings – Litvinenko (ed)
© 2019 World Petroleum Council, London, UK, ISBN 978-0-367-34668-3

Long-term development of Russian refineries based on the most probable scenarios of energy markets evolution

G. Oganian & B. Kharchenko
JSC Gazpromneft Moscow Refinery, Moscow, Russian Federation

ABSTRACT: Existing Russian oil refinery industry limited with strategy development until 2025-2035. During the years of modernization,there will be inevitable changes in the energy markets,especially with dynamically developing sources of alternative energy. New revamping of refinery plants will be required to face market demand. Nevertheless, existing long-term outlooks in Oil and Gas field focused on changes in energy market.In this paper, we suggest approach and research of future efficient refinery configuration based on expected market scenarios. This is done through systematization key factors essential for the development of an oil refinery, modeling scenarios of energy market evolutions and refinery configurations, selection of configurations to developed scenarios. Based on this study it is possible to select refinery vector of development in long-term prospects.

1 INTRODUCTION

The modern Russian oil refining industry is going through a stage of bulk modernization. Largest domestic players have implemented strategic development programs until 2025–2035. Gazpromneft is one of the leading Global Oil companies and its Moscow Oil Refinery has implemented the intensive development program until 2025 accompanied by a selection of further development model up to 2035. As the Moscow Refinery can be considered as a benchmark at least in the Eastern and Central Europe it is assumed that in general there is no well-detailed development program going further then 2025-2035.

Then, most likely, after extensive modernizations there will take place a phase of about 10 years of stable and profitable exploitation. In the meantime, according to the industrial outlooks, the share of non-traditional petroleum fuels will increase significantly (up to 50%) by the middle of the 21st century. These changes, in combination with the likely increase in energy efficiency of vehicles and the growing popularity of efficient small cars, of course, will affect the oil refining industry. These and other markers indicate that the structure of demand for petroleum products will change, which will inevitably trigger new phase of revamping of existing refinery schemes.

However, even the most long-term forecasts do not give well-defined recommendations for further development of oil refineries. In addition, in a marketplace where alternative sources of energy are rapidly developing, the challenge comes from a different angle: traditional fuels will lose dominance, so it is necessary to have an understanding of the future market and plan the vector of development of oil refining companies in the long term.

Thus, the objective of the research is to develop methodology of long-term projection of energy markets development impact on oil refining industry and development of some specific recommendations for the future development of domestic refineries. The planning time-frame until 2050 was selected, since energy consumption in the modern world is restructuring dynamically, and long-term refinery development strategies does not go beyond 2035. At the same time, developing an outlook with such a precision for more than 30-35 years seems inappropriate, since changes on such a far horizon are hard to predict in the context of a particular sector of the oil and gas industry.

2 LITERATURE REVIEW

Despite wide range of existing papers on different outlooks of future oil and gas market, there are no papers discussing prospects of refinery industry, especially approximate set of units to fit market needs from one side and maximize margin from another side. The existing literature and outlooksof oil and gas companies and agencies (among them Outlook of development of International and Russian energy until 2040, Institute of Energy research of The Russian Academic of Science; International Energy Forum in partnership with Duke University, 2014; BP Energy outlooks 2035; Exxon Mobil, The Outlook for Energy: A View to 2040; OPEC, World Oil Outlook; Shell Energy scenarios 2050; StatOil Energy perspectives: Lon-term macro and market outlook; International Energy Agency, World Energy Outlooks: Global energy trends to 2040; Paul Scherrer Institute, World Energy Scenarious, Composing energy futures to 2050; International Energy Agency, Winners, loses and unintended consequences in the outlook for oil products demand, Tae-Yoon Kim, 2019) focuses on future market, aiming on prediction of demand for energy commodities. Since most of this outlooks and forecasts does not contain any specific information on technical side of refining industry and focus mainly on future energy market, this paper offersa new approach to making a specific projection of refinery industry.

3 METHODOLOGY AND MODELING

Methodology of this research can be broken down for several main steps, which are:

1. Studying of key industrial forecasts of the global energy business development and forming the key scenarios;
2. Detection of key factors having the most significant influence on refining business;
3. Modeling and economic assessment of various key refinery configurations in terms of elaborated scenarios.

First step of this methodology is analysis and systematization of key industrial forecasts for the global and Russian development of the fuel and energy business, designing of key scenarios:

- Russia's role in the future of the global energy system;
- Analysis of global primary energy consumption forecast data;
- Analysis of outlooks data on dynamics and structure of consumption of primary energy sources by the transportation sector;
- Analysis of outlooks data on dynamics and structure of consumption of primary energy sources by the petrochemical industry;
- Analysis of outlooks on the dynamics and structure of global oil and gas production;
- Analysis of outlooks for the development of the structure and depth of the global oil refining industry;
- Analysis of outlooks for the development of alternative energy.

Analysis of macroeconomic and microeconomic aspects of market conditions that have the most significant impact on oil refineries:

- Identification of the most significant factors affecting the oil refining industry;
- Methods for determining the cost of construction and operation and profitability of oil refineries;
- Development of a methodology for assessing the economic efficiency of various configurations of refineries for long-term planning.

Modeling:

- Digitization of factors essential for the development of an oil refinery;
- Building models of refineries;
- Evaluation and comparative analysis of the economic efficiency of refinery configurations under scenario conditions.

In order to analyze the efficiency of oil refining in various conjuncts of the forecasting market, due to the impossibility of accurately forecasting the state of the global economy in 2050, it was necessary to make several assumptions. This study does not take into account possible fluctuations in the markets and world energy systems, revolutionary leaps in the development of industry, drastic changes in the vectors of the evolution of world energy production and energy consumption. Moreover, the model, in principle, does not consider the system in dynamics.

Thus, based on the available data and hypotheses, a series of linear forecasts of the most likely scenarios for the development of the world energy system has been developed. Based on these forecasts, only the state at the time of 2050is analyzed.

To determine the values of the key parameters of the economic efficiency of oil refineries and the formation of scenarios for the development of the domestic oil refining industry, it is necessary to determine the parameters for the calculation and comparative analysis of the economic efficiency of various refineries configurations. As such, parameters were taken:

- The price of oil;
- The price of petroleum products;
- The price of auxiliary semi-products;
- The price of petrochemical products;
- dollar rate;
- Inflation rate.

All other parameters will be calculated based on the above parameters, as well as basic information obtained from various sources.

In this study, the development of scenarios for the development of the oil refining industry is aimed at determining the prices of petroleum products. Prices for oil products are determined for each scenario individually based on oil prices, demand and supply dynamics and current prices.

Building a model of an oil refinery consists of three main steps:

1. Construction of material balances of all plant installations. This part of the model allows to determine the products of the refinery before mixing the various streams into marketable products. In most cases, the dependencies of plant outputs on incoming raw materials are assumed to be linear. Nevertheless, some key factors, such as restrictions on secondary raw materials, hydrogen balance, etc., are taken into account during the construction of the model.
2. Blending. The so-called blending or compounding model is aimed to selectthe most optimal combinations of finished products, taking into account key specifications such as saturated vapor pressure, octane number and sulfur content for gasolines, density and viscosity for fuel oil, etc.
3. Economic evaluation and comparison of various configurations of oil refineries in various economic conditions and under certain market conditions.

4 DEVELOPMENT OF INDUSTRIAL SCENARIOS

For simplicity of analysis, all major products were combined into five groups based on the similarity of fractional composition, physical and chemical properties, and pricing principles:

1. Gasoline. This group of petroleum products includes gasolines with various octane numbers.
2. Distillates. This group of petroleum products includes jet fuel, diesel fuel for winter and interseasonal, marine low-viscosity fuel.
3. Raw materials of petrochemistry. This group includes polypropylene, aromatics and C4 processed products.
4. Heavy oil products. This group includes fuel oil, bitumen, coke. Dynamics of prices and demand for this group of petroleum products is not taken into account in this model, since their impact on the profitability of the refinery is insignificant.

5. The rest. This group includes all petroleum products not accounted for in the previous four groups and the dynamics of demand and prices for this group are not taken into account in the model, since their quantity, output and weight in the structure of the refinery's income are insignificant.

Based on a study of existing forecasts of the development of the domestic and world economy and energy systems, the following scenarios were developed:

1. Balanced (base): proportional growth in demand for all petroleum products.
2. Growth of petrochemistry, slowdown of gasoline and distillates: slowdown in the growth rate of demand for motor fuels, growth of consumption of petrochemical products;
3. The growth of petrochemistry, gasoline stagnation: an increase in the growth rate of demand for petrochemical products, a stagnation of gasoline demand; demand for distillates at the base level;
4. The growth of petrochemistry and distillates: an increase in the growth rate of demand for distillates and petrochemical raw materials; gasoline demand at the base level;
5. Transition to gas and alternative sources: reduction of oil production and gradual abandonment of oil as a raw material for the production of fuels and other consumer goods, stagnation in demand for all petroleum products.

Thus, summarizing the developed scenarios, we can conclude on the following hypotheses:

- The probability of a decline in demand and, as a result, a fall in prices (as compared to current ones) for any of the major groups of petroleum products is insignificant, and therefore not reflected in any of the scenarios;
- The probability of accelerating the growth rate of demand for the petrochemical group of petroleum products is estimated highly, therefore, reflected in three of the five scenarios;
- The probability of accelerating gasoline demand growth is assumed to be marginally low, the most optimistic option for gasoline is maintaining prices in accordance with the baseline scenario.

Growth in demand (consumption) in all scenarios, except for the base one, implies an acceleration in the growth rate of demand, which leads to an increase in prices for this group of petroleum products. In this scenario, a coefficient of 1.25 is applied to the base prices, which means a 25% increase in prices for these petroleum products as compared to the baseline scenario.

Slowdown is understood to mean a slowdown in growth rates and, as a result, a 25% reduction in prices for a given petroleum product or a group of petroleum products relative to prices in the baseline scenario.

Stagnation means stopping the growth of demand for this product or group of petroleum products, that is, demand remains at the current level, and prices are considered to be 2 times lower than the baseline.

If nothing is said about the dynamics of demand for petroleum products or a group of petroleum products, this means that their prices remain at the basic level.

The coefficients of 1.25, 0.75, and 0.5, which take into account growth, deceleration, and stagnation, respectively, are deliberately taken in oversized and unlikely sizes in order to increase the visibility of the model and simplify the comparative analysis of various scenarios and configurations. These coefficients are the cause of artificially high or low performance indicators of configurations of oil refineries under a number of scenarios for the development of the oil refining industry.In fact, the percentage of price changes depending on the dynamics of demand will be significantly lower.

As for the price of oil, its main drivers, like any product, are supply and demand. However, since it is the most traded commodity in the world, the speculative factor plays an important role in the formation of oil prices. That is, it is difficult to predict the price of oil in such a long term.Nevertheless, based on the Energy Information Administration (EIA) chart, it can be concluded that the oil price quoted for the 2018 dollar varied in the corridor from $ 20 to $ 40 per barrel for almost 40 years (from 1966 to 2004) . Then there was a period of growth of the world economy and oil consumption, which ended with the world economic crises of 2008 and 2014. During this period of time, there is a high volatility in oil prices, which, recently, is being balanced at the level of 40-50 dollars per barrel.

Since inflation in the United States in June 2016 to December 2014 (basic when calculating the real price of EIA) was 1.9%, the average price in June 2016 (about \$ 48/bbl) is equivalent to \$ 47/bbl in prices of 2014. Thus, if the price of oil is fixed in the corridor of 40-60 \$/bbl for a sufficiently long period of time, this will mean that it has increased compared to the previous long historical period of relatively stable oil prices, which fits into the total logic of development of industry and the world economy.

Thus, in this study, the base oil price in 2050 was adopted at \$ 50/bbl, regardless of the scenario. Nevertheless, in order to reflect other hypothetically possible, albeit unlikely scenarios of oil price dynamics and to form an exhaustive picture of possible scenarios for the development of the domestic oil refining industry, all models were also calculated at prices of 20, 100, 150 and \$ 200/bbl. That is, within each scenario, options are considered based on the following oil prices: \$ 20/bbl, \$ 50/bbl, \$ 100/bbl, \$ 150/bbl and \$ 200/bbl.

The notional and variable costs are calculated on the basis of data obtained in the Gazprom Neft group of companies and are reflected in the form of operating expenses per ton of processed raw materials of each plant.

The ruble exchange rate against the dollar was adopted at the level of 85 rubles/dollar, the forecast was made on the basis of a logarithmic trend line using the dynamics of the dollar exchange rate for the period from January 1998 to July 2018.

All parameters are taken into account in statics, in the forecast level for 2040. Prediction of the dynamics and fluctuations of various factors was not carried out. Thus, the hypothesis was adopted in the paper that the development of the world energy system will take place in an evolutionary way, without significant jumps and fluctuations.

The model indirectly takes into account in the prices of oil and petroleum products various factors influencing the state, such as excise taxes and export duties on petroleum products and MET.In addition, the model does not take into account semi-fixed costs, since they are assumed to be unchanged in all configurations of refineries and scenarios for the development of the refining industry and, as a result, do not play a significant role in determining the development vector of refineries. Environmental aspects taken in the current settings. It is assumed that the modeled plants satisfy all requirements in terms of product quality and environmental performance.

5 REFINERY CONFIGURATIONS

Mismatch between refinery configurations and product demand stimulates the discovering of the most optimal combination of process units. Future refinery model identified by comparative analysis of various configurations of oil refineries, five optimal models were built, taking into account advanced technologies and new refineries patterns. The base case is the configuration that will prevail after the modernization of the Russian refineries, whichis completed by 2025-2-30 years. The configuration isbased on the model of JSC "Gazpromneft-MNPZ" or in public language Moscow oil refinery.

Base case

This version of the refinery can be referred to a group of fuel orientated plants. Such model includes atmospheric and vacuum distillation units, light naphtha isomerization and catalytic reforming units, kerosene and diesel hydrotreatment, fluid catalytic cracking unit,FCC gasoline selective hydrogenation unit, hydrocracking unit, delayed coking unit and bitumen production.

Diesel oriented case

This plant presupposes to operate the same units in general with some changes directed to middle distillates production. Fluid catalytic cracking and cracking gasoline selective hydrogenation are excluded. All vacuum gas oil is routedto hydrocracking unit of VGO. Thus, such refinery probably will required modernization of units duty. No additional technology is required.

Fuel-petrochemical case

This refinery presupposes involvement of polypropylene production unit. The main capacity of VGOis processedat fluid catalytic cracking for the propylene obtaining. Butsince this refinery, besides petrochemistry products and gasoline pool have to produce middle distillates products, VGO is also processed at hydrocracking. C4 fractionis routedto obtain petrochemical products such polybutylene, maleineangedrite.

Diesel-petrochemical oriented refinery

In this configuration, as compared with the fuel- petrochemical refinery, the complex for the production of aromatics, steam cracker unit of gasoline is added, Steam cracker is used to obtain ethylene and propylene fraction for further polymerization. C4 fraction of SC is routed to obtain synthetic rubber. Since the main purposes are diesel and petrochemical products pool, FCC unit not more required.

Gasoline-petrochemical oriented refinery

Since the main fuel pool is gasoline, this configuration of a refinery excludes the installation of hydrocracking unit of vacuum gas oil. Vacuum gasoil now is processed at fluid catalytic cracking to cover theses necessary purposes. Also installation production of octane esters, alkylation and complex aromatics production is added to keep balance between gasoline and petrochemical products.

Petrochemical refinery

Everything in this refinery is routed to maximize petrochemical pool. Gasoline cuts are processed at steam cracker to obtain ethylene, propylene. Fluid catalytic cracking unit is used to extract C3 and C4 cuts to obtain polypropylene, synthetic rubber, maleic anhydride, and polybutylene.

6 CONCLUSION

Development of future refinery industry in conditions of transitional energy markets where balance of traditional fuels can significantly shift is one of the key issues to fit markets needs and maintain business efficiency. Therefore, the development of future refinery configuration vision is of significant oil companies to face these challenges.

There are several possible key scenarios of energy markets evolution: increase in demand for distillates and petrochemical raw materials, stagnant gasoline demand; reduction of oil production and the gradual abandonment of oil as a raw material for the production of fuels and other consumer products, stagnating demand for all petroleum products.

The study of economic efficiency of profitable refinery in terms of market scenarios is based on the following factors: the price of oil, the price of petroleum products, the price of auxiliary semi-products, the price of petrochemical products, dollar rate, inflation rate (2050 by 2018). Models of typical oil refineries with various pools of products were developed: as the base case fuel oriented plant is considered, diesel oriented plant, fuel-petrochemical, diesel-petrochemical oriented refinery, gasoline-petrochemical oriented and petrochemical refinery. The research shows that in the most studied scenarios of energy market development, the diesel-petrochemical oriented configuration proved to be economically viable.

Thus, this study can be used as a basis for long-term forecasting and selection of Russian refineries' development vision. Moreover, the implemented methodology can be adopted for similar kind of assessment of refining industry development in other regions.

At current moment, the study is going on to have a better understanding of set of units that will be required for the refinery industry.

ACKNOWLEDGEMENTS

The studywas carried out with the support of management of JSC Gazpromneft Moscow Refinery, mainly its prospective development teamand engineering modeling team, and Russian

State University of Management, mainly its department of scientific activities management and Department of Economics and Management in the fuel and energy complex.

REFERENCES

Tae-Yoon Kim, Winners, loses and unintended consequences in the outlook for oil products demand, International Energy Agency, 2019

Outlook of development of International and Russian energy until 2040, Institute of Energy research of The Russian Academic of Science, 2014

Makhankov E.S., The newest factors of forming global strategies of vertically integrated oil companies, Proc. Scientific Proceedings of the Siberian Economic Journal. - Vol. 2. - Novosibirsk, 2016

Guseinov B.M., Analysis of the impact of macroeconomic and production among the factors affecting the cost of vertically integrated oil and gas of Companies // Problems of the economy and management of oil and gas, 2012

Annual Energy Outlook – 2015: Executive Summary –http://www.eia.gov/forecasts/aeo/executive_summary.cfm

BP Energy Outlook 2030. – London, January 2011.

BP Statistical Review of World Energy. June 2015.

Chua, J. Financial Performance of the U.S. Oil and Gas Industry: 1980-1990, Financial Markets, Institutions & Instruments. –1994. – V.3. – № 1.

Corporate Power Facts and Stats [Electronic resource] – http://www.globalissues.org/TradeRelated/Corporations/Facts.asp

Doing Business 2014: Doing Business in a More Transparent World. –Wash., DC: IBRD/WB, 2014.

Doran, P. B. Collective Energy Security: The New Approach for Europe, The Journal of Energy Security. – 19 Febr. 2009. –http://www.ensec.org/index.php?option=com_content&view=article&id=177:collective-energy-security-a-new-approach-for-europe&catid=92:issuecontent&Itemid=341.

Energy 2020. A strategy for competitive, sustainable and secure energy: Communication from the Commission to the European Parliament, the Council, theEuropean Economic and Social Committee and the Committee of the Regions: Doc.COM (2010) 639 final. Brussels.

Energy Balances of Non-OECD Countries: 2014/2015: 2015 Edition InternationalEnergy Agency, Paris, 2015.

Energy Independence and Security Act of 2007 -http://www.gpo.gov/fdsys/pkg/BILLS-110hr6enr/pdf/BILLS-110hr6enr.pdf.

Energy Policies of IEA Countries: IEA Energy Policies Review, The EuropeanUnion: 2013, International Energy Agency, Paris, 2015

Energy Policy Act of 1992 (EPAct) -http://www.eia.gov/renewable/afv/archive/index.cfm

EU. Energy Scenarios up to 2030, Brussels, 2006.

Global Competitiveness Report 2014–2015. World Economic Forum,2014.

Hoyos C., The new Seven Sisters: oil and gas giants dwarf western rivals//Financial Times. – March 12, 2007– Mode of access:http://www.ft.com/cms/s/2/471ae1b8-d001-11db-94cb-000b5df10621.html#axzz3u1txkYn5

International Energy Outlook 2015.

International Trade Statistics 2014. – Geneva: WTO, 2014.

Key World Energy Statistics. – International Energy Agency, 2015.

National Energy Conservation Policy Act (NECPA) -http://www.gpo.gov/fdsys/pkg/STATUTE-92/pdf/STATUTE-92-Pg3206.pdf.

Oil Market Report. International Energy Agency. – 13.03.2015– Mode of access:https://www.iea.org/media/omrreports/fullissues/2015-03-13.pdf

Short-Term Energy and Winter Fuels Outlook (STEO), U. S. EnergyInformation Administration (EIA). URL -http://www.eia.gov/forecasts/steo/pdf/steo_full.pdf.

Statistical Review of World Energy 2013. http://www.bp.com/en/global/corporate/about-bp/energyeconomics/statistical-review-of-world-energy-2013.html.

The World in 2050: Beyond the BRICs: emerging market growth prospects.– PwC, March 2008.

The World in 2050: Will the shift in global economic power continue? –PwC, February 2015. – 44 p.

US Energy Information Administration (EIA): Annual Energy Outlook2014.; U. S. Energy Information Administration (EIA). http://www.eia.gov/forecasts/aeo/pdf/0383(2014).pdf.

VYGON Consulting, 2015, http://vygon.consulting/upload/iblock/26f/vygon_consulting_oilprices.pdf

World Development Indicators 2015. The World Bank, Washington.2015.

Youth Technical Sessions Proceedings – Litvinenko (ed)
© 2019 World Petroleum Council, London, UK, ISBN 978-0-367-34668-3

Carbon offsetting strategies for supply-side responsiveness

S. Bozok
Master's of Science, London School of Economics, London, UK

ABSTRACT: Conventional customers with hydrocarbon-induced demand are increasingly gaining carbon footprint awareness. Lower carbon intensity and environmental considerations are therefore high in demand. As demand-side paradigms change, it is worth of research how energy companies, retails, manufacturers etc. adopt strategies in compliance and voluntary carbon offset markets for the emissions they cannot manage to reduce. Employment of carbon offset credits with goods and services such as natural gas, power and water now also brings new additionalities such as biodiversity, air quality and reduced poverty. This paper first demonstrates the degree of complementariness between compliance and voluntary carbon markets. Then it focuses on demand and supply side to map out the kind of strategies that can be adopted in line with the design of those markets. Finally, it discusses the new developments to watch that might affect demand and supply for carbon abatement.

1 INTRODUCTION

The decision to withdraw of the United States (U.S.) from the Paris Agreement by 2020 is only of the uncertainties surrounding the concept of decarbonisation. Nonetheless, given that the U.S. has rather state-level compliance markets and that the federal withdrawal would not significantly draw a setback to them, the withdrawal is even less of a concern for the corporations and individuals offsetting their carbon footprint.

Offsetting emissions can be defined as purchasing credits equivalent of the metric tonne of carbon dioxide of the projects that reduce GHG emissions in the atmosphere such as clean cookstoves, afforestation or reforestation and renewable energy generation, thus compensating for a carbon-intensive activity by paying for carbon abatement elsewhere. These carbon offsetting credits can be traded either as primary or secondary products in compliance carbon markets where the government requires emissions to either be reduced physically or by purchasing offset credits, or alternatively in voluntary carbon markets where buyers and sellers trade according to their volition (Hamrick, K. & Gallant, M. 2018).

Despite the U.S. withdrawal, the Paris Agreement which is a commitment of 195 signatories to reduce carbon emissions galvanised both supply and demand to products that emit less greenhouse gas (GHG) emission and offset the emissions they cannot abate. It is imperative to query how the sectors such as oil and gas, refining, finance and fast-moving consumer good (FMCG) replies this momentum. Firstly, compliance and voluntary carbon markets be assessed by volumes size, prices, standards, and project subjects. Secondly, supply and demand properties will be analysed to finally set out different offsetting strategies with different sector cases. This paper will conclude with indicating new developments to watch for carbon offset markets.

2 CARBON OFFSET MARKET FUNDAMENTALS

The 1997 Kyoto Protocol established the carbon markets by introducing the flexibility mechanisms to reduce GHG emissions. Whilst the emissions trading schemes (ETS) such as the

European Union ETS, the California Cap-and-Trade Programme and New South Wales GHG Reduction Scheme were established as the "compliance markets" in the Annex I parties, the project-based mechanisms, namely the Joint Implementation (JI) and Clean Development Mechanism (CDM), set the foundation of the "voluntary markets", the latter being allowed only in the non-Annex 1 parties. The ensuing 2015 Paris Agreement will regulate the interaction between compliance and voluntary markets for the 2020 Post-Kyoto era.

2.1 *Compliance carbon markets (Cap-and-trade)*

In compliance markets, the sectors such as electricity generation, cement manufacturing and refinery are regulated in a way that the aggregate GHG emissions from the regulated sectors cannot exceed the cap decided by the government, which is set to reduce periodically over the ETS lifetime. Emission allowances to the regulated installations can be allocated freely or via auctions. It's at the government's regulatory discretion how to allocate allowances, decrease the cap and whether allow offsetting, if so to what degree. For instance, California Cap-and-Trade Programme adopts a combination of free allowances and auctions, decreases the cap by 3% annually from 2015 to 2020, and allows only 8% of the allowance obligations to be offset by carbon credits. (Hein, L et. Al. 2017) Likewise, the EU ETS allows only CDM and JI project offset credits to be deducted from emission obligations, whose limit is set on installation level (Use of international credits). The installations that exceed their allowed emission levels purchase allowances or offset credits from those that have allowance surplus, resulting from either an emission abatement improvement in their facilities not exhausting all their allotted allowances. This regulated commerce results in relatively consistent price levels (Hamrick, K. & Gallant, M. 2018) and only unregulated sectors can produce offset credits, which are very minimal (Hamrick, K. & Gallant, M. 2017).

Whereas California and New Zealand allow credits from afforestation or deforestation projects, with geography restrictions, the EU ETS does not allow forest projects to be used to meet obligations (Hein, L et. Al. 2017).

2.2 *Voluntary carbon markets (Baseline-and-credit)*

In voluntary carbo markets, corporations and individuals can purchase and sell credits produced by carbon offset projects even in the absence of compliance markets, often motivated by sustainability purposes, social responsibility programmes, or shareholder pressure. Voluntary carbon projects have prevented 437.1 million tonnes of carbon dioxide equivalent ($MtCO_2e$) to be exhausted to the atmosphere until now. (Hamrick, K. & Gallant, M. 2018). The amount of voluntary carbon offset transactions was 63.4 $MtCO_2e$ with a total market value of $191.3 million (m) (Hamrick, K. & Gallant, M. 2017). Voluntary markets are in many legislation predecessors of compliance markets, because it usually performs as "pre-compliance" training and preparing market actors for the next phase. For that reason, transaction volume and market value can fluctuate. For instance, California compliance market incorporated many voluntary projects when it was created, and it resulted in reduction of voluntary market transaction decrease (Hamrick, K. & Gallant, M. 2017).

The prices differ based on location, project type and standards used and unlike the compliance markets, there is not a cap on the amount of supply, which results in excess offset supply. The rigorous quantifying methodologies as well as monitoring, reporting and verification (MRV) of the projects turned voluntary markets into a testing ground of innovative projects that might eventually evolve into compliance market tools. (Hamrick, K. & Gallant, M. 2018). To ensure the "additionality" principle, net emission abatement incurring from the project compared to business as usual (BAU) or baseline emissions, credit issuing bodies adopt different standards such as CDM, Gold Standard, Voluntary Carbon Standard (VCS), Plan Vivo and the Climate, Community and Biodiversity (CCB), although their additionalities and methodology qualities (permanence, double counting, leakage, verification and transparency) had received much criticism in the past. (Voluntary Carbon Offsets, 2007).

Whilst 57.3 $MtCO_2e$ of offsets were transacted in 2016, 58% (33.234 $MtCO_2e$) of that was VCS-certified. It is followed by 17% of Gold Standard, 8% of CDM, 8% of Climate Action Reserve, 4% ISO-14064, 3% of American Carbon Registry and 0.6% of Plan Vivo certificates. (Hamrick, K. & Gallant, M. 2017)

3 CORPORATE OFFSETTING STRATEGIES

According to the survey conducted for the Ecosystem Marketplace 2017 Report, the respondents indicated having sold 63.4 $MtCO_2e$ in 2016, yet have a supply surplus of 56.2 $MtCO_2e$ in their portfolio. Because of this discrepancy, offset market can be identified as a "buyers' market", where buyers prevail in setting the product designs (Hamrick, K. & Gallant, M. 2017). However, it might also cause that offset providers remain vigilant in producing to avoid selling at a cheap price, then contract supply and elevates prices in national level.

3.1 *Responding the Demand and Supply-Side Considerations*

The properties of compliance and voluntary market products must be taken into account jointly when setting a corporate decarbonisation strategy. The GHG Protocol, which is created by the World Resources Institute and the World Business Council for the Sustainable Development, categorises GHG emissions under three categories (Scope 1, Scope 2 and Scope 3) to measure the carbon footprint of a company (http://www.ghgprotocol.org.). Whereas Scope 1 defines the emissions under direct control and Scope 2 indirect emissions associated with the energy use, Scope 3 defines the indirect emissions with the use of sold product or transportation Compliance markets usually cover Scope 1 emissions which were reported as 5.218 $MtCO_2e$ in 2014 to the Carbon Disclosure Project (CDP), a voluntary-based disclosure platform for companies' carbon footprint; however, Scope 3 emissions which are traded in voluntary markets were reported to be almost 3.5 times (18.118 $MtCO_2e$) greater than that. (Goldstein, A. 2016). Thus, this paper argues that there is a significant degree of complementariness between compliance and voluntary markets to internalise all externalities of GHG emissions.

When reducing carbon footprint, a company should first seek improvement in its overall operations to reduce GHG emissions. If that does not suffice to reach the target, then buying offset credits, and allowances as well in compliance market case) should be pursued. The more an installation abates emissions, the more income it alternatively generates from its reduction surplus (Carbon Trust, 2006). Abating beyond regulation, thus originating offsets turn the offset buyer into an offset provider. However, if the company has no interest in making profitable sales out of its offset surplus, but rather uses them to comply with more rigorous internal emission targets, then this phenomenon is described as "insetting" (Goldstein, A. 2016).

Although purchasing offset credits in voluntary markets were criticised as green-washing and postponing material emissions in corporate operations, there is a correlation between offset purchases of a company and its recognition of climate change and the risks related (Goldstein, A. 2016). Therefore, whereas compliance market buyers are incentivised by low offset prices to fulfil their obligations, voluntary market buyer are incentivised by additional features of an offset project such as reputation, standard, proximity to its supply chain locations, rural preservation and support for indigenous people. Therefore, voluntary compliance markets are more similar to real estate markets where prices of similar houses differ in difference locations than compliance markets where prices reflect the least costly offsetting solution, similar to commodity markets (Hamrick, K. & Gallant, M. 2017).

Setting a comprehensive offsetting strategy requires that the corporations assess all lifecycle emissions from supply to sales, extending over a long period of time given that the process will include as far as convincing procurement contractors to convert their raw materials to less carbon intensive ones (Goldstein, A. 2016).

3.2 *Sectoral cases*

Buyers in compliance carbon market are of the regulated industries such as chemicals as power generation, whereas buyers in voluntary carbon market are of the competitive industries such as finance, FMCG, airlines, retail, technology and even pharmaceuticals. (Goldstein, A. 2016). Although the majority of the offset supply is originated from Asia, demand in developed geographies such as Europe and the U.S. have a preference for domestic projects (Hein, L et. Al. 2017), which is not surprising given that many of the offsetting projects carry sustainable development additionalities, despite the fact that the GHG abatement in the atmosphere is purely global regardless of location. Therefore, many offset retailers offer products combining United Nations' (UN) Sustainable Development Goals (SDGs) such as ending hunger, protecting marine life and energy access provision (Hamrick, K. & Gallant, M. 2018).

Some of the projects boosts credibility of the buyer company and creates closer relationship with the supply location. For instance, Starbucks initiated an agroforestry offset project for its local coffee farmers, from which they can generate additional income. Similarly, Hershey reduces the fuel expenses by creating a biomass project to produce energy from the macadamia shell leftovers from chocolate production (Goldstein, A. 2016).

In addition, consumers opt in for projects whose GHG abatement is easy to comprehend (e.g. wind power plant projects), yet still authentic and charismatic (e.g. forest projects), ready to bear the incurring premium cost per credit. Nonetheless, regulations failed to actively incorporate Reducing Emissions from Deforestation and Degradation (REDD+) to the markets thus far (Kollmuss, A., Zink, H. & Polycarp, C. 2008). Although CDM includes forestry projects, measuring the additionality criteria was much debated and externalities such as displacing indigenous people or mono-culture plantation reduced the attractiveness of forestry projects. California Cap-and-Trade Programme allows American forest projects as well as those from some regions of Brazil and Indonesia; however, the process is rather very complex for timber businesses and small-scale landowners (Kelly, E.C. & Schmitz, M.B. 2016). EU ETS prohibits forestry emission sinks to be used because of uncertainty in quantifying emissions sequestrated by trees. (Voluntary Carbon Offsets, 2007). Tim Participacões S.A, Barclays Africa, Royal Wessanen, SURA Peru and Brambles are among the corporations which have reported to purchasing REDD+ projects in the past (https://cdp.net/en).

Bundling hydrocarbon products (e.g. natural gas and diesel fuels) with carbon offset credits was recently classified as a new "clean product" trend in energy marketing. Offsetting fuel emissions that cannot be reduced is attractive, because the transition from internal combustion engine vehicles to electric ones will may take a long time, requiring transitional products such as carbon-neutral diesel, natural gas and even power. When reflecting the cost of offsetting to the customers, the marketing strategy would prove more impactful should it depends on the fact that electric vehicle manufacturing can be very energy intensive with battery waste incurring at the end (Hein, L et. Al. 2017).

Apart from the climate-change induced preferences from demand side, many companies are directly concerned about the intrinsic and physical danger of the climate change. The 2016 Buyers' Report of the Ecosystem Marketplace highlights that Puma is concerned about the negative effect of increased temperature on cotton farms, Google and Microsoft on their cooling systems and Delta Airlines on aircraft weights. Perhaps the latter example is worth further elaborating on aviation industy (Goldstein, A. 2016).

EU ETS regulates domestic aviation, yet international aviation has never been covered by neither the Kyoto Protocol, nor the Paris Agreement. The International Civil Aviation Organisation (ICAO) made steps to regulate international aviation emissions, which would massively increase demand for offset credits around the world. Having recently agreed on emission offset criteria for the Carbon Offsetting and Reduction System for International Aviation (CORSIA), ICAO will contain all countries in three phases by 2035 (1st phase in 2012-2023, 2nd phase in 2024-2026, 3rd phase in 2027-2035). CORSIA is forecast to generate an offset demand of 1.6 - 3.7 $MtCO_2e$ by 2035 (Sean Healy, CORSIA: Quantification of the Offset Demand (Healy, S. 2017), which would prove to be a game changer in offset trade since many voluntary offset customers like Delta Airlines will have to change their offsetting strategies.

4 CONCLUSION

Carbon intensity of goods and services is under increasing scrutiny and energy sector is not exempt from it. Offsetting is only one element of a comprehensive climate change strategy and not a substitute of emission reduction, but a complement to it. It is observed that the companies that offset beyond their obligations implements emission reduction measures at a higher rate than those that do not offset (Goldstein, A. 2016). Companies can adopt a portfolio of various offset projects from different locations, topics, prices and standards to distribute the risk associated.

The voluntary markets will continue to be natural buyers markets so long as the prices remain low due to weak demand and supply surplus. This trend sets a suitable environment for opportunistic offset buyers to frontload before markets contracts. The CORSIA regulations are not the only trend market participants should watch. As the uncertainty looms over the "Post-Paris era", the EU Commission also seek to internalise via the Effort Sharing Regulation (ESR) the projects that were left outside the EU ETS such as the forestry activities called "Land Use, Land-Use Change and Forestry" (LULUCF). The ESR will become determinant over the fate of many voluntary offset projects after 2020, because it will impose national level abatement targets and restrictions to the project pool to achieve those (Council Regulation, 2018). The final regulation to watch is Climate Neutral Now initiative that is set under the UN Climate Change Secretariat to replace CDM. Although it is endorsed by companies such as Microsoft, Adidas, M&S and Sony, it is uncertain if it will have the same limited demand in Post-Paris era as its predecessor CDM in Post-Kyoto era (Hein, L et. Al. 2017).

This paper argues that offset trading will remain a pivotal flexibility in decarbonisation strategies. Alongside the above market developments, demand side developments such as the CDP, the Science-Based Target initiative and renewable energy certificates trading are worth further research in order to better understand the overall concept of carbon neutrality.

REFERENCES

Brinkman, M. 2010. *A new look at carbon offsets*: a report of McKinsey & Company.

Bushnell, J.B. 2010. *The Economics of Carbon Offsets*. National Bureau of Economic Research.

Carbon Offset Project Types and Categories (2018). *Forest Trends' Ecosystem Marketplace*, viewed 09.03.2019, https://www.forest-trends.org/wp-content/uploads/2018/09/Factsheet_Carbon-Offset-Project-Types-and-Categories.pdf

Council Regulation 2018/1999 on the Governance of the Energy Union and Climate Action, 2018 OJ L 328.

Gold Standard Annual Report 2017. *Gold Standard*, 2018.

Gold Standard Market Report 2017. *Gold Standard*, 2018.

Goldstein, A. 2016. *Taking Stock of the Role of Offsets in Corporate Carbon Strategies*: a report of Forest Trends' Ecosystem Marketplace.

Gössling, S. et. al. 2007. Voluntary Carbon Offsetting Schemes for Aviation: Efficiency, Credibility and Sustainable Tourism. *Journal of Sustainable Tourism* (15) 3: 223–248.

Grainger C.A. & Kolstad C.D. 2009. *Who Pays a Price on Carbon?* National Bureau of Economic Research.

Hamrick, K. & Gallant, M. 2017. *Unlocking Potential: State of the Voluntary Carbon Markets 2917*: a report of Forest Trends' Ecosystem Marketplace.

Hamrick, K. & Gallant, M. 2018. *Voluntary Carbon Markets Insights: 2018 Outlook and First Quarter Trends*: a report of Forest Trends' Ecosystem Marketplace.

Healy, S. 2017. *CORSIA: Quantification of the Offset Demand*: a report of Institute for Applied Ecology.

Hein, L et. Al. 2017. *Hein, L et. Al. 2017*: a report of Concawe.

Kelly, E.C. & Schmitz, M.B. 2016. Forest offsets and the California compliance market: Bringing an abstract ecosystem good to market. *Geoforum* (75): 99–109.

Kollmuss, A., Zink, H. & Polycarp, C. 2008. *Making Sense of the Voluntary Carbon Market: A Comparison of Carbon Offset Standards*: a report of the WWF Germany.

The Carbon Trust three state approach to developing a robust offsetting strategy. Carbon Trust, 2006.

Voluntary Carbon Offsets. International Energy Agency, 2007.

Use of international credits. (n.d.). viewed 09.03.2019, https://ec.europa.eu/clima/policies/ets/credits_en

Youth Technical Sessions Proceedings – Litvinenko (ed)
© 2019 World Petroleum Council, London, UK, ISBN 978-0-367-34668-3

Russian gas market reform: Evaluation of the market efficiency and the Saint Petersburg stock exchange gas market as a new price indicator instead of regulated low price border

A.S. Talipova
Higher School of Economics, National Research University

ABSTRACT: According to the current legislation, after 1st July 2019 Russia should refuse to regulate wholesale gas prices and move to regulation of only transportation tariffs. The reform, which is positioned by the Federal Antimonopoly Service (FAS) as the first stage of deregulation of the domestic gas market, should, therefore, lead to a decrease in gas prices across the country by an average of 5-10% according to FAS estimations. In addition, since 2014, gas again began to be sold at Saint Petersburg Stock Exchange. Consequently, the exchange price for gas is suggested to become a new price indicator instead of high and low price borders. In this paper the gas market in Russia is explores and, in particular, the exchange gas market which started operating again in 2014 after it was closed in 2008. In connection with its appearance, experts disagree about whether it will be able to become a price indicator in place of the existing regime for regulating the lower and upper border of prices. OJC Gazprom, the Russian gas monopolist, insists that the exchange price may becomes an indicator and price limits can be abolished. In this paper, the volatility of the gas market has been studied since the beginning of 2017 using the VaR (Vector Auto Regression). The basic hypothesis accepted is that stock price can be assumed as new indicator. The main result achieved is that all tests strengthen our hypothesis, that existing volatility and particularly with season trends looks suspicious with no evidence why that happens.

1 INTRODUCTION

Natural gas is one of the leading resource in the energy balance of Russia, which plays a significant role as a non-renewable resource for related industries: about 65% of natural gas is used for electricity (about two-thirds of the electricity and heat energy in Russia is produced from gas) and heat, about 10% is consumed by the households, more than 15% in heavy industry, 10% goes to gas transportation needs.

During the past decade and especially in the period 2007-2011, Russia experienced a significant increase in gas prices. According to Federal State Statistics Service, from 2005 to 2015, the wholesale price for natural gas for industrial organizations in Russia increased 3.2 times, amounting to 3,642 rubles per thousand m cub in 2015, which in turn affected the prices for electricity, heat and industry as well. A sharp change in prices with a possible change in the pricing method, may lead to adverse socioeconomic consequences. However, the current pricing model also entails a number of possible negative effects, which include a lack of investment, a shortage of gas transportation capacities in a number of regions, insufficient activity of independent gas producers.

According to the current legislation, after 2018 Russia should refuse to regulate wholesale gas prices and move to regulation of only transportation tariffs. Until 2018, it is supposed to use a formula that determines the gradual achievement of equal returns on gas supplies to the external and internal markets. In this regard, in early 2016, Russian Federal Antimonopoly Service (FAS) announced a pilot project to liberalize natural gas prices in the domestic market in three regions in 2017, which supposes the abolition of the lower boundary of the wholesale price of gas for industry, while prices for the households are supposed to stay constant. The reform, which is positioned by the FAS as the first stage of deregulation of the domestic gas market, should therefore lead to a decrease in gas prices across the country by an average of 5-10% according to FAS estimations. In addition, since 2014, gas again began to be sold at Saint Petersburg Stock Exchange. The main argument of the FAS in favor of abolishing the lower price border of regulation is the difference between exchange and long-term contracts, which in some Russian regions reaches 10-12%. Consequently, exchange price for gas is suggested to become a new price indicator instead of high and low price borders.

The issues of fixing the exchange price as a new indicator of pricing cause rather contradictory opinions among the majority of experts today:

- More than 60% of all gas at the exchange is sold by the companies of OJC Gazprom, which indicates the possibility of price manipulation;
- The volumes of gas sold to the exchange are very small (do not exceed 3 billion cubic meters per month), while domestic gas consumption averages 450 billion cubic meters per year;
- The gas transportation system still remains to be the property of OJC Gazprom and, despite the non-discriminatory access, for the group of Gazprom companies, in fact the transportation tariff remains inside the company, which makes the position of other producers unequal.

In this regard, the purpose of this paper is to study the theoretically and empirically domestic gas market in Russia and, in particular, the exchange gas market in the period from the beginning of 2016 to determine the appropriateness of using the exchange price as a price indicator and the transition to market deregulation.

2 LITERATURE REVIEW

Over the last years, a numerous significant changes took place in gas industry throughout the world which raised the issue of policy regulation. Being regional by nature, gas markets in the past decades went towards deregulation of natural monopolies by splitting the transportation system from production. The first market where this process could be clearly observed and the outputs are already obtained is the US market. The main attention however there was paid to the pricing aspect and to the concept of shadow price particularly, which described by (Thompson, 1985). He insisted that disregarded aspect of shadow price in modelling the effect of monopoly deregulation (American Gas Association, 1983; U.S. Department of Energy, 1982; Woods and Vidas, 1983) may lead to huge underestimations of negative effect from monopolies. Therefore, the object of regulation in the Natural Gas Policy Act of 1978 (NGPA) were wellhead prices or the high-cost gas in other words. On the contrary, (The National Regulatory Research Institute, 1983) investigated the problems of rapidly raised prices after partial decontrol. According to them, much of the controversy can be traced to gas market distortions caused by partial regulation; that is, the gas market is now neither fully regulated from wellhead to burner. tip nor fully deregulated (The National Regulatory Research Institute, 1983). Later works, after the NGPA ultimately finished the process of gas market deregulation, the works discussed success of the state policy (Rosput, 1992) and show the direct dependence of spot and free gas market appearance with the deregulation policy (Vany and Walls, 1994).

The fundamental problems as the central problem of identifying the optimal way to regulate the industry were studied in a numerous works. According to the (Gordon et. al., 2003), the fundamental problem to identify and regulate natural gas monopoly is subadditivity (Berg and Tschirhart, 1988). Supply and demand balance in free market conditions is described by (Wilkinson, 1983). There are two fundamental problems for the competition in gas industry: high transportation costs and regional differences. Therefore, since large markets often have send outs well above the capacity of a single large-diameter pipeline, the duplication or looping of pipelines is common. Where this condition exists, the defining characteristic of natural monopolies - declining costs with scale of operation - may no longer apply (Jensen, 1992).

3 METHODOLOGY

The study process is under three main theories: incomplete contracts, market efficiency and manipulation, pricing on imperfect markets. To analyze the problem, econometric modeling and qualitative descriptive approaches were used. Preliminary, statistical analysis for the data was completed; three econometric models were developed.

The primary benefit of the market is that it allows market participants to hedge the risk of price affecting future purchases or sales. However, the price signals generate futures market must accurately reflect the future spot price of natural gas producers and suppliers are to hedge the risk of future price change primary concern is the extent to which futures market prices accurately the available information which bears on the future spot price. Futures are generally said to be efficient when the futures price today contents relevant information to forecast tomorrow's spot price. Fama (1970) weak-form efficiency to mean that the futures price will be an unbiased predictor of the future spot price (Walls, 1995).

Overall theoretical approach, methodology and process is widely known and described. We used (Walls, 1995) and IMF recommendations.

We include it here to enhance the clarity of the estimations. We start with a futures pricing model in which the price of a futures contract for a commodity is equal to the discounted value of the expected spot price (IMF, 2011):

$$F(t,T) = e^{-\rho(T-t)} E^f(t)[S(T)] \tag{1}$$

Where F(t,T) denotes the futures price in period t with delivery in period T, $E^f(t)[S(T)]$ is the expectation in the futures market in period t of the spot price in period T, and ρ is the continuously compounded rational expectations risk premium. (In a market without limits to arbitrage, expectations in spot and futures markets will be the same, but we retain this notation for a useful derivation below.) (1) states that a position that requires no upfront investment should deliver an expected return equal to the risk premium of the investor. Taking logs of (1) obtains a linear relationship between the futures price and expected spot price. From this point on, we will consider a forecast length of k = T - t so that:

$$f_{t,t+k} = -\rho k + E^f S_{t+k} + 1/2var(s) \tag{2}$$

In (2), f is the log futures price, ρ is the (assumed constant) risk premium scaled to the length of the forecast horizon k, $E^f S_{t+k}$ is the period t expectation in the futures market of the log spot price k periods ahead, and $1/2var(s)$ is Jensen's inequality correction. Henceforth, we will ignore Jensen's term, although it can equivalently be included in the constant term ρk in the case of homoscedastic variance, an assumption that we maintain in this paper. Subtracting the current log spot price S_t from both sides of (2) gives an expression which states that the current spread between the futures price and spot price (the basis) is equal to the expected change in the spot price for the period until delivery less the risk premium.

$$f_{t,t+k} - S_t = -\rho k + E_t^f(S_{t+k} - S_t) \tag{3}$$

Based on (3), one approach to testing for market efficiency is to estimate the following regression.

$$S_{t+k} - S_t = \alpha + \beta(f_{t,t+k} - S_t) + \varepsilon_{t+k} \tag{4}$$

The parameter α can be interpreted as the constant component of the risk premium. If the basis provides an unbiased forecast of the future spot price then α = 0, β=1 and ε_{t+k} has a conditional mean of zero. This regression is typically estimated in the market efficiency literature, including for commodities (e.g., Chinn and Coibion, 2010 and Reeve and Vigfusson, 2011).

4 CURRENT MODEL OF GAS PRICING ON RUSSIAN DOMESTIC MARKET

The system of price rates, for which payments are made for gas supply, service, transportation or acquisition to industrial companies include the following:

- the price of gas (wellhead price);
- payment for distribution services;
- tariffs for gas transportation services;
- special surcharge to the tariff for the transportation of gas through gas distribution grid;
- retail gas prices for households are set by the local administrations.

The Russian domestic gas market is divided into a regulated and unregulated sector. OJC Gazprom is the main gas supplier in the regulated sector, while in the unregulated sector the main suppliers are independent gas producing and oil companies (Novatek, Rosneft etc.). In this case, the dominant position is occupied by the regulated segment of the market. The government regulates:

- wholesale gas prices, for which OJC Gazprom and its affiliate companies sell gas on the domestic market.
- tariffs for gas transportation services through distribution pipelines for independent producers, and for transportation of gas through gas distribution grids;
- payments for distribution services.

Gas from independent producers is sold at unregulated prices, which together satisfy about a third of Russian demand (about 120 billion cubic meters per year).

In the autumn of 2014 the exchange trade in gas began on Saint-Petersburg Stock Exchange, the prices within which are also unregulated. In addition, regulatory system sets differentiated tariffs for transporting gas through Russia's distribution pipelines for independent gas suppliers, while tariffs for gas distribution and distribution organizations are regulated by the state. All the system, that exist today underwent huge transformation and still under active development.

The pricing system for natural gas on the wholesale market has undergone a number of changes throughout the regulatory period, but until 2007 no critical changes in regulation have been observed. Since 2000, at the legislative level, two main priorities have been formulated, on which it was planned to build a future system of pricing for natural gas.

1. As a first priority for changing the pricing system, a transition to the liberalization of wholesale gas prices was set except transportation tariff. This priority was formally designated in 2000, but for the first time the terms of such a transition were determined only in 2010 (initially - by 2015 and to date, the period was postponed till 2018).
2. As a second priority in 2007, the need was set for a two-stage transition from the prices of the domestic market, formed within the limits set by the Government of the Russian Federation for the growth of average gas prices, to prices of equal yield between the external and internal markets. It was assumed that an equal yield will be provided in comparison with European markets (less transportation costs and export duties, that is, netback prices will be formed).

However, setting priorities for the transition to equal profitability of gas prices in 2007 was due to a number of factors that have largely lost their relevance at the present time.

- First, in the years 2007-2010. The expectations for a relatively rapid increase in gas demand in external gas markets (and the reluctance of the producer to sell gas at relatively lower prices in the domestic market) were quite high.
- Second, in the late 1990-th - early 2000-th Russia pursued a policy of curbing the growth of prices for natural gas, which in many respects caused the negative profitability of Gazprom's sales in the domestic market in the mid- 2000s. However, as a result of the annual and inflationary increase in gas prices from 2007 to 2011 (by 15-25% annually for both industrial consumers and the population) the problem of low profitability in the domestic market was actually solved.

Thus, according to Gazprom estimation, the domestic sales profitability has been positive since 2009. In addition, the situation on foreign markets has changed - the demand for gas in

European markets has grown more slowly than previously expected. There have also been changes in the domestic market - there has been a significant increase in the role of independent gas suppliers in Russia (their share in domestic gas sales increased from 14% in 2007 to 36% in 2013), especially among sales to industrial consumers and generating companies, that led to increased competition in the domestic market.

In connection with the emergence of other basic prerequisites, a well-grounded question emerged about whether the transition to equitable markets with the European market is the benchmark of the market price for the Russian domestic market, whether this indicator is too high. According to rough estimates, in order to achieve equal returns, it will be necessary to significantly increase gas prices in the domestic market.

In 2013-2015, several studies were published showing that a further annual increase in domestic gas prices at the same high rates as previously occurred poses significant risks for Russian industry (for example, the Ministry of Economic Development of Russia, Institute for Energy Studies of Russian Academy of Sciences). According to the Ministry of Economic Development of Russia, in order to approach the prices of equal yield, the indexation of the price for gas in 2013-2015 should be at least 15% per year. This, in turn, will lead to an increase in prices for gas and electricity over the same period by 1.4-1.5 times. At present, the growth rates of gas prices in the wholesale market for 2014-2017 have been reduced to less than 5-10% per year (Figure 1). This is a sign that there is a temporary or final abandonment of the logic of equally profitable prices as an indicator of an unregulated price for natural gas in Russia. Therefore, an exchange price today is accounted as a new price indicator. However, it may cause difficulties and inequalities of market players as presented below by which this price can't be accounted as a price indicator.

5 EXCHANGE PRICE AS A NEW PRICE INDICATOR. DISCUSSION.

The analysis accordingly to the above described methodology shows that there is a cointegration in the futures and spot markets. In other words, mathematically we see at least semi strong market efficiency.

The largest players on the stock exchange are the companies of Gazprom group and OJC NOVATEK, as an independent producer, and brokerage companies. Gazprom Group companies account for at least 60% of the total traded gas. The structure of participants from other industries, then it is dominated by generating companies and units of companies from the branches of metallurgy. Of all the listed market participants, the largest amount is generated by generating companies.

But even so, the total volume of buyers of various industries of about 20 billion cubic meters annually is insignificant with the size of the domestic market at more than 400 billion cubic meters. This volume also can not be called essential for market manipulation opportunities. However, Gazprom may become a possible manipulator, since it has an overwhelming market volume in terms of liquidity indicators.

However, a very important question arises here: how to determine whether the established price on the exchange is the object of manipulation? With what it is in this case to compare? As a rule, on world stock exchanges, if the deviation of the transaction price from the prevailing average price on the exchange exceeds the total volatility of this market segment over the past year, then such price may be subject to investigation by authorized bodies (for example, antimonopoly bodies). But when most of the deals are just for a participant who can manipulate the price, then any comparative possibilities disappear. On the other hand, it is possible to compare it with over-the-counter contracts, but here the question arises. And how well is the over-the-counter price of contracts justified so that it can be compared with an exchange one? And how much in this case we allow the level of deviations from the over-the-counter price, so that the stock exchange is not perceived as manipulative. If we assume that the price of a well is economically justified, just like the transport tariff, then discounts remain for large buyers, with which it is possible to compare exchange prices. But the latter, in turn, are discounts to the minimum regulated price of Gazprom, which again is a reason for doubting the validity of the price from which independent manufacturers and discounts are provided.

From this, we see a kind of vicious circle in which Gazprom can set a kind of "monopolistically" low price on the exchange, the reasonableness of which can not be confirmed, and which in the case of comparison with the prices of independent producers, equal to Gazprom's minimum over-the-counter price, taking into account the average discount, can say that they can provide large discounts equal to the difference between exchange and over-the-counter prices. However, it is clear that Gazprom can set such a price, based on its own production and technological characteristics (lower cost and the scale effect), rather than economically sound pricing calculations. In turn, independent producers either will be forced to sell gas at the well, or give large (notice unreasonable) discounts on over-the-counter contracts, based on the difference between the exchange and over-the-counter prices, or to reduce the price of net-back to stock exchange. In all cases, not only independent gas producers are losing, but also PJSC Gazprom, while winning (perhaps, but not the fact) in volumes. Below is an analysis of losses of PJSC Gazprom under the condition of the exchange price as an indicator of determining the price of the well or discounts on over-the-counter contracts.

6 CONCLUSION

Thus, with the launch of exchange trades on gas, it is no longer necessary to argue about discounts and more flexible tariffs: Gazprom and other market participants will now be able to sell gas on relatively equal terms. We note one more important feature related to the bias of exchange quotations as an indicator of prices. As stated by the Ministry of Energy, until the end of 2016, the liquidity of the exchange should reach 35 billion cubic meters of gas per year. At the moment, we omit consideration of the reasons why this happened, but we will analyze the technical side of this situation. So, Gazprom has the technical ability to sell through the exchange up to 30 billion cubic meters of gas per year. However, it is unlikely that Gazprom will be allowed to sell as much gas through the exchange. In addition, gas trading on the exchange is conditioned by a number of restrictions: the total trading volume on the stock exchange can not exceed 35 billion cubic meters per year, while Gazprom can not sell more gas than all independent producers put together, that is, not more than 17.5 billion cubic meters of gas per year. The last restriction, at least according to the Ministry of Energy, was introduced so that the monopolist could not manipulate the exchange price. And restrictions on exchange trades are introduced to preserve OTC trade, which by definition is absurd.

Finally, we conclude, that the development of the Saint Petersburg Stock Exchange is vital and necessary in building an efficient competitive market. However this process should be followed by the equal conditions for all participants, increasing the number of buyers and brokers as well as a very careful analysis of monopolist position.

REFERENCES

Abosedra, S., Elkhal, K., Al-khateeb, F., 2006. Natural Gas Futures Market: An Assessment Of Recent Data 4, 65–70. https://doi.org/10.19030/jber.v4i11.2715

Aggarwal, R.K., Wu, G., 2003. Stock Market Manipulation - Theory and Evidence. Ssrn. https://doi.org/10.2139/ssrn.474582

Arano, K.G., Blair, B.F., 2008. An ex-post welfare analysis of natural gas regulation in the industrial sector. Energy Econ. 30, 789–806. https://doi.org/10.1016/j.eneco.2007.07.003

Asche, F., Misund, B., Sikveland, M., 2013. The relationship between spot and contract gas prices in Europe. Energy Econ. 38, 212–217.

Berg, S., & Tschirhart, J. (1989). Natural Monopoly Regulation: Principles and Practice. Cambridge Surveys of Economic Literature.

Burns, P., Weyman-Jones, T.G., 1998. Is the gas supply market a natural monopoly? Econometric evidence from the British Gas regions. Energy Econ. 20, 223–232. https://doi.org/10.1016/S0140-9883(97)00006-6

Chinn, Menzie D., and Coibion, Olivier, 2010, "The Predictive Content of Commodity Futures," NBER working paper 15830.

Comerton-Forde, C., Putniņš, T.J., 2011. Measuring closing price manipulation. J. Financ. Intermediation 20, 135–158. https://doi.org/10.1016/j.jfi.2010.03.003

Dwyer, G.P., 2015. The Johansen Tests for Cointegration.
Energy Resource Analysis Model, A.G.A. - TERA 83-1 (American Gas Association, Arlington, Va., Feb. 8, 1983).
Gordon, D. V., Gunsch, K., & Pawluk, C. V. (2003). A natural monopoly in natural gas transmission. Energy Economics, 25(5),473–485. https://doi.org/10.1016/S0140-9883(03)00057-4
Grigoryev, Y., 2007. Today or not today: Deregulating the Russian gas sector. Energy Policy 35, 3036–3045. https://doi.org/10.1016/j.enpol.2006.07.014
Haas, R., Glachant, J., Keseric, N., Perez, Y., 2006. Competition in the continental European electricity market: despair or work in progress?, in: Sioshansi, F., Pfaffenberger, W. (Eds.), Electricity Market Reform: An International Perspective.
Hawdon, D., Stevens, N., 2001. Regulatory reform of the UK gas market: The sase of the storage auction. Fisc. Stud. 22, 217–232. https://doi.org/10.1111/j.1475-5890.2001.tb00040.x
Henderson, J., Mitrova, T., Heather, P., Orlova, E., Sergeeva, Z., 2018. The SPIMEX Gas Exchange: Russian Gas Trading Possibilities (No. NG), 126. https://doi.org/10.26889/9781784671013
Herbert, J.H., Kreil, E., 1996. US natural gas markets. Energy Policy 24, 1–5. https://doi.org/10.1016/0301-4215(95)00145-X
Hillion, P., Suominen, M., 2004. The manipulation of closing prices. J. Financ. Mark. 7, 351–375. https://doi.org/10.1016/j.finmar.2004.04.002
Jensen, J.T., 1992. Open access-the new market approach to natural gas policy. Energy Policy 1005–1014.
Johansen, S., 1988. Statistical Analysis of Cointegration Vectors. J. Econ. Dyn. Control 12, 231–254.
Johansen, S., 1991. Estimation and Hypothesis Testing of Cointegration Vectors in Gaussian Vector Autoregressive Models. Econometrica 59, 1551–1580.
Joskow, P., 2008. Electricity Market Liberalization: Lessons Learned. Energy J. Special Is, 9–42. https://doi.org/10.5547/ISSN0195-6574-EJ-Vol29-NoSI2-3
Joskow, P.L., 2006. Markets for Power in the United States 1–36.
Lunden, L.P., Fjaertoft, D., Overland, I., Prachakova, A., 2013. Gazprom vs. other Russian gas producers: The evolution of the Russian gas sector. Energy Policy. https://doi.org/10.1016/j.enpol.2013.06.055
Nosić, A., Karasalihović Sedlar, D., Jukić, L., 2017. Oil and Gas Futures and Options Market. Rud. Zb. 32, 45–54. https://doi.org/10.17794/rgn.2017.4.5
Orlov, A., 2015. An assessment of optimal gas pricing in Russia: A CGE approach. Energy Econ. https://doi.org/10.1016/j.eneco.2015.03.013
Panagiotidis, T., Rutledge, E., 2007. Oil and gas markets in the UK: Evidence from a cointegrating approach. Energy Econ. https://doi.org/10.1016/j.eneco.2006.10.013
Park, H., Mjelde, J.W., Bessler, D.A., 2008. Price interactions and discovery among natural gas spot markets in North America. Energy Policy. https://doi.org/10.1016/j.enpol.2007.09.012
Pierce, R.J., 1983. Reconsidering the Roles of Regulation and Competition in the Natural Gas Industry. Harv. Law Rev. 97, 345–385. https://doi.org/10.2307/1340851
Pindyck, R.S., 2004. Volatility in natural gas and oil markets. J. Energy Dev. 30, 1–19.
Reeve, Trevor A. and Robert J. Vigfusson, 2011, "Evaluating the Forecasting Performance of Commodity Futures Prices," International Finance Discussion Papers, Number 1025.
Reichsfeld, D., Roache, S., 2011. Do Commodity Futures Help Forecast Spot Prices? IMF Work. Pap. 11, 1. https://doi.org/10.5089/9781463923891.001
T.J. Woods and E.H. Vidas, The GRI Hydrocarbon Model (Gas Research Institute, Chicago, IU., 1983).
Talipova, A., 2018. Recently Announced Reforms in the Russian Domestic Gas Market: Impact on the Market Players' Value in New Realities of International Gas Trading, in: Paper Accepted for SPE Russian Petroleum Technology Conference. Society of Petroleum Engineers, Moscow, Russia. https://doi.org/10.2118/191647-18RPTC-MS
Thompson, R. G., & Singleton, F. D. (1985). Shadow Price Principles Applied to Regulated Pricing of Natural gas. Annals of Operations Research, 2, 285–316.
U.S. Department of Energy, Energy Information Administration, Annual Energy Outlook 1982 - Supplement (Washington, D.C.).
Vany, A.De, & Walls, W. D. (1994). Network connectivity and price convergence: Gas pipeline deregulation. Research in Transportation Economics. https://doi.org/10.1016/S0739-8859(09)80003-6
Vany, A.De, Walls, W.D., 1994. Network connectivity and price convergence: Gas pipeline deregulation. Res. Transp. Econ. https://doi.org/10.1016/S0739-8859(09)80003-6
Von Hirschhausen, C., Engerer, H., 1998. Post-Soviet gas sector restructuring in the CIS: a political economy approach. Energy Policy 26, 1113–1123. https://doi.org/10.1016/S0301-4215(98)00062-7
Walls, W.D., 1995. An Econometric Analysis of the Market for Natural Gas Futures. Energy J. 16, 71–83.
Wilkinson, J. W. (1983). The Supply, Demand, and Average Price of Natural Gas under Free-Market Conditions. International Association for Energy Economics, 4(1),99–123. Retrieved from https://www.jstor.org/stable/41321589

Field Geology and Geophysics. Innovative approach

Youth Technical Sessions Proceedings – Litvinenko (ed)
© 2019 World Petroleum Council, London, UK, ISBN 978-0-367-34668-3

A new method of hydrocarbon detection with seismic data in tight reservoirs

R. Jiang
College of Geosciences, China University of Petroleum, Beijing, PR China
Petrochina Company Limited, Research Institute of Petroleum Exploration and Development, Beijing, PR China

C. Liu
College of Geosciences, China University of Petroleum, Beijing, PR China

ABSTRACT: The conventional AVO attributes have the ambiguity to detect hydrocarbon especially in the tight reservoir for its extremely low porosity. The aim of this paper is to find a best sensitive attribute to pore fluid in tight reservoir to reduce the uncertainty of hydrocarbon prediction via the integration of Monte Carlo simulation and AVO attributes. It gives the key step of method and applied the method to predict gas reservoir from Sichuan Basin of China, the test result shows that it can provide more reliable evidence of suggestion for development wells.

1 INTRODUCTION

Tight reservoirs are widely distributed in China, and the burial depth of tight reservoirs is deeper than 3000m, so the cost of drilling is expensive. It is of crucial importance to assess a few high gas production well locations in a predrill situation, but the low porosity of tight reservoir causes challenges in detecting gas and brine with the conventional AVO attribute such as intercept(P) and gradient(G) due to slight AVO response (Yonggen Li, 2008), which are contributed to the nonuniqueness of interpretation with seismic data. We develop a way to find a new seismic attribute which can improve the successful rate of high gas production wells.

2 METHOD

Based on the analysis of statistical rock physics of well data, Monte Carlo simulation, and AVO modeling, we find the most sensitive attribute to fluid by combing P and G of AVO attributes to reduce the ambiguity of seismic data interpretation in tight reservoir.

2.1 *Workflow*

Figure 1 shows the workflow of the method. Firstly Monte Carlo simulation is utilized to simulate the characteristic of elastic parameters of P-velocity, S-velocity and density in various states of reservoir subsequent to statistical rock physics analysis of the known well data, and simulation data's characteristic is similar to the known well data, and the linear or non-linear relationship of each two parameters should be considered. And then we develop an AVO modeling (Ross, 2000. Yongyi, 2007.) from Zoeppritz equation to obtain the AVO attribute of P and G, and then make a crossplot of P and G. Eventually we reconstruct the most sensitive attribute to fluid in tight reservoir via the axis rotation of P and G attribute subsequent to the seeking of rotation angles.

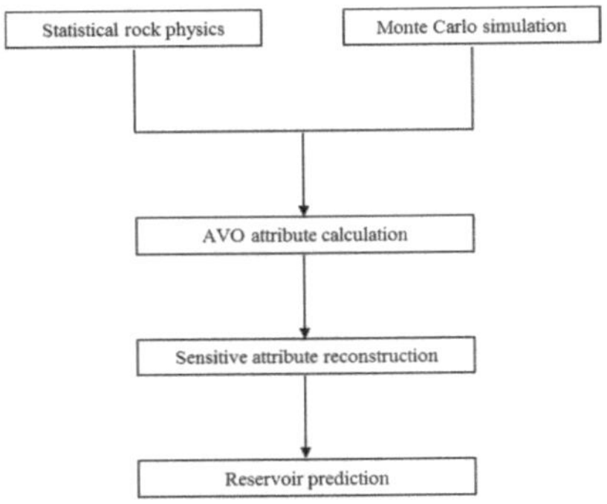

Figure 1. Workflow

2.2 *Monte Carlo simulation*

Well log data provide information about petrophysical and elastic properties of the subsurface but do not necessarily cover all possible scenarios of interest that we could encounter away from well control. To account for plausible variations in the subsurface, we can stretch a gas saturation interval or increase or reduce the gas saturation in the reservoir. Our goal is to perturb these properties in order to describe geological scenarios not represented in the well. The most common technique for achieving this goal is the Monte Carlo simulation (Jack Dvorkin, 2014. Per Avseth, 2005). The method can be divided into four steps: (1) assume a geologically plausible probability distribution of a target rock property (e.g., porosity) that covers ranges not present in the well; (2) randomly sample the property from this distribution; (3) use a deterministic transform between rock properties and its elastic attributes, which could be a theoretical rock physics model or simply a relevant empirical trend and compute the elastic properties for each sampled rock property value; and (4) use this set of computed elastic properties to create a probability distribution of Vp, Vs, and density. Figure 2 (gas reservoir) and Figure 3 (tight reservoir) show the distribution of input well data and simulation data in different states of reservoir, and these figures illustrate the distribution shape of simulation data is similar with the input well data, but the simulation data has more data points than the input data, so it can cover more variations of the reservoir.

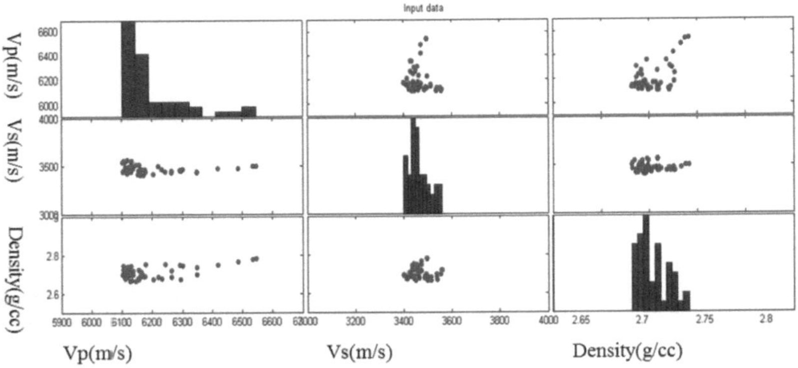

(a)

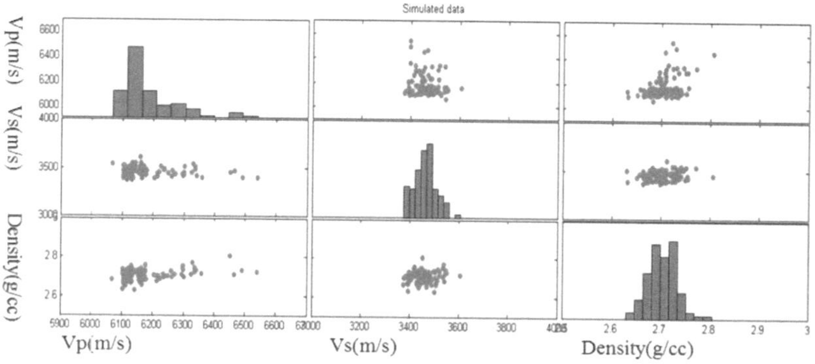

(b)

Figure 2. Monte Carlo simulation of gas reservoir. (a): input data; (b): simulation data

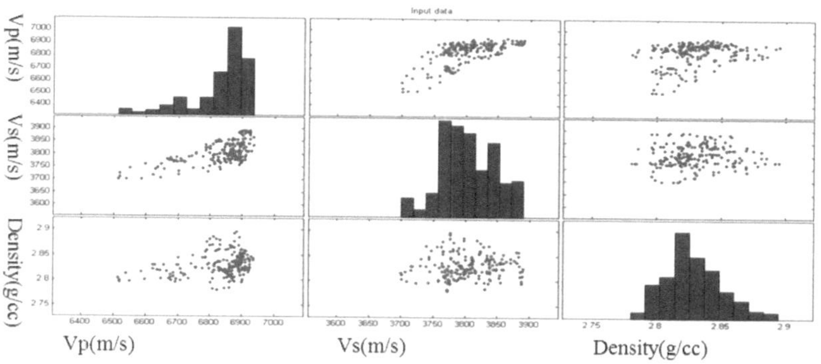

(a)

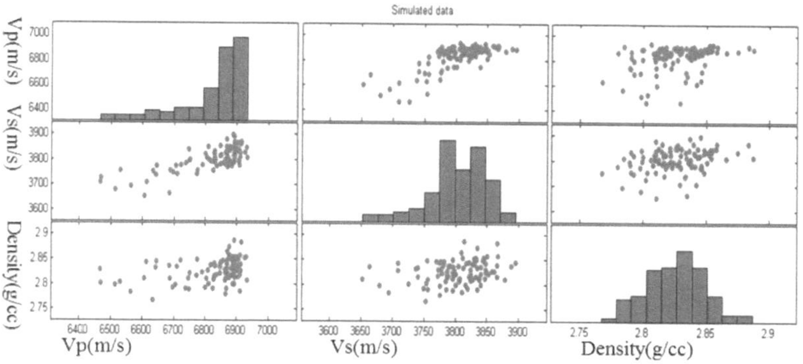

(b)

Figure 3. Monte Carlo simulation of gas reservoir. (a): input data; (b): simulation data

2.3 *Sensitive seismic attribute reconstruction*

Forward modeling of seismic reflections at an interface between two elastic halfspaces is a traditional way of setting expectations for the character of seismic traces between the overburden rock and different states of reservoirs (Figure 4). To conduct such computations, the elastic properties of both half-spaces are required. The elastic parameters of overburden rock are set to a constant obtained by averaging the parameters of the overburden rock, and the elastic parameters of gas or tight reservoir were acquired by Monte Carlo simulation. Additionally the AVO attribute of P and G can be calculated using Shuey's (1985) approximation of the Zoeppritz (1919) equations, and then crossplot P and G (Figure5.a), finally rotate the axis to merge the P and G attributes to a new attribute named WS (figure 5.b). This figure shows that it is much easier to discriminate the gas and tight reservoir with the WS attribute. The axis rotation is actually the linear transform of X and Y axis, and WS is derived by the Equation 4.

$$R_{PP} = \mathrm{P} + \mathrm{G} \cdot sin^2\theta \tag{1}$$

$$P = \frac{1}{2}\left(\frac{\Delta V_P}{V_P} + \frac{\Delta\rho}{\rho}\right) \tag{2}$$

$$G = \frac{1}{2}\left(\frac{\Delta V_P}{V_P} - \frac{\Delta\rho}{\rho} - 2\frac{\Delta V_S}{V_S}\right) \tag{3}$$

Where R_{PP} is the P-P wave reflections coefficient at different angles, P and G are intercept and gradient of AVO attributes respectively, ΔV_P, ΔV_S, $\Delta\rho$ are the difference of P-Velocity, S-Velocity, and density of both side of the interface respectively, V_P, V_S, ρ are the average of P-Velocity, S-Velocity, and density of both sides of the interface respectively.

$$\mathrm{WS} = 4.1826\Delta\mathrm{P} - 0.2464 - \mathrm{G} \tag{4}$$

Where WS is a new attribute from P and G

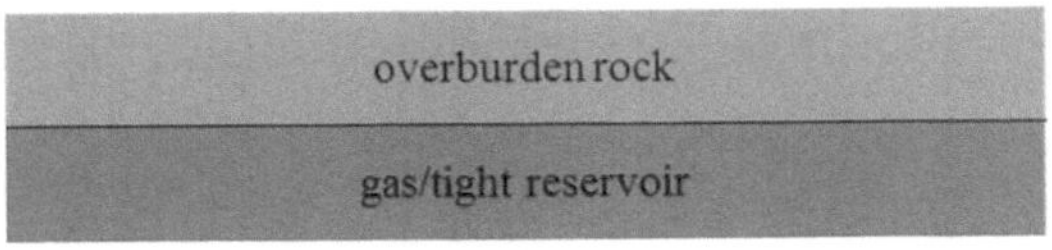

Figure 4. halfspaces model diagram

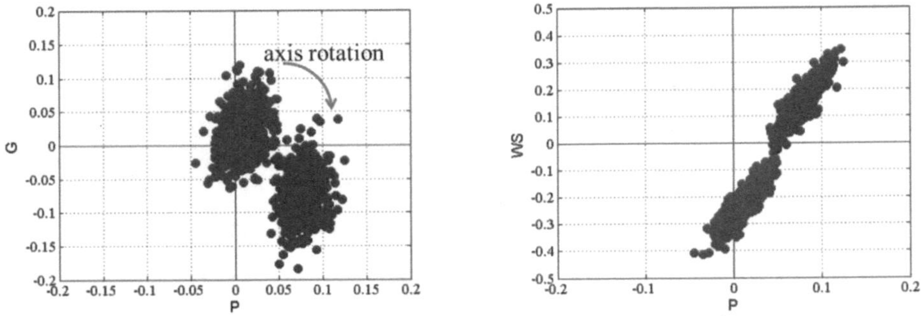

(a) (b)

Figure 5. P vs. G cross-plot (a) and P vs. WS cross-plot (b)

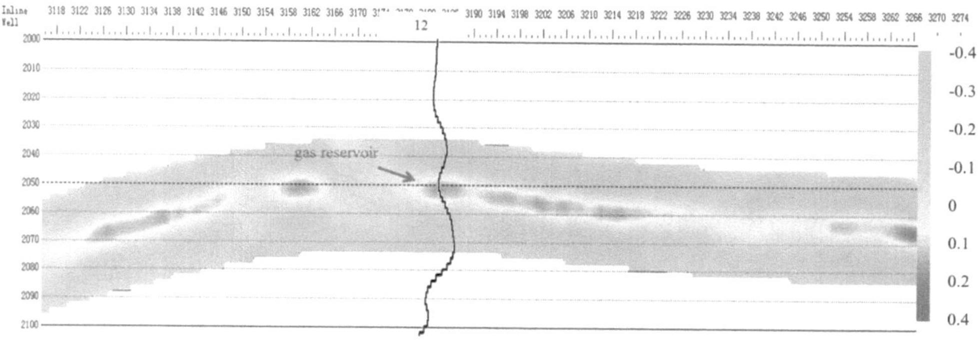

Figure 6. WS attribute section of the project

3 CASE STUDY

3.1 *Geological background of the project*

The Longwangmiao Fm gas reservoir in the Moxi area, Sichuan Basin in China, is a super giant monoblock marine carbonate gas reservoir with its single size being the largest in China (Guangrong Zhang, 2017. Chunchun Xu, 2014. Xizhe Li, 2017). The key to drill high production gas wells is to identify effective reservoir accurately whose porosity is extremely low, ranging from 2-6%, so it is crucial to find the most sensitive attribute to porosity to improve the rate of high production well (Hua T., 2012).

3.2 *Result analysis*

Comparing with conventional AVO attributes, the case study shows that the new attribute obtains the more consistent result with drilling result. Compared with G (Figure 7.a) attribute, WS attribute (Figure 7.b) is much more consistent with the result of well testing, and it can describe the lateral distribution of gas reservoir (Figure 6). The magenta bar indicates the gas volume, and the blue one indicates the water volume. From the Figure 5.b, most of high gas production wells are located in the negative anomaly of WS attribute.

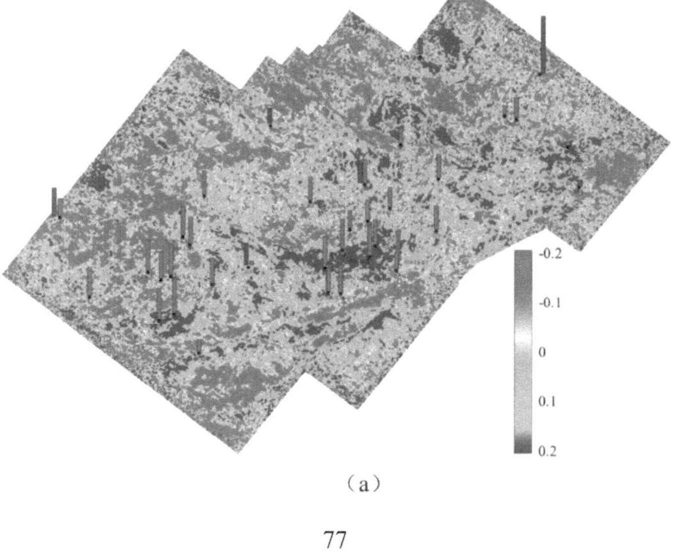

(a)

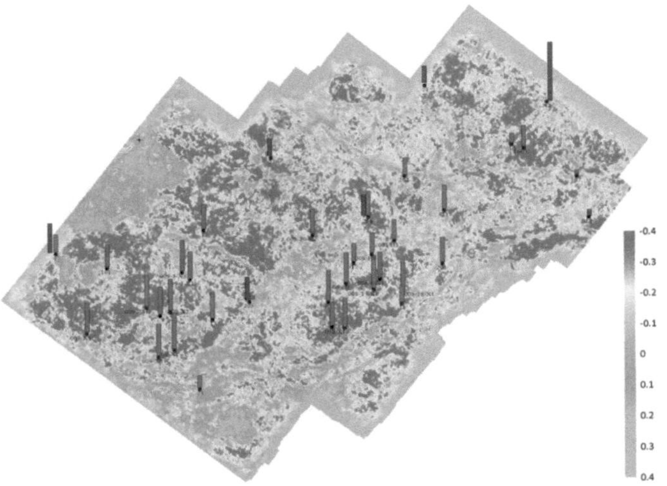

(b)

Figure 7. Hydrocarbon detection with G (a) and WS attribute (b) (The yellow and red background colors predict the later distribution of gas reservoir with attribute, the overlay magenta and blue bar chart represent gas and water of test production volume.)

4 CONCLUSION

The method presented in this paper is proved to be valid in detecting hydrocarbon with seismic data in tight reservoirs, and the new attribute result varied from elastic parameters inversion is not easily affected by seismic noise, and it combines the advantages of P and G, and the fusion attribute has the same anti-noise ability with P and G which are fit by CRP gathers. This method can be of great value to provide a cost-efficient way to increase the probability of successfully choosing the drilling locations. It provides a convenient and qualitative way to generate hydrocarbon distribution maps from 3-D prestack seismic data that can be used as input for well planning in hydrocarbon exploration and reservoir development. Such maps can furthermore serve as constraints in reservoir simulation and production forecasting.

Above all, it provides us an innovative way to produce a new attribute by axis rotation, it can not only be applied to the P and G attribute, but also to other attributes, so in the future, we can optimize the most two sensitive attributes to obtain a new attribute to improve the rate of high gas production wells.

REFERENCES

Avseth, P., Mukerji, T., & Mavko, G. 2010. *Quantitative seismic interpretation: Applying rock physics tools to reduce interpretation risk*. Cambridge university press.

Chunchun Xu, Ping Shen, Yueming Yang, etc. 2014. Accumulation conditions and enrichment patterns of natural gas in the lower Cambrian Longwangmiao fm reservoirs of the leshan-longnüsi palaeohigh, Sichuan basin. *Natural Gas Industry B*, 1(1),51-57.

Dvorkin, J., Gutierrez, M. A., & Grana, D. 2014. *Seismic reflections of rock properties*. Cambridge University Press

GuangRong Zhang, Qi Liao, & Yi Yu, etc. 2017. Seismic prediction on the favorable efficient development areas of the Longwangmiao Fm gas reservoir in the Gaoshiti–Moxi area, Sichuan Basin. *Natural Gas Industry B*, 4(3),203-211.

Hua, T., Shuichang, Z., Shaobo, L., Xingzhi, M., & Hong, Z. 2012. Parameter optimization of tight reservoir porosity determination. *Petroleum Geology & Experiment*, 34(3),334-339.

Ross, & Christopher. P. 2000. Effective AVO Crossplot modeling: a tutorial. *GEOPHYSICS*, 65 (3),700-711.

Shuey, R. T. 1985. A simplification of the Zoeppritz equations, *Geophysics*: 50, 619–624, doi: 10.1190/1.1441936.
Yongyi, LI., Jonathan, Downton., Yong, XU. 2007. Practical aspects of AVO modeling. *The Leading Edge*, 26.3: 295-311.
Xizhe LI, Zhenhua GUO, Yujin WAN, etc. 2017. Geological characteristics and development strategies for Cambrian Longwangmiao Formation gas reservoir in Anyue gas field, Sichuan Basin, SW China. *Petroleum Exploration and Development*, 44(3): 398-406.
Yonggen Li, Shengfeng Xu. 2008. Application of Seismic Rock Physics and Forward Simulation in Predicting Tight Sandstone Reservoirs. *Journal of Oil and Gas Technology*, 30(6),79-83.
Zoeppritz, K. 1919. Erdbebenwellen VIIIB, *On the reflection and propagation of seismic waves*, Gottinger Nachrichten: I, 66–84.

Youth Technical Sessions Proceedings – Litvinenko (ed)
© 2019 World Petroleum Council, London, UK, ISBN 978-0-367-34668-3

Application of machine learning techniques to estimate petrophysical properties

D.A. Balashov & D.S. Grigorev
Gazpromneft STC, Saint Petersburg, Russia

ABSTRACT: Importance of well-logging interpretation in reservoir characterization cannot be overestimated. However, it is not always possible to conduct a logging operation with a complete set of tools as some of them are not available at the field or economically expensive. Moreover, relationships used to evaluate petrophysical properties require prior knowledge of rock matrix properties which inevitably introduce a bias into the result of interpretation. The goal of this work is to build a general mathematical model based on the machine learning algorithms which enables to obtain porosity and permeability values accurately utilizing only well log data. The potential application of machine learning to this problem in comparison to the traditional well-logging interpretation is also explored. It is worth to note that the research is based on the actual information from several oil wells in Western Siberia.

1 INTRODUCTION

Well-logging interpretation plays a crucial role in determining plenty of formation properties such as porosity, permeability, water saturation and others just to name a few. Analysis of logs behavior helps to recognize intervals of interest, which may contain and produce hydrocarbons. Subsurface formations contain prolific amount of valuable data and require implementing operations with bunch of logging tools to extract it fully. The quantity of information and the quality of its processing exert strong influence on the key oil field characteristics such as reserves volume, potential productivity of wells, etc. All the parameters described above are highly important to form an opinion about the reasonability and possible strategy of future field development. At the same time, there is no general rule of reservoir characterization techniques, which gives an opportunity to formulate the universal approach that eases the work process of petrophysicists. This paper is narrowed to the exploration of the algorithms of facies classification and effective porosity and absolute permeability evaluation.

2 OBJECT FOR THE ANALYSIS

One of the oil fields in Tomsk region, Russian Federation, is used that is at the exploration stage. Eight exploration wells were available for the examination, which are spread over the field, and penetrates the reservoirs at different absolute depths with variable saturation. The interval of an interest is found in the Upper Jurassic sediments in Upper Vasuganskaya sub-formation.

The field contains three reservoirs, which are U1, U2 and U3 in the down direction. Core was recovered and well logs were measured in each well what allowed the sedimentological and petrophysical interpretation execution. Moreover, machine-learning algorithms require features for training, which are accessible in all wells used. Unfortunately, different logs had been measured in wells. Therefore, the best possible case was to filter wells and logs and remain with four wells and eight logs what is still sufficient for the work.

3 SEDIMENTOLOGICAL CONCEPT

The field consists of three major reservoirs with different complexity and compartmentalization. Recovered core was analyzed for the sedimentological features with electro-facial analysis simultaneously what led to the sedimentological concept for an each particular layer.

The lower reservoir, according to the authors' concept, was formed in the delta-front conditions. The coarsening upwards succession from core is reflected in the decreasing SP values to the upper section of the reservoir. All these features allowed authors understanding reservoir properties, which are better petrophysical properties in the upper section of the layer that continuously deteriorate down to the bottom. Thus, the upper section is referred to the mouth bar while the lower section is the transition zone of delta front between the prodelta and the mouth bar (Walker 1976).

The second layer was formed in the delta plain sedimentological conditions when the sea regression still was predominant. Two main facies are found by the core analysis and form of SP log which are crevasse splays and channel fill facies. Channel fill facies is characterized by the fining upwards succession and it may be identified by increasing values of SP log (Baraboshkin 2007). Due to the limited lateral size of splays, the second layer is composed of lenticular sand bodies mostly.

The upper layer was formed in another sequence when the regression slowed down and sedimentary supply was limited. The delta front and channel fill facies are the majors for the upper layer.

4 PETROPHYSICAL INTERPRETATION

Performed petrophysical interpretation may be called conventional since the authors followed the basic steps of an interpretation except for the permeability calculation (Korovin et al. 2014). Thus, the first step was to perform lithological interpretation, which was based on SP and GR logs mostly due to the clastic succession. Next step was to estimate the shaliness of the reservoir using GR log and Clavier correction to obtain the effective porosity. Neutron logs, acoustic log and α-SP were basic for porosity calculation when the matrix and drilling mud properties are reference data. The confirmation of the correct porosity estimation is core matching which is core porosity comparison to the calculated values by cross-plotting. Neutron log was also normalized with help of core data to convert neutron index into neutron porosity. Then, water saturation calculation was performed when LLD resistivity was used as a reservoir property, water resistivity was estimated by known salinity and elements presence. "a", "m" and "n" values are estimated by core measurements which are formation factor and resistivity index with porosity and water saturation relationships.

As for permeability estimation, hydraulic flow unit (HFU) theory was integrated in facial analysis to better obtain the facial margins. Thus, flow zone index (FZI) probability distribution function is constructed for each layer to get the number of HFUs. Since the interpretation of pdf plot was controversial, the optimal number of HFU was found by iterative increase in number of HFU with absolute error calculation between core permeability and permeability obtained by HFU theory. The optimal value is such where the "elbow" in error decrease is found (Svirsky et al. 2004). As a result, two HFU are optimal for layers U1 and U2 while the HFU theory was not used for the layer U3 because core data are found only at the upper section of the reservoir that is a mouth bar. Consequently, it was not possible to use more than one correlation. The HFU units converged to the facial analysis and facial margins were refined with HFU theory, especially, where electro-facial interpretation was controversial. The resulting correlations for each facies are given in Table 1. The final variant of the well logs interpretation for one of the wells is shown on the Figure 1.

Table 1. Permeability-porosity correlations

Type	Layer	Facies	Correlation	R
Facies 1	U1	Delta Front	$k = 0.0409 \cdot e^{34.639 \cdot \emptyset}$	0.84
Facies 2	U1	Channel Fill	$k = 0.0001 \cdot e^{63.577 \cdot \emptyset}$	0.81
Facies 3	U2	Crevasse Splay	$k = 0.1041 \cdot e^{25.193 \cdot \emptyset}$	0.88
Facies 4	U2	Channel Fill	$k = 0.0405 \cdot e^{19.323 \cdot \emptyset}$	0.82
Facies 5	U3	Delta Front	$k = 0.001 \cdot e^{53.474 \cdot \emptyset}$	0.84

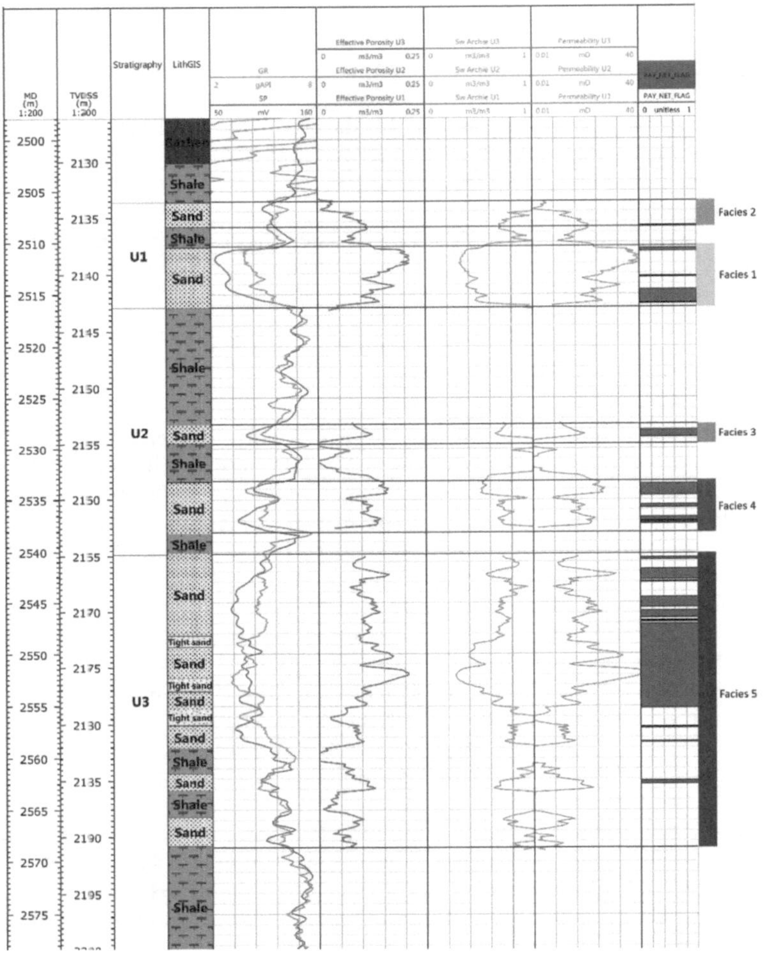

Figure 1. Petrophysical interpretation

5 EXPLORATORY DATA ANALYSIS

As porosity and permeability values of target intervals were determined, and main five facies were defined, exploratory analysis of data should be conducted. This is a preliminary step before machine learning models construction where input parameters are visualized in order to get more information about anomalies, variables distribution and degree of correlation between them. By taking into consideration all patterns amongst the variables, regression models can be

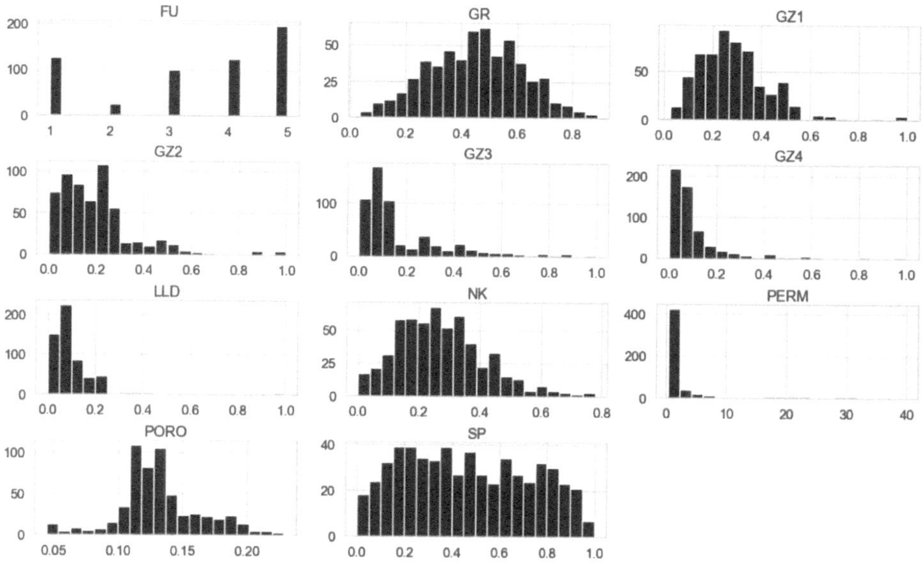

Figure 2. Histogram plots of input variables.

built with more insights and a range of specific questions answered (Behrens 1997). It should be noted that all data set in this work is analyzed graphically as a whole without separation between input and output variables. Besides, the choice of variables included into the training data set is dictated by the field experience and the knowledge in the research area.

One of the main tools to understand data in a graphical form is to build histogram plots of each parameter (Zhong et al. 2015). Figure 2 provides a bunch of well logs, porosity and permeability histogram plots. It should be highlighted that flow units have almost uniform distribution, and porosity and permeability values have similar distributions to normal and log normal distributions respectively. Moreover, it is seen from the plots that permeability values are in the range from 0.3 to 40 millidarcies. The low limit is dictated by cut-off value to divide the interval into reservoir and non-reservoir intervals. All well logs values are normalized in the range from 0 to 1 as a step of data preprocessing.

6 MACHINE LEARNING

The further step of the predictive model construction is to determine what kind of tasks should be solved, and several suitable algorithms should be chosen in order to obtain the most accurate and robust predictor. Nowadays, the advent of machine learning techniques provides vast opportunities to experiment for data scientists and engineers, so the choice in this paper has been limited to the popular ones, such as k-nearest neighbors, random forests and gradient boosting machine that will be discussed in detail further. Exactly these mathematical models are employed to cope with target parameter prediction or classification of objects, and the known answers are given in advance. These tasks are called regression and classification, and they are both related to the supervised machine learning class where the output values are already provided.

The total approach is divided into two steps. Firstly, whereas the intervals in each well have already been interpreted into five geological faciesthe classification task can be solved. It implies that the predictive model can be built to split a new interval in a new well into several groups in accordance to the performed facies recognition. Secondly, well log and facies classification data is used to build the regression model which can successively predict porosity and permeability values. The results show that facies classification included into the final regression model sufficiently increases its predictive power.

First of all, it should be noted that there is no predominant class amongst the facies groups. It gives some freedom to use the simplest metrics to evaluate the quality of the algorithm such as accuracy score which counts the proportion of right answers to the whole set of answers. As the metric is chosen few words should be said about each machine learning technique mentioned above.

The k-nearest neighbors algorithm can be characterized as a universal and simple one as it is used to deal with both classification and regression tasks. The former problem is solved by the assignment of that class which the nearest neighbors in the feature space have. The latter has similar decision but instead of class voting amongst the neighbors the average value is allocated to the target object. The number of neighbors is chosen by the engineer at the training step and plays a role of hyperparameter to tune the predictive power of the classifier. In this paper, the distance between objects in multi-dimensional space is measured by Euclidean distance as it is wide-spread measure, and the input data contains only continuous variables.

The another algorithmused in the research is gradient boosting which is also successfully applied to both classification and regression tasks. The idea of gradient boosting decision trees is to incorporate numerous 'weak' classifiers in one ensemble where 'weak' implies that its predictive power slightly exceeds the random guessing (Friedman et al. 2001). The key benefits of this approach are its relatively high accuracy of predictions, flexible tuning of model hyperparameters and tolerance to the data sets with missing values.

The following step is to train the model on the training set and check the quality of the model on the test set. The idea of such split is to get unbiased estimate of the model's predictive ability on the datawhich is not previously used to construct the model. In this research, split of the initial set to the training and test ones is made in proportion 4:1 in accordance to the commonly used practice. Figure 3 presents the accuracy score for both methods on the test set. It can be concluded that the gradient boosting algorithm shows more promising results predicting correctly 93 objects out of 100.

As facies are distributed along the intervals in all wells with aid of already build classifier, the second task of machine learning can now be solved where the regression model should be made in order to forecast continuous data (porosity and permeability). The advanced machine learning algorithms, such as random forests and gradient boosting are utilized in this part of the study.

Random forest is related to ensemble methods where each tree is built via bagging algorithm employed to diminish variance and prevent overfitting (Friedman et al. 2001). Firstly, the subset of variables for each tree is chosen randomly with replacement. Secondly, subset of features at each node of the tree is chosen where the best split is based on the specific criteria (variance reduction, for example). Finally, the average prediction is obtained amongst all

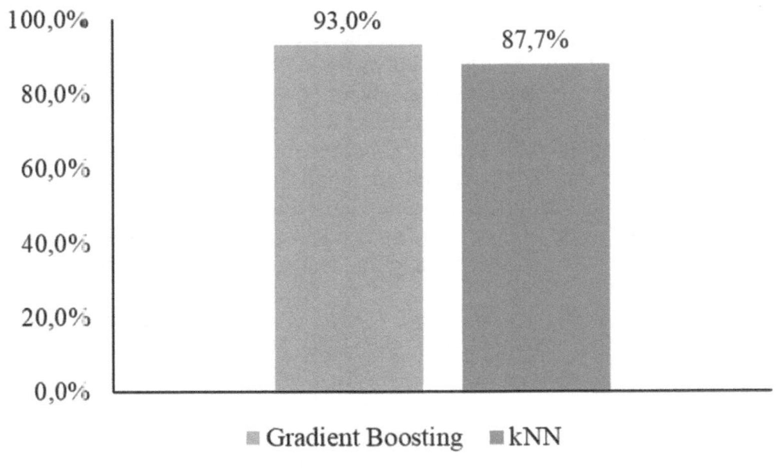

Figure 3. The comparison of two classifiers on the test set.

decision trees in the ensemble. The advantages of this approach are the ability to process large amounts of data and efficiently manage both categorical and continuous variables.

Firstly, porosity values are predicted having the information from well logs and facies classification. Again, the data is split into two subsets in proportion 4:1. The quality of the model is measured by mean absolute error (MAE)calculated in Equation 1 and coefficient of determination (R^2). Table 2 presents the comparison of two algorithms. Clearly, gradient boosting slightly outperforms random forest by both criteria.

$$MAE = \frac{\sum_{i=1}^{n} |y_i - y|}{n} \quad (1)$$

where y_i - actual value; y – predicted value, n – number of objects.

Another approach to estimate the quality of the model is to perform cross-validation, which potentially helps to avoid overfitting i.e. strong ability of the model to repeat the fluctuations of the training data, but gives poor predictions on the independent set. The procedure of cross-validation implies that each $1/n$ of the groups is set aside while the model is trained on the remaining $(n\text{-}1)/n$. This approach is commonly used by scientists to similar problems (Lolon et al. 2016; Schuetter et al. 2015; Wang & Chen 2016). In this work, n =10 which means that 10-fold cross-validation is performed (i.e. 10 regression models are built, and accuracy is measured by averaging the results). Figure 4 presents the comparison of actual values of porosity and predictions made by gradient boosting model.

The final step is to build the model able to predict permeability values. The procedure of data preparation is repeated,and the results of both algorithms are presented in Table 3. The distinct feature of this step is that the model is trained to forecast logarithm of permeability as it is commonly accepted that relationship between porosity and permeability (logarithm) can

Table 2. The results of porosity forecast on the test set

Machine learning algorithm	Accuracy metrics	
	R^2	Mean Absolute Error
Random Forest	0.885	0.0073
Gradient Boosting	0.895	0.0068

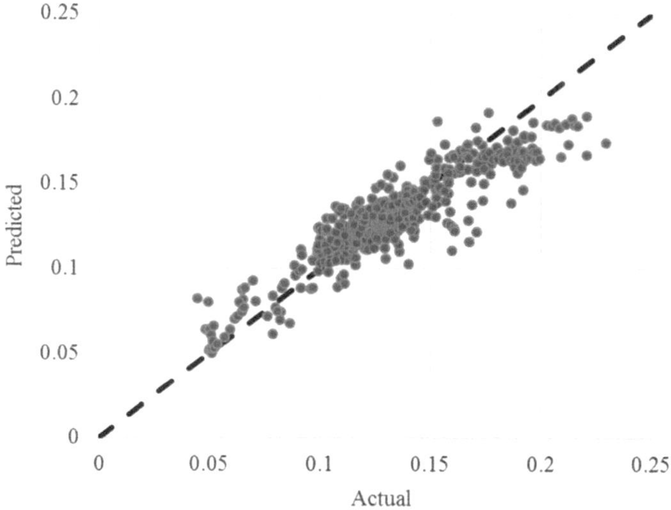

Figure 4. The comparison of actual and predicted porosity values (MAE = 0.0104; R^2 = 0.806)

Table 3. The results of permeability (logarithm) forecast on the test set

Machine learning algorithm	Accuracy metrics	
	R^2	Mean Absolute Error
Random Forest	0.988	0.036
Gradient Boosting	0.992	0.0348

be approximated by linear correlation.Porosity is added to the initial data set as it is expected to improve the accuracy of predictions dramatically. The gradient boosting model demonstrates more accurate result than random forest by both metrics.

The procedure of 10-fold cross-validation gives less promising results: MAE = 0.138 and R^2 = 0.89. These results appear to be more reliable and reflect the actual predictive power of the regression model.

7 CONCLUSIONS

Undoubtedly, well-logging operations will maintain the position on the market as acost-effective and reliable instrument to get information of high quality about the reservoir. The advent of machine learning and data analysis offers novel approaches to the problem of petrophysical interpretation. In this study, two tasks of supervised machine learning applied to reservoir characterization are solved. Firstly, the solution to classification task is provided where the independent data has been classified into five facies with accuracy of 93%. These flow units then are given to the input of regression models which have been trained to forecast successively porosity and permeability values. The final accuracy is accepted to be MAE = 0.0104; 0.138 and R^2 = 0.81; 0.89 for porosity and permeability best predictors respectively. The main advantages of the final models are its high accuracy and robustness, and practical ease of use. The latter implies that only predetermine set of common well logs is required, and there is no need to tune the model for each specific case.

REFERENCES

Baraboshkin, E. 2007. Practical sedimentology (terrigenous reservoirs). Tomsk: Tomsk Polytechnic University.

Behrens, J. T. (1997). Principles and procedures of exploratory data analysis. *Psychological Methods*2(2): 131.

Friedman, J., Hastie T., and Tibshirani R. 2001. The elements of statistical learning. Vol. 1. No. 10. New York: Springer series in statistics.

Korovin, M. & Merkulov, V. 2014. Formation evaluation and well-test analysis for complex interpretation of reservoir permeability distribution. In IOP Conference Series: Earth and Environmental Science (Vol. 21, No. 1, p. 012017). IOP Publishing.

Lolon, E., Hamidieh, K., Weijers, L., Mayerhofer, M., Melcher, H., & Oduba, O. 2016. Evaluating the Relationship Between Well Parameters and Production Using Multivariate Statistical Models: A Middle Bakken and Three Forks Case History. Society of Petroleum Engineers. doi:10.2118/179171-MS

Schuetter, J., Mishra, S., Zhong, M., & LaFollette, R. 2015. Data Analytics for Production Optimization in Unconventional Reservoirs. Unconventional Resources Technology Conference. doi:10.15530/URTEC-2015-2167005

Svirsky, D., Ryazanov, A., Pankov, M., Corbett, P. & Posysoev, A. 2004. Hydraulic Flow Units Resolve Reservoir Description Challenges in a Siberian Oil Field. Society of Petroleum Engineers. doi:10.2118/87056-MS

Walker, R. 1976. Facies Models 2. Turbidites and associated coarse clastic deposits. Geoscience Canada, 3(1).

Wang, S. & Chen, S. 2016. A Comprehensive Evaluation of Well Completion and Production Performance in Bakken Shale Using Data-Driven Approaches. Society of Petroleum Engineers. doi:10.2118/181803-MS

Zhong, M., Schuetter, J., Mishra, S., & LaFollette, R. 2015. Do Data Mining Methods Matter?: A Wolfcamp Shale Case Study. Society of Petroleum Engineers. doi:10.2118/173334-MS

Youth Technical Sessions Proceedings – Litvinenko (ed)
© 2019 World Petroleum Council, London, UK, ISBN 978-0-367-34668-3

Experimental investigation of CO_2-brine-rock interaction at in-situ reservoir conditions and its implication for tight oil EOR

S.T. Wu, X.F. Zhai, H.W. Yu, Z.C. Yu & Z. Yang
CNPC Research Institute of Petroleum Exploration and Development (RIPED), Beijing, China

ABSTRACT: The effects of supercritical CO_2 on the reservoir property changes of the Chang 7 Member tight sandstones of the Yanchang Formation in the Ordos Basin, China, was investigated using CO_2-brine flow-through experiments and in conjunction with a detailed characterization of the rocks and fluids before and after the experiments using X-ray diffraction mineral analysis, 3D Computed Tomography (CT) scanning, high-resolution scanning electron microscopy (SEM), and ICP-AES. An in situ 3D porosity-mineral comparison model has been established in the paper. Minerals such as potassium (K) feldspar, albite and calcite are variably dissolved. The dissolution, migration and re-precipitation of clay minerals are common, which is characterized by the dissolution of chlorite and the re-precipitation of kaolinite, calcium montmorillonite, and sodium montmorillonite. The original chlorite-kaolinite framework was destroyed. It is the precipitation and re-migration of clay particles (e.g., kaolinites and smectites) released by the dissolution, which have been transported in the fluid flow path and accumulated at pore throats, that result in the changes of the pore system and physical properties. The trend of physical property evolution varies among the studied samples, and the porosity is reduced by 4.5%~12.8% after the experiment except an increase of 13.7% in one sample. The permeability reduction is 33.3%~81.2%, which is positively related to the clay mineral content. The influence of the type and content of clay minerals on the porosity is in conclusive and needs to be further studied.

1 INTRODUCTION

The application of CO_2 in EOR and CCUS programs has become a worldwide concern for its significance in environmental protection and its substantial economic returns (Bredesen *et al.*, 2004; Shukla *et al.*, 2010; Sorensen and Hamling, 2016; Moodiea *et al.*, 2019). As is known to all, the great success of tight oil in North America has changed the global energy landscape, yet how to enhance the low oil recovery (only 5%~8%) is a great challenge. The application of CO_2 in tight oil EOR has just commenced, and a large number of scientific problems have to be solved, in which, the reaction between CO_2, brine and minerals is one of the PRIORITIES.

Although previous studies have investigated the CO_2-brine-rock interaction, yet some important problems are still un-answered. Firstly, the influence of CO_2-brine-rock interaction on reservoir physical properties is still controversial at present. Yu et al. (2012), and Abdoulghafour et al. (2016) found that CO_2-brine-rock interaction caused significant decrease in core porosity, pore volume and permeability through experiment; However, Yang et al. (2010) believed that CO_2-brine-rock interaction could improve reservoir physical properties via CT scanning. Dávila et al (2016) and Al-Zaidia et al (2018) discussed the effects of CO_2 storage time and fluid velocity on reservoir porosity and permeability, respectively. Secondly, the published literatures haven't covered CO_2-brine-rock interaction in tight sandstone. The previous study objects had porosity generally >18%, and permeability >10 mD (Yu et al., 2012; Dávila et al., 2016). Thirdly, most researchers used XRD and SEM to study the dissolution and precipitation of minerals, but lacked the fine comparison of the same mineral and the same location before and after experiments. The adopted CT resolution was usually low (Yu et al., 2012; Andrew et al., 2014; Dávila et al., 2016), thus it is difficult to compare micro- and nano-scaled pore structures.

This article aims at investigating more closely in the laboratory the CO_2 injection and storage scenarios and providing references for enhancing recovery by CO_2 injection and for effective CO_2 storage in tight sandstones. To this end, CO_2 injection experiments are conducted on brine-saturated reservoir rocks with different microstructure and mineral content. The Triassic Chang 7 tight sandstone of the Ordos Basin was studied, and X-ray diffraction mineral analysis, 3D Computed Tomography (CT) scanning, high-resolution scanning electron microscopy (SEM), quantitative evaluation of minerals with SEM (QEMSCAN) and ICP-MS were used to determine the evolution of mineral composition and pore structure, as well as the changes of reservoir properties.

2 SAMPLES AND METHOD

The Triassic Chang 7 tight sandstones in the Ordos Basin, the most successful tight oil E&P targets in China, have been selected as the focus of the research. Three core samples were used in the experiments, which are numbered A-173, H-25, and Y-46. All the core samples were 2.54 cm in diameter (D) and the total length (L) of the core was 14.5 cm. The average porosity is 8.56%, and air permeability is only 0.05mD. CO_2-brine-rock interaction is set at in-situ reservoir conditions (78°C, 24MPa). An ionic chemical analysis of the outlet solution, combined with high resolution CT (1.0 micron/pixel), SEM, QEMSCAN and bulk-rock XRD analyses of the core pre- and post-experiment, has been adopted. The core sample preparation is shown in Figure 1.

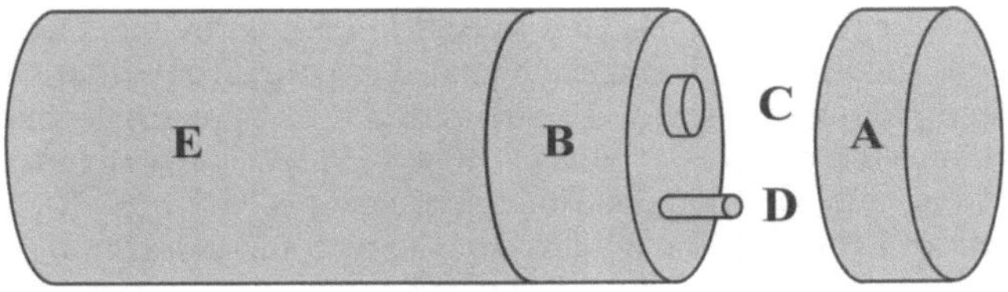

Figure 1. Sketch graph of sample preparation and assembling. Part A and B were used for mineral analysis by X-ray diffraction before and after the experiment; From part B, C and D cylindrical plugs with diameters of 1 cm and 2 mm were drilled respectively. Part C was used for QEMSCAN analysis before and after the experiment. Part D was used for 3D CT scanning analysis before and after the experiment. During the experiment, part C and part D needed to enter part B completely to ensure that the samples were not destroyed under high pressure. Part E was used for measuring porosity and permeability before and after the experiment.

3 RESULTS

3.1 *Changes in water chemistry*

During the experiment, we collected 129.4 ml liquid in total, corresponding to 16.9 times of the total pore volume, in a displacement time of 854 hours. The pH of the original formation water was 7.034. After 36 hours of CO_2 displacement, the water's pH has reduced to 5.387. During the whole experiment, the pH value fluctuated between 5.3 and 6 (Figure 2-A), meaning that the fluid environment was acidic overall.

Different ions were found to have different concentrations (Figure 2). As the interaction progressed, the different ions exhibited differences in their trends. The concentrations of cations including K^+, Na^+, Ca^{2+}, and Mg^{2+} increased in the early stage of the experiment. However, as the interaction continued, the trends in ion concentrations changed. K^+ and Ca^{2+} exhibited similar trends, peaking in the early stage of the experiment and then decreasing gradually before finally stabilizing at about 20 mg/L and 300 mg/L, respectively (Figure 2-B, C). In contrast, Na^+ and Mg^{2+} concentrations stabilized at peak with little change and reached

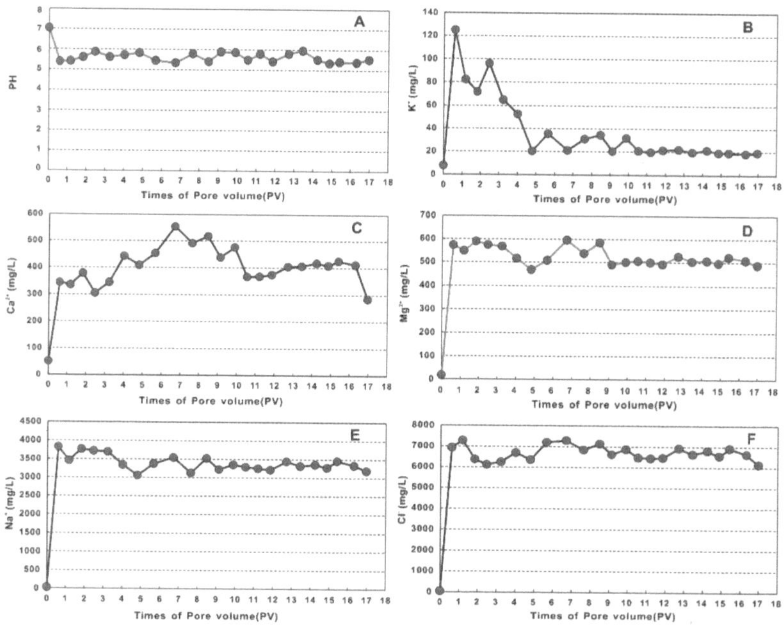

Figure 2. Curves of typical ions and pH values during the experiment. A-pH values, B-K^+; C-Ca^{2+}; D-Mg^{2+}; E-Na^+; F-Cl^-.

3200 mg/L and 500 mg/L, respectively (Figure 2-D, E). The concentrations of Cl^- and $SO4^{2-}$ increased rapidly and peaked at the beginning of the experiment, and then experienced short-term decline when the displacement CO_2 reached three times of pore volume, finally stabilizing when the displacement CO_2 reached eight times of pore volume (Figure 2-F).

3.2 *Mineralogical changes*

SEM images indicate that non-clay minerals (i.e., potassium feldspar, albite and calcite) are variously dissoluted (Figure 3-A_1-A_3, B_1-B_3). The dissolution occurres along the cleavage surfaces of potassium feldspar, biotite and other minerals, or along the existing pore space, and further enlargement of the existent reservoir space. At the same time, a special mineral, halite, on the surface of albite is also dissolved to form new reservoir space (Figure 3-B_1-B_3), which can explain why the concentration of Cl^- increases sharply. Quartz, relatively stable, has little dissolution. Generally, the variation characteristics of non-clay minerals are in good agreement with the analysis results of X-ray diffraction.

SEM reveals that Chang 7 tight sandstone in the Ordos Basin has the multi-stage growth of clay minerals, mainly presenting in the forms of chlorite lining growth-kaolinite filling (Figure 3-C_1-C_3), and illite-montmorillonite combination (Figure 3-D_1-D_3). The changes in clay minerals before and after the experiment are more complex: (1) the combination structure of clay minerals is destroyed. The most obvious feature is that chlorite and kaolinite are dissolved by acid solution, and the original multi-stage growth structure is broken, forming dissolution pores. (2) In the process of CO_2 supercritical fluid flow, clay particles (such as kaolinite and montmorillonite) would migrate and precipitate again; the location of secondary precipitation may be the original pores and throats, or the new pores formed by dissolution (Figure 3-A_2, B_2, C_2 and D_2). In the acidic environment, we also observed the re-growth of kaolinite. For example, in the dissolution pores of albite, it can be seen that the distribution of kaolinite after the experiment is obviously beyond the range before the experiment (Figure 3-D_1-D_3). Apparently, the evolution of clay minerals is more diverse than that of non-clay minerals, thus we should play special attention to this in the evaluation of pore structure and reservoir performance.

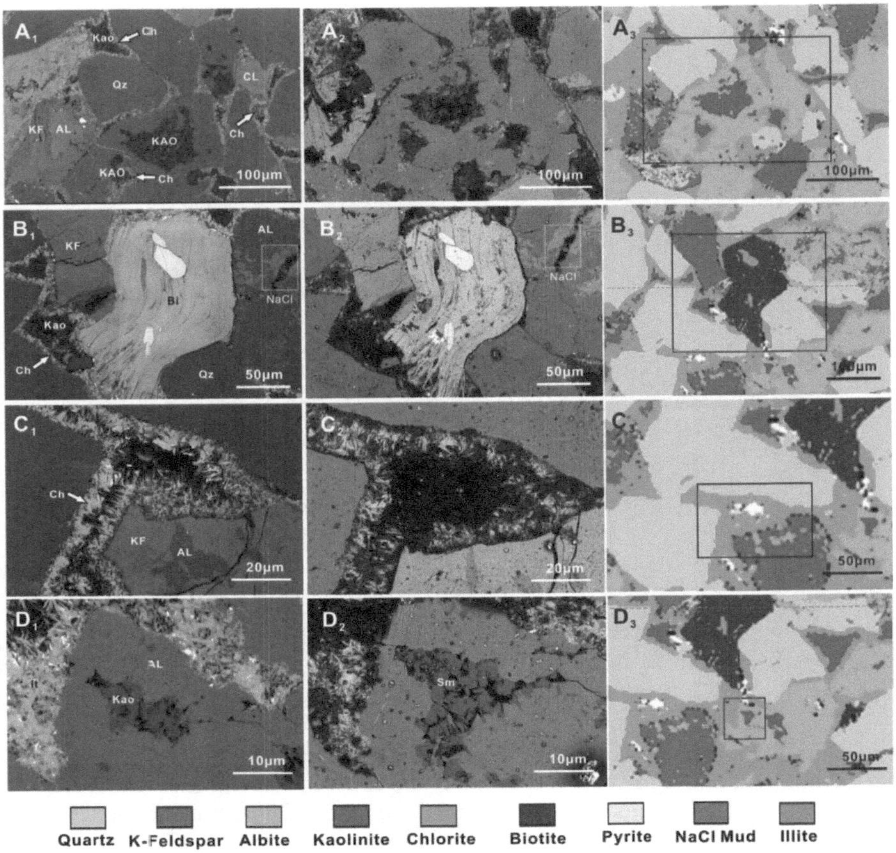

Figure 3. Comparison of in situ minerals before and after the experiment using SEM and QEMSCAN analysis. KF- Potassium feldspar, AL-Albite, CL-Calcite, Qz-Quartz, Ch-chlorite, KAO-Kaolinite, Bi-Biotite. NaCl dissolution was marked with rectangle in white lines.

3.3 *Changes in porosity and permeability*

Figure 4 shows the change of pore structure before and after CO_2-brine -rock interaction. It can be seen that different regions of the sample have obvious differences in pore variation characteristics. Some pores reduce in size (black solid line boxes in Figure 4), while some pores increase in size (white dotted line boxes in Figure 4). The volume change of minerals and the resulting pore structure can be seen in the areas of the white solid boxes in Figure 4. For instance, the re-growth and precipitation of clay minerals can be seen in position D in Figure 4-A_1 and Figure 4-A_2.

The complex effect of CO_2-brine-rock interaction on pore structure is also reflected in the results of physical property testing. The porosity of No.A-173 and No.Y-46 samples decrease, while that of No.H-25 sample increases. The change trend of permeability is basically the same as that of porosity, but the change is more obvious (Figure 5). The porosity of No.A-173 decreased from 11.2% before the experiment to 10.7% after the experiment, with a decrease ratio of 4.5%; while its permeability decreased from 0.028 mD before the experiment to 0.009 mD after the experiment, with a decrease ratio of 60.7%. The porosity of No.Y-46 sample dropped from 9.4% before the experiment to 8.2% after the experiment, with a decrease ratio of 12.8%; while its permeability dropped from 0.117 mD before the experiment to 0.022 mD after the experiment, with a decrease ratio of 81.2%. The porosity of No.H-25 sample increased from 5.1% before the experiment to 5.8% after the experiment, with an increase

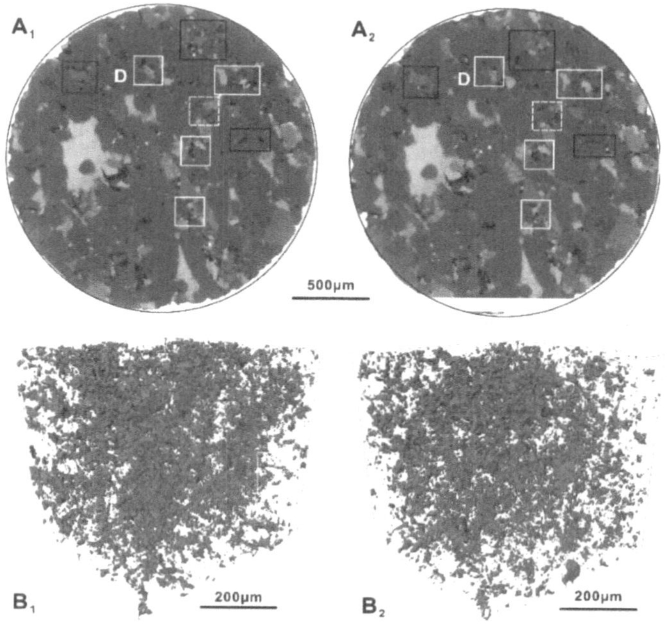

Figure 4. Comparison of CT analysis results of No.Y-46 sample before and after experiment.

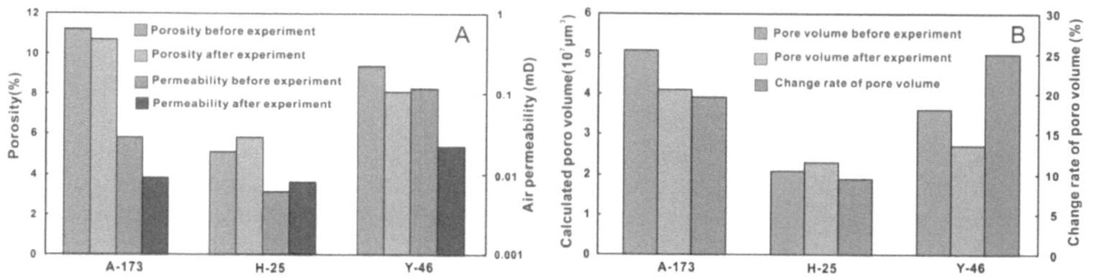

Figure 5. Measured porosity, permeability before and after the experiment and calculation of pore volume change based on CT model

ratio of 13.7%; while its permeability increased from 0.006 mD before the experiment to 0.008 mD after the experiment, keeping basically the same. This means that although the porosity of No.H-25 sample increased to a certain extent, its permeability has not fundamentally improved (Figure 5). CT data also support the same point.

4 DISCUSSION

4.1 *CO_2–brine–rock interaction in the Chang 7 tight sandstones*

Based on our results, changes in ion concentration and mineral transformation in the Chang 7 tight sandstone generally corresponded with each other, and mineral evolution including the dissolution, migration, and re-precipitation of clay minerals and the dissolution of feldspars and calcites, had remarkable effects on pore structure and seepage capacity. Generally, Ca^{2+} is affected by the dissolution and secondary precipitation of calcium-bearing minerals such as calcite and calcium montmorillonite. During the early stage of the experiment, acidic water entered the system and diluted the original water, resulting in a reduction in pH, the dissolution of calcite, and an increase in Ca^{2+} concentration. As the experiment continued, Ca^{2+} concentration continued to increase and combined with other ions to form calcium montmorillonite, leading to a decrease in Ca^{2+} concentration. Mg^{2+} is affected mainly by the dissolution and secondary precipitation of magnesium-bearing minerals such as chlorite and montmorillonite. During the early stage of the experiment, the concentration of Mg^{2+} increased with the dissolution of chlorite. During the late stage of the experiment, chlorite was being dissolved continuously, and montmorillonite was precipitated. There was only a slight change in Mg^{2+} concentration through the experiment because Mg^{2+} was being both produced and consumed.

Based on these results, we constructed a CO_2–brine–mineral interaction model for the Chang 7 tight sandstone. After the injection of $scCO_2$-brine into the rocks, the fluid environment gradually changed from neutral to acidic, resulting in the dissolution and re-precipitation of unstable minerals. Potassium feldspar, albite, calcite and chlorite were dissolved to form new pore spaces. Meanwhile, clay minerals (i.e., kaolinite, calcium montmorillonite, and sodium montmorillonite) were precipitated in both original and new generated pore space, leading to a decrease in pore volume. In addition, through acidic fluid flow, fine particles of clay minerals (i.e., kaolinites) evidently migrated and clustered to occupy smaller pores and throats.

4.2 *The main controls on the performance of tight sandstone reservoirs*

The $scCO_2$–brine–mineral interaction is a relatively complex process involving the synergistic interaction of $scCO_2$ fluid, formation water, and rock minerals under high temperature and high pressure. Thus, the changes of reservoir performance, i.e. porosity and permeability pre- and post-experiment are complex and difficult to predict. The results are largely depended on various factors, including external factors and internal factors. External factors refer to experimental parameters, such as temperature, pressure, experimental duration, and $scCO_2$ injection velocity. Considering the regional particularity of temperature and pressure, most researchers have focused on the influence of experimental duration and fluid injection rate on reservoir porosity and permeability (Shedid *et al.*, 2013; Garcia-Rios *et al.*, 2015; Dávila *et al.*, 2016). Generally, with the extension of experimental time, the porosity will increase (Dávila *et al.*, 2017). Internal factors, mainly those related to mineralogy and pore system are the key to understand the whole evolution process. The dissolution and secondary precipitation of unstable minerals and clay minerals in acidic environments (such as feldspar and calcite) have attracted close interest from researchers (Yu *et al.*, 2012; Dávila *et al.*, 2016, 2017; Pimienta *et al.*, 2017). By comparing the mineral compositions of three samples, we found a positive relationship between the changes of total clay mineral content and the changes of physical properties. The clay mineral contents of A-173 and Y-46 were greater than 27%, whereas that of H-25 was only 12.9%. Higher clay mineral content implies that more clay mineral re-precipitation occurred during the experiments. In particular, the likelihood and intensity of particle migration were evidently higher for tight sandstones with more clay minerals and thus the blocking possibility of small pores and throats increased sharply. Consequently, the porosity and permeability decreased in value. The experimental results show that the relative reductions in permeability of A-173 and Y-46 were greater than 60%–80%. Although the porosity of H-25 increased, its permeability remained almost unchanged owing to the migration and blockage of clay minerals. Therefore, it can be inferred that higher clay mineral contents result in greater porosity and permeability reduction. Future research should be focused on the influence of clay mineral type and content on reservoir physical properties in the processes involved in CO_2–brine flow-through experiments.

5 CONCLUSIONS

(1) In this study, CO_2–brine flow-through experiment was conducted on Upper Triassic Chang 7 tight sandstones from Xin'anbian area, Ordos Basin, China. Xin'anbian area is one of the most successful tight oil exploration targets in China. An *in situ* 3D porosity-mineral

comparison model has been established with the application of X-ray diffraction mineral analysis, 3D Computed Tomography (CT) scanning, high-resolution scanning electron microscopy (SEM), and ICP-AES.

(2) The evolution trends observed for different minerals varied greatly during the experiment. Potassium feldspar, albite, and calcite were the most soluble components. The original chlorite-kaolinite framework was destroyed, and the re-precipitation of kaolinite, calcium montmorillonite, and sodium montmorillonite had a strong impact on pore structure, resulting in the decrease of both porosity and permeability.

(3) The synergistic reaction of scCO_2, brine and minerals at in-situ reservoir conditions has changed the nano-scaled pore systems, and resulted in the different physical properties evolution. Our results revealed that higher clay contents are associated with a greater reduction in the porosity and permeability after the CO_2-brine flow-through experiment. The influence of clay mineral type and content on these physical properties should be a focus of further research. The results provide new insights into CO_2 trapping mechanisms in tight oil reservoirs, and into the potential formation damage that may result from massive injections of CO_2 into reservoirs during EOR and Carbon Capture, Utilization and Storage (CCUS) programs.

ACKNOWLEDGEMENT

This study was supported by the National Key Basic Research Program-973 Project (Grant No. 2014CB239000), the National Science and Technology Major Project of China (Grant No. 2017ZX05001) and the CNPC Science and Technology Project (Grant No. 2016B-03).

REFERENCES

Abdoulghafour, H., Gouze, P., Luquot, L., Leprovost R., 2016. Characterization and modeling of the alteration of fractured class-G Portland cement during flow of CO_2-rich brine. *Int. J. Greenh. Gas Control* 48, 155-170.

Al-Zaidia, E., Fan, X.F., Edlmannb, K., 2018. Supercritical CO_2 behaviour during water displacement in a sandstone core sample. *Int. J. Greenh. Gas Control* 79, 200-211.

Andrew, M., Bijeljic, B., Blunt, M.J., 2014. Pore-scale imaging of trapped supercritical carbon dioxidein sandstones and carbonates, *Int. J. Greenh. Gas Control* 22,1-14.

Bredesen, R., Jordal, K., Bolland, O., 2004. High-temperature membranes in power generation with CO_2 capture. *Chem. Eng. Process* 9, 1129-58.

Dávila, G., Luquot, L., Josep, M.S., Jordi, C., 2016. Interaction between a fractured marl caprock and CO_2-rich sulfate solution under supercritical CO_2 conditions. *Int. J. Greenh. Gas Control* 48, 105-119.

Dávila, G., Cama, J., Luquot, L., Soler, J.M., Ayora, C., 2017. Experimental and modeling study of the interaction between a crushed marl caprock and CO_2-rich solutions under different pressure and temperature conditions. *Chem. Geology* 448, 26-42.

Garcia-Rios, M., Luquot, L., Soler, J.M., Cama, J., 2015. Influence of the flow rate on dissolution and precipitation features during percolation of CO_2-rich sulfate solutions through fractured limestone samples. *Chem. Geol.* 414, 95-108.

Moodiea, N., Ampomahb W., Jia, W., Heathc J., Phersona, B., 2019. Assignment and calibration of relative permeability by hydrostratigraphic units for multiphase flow analysis, case study: CO_2-EOR operations at the Farnsworth Unit, Texas, *Int. J. Greenh. Gas Control* 81, 103-114.

Pimienta, L., Esteban, L., Sarout, J., Liu, K.Y., Dautriat, J., Piane, C.D., Clennell, M.B., 2017. Supercritical CO_2 injection and residence time in fluid-saturated rocks: Evidence for calcite dissolution and effects on rock integrity. *Int. J. Greenh. Gas Control* 67, 31-48.

Shedid, S.A., Salem, A.M., 2013. Experimental investigations of CO_2 solubility and variations in petrophysical properties due to CO_2 storage in carbonate reservoir rocks. *SPE164632*.

Yang, T., Yang, D., Kang, Z.Q., 2010. Influence of super critical CO_2 injection on coal permeability. *Coal Scie. Tech.* 4, 108-111.

Yu, Z.C., Liu, L., Yang, S.Y., Li, S., Yang, Y.Z., 2012. An experimental study of CO_2-brine-rock interaction at in situ pressure-temperature reservoir conditions. *Chemical Geology* 326, 88-101.

Youth Technical Sessions Proceedings – Litvinenko (ed)
© 2019 World Petroleum Council, London, UK, ISBN 978-0-367-34668-3

Utilizing of the value of information and risk adjusted VOI for making better exploration decisions

E.V. Vekshina
"Gazprom EP International B.V."

ABSTRACT: Subsurface underappraisal programs are aimed at reduction of uncertainties which are peculiar to all prospective projects especially in new geological regions and sedimentary basins. Currently exploration and production companies apply different methods which treat a range of uncertainties by applying present value discounting methodic, probability discounting and value of information approaches. Unfortunately, these methods do not take into account exploration and production company's attitude to financial risk of losing money from wrong decisions. Subsequently incorrect estimation and treatment of existing uncertainty can lead to application of suboptimal development programs or even to opportunity loss. This paper describes a combination of the Value of Information approach with Risk preference theory by application Risk Aversion coefficient to estimate Risk Adjusted Value of Information of the proposed underappraisal program. Moreover, two different cases are considered. These cases are described by equal magnitude of uncertainty while the risk in case 1 is lower than risk in case 2. It is shown below how project risk level influences the Value of Information and thus decision to be taken by the exploration company. This paper reviews the approach of the most suitable exploration well placement by taking into account not only monetary values but also a number of subsurface and abovesurface parameters. It is shown that such multi-variable chart can be a very useful and applicable tool to substantiate and improve decision-making process.

1 INTRODUCTION

Reservoir uncertainty is inherent to all hydrocarbon prospects and is one of the most influencing factors when companies making decision whether to invest in the project or underappraisal programs such as acquisition of additional seismic information, drilling of additional exploration well, coring, well logging, run special core analysis (SCAL) etc. Management of exploration and production companies make final investment decisions taking into account risks and uncertainties basically related only to reservoir [1]. It seems practically possible because many companies simply do not know their attitudes to risk.

Further, such uncertainties give rise to development related risks that can have negative impact on final economic performance of the project or underappraisal program. However, making decision to invest into development or underappraisal often does not take into account the financial consequences for the company when risk realizes [2].

Basically majority of companies evaluate projects applying either or two methods mentioned below:

- Present Value discounting method
- Expected Net Present Value calculation and VOI estimation

Present value discounting method takes into account the time-value of money by application of discounting factor. Expected Net Present Value approach account for the chances of each possible Net Present Values and results in an Expected Net Present Value (ENPV). Using ENPV a Value of Information (VOI) method can be applied to evaluate the outcome of different scenarios. This paper will focus on the evaluation of the necessity to drill additional exploration well to reduce reservoir uncertainty and to estimate the influence of the information gathered from this well on development scenario. Combining decision tree analysis with probabilities of each NPV outcome a Value of Information approach which ties reservoir uncertainty to potential economic consequences [3].

While ENPV and VOI approaches weighs financial consequences by their probabilities and compare the utility of additional information in monetary equivalent, they do not take into account decision-maker's reaction to risk. This paper discusses a model that uses utility theory to calculate Risk Adjusted value by combining ENPV calculation with and assessment of the Risk Tolerance (RT) of the Exploration company.

In addition, if company can evaluate risk aversion coefficient (r) it becomes able to calculate Optimum Working Interest (OWI) in the project but its estimation is not a key issue of this paper.

Practical application of the proposed method can help exploration company to better understand should it reduce uncertainty by investing additional money on underappraisal programs and to reflect the financial consequences of the investment decision of the company.

2 FIELD DESCRIPTION

As a basis for the study the transboundary field has been taken. It is an offshore sour gas field which has been discovered in 1966 year by drilling of exploration well. In addition, prior drilling of exploration well a 2D and 3D seismic study were completed (Figure 1). Water depth varies between 45 and 60 meters while depth of targeted reservoirs varies from 3100 to 3700 meters.

The field is presented by anticline four-way dip structure formed as a result of salt diapirism observed during Permian time. Targeted reservoir intervals are comprised of thick carbonate structures of different lithology types with observable macro and micro fracturing across all reservoir section (Figure 2). A standard procedure was followed while creating geological model.

Reservoir engineering data obtained from existed exploration wells shows that gas-water contact has not been encountered during drilling. Thus, first exploration well that was drilled in 1966 did not reach targeted reservoir layers. Second well was drilled through the crest of the structure in 1967, observed gas but did not encounter gas-water contact as well.

In order to estimate depth of gas-water contact and its possible variability across the reservoir all available data, including DST, pressure tests and XPT data were analyzed. In total 12 tests were conducted 2 of which were dry and 10 encountered gas. Based on the results a gradient of 0,136 psi/ft was calculated (Figure 4). No water gradient was observed during studies thus it is assumed that it is located lower than tested intervals.

Uncertainty analysis was performed based on the static model using standard Monte-Carlo approach. Main uncertainties include GWC, porosity, water saturation, net-to-gross ratio, gas formation volume factor were evaluated. The uncertainty ranges for each of these parameters are taken according to geological concept and all available data [4]. A systematic and comprehensive analysis should be applied to define a type of distribution (triangular, normal, lognormal, etc), the ranges of variables and necessary number of realizations. These considerations must have a strong bound with geology and feasibility of assumptions [5]. As an output

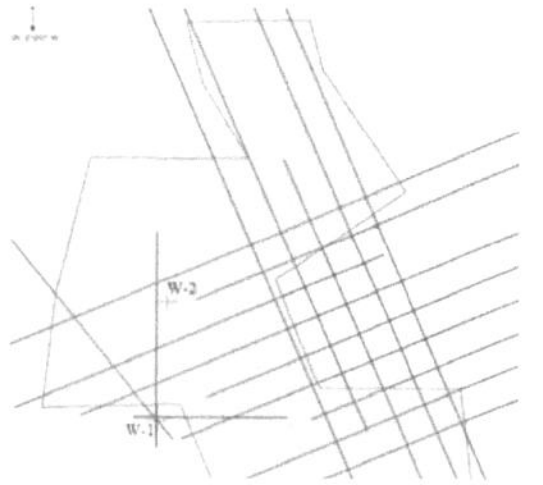

Figure 1. Initially available information on the Field

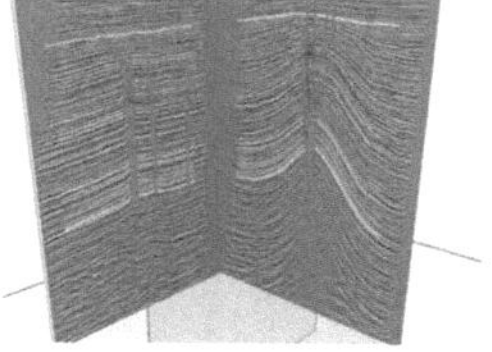

Figure 2. Initially available information on the Field

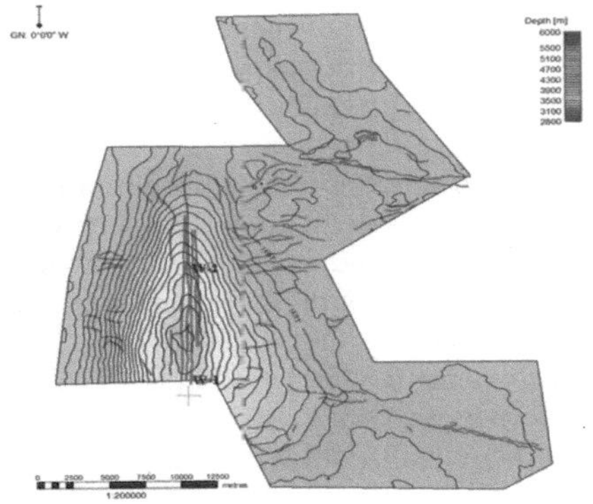

Figure 3. Depth converted Top of one of the targeted reservoirs

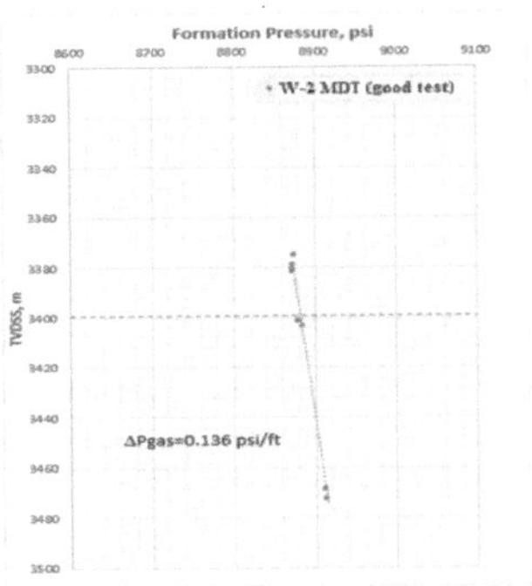

Figure 4. Results of well tests data interpretation

parameter GIIP values were chosen to allow us to identify which of the uncertainties materially (through gas production volumes and monetary values) impact exploration company's investment decision.

To determine possible GWC positions all available logging data, geological maps and well tests were reviewed. A Gas-Down-To was estimated at 3700 m TVDSS. According to the widely applied practices, lowest closing contour was applied as a maximum GWC position. Corresponding distances of new exploration well were determined based on top structure maps and possible GWC positions (Tables 1, 2, 3).

The process of modelling is repeated for 100 realizations, giving a probability distribution function of the output - gas initially in place volumes [6]. Figure 5 illustrates to which extent each parameter influences gas in place volume calculation and gas water contact has the highest impact. The results of GIIP ranges are presented in Table 3 and Figure 6.

It's important to discuss the probability distribution of GWC taken as an input to uncertainty analysis. The reasons for lognormal type of distribution is that between minimum and

Table 1. GWC positions

Case	GWC position, m TVDSS	Distance from crest, km
Low	-3700	10
Mid	-4000	13
High	-4440	17

Table 2. Top view of GWC positions for estimated reservoir

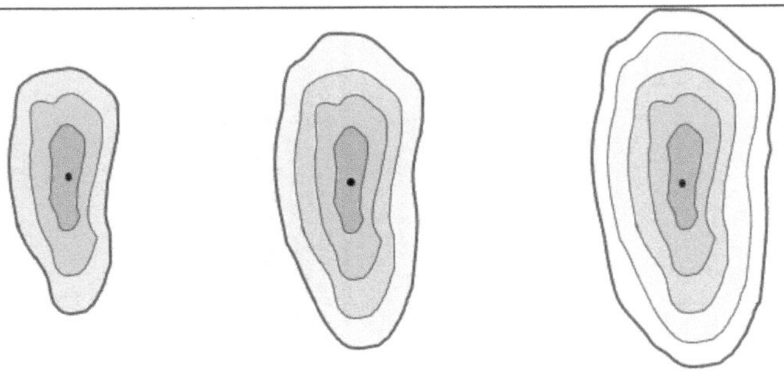

GWC @ -3700	GWC @ -4000	GWC @ -4440

Table 3. GIIP estimation

Case	GIIP, Tcf
Low	8.4
Mid	12.8
High	19.1

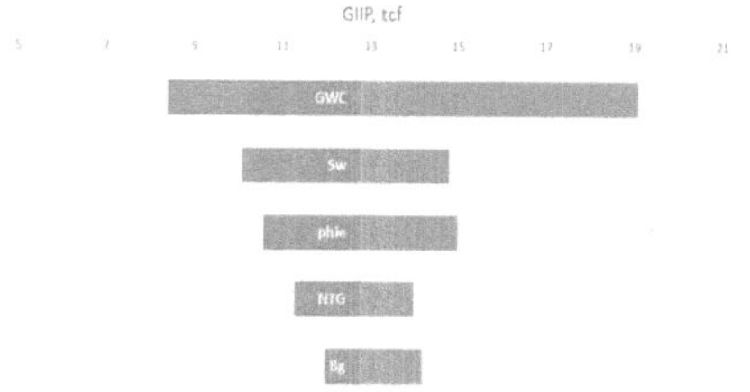

Figure 5. Tornado chart for gas in place uncertainty analysis

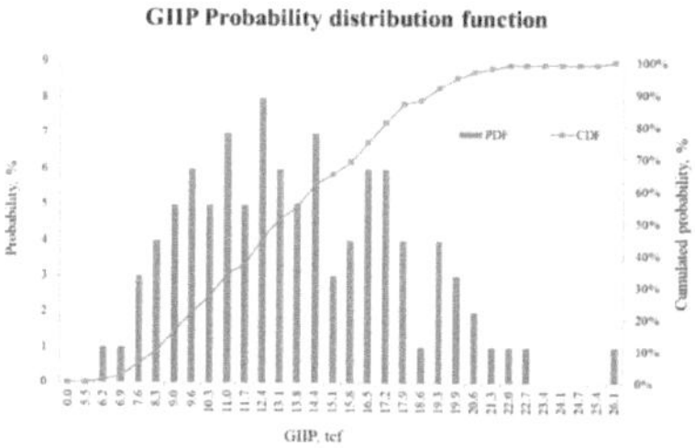

Figure 6. Probability distribution function of gas initially in place

maximum depths more likely GWC will lie closer to minimum value (gas-down-to) rather than maximum (lowest closed contour) since the first one is hard data, proved by well data and DST tests, while the second one is an expectation without any solid confirmation which was taken as a lowest closed contour in accordance with extrapolated 3D seismic map on the territory were no seismic data is available.

3 PROBABILITY FUNCTION DISCRETIZATION

Probability distribution function (PDF, figure 6) defines the picture about our beliefs regarding the outcomes. It can be described by its expected value (or average weighted value):

$$E[X] = \sum_{i=1}^{n} p_n * x_n, \text{ where}$$

p – probability of the outcome, x – absolute value of outcome

However being a measure of the center of a probability distribution, the expected value would not, by itself, provide enough information about variability of the distribution and thus, not giving the full picture. Dispersion is most commonly to be defined by standard deviation:

$$\sigma(X) = \sqrt{Var(X)}; \quad Var[X] = \sum_{i=1}^{n} p_n(x_n - \mu), \text{ where}$$

Var(X) – variance, σ(X) – standard deviation, μ - expected value

For further analysis and decision tree procedure, it's necessary to work out probabilities for each GIIP outcome considered along the branches of a tree. An essential step is to convert probability distribution derived from the uncertainty analysis to 3-points probability distribution, or in other words, to discretize the given PDF. For comparing the results of different discretization procedures we can use both expected value and standard deviation for original PDF and discretized 3-points distribution.

For that aim we use three approaches in this paper:

- Swanson-Megill method, assuming percentile P10 GIIP has 30% of probability, P50 = 40%, P90 = 30%;

Table 4. Probabilities discretization by Swanson and Megill

Swanson-Megill approach (1) Percentile	Value	Probability
10	479	0.3
50	728	0.4
90	1086	0.3
Expected value		761
Standard deviation		237

Table 5. Probabilities discretization by Pearson-Tukey

Pearson-Tukey approach (2)		
Percentile	Value	Probability
5	440	0.185
50	728	0.63
95	1151	0.185
Expected value		753
Standard deviation		219

- Pearson-Tukey method is quite similar, but the probabilities are defined as follows: percentile P5 – 18,5%, P50 has 63% probability, P95 – 18,5%.
- Proposed method by choosing the probabilities to meet the condition: expected value and standard deviation for original PDF and discretized PDF are the same

Results of 1st and 2nd methods are given in tables (4,5).

Expected value for original PDF is 782, standard deviation – 229. Expected values from both 1st band 2nd methods have a small difference with original (782).

For better results we propose to capture the probabilities of P10, P50, P90 estimates to fit the expected value and standard deviation of original PDF:

Table 6. Proposed probabilities discretization

Proposed approach (3)		
P	Value	Probability
10	479	0.235
50	728	0.450
90	1086	0.315
Expected value		782
Standard deviation		229

GIIP probability values are now applicable to use in decision tree with three outcomes (figure 6).

4 CASE DESCRIPTION AND ESTABLISHING OF PROBLEMATICS

Suppose that exploration company is at the point of making a decision to drill an exploration well. The essential question in field exploration is whether the expected reduction in uncertainty is worth the cost of achieving the information. The primary goal is to make a decision whether to drill an exploration well or not, and to plan a best location according to available data and GIIP estimates.

Before going through the calculation procedure in detail, we first consider a decision tree for proper calculation of the Value of Information (Figure 1).

The "no drilling' case assumes no exploration well is drilled, thus, the production wells drilling starts just after making final investment decision. Due to the high uncertainty and dispersion of GIIP estimation exploration company has to deal with the risk to over- or/and underestimate reserves and thus develop suboptimal development program and surface facilities object resulting in failure of the estimation of capital expenditures. The fact that uncertainty is very high and taking into account abovementioned discrepancies exploration company can not be sure that investing in this project is worth.

From the other point, if company considers drilling additional exploration well, there is an uncertainty to encounter gas-water contact by the well. If the well encountered GWC it will be

Table 7. Corresponding new exploration well positions based on the GWC estimates

GWC @ -3700 Distance to new well 10 km	GWC @ -4000 Distance to new well 13 km	GWC @ -4440 Distance to new well 17 km

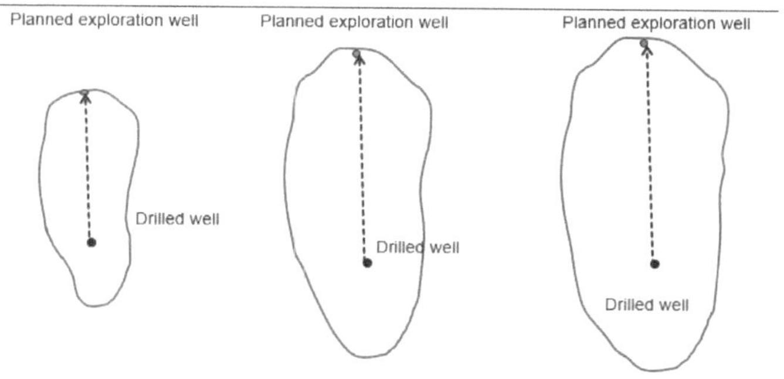

considered to achieve perfect information, hence, decreasing an uncertainty in GIIP and having a chance to correctly estimate development program, predict production profiles, project cash flows and to elaborate surface facilities objects diminishing capital expenditures and maximizing the value of the project. Worst possible case is to drill a well into water zone outside the gas saturated part of the reservoir, thus, not being able to remove uncertainty. In this case, the uncertainty ranges will be updated based on new well information and new development program and surface facilities will rely on new estimate, but still, remaining a risk of excessive CAPEX or insufficient capacity of the system.

A tentative cost estimation of drilling of exploration well was done on the basis of information from the investigated region where studied field is located. Information on typical well construction, drilling time, daily drilling rig rates were calculated and resulted that the cost of exploration well is 140 US$ MM.

5 DECISION TREE AND VOI WITHOUT RISK ESTIMATION

Basically all exploration companies use decision tree as an analytical tool that includes choices, risks, objectives, monetary gains and probabilities of each possible outcome. For this paper a decision tree with 12 cases was elaborated. Each case is described by its probability and related Net Present Value. It is worth to mention that three different well locations were considered: 10, 13 and 17 km from the crest. It is shown in the decision three by cases 1 to 9 (Figure 1). The lower part of the decision tree implies that exploration company rejects the opportunity to drill an exploration well. Assigned probabilities were taken from discrete probability distribution function. An economic model was created to reveal the fiscal regime of the government and to estimate the share of exploration company. The results are presented in the table below.

The next step was calculation of EMV for both options:

Option 1. To drill additional exploration well
Option 2. No additional drilling

$$EMV = \sum_{i=1}^{n} p_n * NPV_n, \text{ where}$$

EMV – Expected Monetary Value, P – Total probability of the case, NPV – Net present Value for each case

After the estimation of EMV for two options, Value of Information gathered from additional exploration well was calculated [7].

$$VOI = EMV_{with\ well} - EMV_{no\ well}$$

Table 8 shows the results of calculations of EMV and related VOI of additional exploration well.

Comparing the cost of drilling an exploration well with information obtained by gathering information from this well it is obvious that decision to drill a well should be taken as VOI is higher that the cost of the well.

Table 8. EMV and VOI estimation

EMV (no well), US$ MM	EMV (with well), US$ MM	VOI, US$ MM
3 571.6	4 340.5	768.9

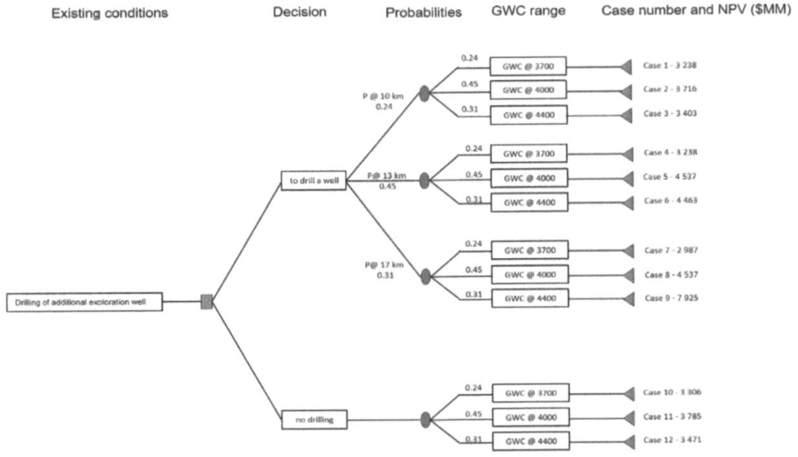

Figure 1. Case 1. Decision tree results

6 RISK ADJUSTED VALUE OF INFORMATION ESTIMATION

While EMV and hence VOI method does not reduce or eliminate uncertainty it can be used as a good tool to decision maker's. Unfortunately both EMV and VOI also do not account for risk attitude of the Exploration company to the upcoming decision and expenditures that are related to reduction of uncertainty [8].

One of possible ways to account for risk activity and company's attitude to risk is to use Preference theory through utility function. Utility function can be defined empirically or analytically by means of interview using linear, logarithmic, exponential or square root function. Both theoretical and practical experiences shown that it is appropriate to use a particular form of function – exponential. For the risk averse decision makers or as in our case exploration campaigns this function has the form:

$$U(x) = 1 - EXP^{\wedge}(-x/R), \text{where}$$

R – risk tolerance which is reciprocal to risk aversion = (1/r), x – evaluation measure or NPV of the outcome

As R increases utility function displays less risk aversion. In fact, with increasing R value the decision maker become risk neutral. In this paper this can be depicted by the increasing of the Risk Adjusted EMV values which in zero aversion level equals EMV (calculated previously).

In his paper on March 1978, John M. Cozzolino derived exponential utility function which is the only function shows constant risk-aversion and may be used to substitute other functions [9].

$$U(x) = (1/r) * (1 - EXP^{\wedge}(-r * x))), \text{where}$$

r – risk aversion coefficient, x – wealth at the end of the period (NPV for each case)

Further Risk Adjusted Value (RAV) formula was derived and applied in the calculations of this paper. For the N outcome venture further RAV formulae was used:

$$RAV = (-1/r) * LN(P1 * EXP^\wedge(-r * NPV(1)) + P2 * EXP^\wedge(-r * NPV(2)) + \ldots P1 * EXP^\wedge(-r * NPV(1))), \text{where}$$

r – risk aversion coefficient, P1,...Pn – probabilities of the outcome, NPV(1),...NPV(n) – net present value of the outcome

Described above risk preference theory does not allow to quantify the risk aversion level that exploration company should use. A "rule of thumb" mentioned in [9] is that risk aversion coefficient is reciprocal to company's exploration investment budget. The results of risk aversion level calculations are shown in Table 9.

By combining risk aversion coefficients taken by the company with estimated probabilities and NPVs for all expected outcomes a number of RAV for the decision to drill a well and not to drill were calculated (Table 10).

On the basis of calculated Risk Adjusted Values it seems possible to re-estimate the Value of Information from the drilling of additional exploration well with taking into account a risk attitude of exploration company on spending money. The resulted Risk Adjusted Value of Information calculations which is the difference between Risk adjusted value with drilling minus Risk adjusted value without drilling are shown in Table 11.

$$RAVOI = RAEMV(\text{drilling}) - RAEMV(\text{no drilling}), \text{ MM US\$}$$

Let's consider project in another region with more harsh and unprogressive fiscal terms. To make comparison viable in this paper author made an assumption that case 2 project of the same magnitude GIIP and the same total probability as in case 1. Due to risk involvement actual NPV for branches has higher degree of dispersion (table 12). For this case standard VoI procedure gives a value of VoI equal to 346. Regarding this fact, company can make a decision to drill a well, since 346 is high than drilling cost (140). But in risky economic environment decision makers should account for company's exploration investment budget and adjust VoI for the well. In case of small investment budget company will have RAVOI less than drilling cost (table 13).

Table 11 shows that Risk Adjusted Value of Information is always higher than the cost of additional exploration well. It means that exploration company can make decision to drill a well disregarding its annual exploration budget. This decision related to the fact that project bears high geologic uncertainty and in parallel low risks (dispersion between NPV between and thus EMV are high).

Table 9. Risk aversion coefficient values

Investment budget, MM US$	Risk aversion coefficient, millionths
100	0,01
200	0,005
500	0,002
1 000	0,001

Table 10. Risk adjusted values calculation

Risk aversion coefficient	Risk adjusted EMV drilling, MM US$	Risk adjusted EMV no drilling, MM US$
0,01	3 644,4	3 424,3
0,005	3 774,9	3 479,8
0,002	4 007,2	3 531,5
0,001	4 151,4	3 551,3

Table 11. RAVOI results low risk project

Risk aversion coefficient	RAVOI, MM US$
0,01	220,03
0,005	295,03
0,002	475,71
0,001	600,14

Table 12. Case 2. High risk project. Decision tree results

Decision	Case	Total probability, fr.unit	GIIP, Tcf	NPV, US$ MM
to drill a well	1	0.05909	8.4	2099
	2	0.10895	12.8	3330
	3	0.07505	19.1	3890
	4	0.10895	8.4	2099
	5	0.20085	12.8	4728
	6	0.13837	19.1	4619
	7	0.07505	8.4	1968
	8	0.13837	12.8	4501
	9	0.09532	19.1	7813
no drilling	10	0.24309	8.4	3207
	11	0.44816	12.8	3891
	12	0.30874	19.1	4021

Table 13. RAVOI results high risk project

Risk aversion coefficient	RAVOI, MM US$
0,01	-7
0,005	4
0,002	105
0,001	206

In opposite Table 13 contains the estimation of Risk Adjusted Value of Information for project with same uncertainty but considerably higher risk. Based on the calculation the exploration company can afford to drill additional exploration well only when annual exploration drilling budget is high enough somewhere between 0,005 and 0,002 risk aversion levels which correspond to 200 and 500 $US MM accordingly. Solving this simple mathematic question and assuming that cost of exploration well costs 140 $US MM the risk aversion of the exploration company should be 0,0041 which corresponds to the minimum required budget of 240 $US MM.

7 WELL LOCATION

In case of making decision on placement of exploration well company can face a lot of issues can arise (Ex. The question of satisfaction of the number of goals: while trying to maximize NPV company can face drilling limitations, or striving to improve and increase reserves estimation can lead to penetration of low quality reservoir) [10].

In order to find the best decision the paper proposes a methodology of ranking the objectives in specific way. At first, definition of the list of objectives (Table 14). The list of objectives consists of measures with different levels: economics, safety, technical requirements and limitations, and other desirable objectives. Although EMV, VoI and decision tree approaches use monetary values in terms of a single economic metric (NPV) which influence the decision, these methods do not require that there be a single metric or that it be an economic one. Therefore, it can be applied to multi-objective decisions e.x. choosing an optimum well location.

Secondly, company should calculate two kind of ranks:

- The first one is to account for dispersion of each parameter for different well location (swing rank).
- The second one is for weight of the objective to the decision making process.

Combining these ranks, company can achieve final numeric weighting parameter for alternatives A, B and C. Each alternative has its own objective value which is scaled from 100

Table 14. Well placement decision matrix

Objectives	Ranks		Weights		Location Values		
					10 km	13 km	17 km
	Swing Rank	Influence Rank	Abs.	Norm.	A	B	C
Risk of encountering water saturated reservoir	80	60	140	0.14	100	50	0
EMV, $mln	60	100	160	0.16	65	60	100
well design complexity	50	40	90	0.09	70	60	40
Reserves added, tcf	100	90	190	0.19	0	50	100
Possibility to involve the well into production	90	80	170	0.17	100	80	0
Possibility to encounter low reservoir quality	70	50	120	0.12	40	55	20
Initial rates	40	70	110	0.11	50	80	35
Total	490	490	980	1.0	59	65	46

(optimistic option) to 0 (pessimistic option). Influence rank of an objective can be either identified as exact, numeric metrics (like economic one) or as a measure of unquantified objective (like risk, safety or complicity level), thus, bringing an expert opinion into the method.

As a conclusion it is worth mention that using weighted rankings, identified in table 14, company can identify resulting weighted parameter that can show the degree of risk and possible outcomes related to different exploration well locations. Given the priorities of each case, through a list of identified criteria, one can choose the best location of a planned well, improving exploration decisions. In that case the most appropriate well position is 13 km from the crest of the structure which corresponds to Mid case.

8 CONCLUSIONS

However uncertainty should be distinguished from risk which is negative consequence of uncertainty. Thus in order to improve decision making process exploration company should take into account not only uncertainties but also risks. Above-mentioned cases are both characterized by high uncertainties but the first bears low risks and the second one can be described by high risks.

Implementation of the value of information approach is suitable in projects with high uncertainty and low risk level while for high risk project VoI is unreliable approach and results in wrong decisions due to the incomplete risk recognition and consideration.

Combination of value of information and risk aversion approaches gives an improved and robust understanding of the decision making process to the exploration company with constrained exploration budget making it feasible to increase the project value.

Proposed procedure for choosing of exploration well placement considers multiple objectives simultaneously, achieving maximum weighted value that can be used to improve and simplify decision making process.

REFERENCE LIST

1. SPE 1811246 Technical Report "Guidance for Decision Quality for Multycompany upstream projects", 2016.
2. Duane McVay, SPE-0215-0072-JP, Industry needs Re-education in Uncertainty Assessment.
3. Reidar B. Bratvold and Steve H. Begg – Making good decisions, 2010.
4. Randy Freeborn, Boyd Russel - Creating More-representative Type Wells.
5. J. S. Devis - Geostatistics, 1990.
6. Jef Caers – Modelling uncertainty in the Earth Sciences, 2011.
7. R. Kumar SPE 96171 – Value of Information for Appraisal of Multiple Dependent Prospects, 2005.
8. James A. MacKay SPE 30043 Utilizing Risk Tolerance To optimize Working Interest.
9. John M. Cozzolino (U. of Pennsylvania) SPE-6632 - A New Method For Measurement And Control Of Exploration Risk, 1978.
10. Willliam J. Haskett SPE 170960-MS – The Myth of sweet spot exploration.

Future of LNG

Youth Technical Sessions Proceedings – Litvinenko (ed)
© 2019 World Petroleum Council, London, UK, ISBN 978-0-367-34668-3

A technological solutions set to ensure the safety of storage of liquefied natural gas

Y.V. Martynenko & V.A. Voronov
Saint-Petersburg Mining University, St. Petersburg, Russia

ABSTRACT: The article offers a set of technological solutions to ensure safety during storage and low-tonnage production of liquefied natural gas (LNG). Tanks are equipped with gas discharge pipelines and safety valves to prevent planned and emergency pressure, through which the excess steam phase is discharged to the system torch or into the atmosphere. The efficiency of gas-equalizing tank strapping has been proven theoretically, including the ejector system, in order to reduce the loss of liquefied natural gas and energy consumption during its safe storage. The most productive aerodynamic scheme of the flow part of the liquid-gas ejector was chosen to significantly reduce the investment in construction and operation of cryogenic tanks.

1 INTRODUCTION

Much attention is given to operations of vapor phase dumping not only due to ecological and fire safety, but also because of the significant losses of the product. However, failure to observe safety regulations can lead to heavy losses, including cases of damages and leaks. Overpressure in the storage tank caused by damaged tools, pressure safety valves or operators' errors is one of the reasons of liquefied gases leakage.

LNG vapor dumping from the tanks through the gas-discharge pipeline and pressure safety valves takes place in the following emergency situations [8]:

- in case of total loss of vacuum in the insulation cavity or destruction of thermal insulation;
- when the pressure regulator jams in the open position;
- when the flowing pressure is exceeded in emergency situations.

Filling tanks which already contain some remains of fuel with a new portion of LNG with density and temperature different from these parameters of the remaining fuel can lead to the formation of stratified macro-layers in the reservoir and to intensive vapor generation. This situation also requires an accident-prevention dumping system.

Cryogenic tanks provide not only emergency, but also periodic gas discharges, such as the discharge of cold vapors during tanks refueling or during technological operations.

For this reason, the pneumatic-hydraulic circuit of cryogenic reservoir should provide an emergency shutdown system , which controls the pressure and level of LNG in the tank.

According to safety rules, dumping of the excess of LNG vapor phase, which becomes burnt in the flare, happens in case of the excess of surplus pressure relatively to nominal by the given number. LNG is discharged into the gas bleeder into the atmosphere directly, if the required pressure reduction does not occur. This leads to significant losses of expensive fuel. In case the reservoir is equipped with treating iron, the separation of vapor phase excess is made by compressors, and that leads to additional energy costs.

2 EJECTOR SYSTEMS FOR VAPOR DUMPING

Pump-ejector systems are suggested as an alternative to existing control and dumping systems. They are included into the treating iron of the reservoir in purpose of operational pressure maintaining by dumping the fuel to separator using the energy of high vapor phase pressure. Such systems are distinguished primarily by simplicity, small capital investments and the absence of the need for additional expenditure of energy from outside.

The choice of the pump jet system depends on the parameters of the exhausted fluid such as pressure, compression ratio, flow rate. The liquid-gas jet is of the greatest interest for cryogenic LNG tanks. The operating principle of this device is based on the following: low-pressure liquid phase comes to mixing chamber because of its vacuum area caused by high-pressure flow of the fluid. As pressure safety valves in LNG reservoir trip when the vapor exceeds the pressure level, there is no need in high-pressure flow.

There is a suggestion to include liquid-gas ejector 3 to the treating iron of cryogenic LNG reservoirs so that the liquid phase would come to the mixing chamber because of the high-pressure flow of vapor phase, which occurs during the emergency or periodical dumping. The two-phase flow after passing liquid-gas ejector is going to the separator 4. After that, gas phase can be used for technological supply of the reservoir park, while LNG (if it fits the required pressure and temperature parameters of the system) can be transported back to reservoir 1. Thus, pumping with the economically efficient fluid jet device can decrease the volume in the reservoir and lessen the excess pressure (Figure 1) [1].

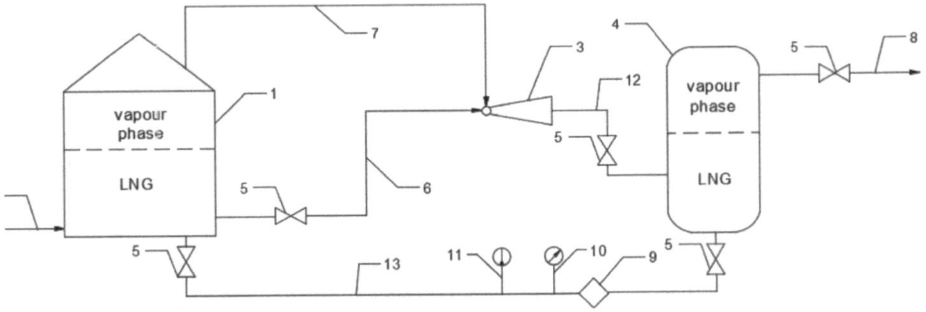

Figure 1. Schematic diagram of the ejector system for LNG vapor dumping.

1 - LNG tank, 2 - LNG pipeline, 3 - liquid and gas ejector, 4 - separation tank, 5 - gate valve, 6 - LNG supply line to the ejector, 7 - steam line to the ejector, 8 - pipeline for technological needs, 9 - a filter, 10 - a manometer, 11 - a thermometer, 12 - a line for supplying a two-phase flow, 13 - LNG piping.

3 METHOD FOR CALCULATING THE DETERMINING PARAMETERS FOR THE EFFICIENCY OF THE APPLICATION OF LIQUID-GAS EJECTOR

The main indicators which show the efficiency of liquid-gas ejector usage are the coefficient of ejection, capacity and efficiency factor depending on the chosen aerodynamic configuration of the device. Scientific works show different effective aerodynamic configurations for the calculation of geometrical design factors of the flow part of ejector, however the calculation of fluidjet devices with cryogenic fuel as an work environment has never been done before [4].

The main indicators of liquid-gas ejector:

- inner displacement rate of liquid, transported by liquid-gas ejector (equation 1):

$$Q_f^{optm} = \frac{z \cdot P_0 \cdot T_f}{(P_g - P_s) \cdot T_0} \cdot Q_{f0} \tag{1}$$

where z - LNG compressibility factor; P_s - saturated vapor pressure of working fluid, MPa, P_g - gas pressure (LNG vapor), МПа; T_f - liquid temperature (LNG), К; Q_{f0} - flow rate of pumped liquid, м^3/u; T_o, P_o - absolute indictors of temperature and pressure, K, MPa.

- reduced working fluid pressure (equation 2):

$$\bar{P} = \frac{P_g - P_s}{P_f - P_s} \tag{2}$$

- coefficient of ejection (inner coefficient of displacement rate (equation 3)):

$$u_{optm} = u_{\max} \cdot \left[1 - \exp(-B_{optm} \cdot \sqrt{\bar{P} - 1})\right] \tag{3}$$

where $u_{\max}$ - maximal coefficient of ejection, depending on the type of aerodynamic system (table 1), B_{optm} - empirical constant, depending on the type of aerodynamic system (table 1).

- coefficient of pressure recovery (equation 4):

$$\psi_{optm} = \psi_{\max} \cdot \left[1 - \exp(-a_{optm} \cdot \sqrt{\bar{P} - 1})\right] \tag{4}$$

where $\psi_{\max}$ - maximal coefficient of pressure recovery, depending on the type of aerodynamic system (table 1),a_{optm} - empirical constant, depending on the type of aerodynamic system (table 1).

- flow rate of working gas flow (equation 5):

$$Q_g = \frac{Q_f^{optm}}{u_{optm}} \tag{5}$$

- gas compression ratio (equation 6):

$$\varepsilon = 1 + \psi_{optm} \cdot (\bar{P} - 1) \tag{6}$$

- mixture pressure on the way out of liquid-gas ejector (equation 7):

$$P_{mix}^{optm} = P_f + \psi_{optm} \cdot (P_g - P_f) \tag{7}$$

- efficiency coefficient (equation 8):

$$\eta = \frac{Q_f \cdot P_f}{Q_g \cdot (P_g - P_{mix})} \cdot \ln\frac{P_{mix}}{P_g} \tag{8}$$

- power, spent on compressing while the ejector is working, with accuracy up to the efficiency coefficient of the pump (equation 9) [8]:

$$N = P_g \cdot Q_g \tag{9}$$

- reduced power spent on compressing (equation 10):

$$\bar{N} = \frac{N}{Q_f \cdot (P_f - P_s)} \tag{10}$$

The difference of the suggested technological scheme is that the working fluid for ejector is gas, because due to its energy LNG is passing to jet device and then makes a two-phase flow. That's why the described calculation method pays attention to pressure and flow rate parameters' changes relatively to gas and liquid.

4 THEORETICAL CALCULATION OF THE CHARACTERISTICS OF LIQUID-GAS EJECTOR AND SELECTION OF THE AERODYNAMIC SCHEME.

The following parameters were determinate for the calculations:

- LNG mixture composition ($CH_4 = 95,5\%$; $C_2H_6 = 2,3\%$; $N_2 = 1,7\%$ и др.) [7];
- LNG compression ratio in the enter to the ejector $z = 0,03671$;
- LNG temperature in the enter to the ejector $T_f = 140K$;
- absolute LNG pressure in the enter to the ejector $P_f = 1,102\, Mpa$;
- vapor pressure in the enter to the ejector (after pressure increase) $P_g = 1,547 MPa$;
- fuel flow rate $Q_{f0} = 4,9 M^3/u$;
- heavy vapor pressure $P_s = 0,6375\, MPa$.

Starting parameters correspond to national standard (GOST R 56851-2016), where thermodynamic LNG properties are listed for different mixture types. There was considered a case which caused the discharge of the gas phase due to temperature increase from 120 K to 140 K and, correspondingly, to the increase of saturated vapor pressure.

The choice of the aerodynamic scheme of the ejector depends on the most effective efficiency factors, the ejection factor and the reduced power. It is recommended to choose the aerodynamic scheme, which provides the highest efficiency at the design value of the compression ratio.

The results of calculating the main parameters, taking into account the coefficients for the optimal regime, are presented in Table 1.

For a clear analysis of the results, the dependences of the mixture pressure at the outlet from the ejector, the reduced power and the efficiency of the ejection coefficient on the four aerodynamic schemes under consideration are plotted (Figure 2).

Thus, the aerodynamic scheme №1 is the most effective, since its geometric characteristics make it possible to obtain the highest values of efficiency, power and pressure of the mixture.

Table 1. The results of calculating the characteristics of the ejector at the optimal regim

Schematic type	u_{optm}	Q_g, м^3/u	φ_{optm}	ε	P_{mix}, MPa	N, MW	N_{reduce}	η, %
1	0,955	0,010	0,146	1,140	1,166	0,016	3,489	15,938
2	1,114	0,009	0,082	1,079	1,138	0,014	2,992	9,871
3	1,401	0,007	0,068	1,06	1,131	0,011	2,380	10,132
4	1,812	0,005	0,045	1,044	1,121	0,008	1,840	8,631

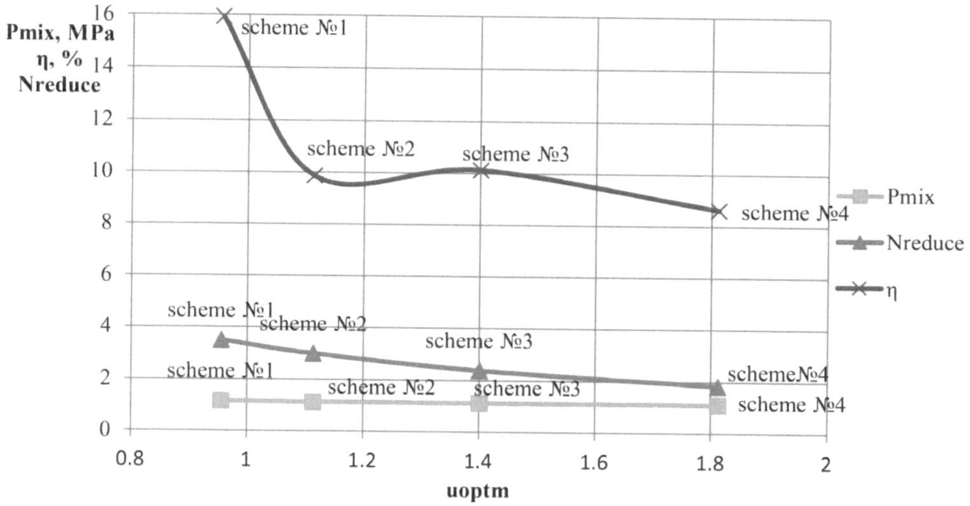

Figure 2. The results of calculating the characteristics of the ejector at the optimal mode.

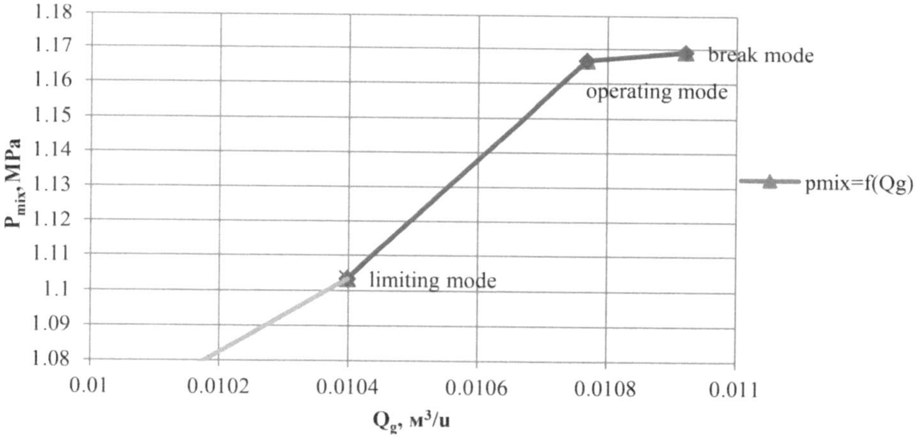

Figure 3. Pressure characteristic for the aerodynamic scheme No.1

According to the chosen scheme, the parameters of the ejector were calculated at the limiting and disruptive operating conditions of the liquid-gas ejector and a pressure characteristic was constructed (Figure 3).

5 CALCULATION OF GEOMETRICAL PARAMETERS

To determine the geometric parameters of the selected ejector circuit, it is necessary to calculate the nozzle diameter (equation 11) and the diameter of the working chamber (equation 12). The remaining dimensions of the flow section are selected according to Table 6 [4].

$$d_0 = \sqrt{\frac{4Q_g}{\pi \cdot \mu_{noz} \cdot i} \cdot \sqrt{\frac{\rho_g}{2(P_g - P_f)}}}, \tag{11}$$

where μ_{noz} - coefficient of nozzle flow rate ($\mu_{noz} = 0,79$ for scheme №1), i - number of nozzles (i=7).

$$d_{wc} = d_0 \cdot \sqrt{i \cdot m} \tag{12}$$

where m - coefficient for scheme 1 (m=3,2).

The initial parameters were used to calculate the geometric dimensions of the flowing part of the liquid-gas ejector. The results are shown in Table 2.

The calculated geometric values correspond to the constructive scheme of the flow part of the ejector, shown in Figure 4.

Table 2. Geometrical dimensions of the flowing part of liquid-gas ejector

d_0,m	a,m	D_0,m	D_1,m	d_2,m	d_{wc},m	l_{noz},m	l_1,m	l_2,m*	l_{diff},m*
0,0075	0,0023	0,1012	0,0237	0,0316	0,0355	0,142	0,112	0,379	0,056

* $l_1 = \frac{d_{wc}-d_2}{2tg\cdot\frac{\gamma_1}{2}}$; $l_{diff} = \frac{2\cdot d_{wc}-d_{2y}}{2tg\cdot\frac{\gamma_{diff}}{2}}$.

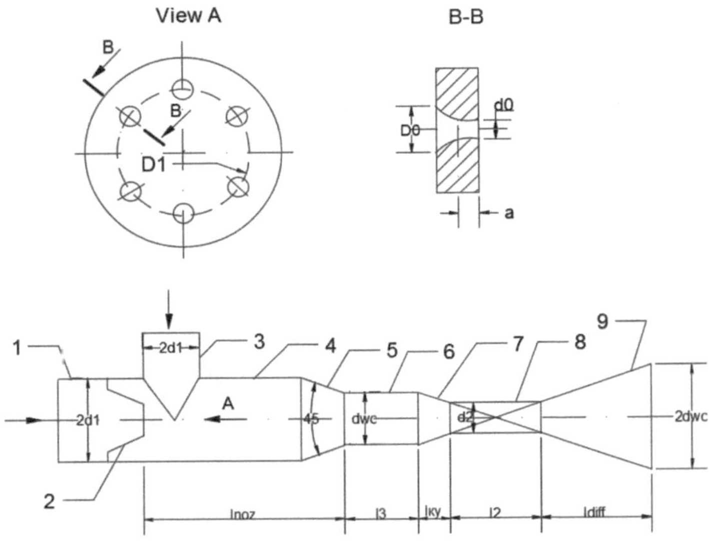

Figure 4. Constructive diagram of the flow part of the ejector
1 – nozzle inlet working gas; 2 – nozzle; 3 – receiving pipe; 4 – the pre-chamber; 5 – confuser; 6 – mixing chamber; 7 – a convergent phase; 8 – cylindrical portion; 9 – diffuser area.

6 CONCLUSIONS

In solving these problems, a complex method of research is used: generalization and analysis of theoretical and experimental works in the field of liquid-gas ejectors, comparison of existing ejector systems, numerical methods for theoretical calculation of the determining parameters of liquid-gas ejectors, synthesis of the data obtained and the existing aerodynamic schemes of ejectors.

The theoretical calculation of the defining indicators was made to substantiate the effectiveness of the use of liquid-gas ejectors, according to which the most efficient aerodynamic scheme was chosen. For a clear analysis of the results, the dependences of the mixture pressure at the outlet from the ejector, the reduced power and the efficiency of the ejection coefficient under consideration are plotted. According to the chosen scheme, the parameters of the ejector were calculated at the limiting and disruptive operating conditions of the liquid-gas ejector and a pressure characteristic was constructed. The calculation of the geometric dimensions of the flow part of the ejector is made and its design scheme is presented.

Thus, as a result of complex work on LNG storage system technological scheme modernization, several key points can be identified:

1. the turboexpander connection in parallel to the reduction unit has an advantage in terms of efficiency over the method of connection in parallel to the entire station;
2. the main gas flow separation on the technological and production flows after the turboexpander allows to reduce the LNG production specific energy consumption due to the flow additional cooling;
3. the use of a gas equalization system in LNG tank farms with an implemented ejector will ensure safety in fuel storage, while reducing capital and operating costs.

REFERENCES

1. Voronov V.A., Martynenko Y.V. 2018. Patent "Method of resetting a vapor from a reservoir of liquefied natural gas" RU 2677022.
2. Voronov V.A., Martynenko Y.V. 2018. Application of liquid-gas ejector in liquefied natural gas storage systems. The book of abstracts "Innovations and prospects of development of mining machinery and electromechanics: IPDME-2018": 120.
3. Voronov V.A., Martynenko Y.V., Nazarova M.N. 2017. Comparative analysis of gas transportation through pipelines in liquid and gas-air conditions, Neftegaz.ru, №10:20-22
4. Donec K.G. 1990. Hydraulic drive jet compressor units. Moscow: Nedra.
5. Morozova N. In. Korshak A. A. 2013. Universal characteristics of liquid-gas ejectors. Oil and Gas business (electronic scientific journal), №6:368-383.
6. Rachevsky B.S. 2009. Liquefied hydrocarbon gases. Moscow: "OIL and GAS".
7. GOST R 56851-2016 "Liquefied natural gas. Thermodynamic properties calculation method". - Introduced 18.01.2016. - Moscow: Standartinform, 2016.
8. SP 240.1311500.2015 "Liquefied natural gas storages. Fire safety requirements". - Introduced 31.08.2015. - Moscow: Tekhexpert, 2016.
9. GOST R 55892-2013. Objects of low-tonnage production and consumption of liquefied natural gas. General technical requirements. - Enter.01-06-2014. - M.: Standartinform, 2014.
10. GOST R 56021-2014. Combustible natural gas LPG. Fuel for internal combustion engines and power plants. Technical conditions. - Enter.01-01-2016. - M.: Techexpert, 2016.

Youth Technical Sessions Proceedings – Litvinenko (ed)
© 2019 World Petroleum Council, London, UK, ISBN 978-0-367-34668-3

Role of LNG in the exporting strategy of Russia

V.A. Fedorova
Gubkin Russian State University of Oil and Gas (National Research University), Moscow, Russia

ABSTRACT: By IEA forecast LNG trade more than doubles, reaching almost 900 Bcm in 2040 up from around 400 Bcm in 2017. Russia has abundant available gas reserves, but its domestic demand is low and opportunities to expand exports are currently limited. Russia faces a difficult task in attempts to diversify its export. Pipeline connections are inflexible, and Russia sells almost 90% of natural gas export volume to European markets, where due to slow population growth and energy efficiency policy gas consumption is declining and simultaneously struggles with political and regulatory pressure. The key question for Russia is what export strategies to implement in an increasing, connecting and competitive global gas market, in which international trade is rapidly moving towards LNG.

1 ROLE OF LNG IN WORLD ENERGY MARKETS

The global energy markets are in constant transformation in accordance with the capabilities and needs of the global community. The current period of change is characterized by a high level of uncertainty, a change in the technological structure, as well as promised ending of hydrocarbon economy and continuing search for a more environmentally friendly alternative energy sources.

According to the International Energy Agency, the global energy demand, despite its slowing growth rates, will increase by 30% by 2040. Big growth in energy consumption will be shown by developing countries in Asia, primarily China and India and the countries of the Middle East, Africa and Latin America.

In accordance with the forecasts of the world's leading energy companies and institutions, the expected growth in primary energy consumption will be approximately between 16,482 million toe and 21,573 million toe. The main energy sources that show growth in all forecast scenarios are natural gas and renewable energy sources (RES). Experts of the International Energy Agency suggest that gas demand will increase by 50% by 2040. Annual growth is expected to be at 1.5%, which is significantly higher than other traditional fuels, but far below the 2.3% growth rate shown in the previous 25 years. Consumption of natural gas will increase worldwide, with the exception of Japan, where today's consumption levels will significantly decrease due to the commissioning of nuclear power generation facilities. The main growth in demand is expected to be in the Middle East, as well as India and China, where most of the natural gas must be imported.

Natural gas has its own advantages, which make it a popular source of energy: no other traditional fuel is so flexible, efficient and suitable for solving various problems. Gas installations are generally cheaper than their coal equivalents. While prices are subject to market trends, a combination of lower emissions and an average lower price encourages consumers to switch to natural gas. When burning, natural gas produces half of the carbon dioxide and only a tenth of the air pollutants produced by burning coal, and therefore has long been considered a relatively "clean" fossil fuel.

Given the increasing competition from coal and renewable energy sources, the natural gas industry will have to work on a high level of efficiency in order to secure markets for itself. Gas demand in industry will grow faster (by 2.1% per year) than in the power generation sector (1.3% per year); the fastest growth (3.4% per year) will be shown by the transport sector, including the usage of LNG for heavy freight and maritime transport (IGU Global Gas Review 2018).

In recent years, approximately 70% of natural gas volumes worldwide are transported to the domestic market, while the remaining 20% cross international borders through a pipeline system, and another 10% is transferred to markets as liquefied natural gas (LNG).

Until 2020 the main increase in production is expected in Australia and the USA, however, after 2020, an increase in supply from a larger number of countries is expected. East Africa will enter the market as an exporter after the development of offshore fields in Mozambique and Tanzania. Egypt will return to the market again, which will contribute to the start of production from the Zohr field. Argentina is also expected to become a major exporter of natural gas with the start of developing promising shale deposits in the Vaca Muerta area. After several years of relatively stable production at the beginning of the 2020s, gas production in Russia will start to grow simultaneously with export via the new pipeline route to China. New projects in North America, Australia, Africa, the Middle East and Russia will help increase the share of LNG in interregional trade from the current 42% to 53% by 2040 (Figure 1)

Natural gas is responsible for a quarter of the global energy balance, 9.8% of which is supplied in the form of LNG. At the moment, the share of LNG in the market has stabilized, but the increase in liquefaction capacity expected by 2020 once again demonstrates the growing demand for this source of fuel. According to forecasts of International Energy Agency, world trade in natural gas will increase from 706 bln m^3 in 2016 to 1229 bln m^3 by 2040. Over 90% of the increase in gas trade over this period will be due to LNG. The share of LNG in the international gas trade will exceed the share of pipeline transportation already after 2025 and will be about 60% by 2040.

Last year, the market for liquefied natural gas showed a record growth, reaching 293 million tons in world trade, an increase of 34 million tons compared with the previous year. The number of suppliers is growing; projects continue to be built around the world - from the United States of America to Australia. LNG is conquering new markets; an increasing number of countries are choosing this way of gas delivery for the needs of their fuel and energy complex. At the same time, supply growth is constrained by the slow pace of commissioning of new liquefaction capacity, while prices remain lower than the cost of LNG in new projects. Different forces affect the market. On the one hand, there is a huge investment in the construction of LNG plants, on the other hand, a decline in the growth rate of natural gas consumption from 2.8% in the first decade of 2000 to 1.3% in the period from 2010 to 2016 (IGU Global Gas Review 2018).

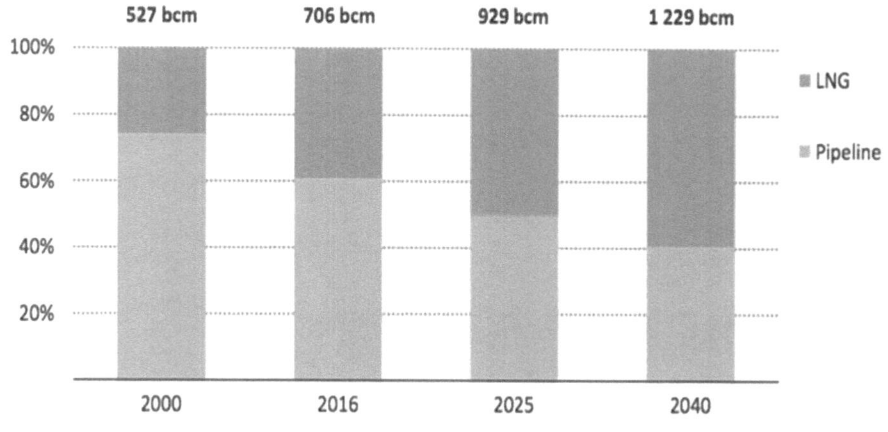

Figure 1. Natural gas trade until 2040 (IEA World Energy Outlook 2018)

The volume of planned liquefaction plants reached a level of 879 million tons by January 2017 in an attempt to rationalize market demand. Most of these projects will not reach FID, since demand remains at a much lower level than the declared volume, especially in the conditions of price underbidding in pipeline supplies: supply from Russia and Norway to Europe, and from the USA to Mexico significantly reduced options for LNG provision for these markets.

Despite the decline in demand growth in the last few years, LNG trade continues to grow, and each country has its own reasons. In the Asia-Pacific market, drivers are, first of all, geographic isolation and low self-sufficiency. Uncertainty of society towards nuclear energy is also a factor supporting the import of LNG. Such countries often face either depletion of reserves or are unable to increase their own gas production in sufficient volume to meet domestic demand. Despite the fact that the UK increased its gas production in 2016 compared to 2015, in the long-term, gas production is decreasing. Additional restrictions on the Groningen field in the Netherlands also reduce gas production in this country. In Kuwait, Thailand and Argentina, the situation is similar: own production is not able to meet the rapidly growing domestic demand, so these countries are forced to enter the LNG market as a consumer.

LNG continues to be used as a tool to ensure energy security, even in countries that have developed imported pipeline infrastructure. European importers, such as France, Italy and Turkey, use LNG to diversify their fuel sources, as well as an insurance in case of disruption of pipeline supplies. In the past few years, concerns about pricing and security of pipeline supplies have pushed Latvia and Poland into entering the LNG market as a consumer.

While the import market concentrates on Asia, the export situation becomes more diverse due to the list of new suppliers in the foreseeable period. Today, about 60% of LNG exports go from Qatar and Australia. During the forecast period, the United States, and then the sub-Saharan African countries, increase exports by 90 bln m^3. Russia increases LNG exports by 60 billion cubic meters. Together, these three regions occupy larger positions in global LNG exports, doubling their share from 23% today to more than 40% by 2040 (Figure 2).

The consumption of natural gas will continue to grow due to the environmental friendliness and cost-effectiveness of this type of fuel. However, the future of gas, as a global energy carrier, will depend on three important criteria - affordability, security of supply and environmental efficiency. The LNG market will continue to grow, and, due to its advantages, it will bypass pipeline transportation in terms of transportation volumes by 2035. According to all of the above, LNG has to secure its own place in the export strategy of the Russian Federation.

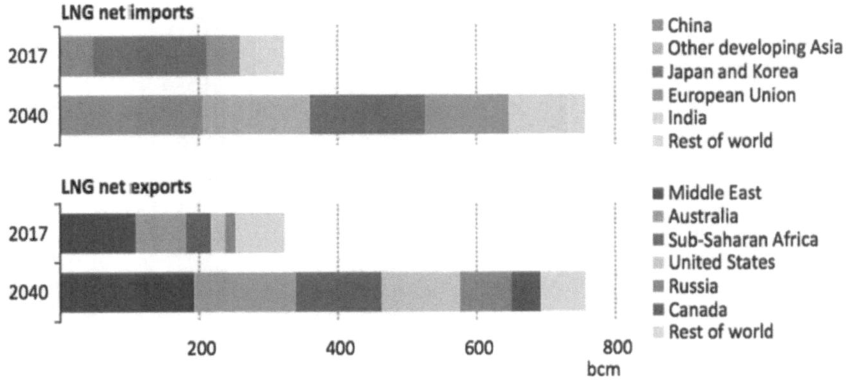

Figure 2. LNG trade by region by 2040 (IEA World Energy Outlook 2018)

2 IMPLEMENTING LNG IN THE EXPORTING STRATEGY OF RUSSIA

Russia is a country with the largest gas deposits. Gas accounts for 64% of country's energy balance, income from energy exports accounts for about 50% of budget revenues, and any supply disruption threatens the inevitable shortage of government spending. In the case of exporting countries, such risks can be hedged in the same way as in importing countries: through diversification of export routes. A relatively high growth in gas usage in industry and at the same time a decrease in population leads to an increase by only 3% by 2040, predicted by the IEA. Electricity and heat production are Russia's largest natural gas consumers (about 25% and 30%, respectively), however, there are government initiatives aimed at improving energy efficiency in these areas, and as a result, natural gas consumption will decrease.

In 2017, Russia exported 231 bln m^3 of gas, of which 81% was transported via pipelines to Europe, 12% to the CIS countries and only 7% to foreign markets in the form of LNG, which once again demonstrates Russia's enormous dependence on Europe as an importer, and the inability to diversify our distribution channels. In 2017 Russia shipped record volumes of gas to Europe, but access to other markets is impossible until the completion of the construction of the Power of Siberia pipeline to China and the planned expansion of LNG capacity. For Russia, the key question is which export strategies should be applied in an increasingly globalizing and competitive global gas market, where trends are emerging - a shift in demand towards the Asia-Pacific region and an increase in the growth of international gas trade in the form of LNG. Russia faces a difficult task trying to diversify its exports. Russia sells almost 90% of its natural gas exports to European markets, where, due to slower population growth and energy efficiency policies, gas consumption is falling and at the same time facing political and regulatory pressure in an attempt to ensure its own energy security.

Expansion of pipeline exports from Russia is expected by 2020 with the launch of Power of Siberia, as well as with the possible decision to build the Altai pipeline (or the Power of Siberia-2). In China, despite a significant increase in domestic production, the growth in demand is ahead of supply, with the result of dependence on imports will increase to more than 40% by 2040. About half of this additional import is satisfied by increasing capacity of pipelines from Russia and other CIS countries, and the rest from will be delivered in the form of LNG.

The increased supply of LNG leads to increased competition between LNG and pipeline gas, especially in Europe and China, the two largest importers of gas. The greater ease of transportation means that pipeline gas has a significant cost advantage over LNG. However, the main limitation of the pipeline import in Europe is concern about dependence on Russia for gas and fear of the 2009 crisis relapse. It is possible to solve this kind of concern in two ways: bypass pipelines construction in order to eliminate transit risks - in this regard, Russia continues to implement further large-scale pipeline projects to Europe, such as the Nord Stream II project with a throughput of 55 billion cubic meters a year and two lines of Turkish Stream (the capacity of each is 15.75 billion cubic meters per year). The second way is the development of LNG trade, which will lead to a more liquid and competitive market, which in turn will ease Europe's tensions with Russia, allowing Gazprom to slightly increase its share in European gas pipeline imports.

According to forecasts, natural gas production in Russia is growing steadily until 2040, retaining its position as the second largest gas producer in the world. Today, almost all production in Russia comes from fields in Western Siberia and the Yamal Peninsula, but the opening of new routes to China leads to the expansion of production in Eastern Siberia and the Far East of Russia. Domestic consumption in Russia remains almost at the same level, which means that production growth must find export markets. How does this LNG exported by various companies coexist with the remaining Gazprom's monopoly on pipeline export is another strategic issue that may ultimately serve as a catalyst for changes in the organization of the Russian gas market.

After the launch of the second, third and fourth lines of Yamal LNG (approximately until the end of 2019), the total capacity of Russian large-capacity LNG production will reach 28.2 million tons, allowing Russia to reach the third place in terms of exports after Qatar and

Australia. However, Russia's future plans for capacity expansion are associated with large uncertainties. PJSC Gazprom has two large-scale projects: the Sakhalin-2 project has been discussing the expansion for many years by building a third line with a capacity of 5.4 million tons. In addition, the Baltic LNG project is being discussed, involving the Unified Gas Supply System as a source of gas. PJSC Rosneft maintains plans for the construction of its own LNG production ("Far Eastern LNG"). The company is also considering the Pechora LNG project on the basis of the Korovinskoe fields, while there is also little certainty on this project.

Among the most likely new productions is the Arctic LNG 2 project (PJSC Novatek) - three lines of 6.6 million tons each, with the resource base of the Morning field of the Gydan Peninsula. The Arctic LNG 2 project involves the construction of an LNG plant on the basis of a gravity type platform (300 m long and 150 m wide), which is expected to reduce capital costs by 30%. The construction of the LNG plant on the platforms will reduce logistics costs and install the main equipment "remotely" - the LNG trains will be built at the Center for the construction of large-scale offshore facilities in Murmansk.

Moreover, based on the existing resource base on the Yamal and Gydan Peninsula, Novatek PJSC is also considering the possibility of implementing the Arctic LNG 1 and Arctic LNG 3 projects, also by 19.8 million tons each, which implies an increase in total capacities of the company's facilities in the region up to 76.8 million tons (including those under construction and existing production), which is comparable to the existing liquefaction volumes of Qatar.

Based on the current situation and statements by company representatives, the possible total LNG capacity in Russia by the early 2030s can be estimated at 60-80 million tons, although it is clear that this is a preliminary estimate. In that situation, it is crucial to implement strategic steps to guarantee investments in future projects and their further profitability.

Russia, which plans to take a strong position in LNG industry, has no other options than to look for new opportunities and implement new strategies:

- Reducing the cost of LNG through the supply chain
  To be cost competitive with coal in Asia, simply based on the marginal economy, a simple analysis of the cost of electricity suggests that gas imports will cost about $ 4 to $ 6 per million BTU - about half the cost today for many contracts for LNG. That means, that reductions of costs must be implied through all components of the supply chain
- More flexible LNG contract
  A key remaining challenge in the LNG sector is the rigidity of existing LNG contracts and market structures. Although the LNG development model as a "virtual pipeline" using long-term contracts and oil indexation has helped create market stability and risk management for sellers and many large buyers, it does not necessarily contribute to a dynamic or flexible market. In 2017, momentum continued to build up around the more liquid and global commodity LNG market. This included an increase in the share of global short-term sales (30% vs. <15% until 2010), a shorter average contract duration (7 years against> 15 years until 2010) and contracts with a lower average volume (0.75 MTPA against> 1, 5 MTPA pre-2010). Cancelation of destination clause in contracts and trading in smaller volumes and shorter periods are crucial for maintaining role of reliable exporter.
- Infrastructure investments
  The speed and nature of gas demand growth, especially in non-OECD countries, depends on the pace at which the necessary supporting infrastructure is created: this is a key source of uncertainty regarding the prospects for natural gas. At the same time, while fighting for the consumer, traders and LNG producers are beginning to invest in the downstream infrastructure of LNG importers in order to create additional gas demand and gain a foothold in new markets, building "special" relationships with customers.
- Export not only products, but also technologies
  The goal of cooperation should not be the resource use of the Russian industrial and scientific base, but the creation of a technology that in the future can give a push to the development of an entire industry, and even be exported as a product.

- Development of new supply routes
  Due to reducing costs through supply chain, it is very important to optimize logistics too. Development of Northern sea route could reduce time of LNG delivery and therefore cut transportation expenditures.

3 CONCLUSION

The obvious shift in Russian energy export policy is associated with the expansion of sales markets towards the Asian region, following the path of oil market, since Rosneft and other companies have contributed to a significant shift in Russian oil exports from Western to Eastern markets. By 2017, 30 percent of the country's oil exports went to Asia, whereas in 2010 this figure was only eight percent. The development of the Eastern Siberia - Pacific Ocean (ESPO) pipeline allowed, first of all, to export oil via a pipeline to China and the Asia-Pacific region.

The LNG market is becoming more complex and flexible, resembling the path made by oil market. Russia, which plans to take a strong position in LNG industry, has no other options than to reduce costs, open new markets (including developing of the domestic market), introduce new technologies and types of contracts, optimize logistics and build strong relationships with importers. Rivalry is high, but only those who can work most efficiently and offer the most attractive conditions to customers will achieve success at the transforming and globalizing market.

REFERENCES

BP Energy Outlook 2019, viewed 03.03.2019, https://www.bp.com/content/dam/bp/business-sites/en/global/corporate/pdfs/energy-economics/energy-outlook/bp-energy-outlook-2019.pdf

EIA – International Energy Outlook 2018, viewed 27.02.2019, https://www.eia.gov/pressroom/presentations/capuano_07242018.pdf

ExxonMobil – Outlook for Energy: *A View to 2040*, viewed 11.03.2019, https://cdn.exxonmobil.com/~/media/global/files/outlook-for-energy/2016/2016-outlook-for-energy.pdf

INES RAS – *World and Russia Energy Outlook 2016*, viewed 11.03.2019, https://www.eriras.ru/files/prognoz-2040.pdf

International Energy Agency – *World Energy Outlook 2018*, viewed 03.03.2019, https://www.iea.org/weo2018/

International Energy Agency – *Special Report Outlook for Producer Economies*, 2019, viewed 03.03.2019, https://www.connaissancedesenergies.org/sites/default/files/pdfptvue/weo_2018_special_-report_outlook_for_producer_economies.pdf

International Gas Union – *Global Gas Review 2018*, viewed 03.03.2019, https://www.igu.org/news/2018-global-gas-report

Skolkovo Energy Centre – *"Global LNG market transformation: ways not to miss the window of opportunities for Russia"*, viewed 07.03.2019, https://energy.skolkovo.ru/downloads/docu-ments/SEneC/News/Russia-on-global-spg-market-Eng.pdf

Shell LNG Outlook 2019, viewed 17.03.2019, https://www.shell.com/energy-and-innovation/natural-gas/liquefied-natural-gas-lng/lng-outlook-2019.html

World Energy Council – *World Energy Scenarios*, viewed 17.03.2019, https://www.worldenergy.org/wp-content/uploads/2016/10/World-Energy-Scenarios-2016_Full-Report.pdf

Youth Technical Sessions Proceedings – Litvinenko (ed)
 ISBN 978-0-367-34668-3

The problem of LNG flow rate determination

S.S. Matushov
Saint-Petersburg Mining university, Saint-Petersburg, Russian Federation

ABSTRACT: The research is being carried out in order to make proposals of innovative methods of LNG (liquefied natural gas) flow rate determination and new devices for its measuring in pipelines. Nowadays systems are not able to provide satisfactory results without any drawbacks. The analysis of existing methods and devices was carried out. Theoretical research and calculations in the aggregate with computer simulations and experimental study on laboratory equipment are being performed taking into account most perspective ideas. The proposed developments tend to raise the economic efficiency of the industry.

1 INTRODUCTION

The importance of process fluids flow rates exact determination is hard to underestimate. The economic efficiency of the industry, as well as the ecological safety of the environment, depend on the exact determination of the fuels flow rate in pipelines. The deficit of pumped over fluids' volumes leads to the loss of manufactures' profits. The problem of an accurate determination of steams' flow rates in pipelines is being solved with the use of highly-precise measurement systems.

This problem does not lose its actuality in case of LNG. The increasing LNG production volumes (figure 1) makes accurate measurement of LNG flow rate even more important.

During the LNG transport in pipelines pressure drops occur in them, which leads to some fuel's part boil off. The LNG transition into gas phase is accompanied with an approximately six hundred times volume increase, which can lead to multiple pressure increase because of the enclosed volume of the pipeline and, as a result, to negative consequences. Because of this the

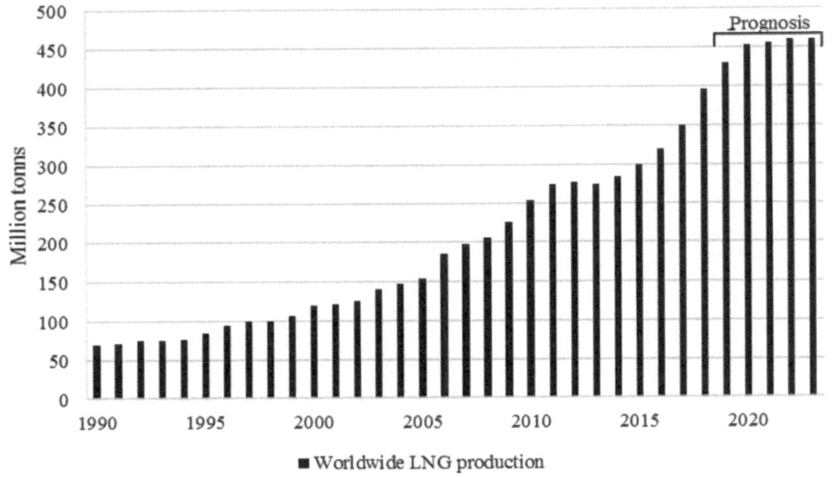

Figure 1. The world LNG production, million tones. (HIS Markit, Company Announcement)

derivation of the evaporated fuel is an obligatory part of the technological process. But even if the two-phase LNG-vapor flow pump off continues in normal mode, it is necessary to account the gas phase flow rate as a part of the total system flow, which is insufficiently provided by currently used LNG flow meters.

2 ANALISYS

The constantly growing LNG turnover demands innovative precise measurement system appliance, which produce minimal errors. Nevertheless, Russian industry is not sufficiently supplied with highly-precise multiphase flows measurement instruments.

Many companies which specialize in progressive cryogenic liquids accounting systems production write about their measuring instruments use for liquefied gases referring to liquid nitrogen, oxygen, carbon dioxide, etc., concealing its efficiency in appliance to LNG or two-phase systems of these substances in general.

A lot of modern gas-liquid mixtures flow rate measurers determine the total mixture flow rate Q_M with the use of Venturi tube, and for the gas phase flow rate portion in general flow rate β (gas portion) determination they use gamma-radiation or electrical properties of fluids, etc. However, densities of liquid and gas phases in the two-phase flow differ significantly, therefor the system tends to stratification into liquid and gas. It results into rapid change in the stream scheme (figure 2) depending on gas portion, and separated phases transfer with different speeds. Thus difficulties with the stream scheme change accounting and its flow rate determination by means of flow pulse meter appear.

The usage of gamma radiation and electrical fluid properties to determine the gas portion in the system's flow is expensive; there are also concerns about these methods harmful effects on human health, which primarily refers gamma radiation. Measurement results based on the electrical properties of fluids get distorted the fluid contains heavy metals and salts.

Currently some developments that do not have such disadvantages exist. Volume double-phase flow meter of gas-liquid mixture and the accompanying multiphasic flow measurement system is one of them. The invention is offered by Kitami H. and Kavaoto H. from Oval Corp., Japan, in patent RU 2507484. It is represented as a replacement of traditional triple-phase oil-gas-water mixture flow measurement system in fields which divide the flow in phases in separators, determine the flow rate of each phase separately and measure portions of each phase in the general flow by sampling.

The method which is widely used currently provides a high accuracy only with the use of large and expensive technical complexes. Meanwhile the invention offered in the patient

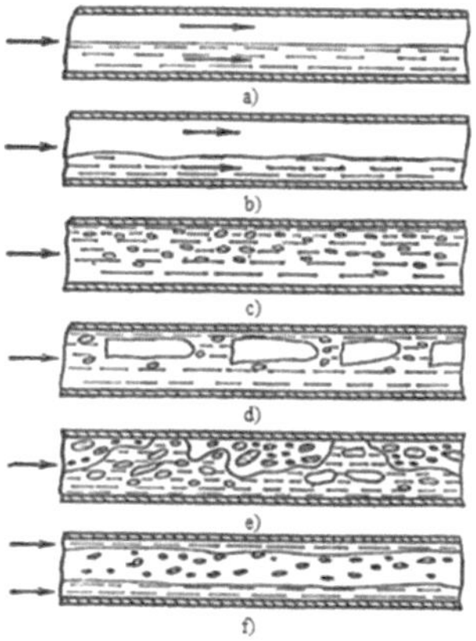

Figure 2. Stream schemes of gas-liquid mixtures flows in pipelines a) Stratified; b) stratified wavy; c) bubbly; d) plugged; e) bubbly-plugged; f) circular

measures liquid and gas flow rates in the general flow without dividing it in separate phases and is much more compact.

The scheme of the described biphasic flow meter is presented in the figure 3.

The flowmeter 1 consists of the pressure measurement block 2 on the entrance into the gas-liquid mixing chamber 5, which mixes liquid and gas in the gas-liquid biphasic flow and creates constant stream scheme inside the flow meter that is convenient for the measurement, the pressure difference measurement block 3 between points before the mixing chamber and after the measurement chamber 7, the temperature measurement block 4 and a pair of rotors 8, which are placed inside the flow meter body 6.

The gas-liquid mixture flow rate and gas and water portions measurement is based on the system of two equations, which represent the functional dependence of, firstly, the general mixture flow on the angular velocity of rotors ω and the gas portion β (equation 1), and secondly, the mixture pressure loss ΔP_M in the measurement system on the mixture's flow rate and the gas portion (equation 2)

$$Q_M = f(\omega, \beta), \quad (1)$$

$$\Delta P_M = f(Q_M, \beta). \quad (2)$$

Let's consider the sequence of the needed parameters of the biphasic gas-liquid flow determination.

During the determination of the rotors angular velocity ω its slowdown due to the increase in the gas portion β in the constant gas-liquid flow Q_M is taken into account. For this reason, the rotation coefficient ε is entered (equation 3)

$$\varepsilon = f(\beta), \quad (3)$$

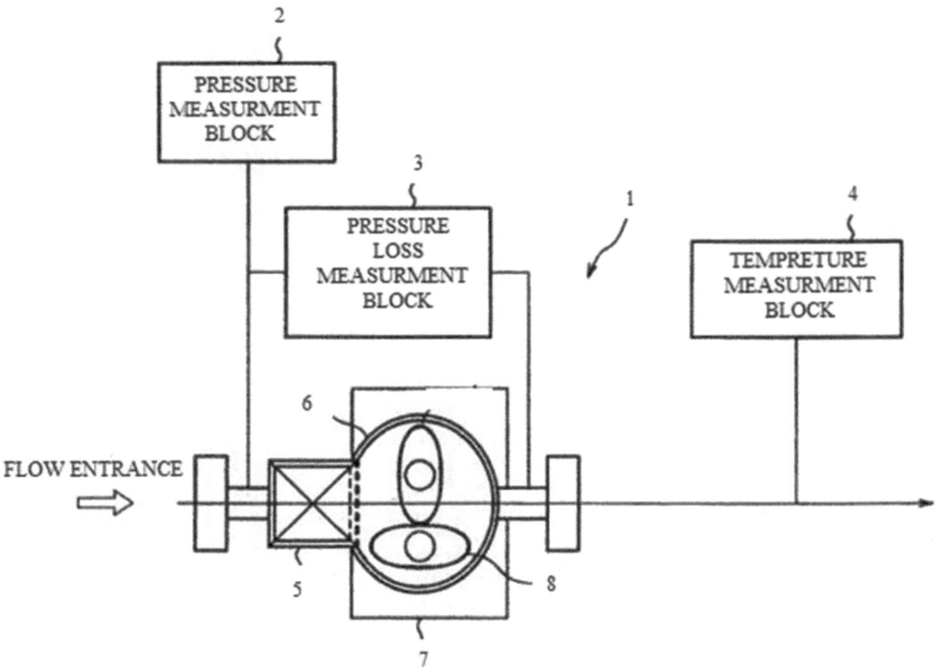

Figure 3. The block-scheme of the volume double-phase flow meter of gas-liquid mixture

which is experimentally determined in advance. Using this coefficient, the general gas-liquid flow rate can be presented the next way

$$Q_M = M_f \cdot \frac{\omega}{\varepsilon}, \tag{4}$$

where M_f – the volume of the liquid, transferring through the flow meter during the one turn of the rotor in case of single-phase liquid flow, l.

Then two pressure losses are considered: ΔP_L, which forms in a single-phase liquid flow, and ΔP_M, which appears and grows gradually with the increase in gas portion β in constant liquid flow. The pressure loss ΔP_M increase in comparison with ΔP_L is presented like

$$\frac{\Delta P_M}{\Delta P_L} = (1-\beta)^{-z}, \tag{5}$$

where z – the pressure loss increase coefficient. It is experimentally determined in advance. The pressure loss of the liquid can be found the next way

$$\Delta P_L = \frac{1}{2} C \rho_L \left[\frac{Q_M(1-\beta)}{A}\right]^2, \tag{6}$$

where C – the coefficient of the mixing and measuring chambers profile resistance both; ρ_L - the density of the liquid, kg/m^3; A – the stream's cross section area on the entrance into the mixing chamber, m^2.

After equation (6) substitution into (5) we get the gas-liquid mixture pressure loss

$$\Delta P_M = \frac{1}{2} C \rho_L (1-\beta)^{2-z} \left[\frac{Q_M}{A}\right]^2. \tag{7}$$

The general mixture flow rate Q_M and the gas portion β can be determined from equations (4) and (7). Liquid and gas flow rates Q_L and Q_G separately can be found from the equations below

$$Q_L = Q_M \cdot (1-\beta), \tag{8}$$

$$Q_G = Q_M \cdot \beta. \tag{9}$$

The described flow meter shows a high accuracy of measurements of biphasic gas-oil flows in comparison with the accuracy of the available flow meters. In the inventors' experiments with gas-saturated oils a relative error for determining the gas fraction β was from (-4) to 2% - and for the liquid and gas phases measurements separately Q_L and Q_G +/- 10%. The measurement range of the invention is wider (1:10 and higher against approximately 1:3 of an obvious multiphasic flow meter), and in the measurement range of the systems flow rate from 5 to 20 m^3/h even the less error +/- 5% is guaranteed.

We offer to take the described device and the calculation method as basis for the research on the biphasic LNG flow rate determination in pipelines. The equal hydrocarbon nature of the gas saturated oil and the LNG confirms the perspectives of the offered way of the research.

At the current stage of the research the possibility of the biphasic flow rate measurement with the laser Doppler flow meter is studied. This device measures the frequency difference, which appears when the light beam reflects from the moving particles of the flow. This phenomenon can be described mathematically the next way

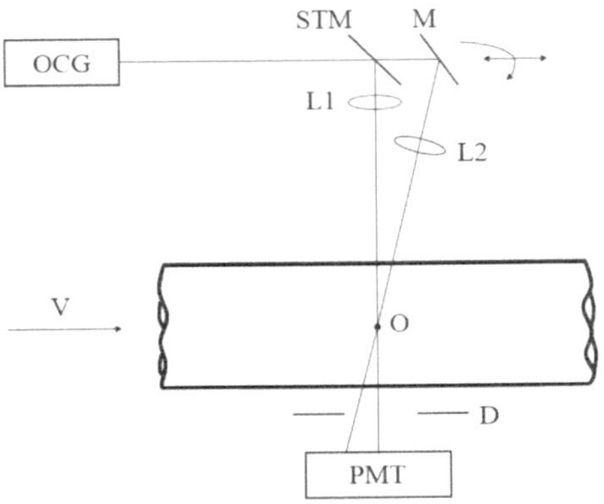

Figure 4. The laser Dcppler flowmeter scheme

$$\lambda = 2\pi \frac{(c - v)}{\omega_0}, \tag{10}$$

where ω_0 – the angular frequency of the source's waves, s^{-1}.

The fluid's velocity can be determined this way.

The principal scheme of the Doppler flow meter is presented in the figure 4.

The measurements are taken the next way. After leaving the optical coherent generator (OCG) the light beam meets the semitransparent mirror (STM) and partially reflects from it, forming the reference beam, which then goes through the lens L1, then through the liquid (perpendicular to its movement direction without a Doppler effect) and passing the diaphragm D gets on the photocathode – photomultiplier tube (PMT). The other part of the beam travels from the OCG through the STM, reflects from the second mirror M, focuses in the point O by the lens L2 (partially dissipated) and forms the working beam, which passes the diaphragm D and gets on the photocathode PMT. The position of the point O can be regulated by the movement of the mirror M.

Currently Doppler flowmeters are used in medical studies of biological liquids circulations in vessels of living organisms, including humane.

3 FORECASTING

At the current stage of the research the possibility of optical, including Doppler, flowmeters use for the LNG flow rate determination in pipelines is studied. To solve the described problem, we need to reach next goals: consider the specialty of optical flow meters work in cryogenic systems; formulate methods of the gas portion in the general flow measurement; carry out needed experiments in the laboratory; make appropriate calculations, analyze them and on the basis of the work done offer a new methods of the LNG flow rate measurement.

4 CONCLUSION

The fact that in current time LNG companies are able to measure only the flow rate of the liquid phase of the system and calculate the volume of the gas phase as a loss of the liquid fuel weighting filled containers, and until recently were measuring shipments by the change in the reservoir liquid level, states the non-nominal actuality of the research.

The upgrade of the equipment and methods of the LNG flow rate determination during its flow in pipelines will raise the economic efficiency of companies and correct many assumptions and inaccuracies in existing techniques of flows' parameters measurement.

REFERENCES

Kavaoto H., Kitami H. 2014. Pat. 2507484 Russian Federation. Volume double-phase flow meter of gas liquid mixture and the multiphasic flow measurement system.

Kremlevski P.P. *Rashodomeri I schetchiki kolichestva veshestv* [Flow and matter amount meters]. Handbook (Volume II): 312-330.

Soloviev V.G., Reut G.I., Reut V.I., Mukhametshina G.F., Pogorelov S.N., Tukhvatullin R.R., Chupaev A.V. Metrological maintenance of liquefied gases flow measurments. *Vestnik Kazanskogo technologicheskogo universiteta* [The Kazan technological university messenger]: 146-150.

Flow Instruments. Available at: http://www.chartindustries.com/Businesses/Flow-Instruments (accessed at 10 November 2018).

Raschodomeri [Flowmeters]. Available at: https://ru.krohne.com/ru/pribory/ (accessed at 11 November 2018).

Youth Technical Sessions Proceedings – Litvinenko (ed)
© 2019 World Petroleum Council, London, UK, ISBN 978-0-367-34668-3

Peculiarities of flow of liquefied natural gas in pipelines

V.A. Voronov
Saint-Petersburg Mining University, Saint-Petersburg, Russian Federation

S.A. Ivanik
Saint-Petersburg Mining University, Saint-Petersburg, Russian Federation

ABSTRACT: The authors of the article consider the features of the flow of liquefied natural gas through cryogenic pipelines, the parameters of dispersed-ring and stratified modes of natural gas transport in the form of gas-liquid mixture, conduct a comparative analysis of pressure losses in the pipeline section for two-phase and single-phase modes of natural gas flow and estimate the limits of possible values of throughput for LNG.

1 INTRODUCTION

The development of the oil and gas industry of the Russian Federation is closely related to the development of the production and use of liquefied gases, with special attention paid throughout the world to the development of liquefied natural gas (LNG) technology (Golovina, 2018). The need for production and consumption of LNG is associated with the conditions under which the construction of gas pipelines is unprofitable or impossible. The main type of LNG transport is sea gas tankers for transshipment terminals. Delivery to consumers is carried out in cryogenic tanks by refrigerator cars. Unloading and filling of all types of LNG transport is carried out at transshipment bases via technological pipelines. In this case, the movement of the product is carried out using cryogenic pumps, supercharged inert gas in the storage or LNG vapors, as well as using hydrostatic pressure generated in the tank (Figure 1).

Filling of cryogenic pipelines is a complex non-stationary process, accompanied by the evaporation of a liquid, pressure pulsation, hydraulic shock, geyser effects (Lensky, 2016). The most dangerous are pressure pulsations, in which the maximum pressure can exceed the pressure of fluid supply to the pipeline 3-5 times. The pipelines are filled each time when loading and unloading tankers or when supplying a stream to a regasification unit. Therefore, the corresponding systems are carried out with the utmost care. Thus, the operating pressure of the pipelines for loading tankers is assumed to be 2.5 MPa, while the pressure of the pumps can be 80-120 m, which corresponds to a pressure of 0.5-1 MPa. Onshore pipelines are supplied with low temperature liquid or vapor circulation systems ensuring their constant maintenance in a frozen state. In order to keep the LNG pumps in a frozen state, small circulation pumps, including vortex pumps, are used (Safonov, 2017).

Currently, LNG pipeline transport is used on relatively short sections, mainly in the process pipelines of plants for the production of liquefied natural gas and terminals for receiving and shipping the product. This fact is caused by the high cost of such pipelines, since their production requires the use of special steel grades and additional insulation coating. In addition, special cryogenic pumps, additional devices are needed to compensate for the unevenness of temperature deformations, to drain the vapor phase formed and to ensure the safety of operation of cryogenic pipelines. Given the significant external technical and economic issues, very little attention is paid to complex internal processes of pumping cryogenic fluid. Despite the fact that the transport of liquefied natural gas through pipelines is carried out mainly through the technological lines of LNG installations, it is necessary to kkep the liquefied gas in the

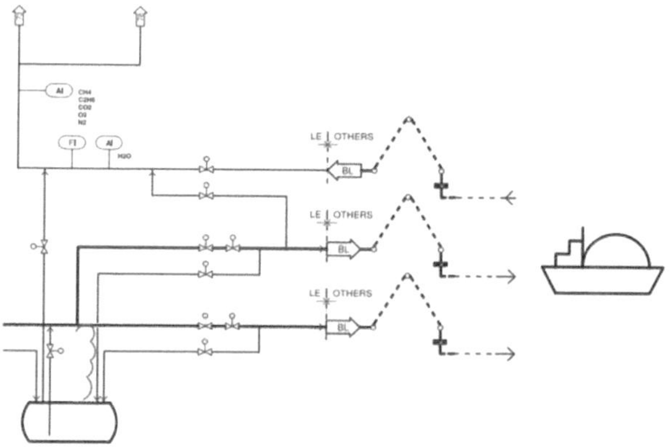

Figure 1. Scheme of tank filling.

pipe in a liquid state. The temperature of the liquefied gas should always remain below the boiling point at working pressure, despite the pressure loss and heat influx. External mooring walls of the discharge complex are located at a distance from the coast and are connected to the onshore facilities by a transition bridge or overpass. Thus, all connections for interaction with the gas carrier, including cryogenic pipelines, are laid on transition bridges or ramps. This solution has proven itself to be safe and secure. An analysis of world experience has shown that the length of the racks built varies from hundreds of meters to ten kilometers. For example, the distance of such an overpass from the storage facility of the LNG plant in the area of the Portovaya compressor station, intended for the uninterrupted gas supply to the Kaliningrad region, is 2.8 km. Designing a cryogenic pipeline of such length is a complex engineering task. At the initial stage of choosing the type of cryogenic pipeline, attention is paid to the main factors affecting the ultimate economic efficiency: the material and the design of the pipeline; thermodynamic efficiency of pipe insulation; operating pressure; experience in using the type of cryogenic pipeline and the readiness of the industry for its production, and even the presence of special conditions for the installation and operation of the pipeline. The following is taken into account when choosing a working pressure in the pipeline. The amount of boiling up vapor generated in the pipeline will be excessive if the transfer pressure is too low. The increase in pressure entails higher requirements for the strength characteristics of the pipeline and an increase in pump power.Existing designs of cryogenic pipelines based on vacuum and bulk types of thermal insulation are not able to provide transportation of LNG in a single-phase state, and technological methods for maintaining LNG in a single-phase state are not feasible with a low-pressure method of LNG transportation. With the joint flow of liquid and gas in horizontal pipes, the type of flow is determined by the volume flow and the properties of the flow itself (Voronov, 2017). The following fluid flow regimes with viscosities less than 0.1 Pa·s and gases approximately equal to the density of air were established experimentally: "foamy" flow, "bubble" flow; "flat" ("smooth") flow; "wave" flow; "piston" flow; "layered" ("ring") flow; "dispersed" ("sprayed") (Figure 2).

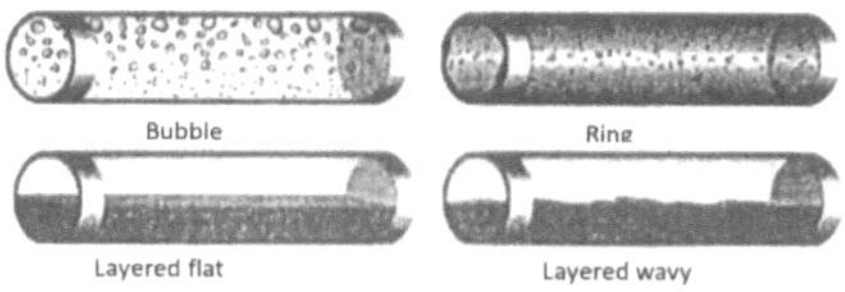

Figure 2. Modes of flow of two-phase fluid in horizontal pipelines.

1. The flat flow (stratified flow) is characterized by the movement of fluid along the bottom of the pipe, and of the gas along the smooth interface. The value of the fluid velocity is less than 0,15 m/c, gas velocity – 0,6-3 m/c.
2. Wave flow is characterized by the formation of waves at the interface. Fluid velocity is less than 0,3 m/c, gas velocity ≈ 4,5 m/c.
3. Bubble flow. Portions of liquid and gas alternately move along the top of the pipe. Fluid velocity is less than ≈ 0,6 m/c, gas velocity ≈ 0,9 m/c.
4. Piston flow. The fast-moving gas periodically picks up a portion of the liquid and forms a foam tube. The velocity of the gas projectile is greater than the average velocity of the fluid. Piston mode is characterized by intense vibrations of pipelines (for example, fittings).
5. The emulsion mode has a homogeneous structure. Gas bubbles have irregular shape.
6. Dispersive ring mode. One of the phases moves in the core of the flow, and the other moves in the film on the wall. This mode is applicable when the pipeline is cooled down with cryogenic liquids and high flow rates (Babenko, 2016).

Two modes were considered during the study for the transport of natural gas in the form of a gas-liquid mixture: dispersed-annular and stratified (Figure 3).

The complexity of the calculations in the dispersed-ring mode lies in the fact that the gas-liquid flow moves in a pulsating mode with a constant frequency of parameter changes (Sitenkov, 2006). Therefore, it is necessary to consider areas of small extent, so that the indicators of hydraulic resistance were the most accurate. It should be determined whether the liquid or gas will be the active dispersion medium at the beginning of the hydraulic calculation. If gas content $\beta > 0,7$, the pipeline is calculated by the formulas, taking into account the compressibility of the medium and the viscosity of the gas. In this case, the liquid phase is distributed along the wall of the pipeline, and the gas in the flow core. If gas content $\beta < 0,5$, the calculation is carried out on the liquid, taking into account its viscosity. The density of the mixture should be used in the calculations. The throughput was calculated using the standard formula with the help of the initial parameters (speed, diameter of the pipeline and a given consumption).

$$Q = \frac{\pi d^2 \upsilon}{4} \tag{1}$$

where υ = speed; d = diameter of the pipeline.

Pressure losses were calculated for two cases of transportation as part of the study of the dispersed-ring mode:

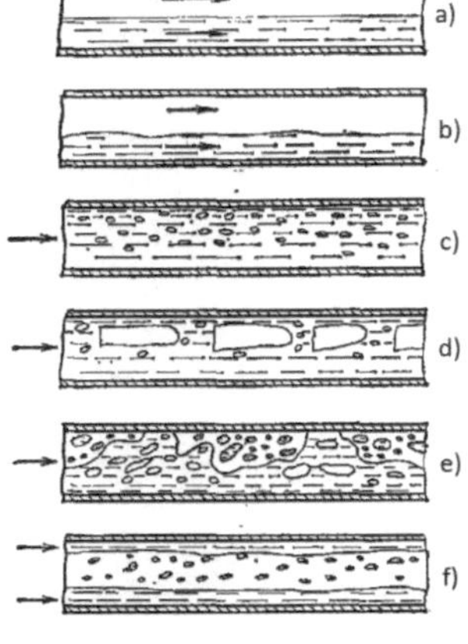

Figure 3. Modes of flow of two-phase flows in horizontal pipelines: a) stratified; b) wave; c) bubbly; d) piston; e) emulsion; e) dispersion ring.

1. Gas consumption Q_g = 1800 m^3/h, the flow rate of the liquid phase Q_g = 700 m^3/h (the active state is gas).
2. Gas consumption Q_g = 700 m^3/h, the flow rate of the liquid phase Q_g = 1800 m^3/h (active state - LNG).

The initial data for the hydraulic calculation were taken experimentally for a visual result. So, the length of the pipeline route is increased to L = 26 km (within 5-50 km), pressure at one end of the pipeline p_1 = 5 MPa, mixture speed υ = 10 m/c, which is characteristic of the stream in question. It is necessary that the maximum field strength exceed the difference in the values of the specific weight of the two phases to create a dispersed-ring mode (formula 2):

$$\frac{6,16\upsilon^2 \cdot \rho mix}{D} > \gamma_l - \gamma_g \qquad (2)$$

If the inequality holds, then it is necessary to determine the calculated diameter of the flow, taking into account the coefficient α for a dispersed system that characterizes the equilibrium position of the field strength and $\gamma_l - \gamma_g$. Thus, the calculated diameter is calculated by the formula 3:

$$D_p = 0,5 \cdot D \cdot (1 + \alpha) \qquad (3)$$

where D = preselected diameter.

Reynolds number value Re determined by formula 4, taking into account the speed of the mixture, the density of the mixture and the viscosity of the active state.

$$\mathrm{Re} = \frac{\upsilon \cdot D \cdot \rho}{\mu} \qquad (4)$$

As a result of calculations the value of the Reynolds criterion is Re > 2300. Thus, the annular flow is a self-similar mode in which the coefficient of hydraulic friction does not take into account the roughness of the pipeline walls and is calculated by the formula 5:

$$\lambda = \frac{0,3164}{\mathrm{Re}^{0,25}} \qquad (5)$$

The calculation of the pressure at the end of the route is calculated by the formula 6:

$$p_2 = \left[p_1^2 + \frac{4200 \cdot \lambda \cdot \rho mix \cdot Q^2 \cdot T_{av} \cdot L}{D_p^5}\right]^{0,5} \qquad (6)$$

where = mixture density; Q = active phase flow, m^3/c; T_{av} = average temperature of the mixture, K.

If inequality (2) is not satisfied, which means that the strength of the boundary layer is not enough to implement the dispersed structure of the flow, then a layered flow is established. This is possible when the value of the mixture speed is υ = 0,5 m/c. In this case, the diameter is calculated by the formula 7:

$$d_p = \left[\frac{1,772 \cdot \upsilon_{mix}^2 \cdot \rho l \cdot D^4}{\gamma_l - \gamma_g}\right]^{0,2} \qquad (7)$$

The calculations performed showed that the pressure loss in the layered flow regime is much higher than in the case of a dispersed-ring flow. Comparing the similar indicators for two cases of high-speed flow, the loss indicators are higher during the liquid active phase. The results of the study are shown in Figure 4.

The pressure loss for single-phase LNG flows and gas-air medium was calculated without changing the initial data. Comparative analysis showed that a dispersion-ring mode have the smallest indicators of hydraulic resistance of the considered methods of transporting natural gas (Figure 5).

Hydraulic calculation showed that a dispersion-ring flow regime has the smallest pressure loss. To design an LNG pipeline, it is necessary to take into account not only the characteristics of the pipe material and insulation, but also the parameters of pumping and external influence. It is required to provide compensation for temperature deformations with decreasing temperature, as well as safety of operation of cryogenic pipelines. LNG consumption exceeds the value of the throughput of natural gas, which represents the feasibility of selecting the optimal values for the parameters of pipeline transport. As an example, we consider the range

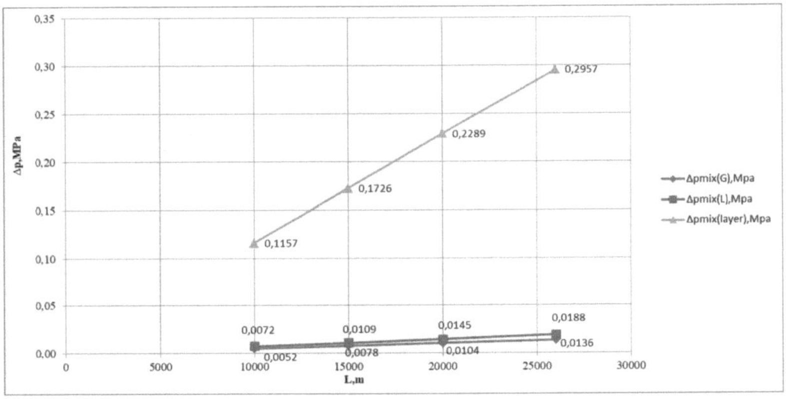

Figure 4. Comparative analysis of pressure loss in the pipeline section with ring and layered flows:$\Delta p_{mix}(G)$ - loss of pressure over the ring flow (the active state is gas); $\Delta p_{mix}(L)$ – pressure loss over the ring flow (active state - LNG); $\Delta p_{mix}(layer)$– pressure loss in case of stratified flow.

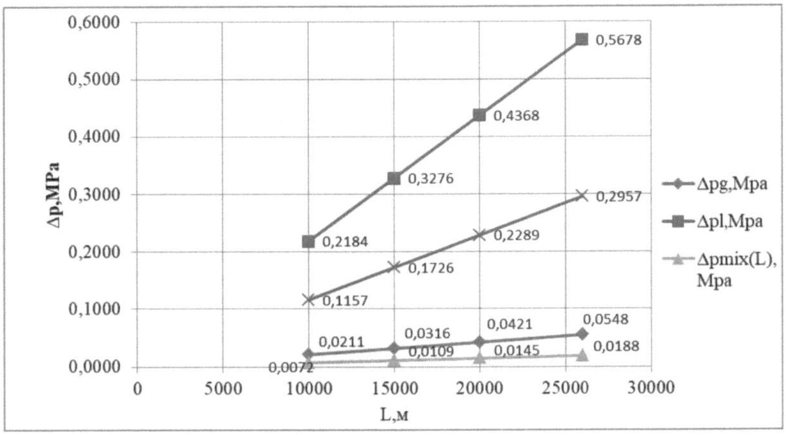

Figure 5. Comparative analysis of pressure loss at the pipeline section for two-phase and single-phase flow modes of natural gas: Δp_g – pressure loss during gas-air transport; Δp_l – LNG transport pressure loss; $\Delta p_{mix}(L)$ – pressure loss during the annular flow (active medium - LNG); $\Delta p_{mix}(layer)$ – pressure loss in case of stratified flow.

of diameters for pumping LNG d = 83-530 mm. Thermal insulation (for example, polyurethane foam or screen-vacuum) is necessary in order to ensure a high level of safety and reliability, which in turn depends on the operating temperature of the LNG, the occurrence and propagation of cracks in the pipe, and temperature deformations. Speed values υ = 0,064 – 3 m/c in accordance with the specified diameters calculated on the basis of thermodynamic parameters (temperature, pressure, compressibility factor). Values of the maximum, minimum and average throughput capacities of the pipeline are shown in the figure on the basis of the calculated speed data (Martynenko, 2017).

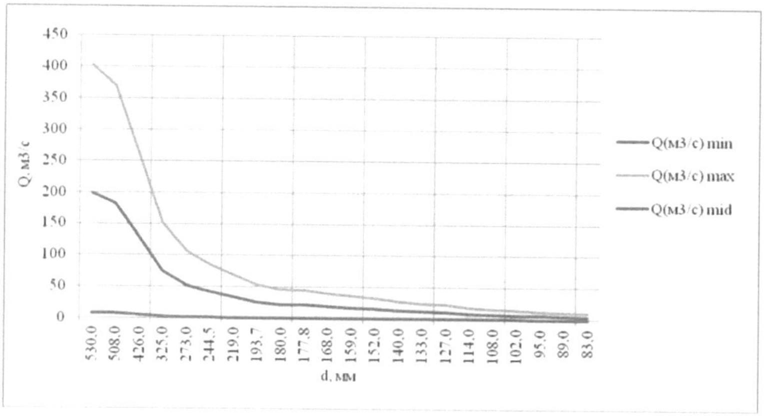

Figure 6. Limits of possible throughput values for LNG

The following known formula was used to determine the speed in accordance with the diameters in question:

$$v = 0,1273 \frac{Qz \cdot T}{d^2 \cdot p} \tag{8}$$

where Q = gas flow through the section, m^3/h; z = compressibility factor; T = gas temperature, K; d = internal diameter of the pipeline, mm; p = absolute gas pressure, MPa.

The features of the partial components are not taken into account in many cases during the hydraulic calculations of liquefied natural gas pipelines (Rachevsky, 2009). It is necessary to determine both the average density of a multicomponent medium and the average viscosity of the flow (and structural viscosity in cases of multiphase flow) (Voronov, 2019). The viscosity of the mixture at a given temperature T (K) of liquefied natural gas is calculated by the formula 9:

$$\lg(\mu_{mix}) = \sum_{i=1}^{n} y_i \cdot \lg(\mu_i) \tag{9}$$

where $\lg(\mu_i)$ = the viscosity of the mixture component, calculated by the formula 10:

$$\lg(\mu_i) = A \cdot \left(\frac{1}{T} - \frac{1}{B}\right) \tag{10}$$

where A and B = empirically determined constants in the viscosity equation (Voronov, 2019).

The vapor pressure at a temperature of 143 K is determined by the formula 11:

$$p_{abs}^{mix} = \sum_{i=1}^{n} z_i \cdot p_i \tag{11}$$

where z_i = molar concentrations of the components; p_i = component pressure is determined by the formula of Antoine:

$$\ln(p) = A - \frac{B}{C + T'} \tag{12}$$

where A, B and C = constants of Antoine's equation (Voronov, 2019).

It is very important to convert mass concentrations to molar and determine the value of the vapor pressure of the mixture for accuracy of calculations.

A distinctive feature of pipeline transport of liquefied gases is the dependence of the transported state on the nature of changes in pressure and temperature along the length of the pipeline. If the pressure in the pipeline falls below the saturation pressure of the liquefied gas at the existing temperature, the liquid will begin to boil, the resulting vapor phase will begin to fill up a portion of the living section of the pipeline. This can lead to a dramatic reduction in pipeline capacity, as well as an emergency situation. Therefore, it is extremely important to carry out a verification calculation at the highest "point" and, as a consequence, the most dangerous one.

The relevance of liquefied natural gas increases every year and is actively developing. Large-capacity LNG plants are built around the world and increase their productivity annually. For example, there are 4 mini-plants for the production of liquefied natural gas in the Leningrad region, the fuel from them is currently delivered to the consumer by means of gas carriers. However, actively developing technologies can make even LNG trunk pipelines economically profitable over short distances already in the coming years.

REFERENCES

Safonov V.S. 2017. Justification of operating parameters of technological pipelines of LNG complexes taking into account industrial safety requirements. Improving the reliability and safety of gas industry facilities. *Moscow, VNIIGAZ*. №1 (29):83-99.

Voronov V.A., Martynenko Ya.V., Nazarova M.N. 2017. Comparative analysis of gas transportation through pipelines in liquid and gas-air states. *Business magazine Neftegaz.RU*. №10:20-23.

Babenko A.V. 2016. Hydraulic calculation of two-phase gas-liquid flows: a modern approach. *Calculations and modeling*: 38-42

Martynenko Ya.V., Voronov V.A. 2017. Comparative analysis of single-phase and two-phase modes of natural gas transportation through pipeline systems. *Yekaterinburg, International Research Journal*, № 04(58),Part 4: 28-34.

Voronov V.A., Ivanik S.A., Shalygin A.V. 2019. Transport of liquefied natural gas. Study Guide. *St. Petersburg, SPMI.*

Golovina E.I., Pashkevich N.V., Tarabarinova T.A. 2018. Problems of reflecting information on subsoil assets in International Financial Reporting Standards. *Academy of Strategic Management Journal*, № 17, V 3: 1 – 9.

Sitenkov V.T. 2006. Hydraulics. Theory and calculation of two-phase systems. *Nizhnevartovsk*. 204 p.

Rachevsky B.S. 2009. LNG. *M., Oil and gas*. 640 p.

Lensky V., Fisenko O. Kuris P. 2016. Underwater LNG pipelines. *Offshore (Russia). Pipelines.* №11:66-71.

Youth Technical Sessions Proceedings – Litvinenko (ed)
© 2019 World Petroleum Council, London, UK, ISBN 978-0-367-34668-3

Vehicles with LNG engines V electric cars: Comparison of ecological an economic characteristics

L.R. Gimadeeva
Gubkin Russian State University of Oil and Gas (National Research University), Moscow, Russia

The high concentration of hazardous emissions in urban air is mainly caused by emissions from vehicles with gasoline and diesel engines. Experts believe that changing to another type of energy is considered a less rational solution than reducing the intensity of movement. Financial investments are also more useful not in introducing new types of energy, but directly in saving energy. Despite these facts, the use of electric cars is considered to be a means for reduction of pollutants in the air, but harmful chemicals released in the production of electricity are not taken into account. They are emitted in production of electricity by not renewable source of energy.

Also, electric cars have the following disadvantages:

1. The use of electric vehicles will greatly increase the use of electricity, which will inevitably lead to additional construction of new power plants.
2. Battery manufacturing is accompanied by releasing of toxic metal compounds.
3. The more expedient solution of the problem can be a changeover to vehicles with LNG engines. An additional plus of using LNG as a gas engine fuel is a small amount of noise from LNG engines. Re-equipment of the machine for switching to LNG is based on the installation of a specialized cryogenic storage tank for this gas, an evaporator unit that uses the heat of exhaust gases.

But large-scale production and exploitation of gas-engine fuel vehicles are limited due to the following reasons:

1. Restriction on the use of a gas cylinder at very low temperatures in winter and very high in summer temperatures.
2. A limited number of gas filling stations in cities.

Concerning the future of LNG production, it is expected to grow. LNG plants will be operating in countries such as the USA, Australia, Iran, Mozambique, Tanzania, Nigeria and Canada in 2040.

Concerning economic characteristics, manufacturers of both type of vehicles promise more economical consumption of fuel in comparison with using gasoline and diesel fuel. Using LNG is more efficient than using GTL technology. Since it has a very high cost of production and low calorific value obtained in the process of diesel fuel.

We compared the environmentally unfriendly emissions of LNG plants for producing the necessary quantities of LNG for driving the distance of 100 kilometers and the emissions of different types of electrical power plants to produce the necessary quantities of electricity needed to drive the electric car for 100 kilometers.

2 models of cars were taken for complex ecological and economic analysis of using LNG as motor fuel and electricity as energy for electric cars. TESLA Model X was taken as electric car. It is required 18 kwt*h of electricity for 100 kilometers distance. Volvo V-90 Bi-fuel was taken as vehicle with LNG engine. It is required 2,8 liters of LNG for 100 kilometers distance. It is shown the comparison of vehicle characteristics in the table 1. It is clear that their technical characteristics are almost the same.

Table 1. Comparison of vehicles characteristics.

Indicator	Volvo V-90 Bi-fuel	Tesla Model X
High speed, km/h	250,0	225,0
Engine power, hp	245,0	329,0
Acceleration, s	6,1	6,2
Range, g	400,0	417,0
CO_2 emissions, g/km	111,0	0,0
required amount of energy for 100 km distance, kwt/h	19,6	18,0

In the research were calculated amount of air pollutants in electricity production by coal and gas consumption, nuclear power plants and hydroelectric power plants. Energy research institute of the Russian academy of science shows 3 scenarios of electricity production and it is supposed that in each of them amount of produce electricity by coal consumption is the largest. It is so because of high prices for natural gas and renewable sources.

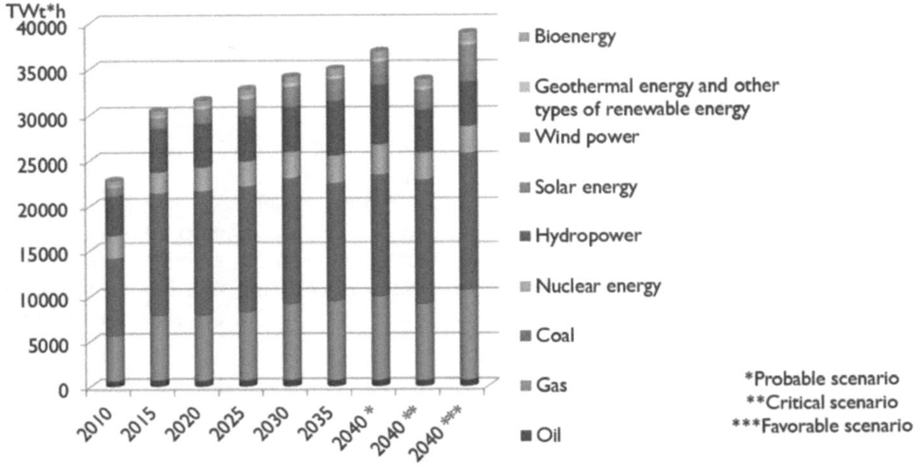

Diagram 1. Electricity production by types for 3 scenarios.

Information about air pollutants was taken from Russian LNG and electricity production companies reports. The amounts of hazardous emissions were calculated for 100 kilometers distance.

Amount of nitrogen dioxide emissions in producing electricity by natural gas combustion is larger than in LNG production. And amount of carbon oxide is larger in producing of LNG.

Amount of carbon oxide, nitrogen dioxide and sulfur dioxide are larger in production of electricity by coal consumption.

Amounts of air pollutants in production of LNG is larger in comparison with production of electricity by nuclear power plants. But it is worth mentioning about radioactive emissions and risks in exploitation, which can lead to disastrous consequences.

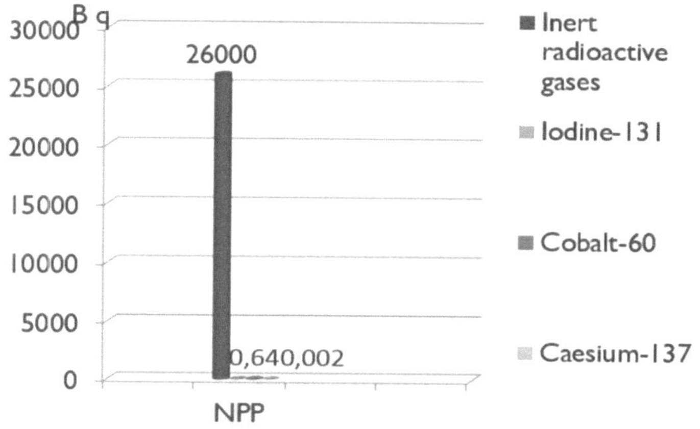

Diagram 2. Radioactive emissions from production of electric energy in nuclear power plant in calculation for 100 km.

Hydropower plant seems to be environmental friendly source of energy, but in a practice, decomposition of natural products of water reservoirs leads to large amount of carbon dioxide emission.

One more problem of using electric cars is their batteries. Nowadays lithium-ion batteries are most commonly used. Using of lithium leads to increase of demand for it. And there is limited amount of lithium resources in the world which cannot satisfy this demand in the future.

But lithium batteries have several advantages:

1. Large service life with small size and low weight
2. Higher energy capacity compared with other types of batteries.
3. The charging process is faster compared to other types of batteries.

Lithium batteries have the following disadvantages:

1. It is impossible to reanimate a fully discharged battery
2. Work in a narrow temperature range
3. High price

As for the recycling of lithium-ion batteries, it is a difficult and dangerous process. Spent lithium-ion batteries belong to the second class of danger. They contain the active element lithium, which can ignite in air with water interacting. Because of this, lithium batteries often cause a fire at various landfills and even waste collection points.

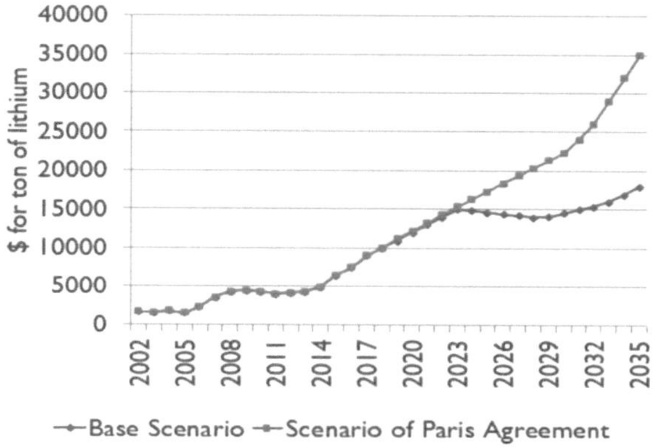

Graph 1. Dynamics of lithium carbonate price increase.

In the research was conducted an economic evaluation for current price situation in Russia. The diagram 3 shows that that in all cases, except for night time in divided tariffs, electricity for electric cars is more expensive than LNG in calculation for 100 kilometers distance.

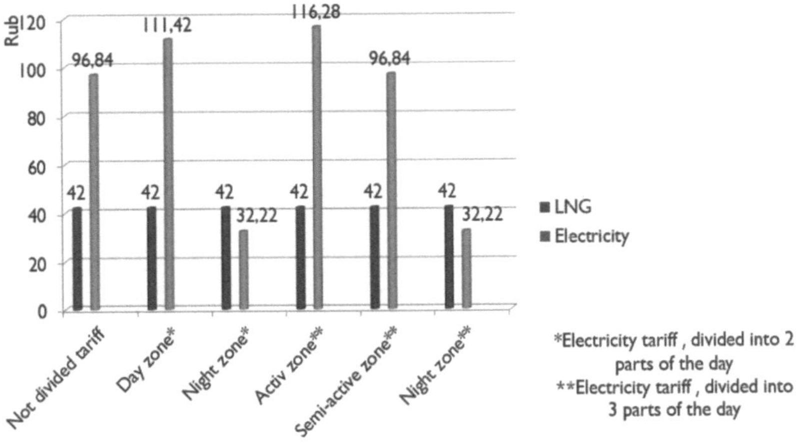

Graph 3. Prices of LNG and electricity in calculation for 100 km.

It was revealed that LNG-engine vehicles are more environmentally friendly and more cost-effective in most cases (depending on the type of electricity tariff).

The main conclusions:

1. It is the almost same amount of required energy for 100 km distance of vehicles with LNG engine and electric car.
2. The largest part of electricity in the world is produced by thermal power stations, nuclear power plants and hydroelectric power plants. alternative energy sources will not become serious competitors of traditional sources in the coming 25 years.
3. CO, SO_2, NO_2, CO_2 are emitted in the electricity production.
4. CO, SO_2, NO_2, CO_2 are emitted in the LNG production too. amount of these pollutants in LNG production in calculation for 100 km is less than in the electricity production.
5. Nowadays extraction of lithium from lithium-ion batteries is impossible, therefore lithium is not a renewable resource which is in limited quantities in the world, so it will be very difficult problem for electric cars manufacturers to satisfy the demand for lithium as component of lithium-ion batteries. this fact gives a chance for vehicles with LNG engines entry into the market in the coming 10 years.
6. Exploitation of vehicle with LNG engine is cheaper than exploitation of electric car in some cases.
7. The potential of natural gas as motor fuel is not fully realized.

Table 2. Comparison of ecological and economic characteristics of vehicles with lng engines and electric cars.

Indicator	Volvo V-90 Bi-fuel	Tesla Model X
Distance, km	100,000	100,000
Energy consumption, Kwt*h	19,600	18,000
CO emissions, g	0,229	3,240
NO_2 emissions, g	0,229	3,024
SO_2 emissions, g	0,002	35,353
CO_2 emissions, g	261,800	287,940
Radioactive emissions, bq	0,000	3023,875
Cost of energy, rub	42,000	80,970

* For Europe only one reported case in 1970.

This research will be of use for PAO «Gazprom» as they forge ahead with their plans of bringing the LNG based vehicles to the market. It is also of great benefits for the national economy as natural gas is readily available in huge quantities and this will lead to more developments in the transport sector.

REFERENCES

1. Makarova, A.A. & Grigorieva, L.M. 2016. Prognosis of energy development in the world and Russia: 180 – 200.
2. Kiselev, I.G. 2014. Using of LNG in energetics and transport area: 25-37.
3. Yupatov, S.K. 2000. Transport of 21st century. Electric cars: 155-163.

Health, Safety, Environment, Quality and Social Impact

Youth Technical Sessions Proceedings – Litvinenko (ed)
© 2019 World Petroleum Council, London, UK, ISBN 978-0-367-34668-3

The normative justification of recultivation process of soils polluted with oil in the agricultural area

N. Belyaeva, N. Antonov & I. Shaidullina
TatNIPIneft Institute – PJSC Tatneft, Bugulma, Russia

ABSTRACT: The oil production facilities of TatNIPIneft Institute – PJSC Tatneft are situated in a region of high agricultural activity, arable lands occupy more than 80% of the area. So, land protection and recultivation of disturbed soils are an important task. Soils are considered polluted, when the concentration of oil products reaches value that leads to negative ecological changes in environment. Oil and oil products disturb ecological state of soils and deform the structure of biocommunity. Permissible level of oil content in soils means that somewhat significant negative influence on surrounding environments – atmosphere, surface and ground water, flora, soil biota – is not observed. This paper presents the normative standard for permissible residual oil content and products of oil transformation in agricultural soils in the Republic of Tatarstan after recultivation and restoration. Methods of investigations of oil polluted soils to achieve the desired goals are also discussed.

1 INTRODUCTION

Absence of real oil pollution assessment in normative documents does not allow to solve the problem of oil polluted soils restoration. Therefore, the legislation of the Russia Federation provides an introduction of regional normative standards for permissible residual oil content in soils and products of its transformation in soils after recultivation and restoration. The document "Interim recommendations for development and implementation of Normative Standards for permissible Residual Oil content in Soils and Products of its Transformation in Soils after Recultivation and Restoration", was approved by Department of Environmental Resources in 2002. For the first time, in this document a concept of normative standards for permissible residual oil content in soils (PROCS standards) was formulated.

By permissible residual oil content in soil is meant oil content and products of oils transformation after recultivation determined with certified methods. This definition excludes possibility of oil entry and products of oil transformation in contiguous areas. Agricultural lands can be used as intended, though with restictions, or in conservation mode. Conservation mode allows to achieve of maximum allowable concentration (MAC) of oil products in soils during self-restoration process without extra special activity. In the Republic of Tatarstan maximum allowable concentration of oil products in soils is 1.5 g/kg. Recultivation activities on oil polluted territories shows that achievement of such a low level of oil content in soil is unreal without extra damage to environment. Obviously, in order to accomplish such strict normatives, it is required to destroy topsoil completely.

The aim of this work is the development PROCS standards in six types of soils in the Republic of Tatarstan: typical chernozem, podzolized chernozem, sod-podzolic, dark gray forest, sod-carbonate leached, sod-carbonate podzolic after recultivation work for agricultural soils.

2 LITERATURE REVIEW

The normatives of oil content in soils should be developed for specific area and for specific type of soils. The existing literature contains data about PROCS standards in different Russian regions and republics. For the first time, in 2004, the PROCS standards were introduced in Khanty-Mansi Autonomous Region, which is a leading Russian region in terms of oil production and, unfortunately, in terms of pipeline breaking accidents (Vershinin 2006). The regional PROCS standards for agricultural, industrial and forest soils were introduced in the Republic of Komi in 2007. The regional PROCS standard for agricultural soils on the Stavropol territory was introdused in 2010.

The investigations of oil polluted soils in the Republic of Tatarstan were initially carried out on chernozems of the trans-Kama region, where the most part of PJSC TATNEFT oilfield facilities are situated. In recent years, the zone of oil production has moved to the regions where the dark grey forest soils prevail. In the Republic of Tatarstan PROCS standards in leached chernozems for agricultural soils is 2.9 g/kg (Shaidullina 2006). The regional normatives for light gray forest and gray forest soils developed in 2011, are 2.0 and 2.8 g/kg respectively.

The review found data about standards of oil content in soils in the USA for various categories (Harmonization of ecological standards 2008): children's playground soils and agricultural soils - 200 mg/kg, temporary and permanent coverage - 2000 mg/kg, forest parks and green areas of the city - 10000 mg/kg. Signal levels of oil pollution have been accepted in Western Europe (Germany, the Netherlands). This necessitates environmental protection measures. Background content of oil typical for country territory is 50 mg/kg. Increased oil content (1000 mg/kg) requires monitoring the pollution dynamics and elimination of the cause of pollution. Soil is considered dirty when oil pollution has reached 5000 mg/kg (Kurakov & Guzev 2003).

3 METHODOLOGY

For laboratory and vegetation experiments we carried out reconnissance survey of oil polluted and background soils on the territory of the Republic of Tatarstan. We carried out laboratory reseach to determine the level of permissible residual oil content in soil. We chose oil polluted territories located in Almetyevsk, Bavly, Bugulma, Aznakaevo areas. As background soils we used unpolluted soils on the same territories. Sample preparation was implemented according to GOST 29269-91 «Soils. General requirements for the fulfilment of analyses». After predrying and removal of the roots of plants, samples of soils were sifted through a Vinkler sieve with mesh size from 1 to 5 mm. These soils for investigation were prepared by mixing oil polluted and pure (background) soils from relevant sites.

4 ANALYSIS

For laboratory investigations we selected soils from recultivation sites after spill of sulphurous oil, because of its higher toxicity. For each type of soils we determined the following parameters: physical and chemical state of soil; soil toxicity by biotesting; microbiological indicators; soil microflora activity; air, water, translocation, general sanitary migration indicators; phytoproductivity using two plants as test objects: spring wheat (Triticum vulgare L.), field pea (Pisum sativum L.).

4.1 *Physical and chemical state of soil*

Physical and chemical state of soil was determined by the following parameters: oil content, grain-size distribution, organic matter, total nutrogen, mobile forms of potassium and phosphorus, water and salt suspensions' pH, water extract, exchangeable sodium, capacity of cation exchange, heavy metals (Cd, Pb, Zn, Ni, Cu).

4.2 *Soil toxicity*

4.2.1 *Determination of soil toxicity by contact method*

Determination of soil toxicity is carried out by germination of seeds (preliminary stage) and by change of the length of seedling roots of higher plants at early stage (final stage). As a test object we used seeds of spring wheat. Germination was at least 95 %. Unsignificant suppression of wheat germination was noticed in oil polluted podzolized chernozem with oil concentration 9.95 g/kg (88%) and 1.22 g/kg (90%), in oil polluted typical chernozem oil concentration was 29.86 g/kg (88%). Interconnection between oil concentration in soils and change of the length of wheat roots was revealed by using contact method.

4.2.2 *Determination of the effect of water extract from the soils on root wheat growth*

Testing of water extracts of investigative soils under analysis showed that oil content in sod-podzolic, dark grey forest, podzolized chernozem did not affect the rate of wheat root growth. IThe increase of oil content in the samples of sod-carbonate podzolic, sod-carbonate leached soils led to decrease of the length of wheat seedling roots. The increase of oil content to 14.9 g/kg in the samples of typical chernozem led to stimulation of length of seedling roots of wheats in water extract from soil.

4.2.3 *Determination of water extract toxicity on the hydrobionts*

For biotesting on hydrobionts Ceriodaphnia affinis and Paramecium caudatum were used as test-objects. Analysis of water extract toxicity did not reveal toxic effect (mortality of test objects was 50 % and more) for all types of soils.

4.3 *Microbiological indicators*

Microbiological research at various levels of soil contamanation included: determination of total number of microorganisms, number of spore-forming and hidrocarbon oxidizing - microorganisms, nitrogen fixer and nitrifier, actinomycetes and micromycetes. As a rule, the total number of microorganisms, as a rule, clearly reflects soil's microbiological activity, rate of organic substance decay and cycle of mineral elements. We can use this indicator for assessment of the level of soil pollution, potential ability to self –restoration, the processes of oil degradation in natural environment and during land recultivation. By active chemical agent concentration is meant concentration, which exerts depressive effect by at least 50 % vs. reference values.

4.4 *Soil microflora activity*

Determination of soil microflora activity include respiration intensity, catalase and urease activity of the soils. The coefficient of microbiological respiration is calculated from the correlation of basal and substrate-induced restoration indicators (Anderson et.al 1993). The determination of the catalase activity is carried out by the titrimetic method (Katznelson 1958). Catalase activity is expressed in mg H_2O_2/g of soil. The determination of urease activity is carried out by the method of soil enzymology (Khaziev 2005). Urease activity is expressed in mkg N/g*h.

4.5 *Air, water, translocation, general sanitary migration indicators of harmfulness*

The assessment of the impact of oil pollution on the surrounding environments of the studied soils is based on the determination of the following indicators of harmfulness: water migration, air migration, translocation and general sanitary migration (Tab. 1).

Table 1. Permissible residual oil content and products of oil transformation according to migration indicators of harmfulness

Soil	Water migration indicator of harmfulness, g/kg	Air migration indicator of harmfulness, g/kg	General sanitary migration indicator of harmfulness, g/kg	Translocation indicator of harmfulness, g/kg
Sod-podzolic	11.4	Not limited	11.4	Not limited
Dark gray forest	9.9	Not limited	5.0	Not limited
Sod-carbonate leached	5.7	Not limited	5.7	Not limited
Sod-carbonate podzolic	5.7	Not limited	5.7	Not limited
Podzolized chernozem	14.9	Not limited	9.9	Not limited
Typical chernozem	14.9	Not limited	14.9	Not limited

4.5.1 *The determination of water migration indicator of harmfulness*

The determination of oil permissible concentration in soils by water migration indicator was carried out comparing the concentrations of oil products in filtrate with maximum allowable concentration of water objects used for fisheries purposes (MAC - 0.05 mg/dm^3). Experiments have shown that in most cases, even at the maximum tested concentrations, oil content in the water extracts from soil was lower then fisheries' MAC.

4.5.2 *The determination of air migration indicator of harmfulness*

Calculation of permissible oil content in soils by migration indicator of harmfulness was not carried out, because the results of gas chromatographic analysis showed that oil content in soils did not exceed the accepted standards for atmospheric air. Concentration of oil products in air was less than 1 mg/m^3.

4.5.3 *The determination of translocation indicator of harmfulness*

The purpose of this investigation was the determination of heavy metals content in grown test plants. These are heavy metals of the first and the second classes of danger - Cd, Zn, Cu, Pb. The dependence between heavy metals content in plants and in oil was absent. Therefore, the translocation of metals in oil was not considered.

4.5.4 *The determination of general sanitary migration indicator of harmfulness*

According to Recommended Practice of MAC Hygienic Justification of Chemical Substances in Soils, strains of coliform bacilli were used to determine general sanitary migration indicator of harmfulness. The strains were cultivated at at 37°C for 24 hours on meat-and-peptone agar. A depressive effect (more than 50 %) of oil products on the growth and development of coliform bacilli was revealed.

4.6 *Plants phytoproductivity in the vegetative experiments*

Vegetative experiments were conducted according to GOST 22030-2009 "Soil Quality. Biological Methods. Chronic Phytotoxicity for Higher Plants". Experimental soils samples with the given parameters of oil content (background soils) were prepared by mixing polluted and pure soils in different proportions. In all soil samples oil content was preliminarily determined.

Table 2. Permissible residual oil content in recultivated soils according to the results of vegetation experiments

Soil	PROCS standards, g/kg field pea	wheat
Sod-podzolic	4.91 ($y=-0.0527x+0.9681$; $R^2=0.91$)	2.04≈2.0 ($y=-0.0579x+0.5371$; $R^2=0.95$)
Dark gray forest	11.61 ($y=-0.036x+0.663$; $R^2=0.95$)	2.76≈2.8 ($y=-0.052x+0.616$; $R^2=0.95$)
Sod-carbonate leached	2.02≈2.0 ($y=-0.0868x+0.7997$; $R^2=0.90$)	2.03 ($y=-0.0313x+0.2951$; $R^2=0.97$)
Sod-carbonate podzolic	2.03 ($y=-0.0953x+0.9073$; $R^2=0.81$)	1.98≈2.0 ($y=-0.0524x+0.547$; $R^2=0.94$)
Podzolized chernozem	2.99 ($y=-0.035x+0.5205$; $R^2=0.91$)	2.96≈3.0 ($y=-0.0227x+0.275$; $R^2=0.84$)
Typical chernozem	5.67 ($y=-0.0239x+0.6135$; $R^2=0.90$)	3.05≈3.1 ($y=-0.0116x+0.1983$; $R^2=0.93$)

We used two types of plants: spring wheat (Triticum vulgare L.), field pea (Pisum sativum L.), as test objects in vegetative experiments. The choice of arable crops was conditioned to the high significance for the Republic of Tatarstan. Vegetative experiment of phytoproductivity was carried out during 42 hours under the certain conditions: artificial lighting using fitolamps with light intensity 5000 Lk and the temperature 23- 26°C. The calculation is based on dry biomass data of wheat and pea yield. We determined the phytomass of constant weight at a temperature of 65°C with an accuracy of 0.0001 g.

The data of phytoproductivity, depending on different concentrations of oil products allowed us to obtain the linear regression equations. This determined the maximum permissible concentrations of oil products that did not have a phytotoxic effect on the test objects (PC_{20}). In this way, oil content which led to loss of phytoproductivity no more then 20 % compared with the control site, considered permissible (Tab. 2). Wheat turned out the most sensitive plant during the vegetation experiments.

5 DISCUSSION

Residual oil concentration that do not affect the surrounding environments judging from the analyzed biological parameters show that phytoproductivity parameters can be used as a limiting nuisance value of residual oil content in recultivated soils (Tab. 3).

It should be noted that spills of oil are often accompanied by invasion of highly mineralized formation waters, which is the cause of soil alkalinization. The latter can be a determining factor of disturbance of grass cover and soil biocenosis, along with oil contamination. Concentrations of chlorides in soils in excess of 0.2 % lead to the marked inhibition of the growth of most herbs. Increase of chlorides' concentration to 1 % put an end to herbaceous cover development altogether. In these cases, polluted sites of soils besides must be recultivated and meliorated. Currently, the normative standards soluble salt content are absent. We offer to establish additional land acceptance requirements in terms of salt content control.

Table 3. The residual concentrations of oil products in the recultivated soils that do not affect adversely surrounding environments

Soil	Respiration intensity coefficient	Microbiological analysis	Soil toxicity	Water extract toxicity on the hydrobionts	Soil microflora activity		Chronic toxicity, wheat, field pea	
	OP, g/kg	OP, g/kg	PC_{50}, g/kg	PC_{10}, g/kg	Catalase, mg H_2O_2/g	Urease, mkg N/g·h	PC_{50}, g/kg	PC_{20}, g/kg
Sod-podzolic	22.83	11.41	6.93	22.83	22.83	11.41	4.9	2.0
Dark gray forest	14.93	4.98	11.43	5.56	14.93	14.93	5.0	2.8
Sod-carbonate leached	11.47	5.74	22.50	4.61	4.56	11.47	4.6	2.0
Sod-carbonate podzolic	5.74	11.47	11.16	9.78	5.74	11.47	4.8	2.0
Podzolized chernozem	4.98	9.95	21.74	14.93	14.93	14.93	5.0	3.0
Typical chernozem	14.93	4.98	15.98	12.00	29.86	29.86	5.0	3.1

6 CONCLUSION

For arable layers of agricultural lands in the Republic of Tatarstan we recommend the following permissible residual concentrations of oil and oil products determined by phytoproductivity: for sod-podzolic, sod-carbonate leached, sod-carbonate podzolic soils – 2.0 g/kg, dark gray forest – 2.8 g/kg, podzolized chernozem – 3.0 g/kg, typical chernozem – 3.1 g/kg.

In our investigations we used sulfurous oil, which is more toxic than Devonian oil. It can be argued that the obtained results are more stringent and can be used for all types of oil produced and transported in the Republic of Tatarstan.

Implementation PROCS standards will facilitate the acceptance of soils after recultivation and preparation of documents on indemnification caused by oil pollution of soils. Also, it will allow land recultivation to meet environmental standards without extra damage to soils and exclude expenses on recultivation to achieve maximum allowable concentration.

REFERENCES

Alehin, V., Emscev,V. & Rogozina E. 1998. Biologic activity and microbiologic recultivation of oil contaminated soils. *Collection of scientific papes*: 95-105. Nizhnevartovsk.

Anderson, T. & Domsch, K. 1993. The metabolic quotient for CO_2 (qCO_2) as a specific activity parameter to assess the effect of environmental conditions, such as pH, on the microbial biomass of forest soils.*Soil Biol. Biochem*: 393-395. Germany.

Hana, D. 2008. Harmonization of ecological standards. *Rationing features of pollutants content in soils in Russia and abroad*: 18. S-Pb.

Ismailov, N. & Pikovski, Yu. 1988. Present state of recultivation methods of oil polluted soils. *Restoration of oil polluted ecosystems*: 42-56. Moscow.

Kurakov, A. & Guzev,V. 2003. Oil contaminated soils: property modification, monitoring and recultivation biotechnology. *Study guide. Oil contamination: control and rehabilitation of ecosystem:* 194.

Katznelson, R. & Ershov V. 1958. Microflora investigation of virgin and cultivated soils of the Karelian ASSR. Soil biological activity of Karelian ASSR. *Microbiology:* 82-88.

Kireeva, N., Kyzahmetov, G., Miftahova, A. & Vodopyanov, V. 2003. Phytotoxicity of anthropogenically polluted soils: 266. Ufa.

Khaziev, F. 2005. The methods of soil enzymology. *Institute of Biology of the Ufa Scientific Center:* 252. Moscow.

Khaziev, F. & Fathiev, F. 1988. Change of chemical characteristics of soil under the influence of oil pollution. *Izv. Penz. gov. pedagog. univ. V.G. Belinskogo*: 102-111. Penza.

Maganov, R., Makarova, M. & Mulyak, V. 2006. Recultivation of oil contaminated soils in the Ysinski region of the Republic of Komi. *Environmental works at oil and gas companies. Part 1*: 208. Syktyvkar.

Regional normative of permissible residual oil content in soils after recultivation and restoration in Khanty-Mansi Autonomous Region – Yugra. *Government Decree of Khanty-Mansi Autonomous Region – Yugra № 466-p, 10.12.*2004.

Regional normative of permissible residual oil content in soils after recultivation and restoration in the Stavropol territory. *The Order of Ministry of Natural Resources and Environment Protection № 468, 20.12.*2010.

Regional normative of permissible residual oil content in soils after recultivation and restoration in the Republic of Tatarstan. *The Order of Ministry of Natural Resources and Environment Protection № 786, 22. 07.*2009.

Regional normative of permissible residual oil content in soils after recultivation and restoration in the Republic of Tatarstan. *The Order of Ministry of Natural Resources and Environment Protection № 303-p, 14.07.*2001.

Shaidullina, I. 2006. Development of Standards of Allowable Concentration of Oil in leached Chernozem in Republic of Tatarstan. *Environ. Radioecol. Appl. Ecol:* 137. Kazan.

Farrahova, I., Gilyazov, M. & Sardiev, P. 2008. Phytotoxicity of oil polluted grey forest soil and character of its self-cleaning in Predkamie region of Republic of Tatarstan. *Kazan State Agrarian University Bulletin:* 119-123.

Vaver, V. 1997. Recultivation of oil polluted soils. *Collection of scientific papes. Biological resourses and environmental management*: 114-135. Nizhnevartovsk.

Vershinin, A., Shagidullin, R., Ignatiev, Y. & Petrov A. 2010. Respiratory activity of oil polluted sod-carbonate soil. *Jornal of ecology and industrial safety, № 3*: 8-9.

Zainulgabidinov, E., Petrov, A. & Yakimova, T. 2010. Features of phytotesting of oil polluted soils. *Jornal of ecology and industrial safety, № 3*: 24-25.

Youth Technical Sessions Proceedings – Litvinenko (ed)
 ISBN 978-0-367-34668-3

Ecological, social and corporate governance within Arctic petroleum project management: Legal aspects

V. Ruzakova
International Institute of Energy Policy and Diplomacy, MGIMO University, Moscow, Russian Federation

ABSTRACT: The aim of the present research is to elicit and systematize the key legal aspects of ESG compliance relevant to the development of Arctic petroleum resources, including matters of ecological security, corporate social responsibility and sustainable governance. The article is an attempt to show that ESG in the Arctic has its own specifics, that both requires for a special approach to the ESG assessment for the projects developed in the region, and can serve as a paragon in many ESG aspects.

1 INTRODUCTION

With the adoption of the 2030 Agenda for Sustainable Development by all United Nations member states, as well as the publication of the report Fiduciary Duty in the 21st Century by the PRI, UNEP and UN Global Compact, which acknowledged ESG assessment as a fiducial duty of investor, it is now both a public duty and a financial interest for a company to integrate best ESG practices in project management. The Arctic region has an outstanding resource potential, and at the same time is characterized by a harsh climate, fragile ecosystems and a stakeholder role of Arctic indigenous peoples. Analysis of ecological, social and corporate governance within Arctic petroleum project management gives therefore a unique opportunity to study a broad range of best practices, as well as to identify areas for further improvement in approaches to project management in a region, where the oil and gas sector is of particular importance to sustainability.

2 LITERATURE REVIEW

Although there has been a lot of scientific research both on socially responsible investment (Renneboog, Ter & Zhang, 2007) and on the ESG criteria (Barber, Morse & Yasuda, 2018; Doyle, 2018) carried out recently, there has been developed no study of SRI or ESG in respect of their Arctic regional specifics.

Therefore, this study will be the first paper to argue about the ESG particularities in reference to the Arctic region, also drawing upon the recent studies of the Arctic environment (Morgunova & Westphal, 2016), the Arctic indigenous peoples (Tysiachniouk, Henry, Lamers & Tatenhove, 2017; Hughes, 2018; Gassiy, 2018) and the Arctic projects of major oil and gas companies (Ashabov, Burzev, Kusnezov & Timonina, 2014) - and others.

3 METHODOLOGY

Comparative legal analysis (analysis of national and international legislation), interdisciplinary analysis (analysis of corporative public accountability, case study, social and ecological impact assessment).

4 ANALYSIS

With the adoption of the 2030 Agenda for Sustainable Development by all United Nations member states, as well as the publication of the report Fiduciary Duty in the 21st Century by the Principles of Responsible Investment (PRI), UNEP and UN Global Compact, which acknowledged ESG assessment as a fiducial duty of investor, it is now both a public duty and a financial interest for a company to integrate best ESG practices in project management. While economic investment criteria generally include profitability, business models, sector analysis and competitiveness, ESG are the extra-financial factors, considered by responsible investor. Environmental criteria allows for the investor to analyse the footprint of the company, for example initiatives of the company in such areas as energy saving and the reduction of pollutant emissions. Under the social criteria the investor examines working conditions, relationships with clients and suppliers and human resources management. Finally, within the "governance" group of criteria it is the corporate governance structure that is examined, with the aim to assess its independency and transparency.

As a result of the unilateral restrictions of the US, EU countries and Norway on investment and technological cooperation in the sphere of deep-water drilling in the Arctic with Russia, Russian Arctic projects are being actively reoriented in the search for foreign investment to other countries: a prominent example here are the Russian Arctic LNG-projects, which attract such major investors as China and Saudi Arabia. A number of fields in the Norwegian continental shelf of the Barents Sea are now open for new licensing rounds, where not only operators, but also investors form the corporate side of the Joint Operating Agreements, and in the United States and Canada, projects for the development of the Arctic transport infrastructure, including pipeline transport are now of great importance and search for investment. Here we can bring the example of the Alaska Highway Pipeline Project which consists in construction of a transboundary gas pipeline that could open the southern markets for Alaskan gas and that was frozen until 2022 due to lack of investment. Consequently, the ESG issue is relevant for Arctic projects, and, moreover, it is most acute because of the specifics of the region's fragile ecosystem, the difficult climate, and the inevitable impact of oil and gas projects on the indigenous communities of the Arctic.

Let's look at the legal basis of the ESG specifics as applied to the Arctic region – both of the legally-binding sources of the "soft law".

4.1 *ESG Legal Basis for the Arctic*

The standards and principles of environmental protection and social responsibility in the Arctic are key sources applicable to the ESG in the Arctic. There fall not only universal agreements (as for example The International Convention for the Prevention of Pollution from Ships of 1973/78,The Convention on Long-range Transboundary Air Pollution of 1979 or The International Convention on the Establishment of an International Fund forCompensation for Oil Pollution Damage, 1992), but also those regional in nature (e.g. the Agreement on Cooperation on Aeronautical and Maritime Search and Rescue in the Arctic 2011, Agreement on Cooperation on Marine Oil Pollution, Preparedness and Response in the Arctic 2013, Agreement on Enhancing International Arctic Scientific Cooperation 2017) – see Table 1.

Of course, the Arctic ESG legal basis, being an integral part of the global legal system, complies with such general legal principles as: the principle of sustainable development, the principle of international environmental cooperation, the principle of responsibility for damage to the environment, the principle of prohibiting military and any other negative impact on the environment (Kukushkina, Shyshkin, 2010).

However, it should be noted here that the ESG approaches required for oil and gas activities in the Arctic are characterized by a number of "industry-specific" principles, such as follows:

- The principle of "precautionary approach": where there is a risk of serious and irreparable damage to the environment, the lack of complete scientific uncertainty should not be the

Table 1. Major sources of international law applicable to the establishment of the ESG criteria for oil and gas Arctic projects

Title	Participants (among the AC members)	Year
Multilateral universal agreements applicable to the ESG in the Arctic (Hludeneva, 2012; Solovyanov, 2011)		
– The International Convention on Civil Liability for Oil Pollution Damage	AC members	1969
– The International Convention on the Establishment of an International Fund for Compensation for Oil Pollution Damage		1971
– Protocol to amend the International Convention on Civil Liability for Oil Pollution Damage		1992
– Protocol to amend the International Convention on the Establishment of an International Fund for Compensation for Oil Pollution Damage		1992
– The Convention on the Prevention of Marine Pollution by Dumping of Wastes and Other Matter ("the London Convention")		1972
– Declaration of the UN Stockholm Conference on the Human Environment		1972
– International Convention for the Prevention of Pollution from Ships		1973/78
– Convention on Civil Liability for Oil Pollution Damage Resulting from Exploration for and Exploitation of Seabed Mineral Resources		1976
– The Convention on Long-Range Transboundary Air Pollution		1979
– UN Convention on the Law of the Sea	the USA does not participate	1982
– International Convention on Oil Pollution Preparedness, Response and Co-operation (OPRC)	the Russian Federation does not participate	1990
– Convention on Biological Diversity	the USA did not ratify	1992
– The United Nations Framework Convention on Climate Change (UNFCCC)	AC members	1992
– Kyoto Protocol to the 1992 United Nations Framework Convention on Climate Change		1997
– Rio de Janeiro Declaration on Environment and Development		1992
– Stockholm Convention on Persistent Organic Pollutants	the USA does not participate	2001
Multilateral regional agreements concerning international legal protection of the Arctic environment		
– Agreement on the Conservation of Polar Bears	the Kingdom of Denmark, Canada, Norway, the Russian Federation, the USA	1973
– Agreement on cooperation within the field of emergency prevention, preparedness and response	BEAC members	2008
	AC members	2011

– Agreement on Cooperation on Aeronautical and Maritime Search and Rescue in the Arctic		
– Agreement on Cooperation on Marine Oil Pollution, Preparedness and Response in the Arctic		2013
– Agreement on Enhancing International Arctic Scientific Cooperation	AC members	2017
Bilateral international agreements applicable to the ESG in the Arctic		
– Memorandum of Understanding and Cooperation in the Prevention and Preservation of Pollution of the Arctic Environment of the Arctic between the Government of the Union of Soviet Socialist Republics and the Government of Canada	the Russian Federation (for the USSR), Canada	1989
– Agreement Between the Government of Canada and the Government of the Russian Federation Concerning Environmental Cooperation	the Russian Federation, Canada	1993
– Memorandum of Understanding between the Government of the Russian Federation and the Government of the Kingdom of Norway on Cooperation in Search and Rescue as well as Warning of Serious Incidents	the Russian Federation, Norway	2000
– Agreement between the Government of the Russian Federation and the Government of the Kingdom of Denmark on Cooperation in the Field of Environmental Protection	the Russian Federation, the Kingdom of Denmark	1992
– Agreement between the Government of the United States of America and the Government of the Union of Soviet Socialist Republics concerning Cooperation in Combating Pollution in the Bering and Chukchi Seas in Emergency Situations	the Russian Federation, the USA	1989
– Agreement between the Kingdom of Denmark and Canada for Cooperation Relating to the Marine Environment	the Kingdom of Denmark, Canada	1983

reason for postponing the adoption of measures to prevent environmental degradation due to their high cost. (The formation of this principle was recorded at the Millennium Summit, where it was stated that the basis for the protection and rational use of all living organisms and natural resources should be put cautious in accordance with the postulates of sustainable development);

- The "polluter pays" principle: the subject whose actions have lead to environmental pollution must bear costs equal to the price of pollution, taking into account public interests and not distorting the process of international trade and investment. Accordingly, states are required to internationalize the costs of environmental protection in the Arctic, while the oil and companies are encouraged to put extra effort into the environmental safety;
- The principle of rational development. It includes such aspects as: the protection of biological diversity; the duty of using the best available technology; the obligation not to transfer, directly or indirectly, the harmful effects or danger of harmful effects from one area of the marine environment to another and not to transform one type of pollution into another.

4.2 *ESG "Soft Law" in the Arctic*

The key international forum directly concerned with environmental protection in the Arctic is the Arctic Council, transformed from the Arctic Environmental Protection Strategy (AEPS), adopted in Rovaniemi in 1991 as part of the First Arctic Ministerial Conference. In addition, in their activities the Council of the Barents/Euro-Arctic Region and the Northern Dimension joint project of the EU, Norway, Russia and Iceland also affect environmental issues in the Arctic.

The Arctic Council is actively working to develop various assessment reports and guidance documents on the sustainable environmental development of the Arctic in the context of the development of the oil and gas regional industry.

The AMAP Arctic Council Working Group has issued "the Assessment 2007: Oil and Gas Activities in the Arctic-Effects and Potential Effects", the PAME Arctic Council Working Group has issued "the Arctic Offshore Oil and Gas Guidelines" in 2009 and "the Arctic Ocean Review Phase II report" in 2013, the EPPR Working Group has produced "the Recommended Practices for Arctic Oil Spill Prevention" as well as "the Summary Report and Recommendations on Prevention of Marine Oil Spill Pollution" in 2013. All these documents contain recommendations on political decisions and administration in order to prevent emergencies and pollution in the Arctic as a result of oil and gas projects on the continental shelf of the Arctic (Arctic Council).

The Arctic Council has already developed a number of guidelines and recommendations, comprehensively devoted to the issues of operating on the Arctic shelf. Among them: the Arctic Marine Shipping Assessment, the Arctic Region Oil Spill Response Resource and Logistic Guide, the Arctic Response Cooperation Guidelines, the Guidelines for Transfer of Refined Oil and Oil Products in Arctic Waters, A Field Guide to Oil Spill Response in Arctic Waters, the Arctic Shoreline Clean-up Assessment Technique Manual, the Environmental Risk Analysis of Arctic Activities, the Circumpolar Map of Resources at Risk from Oil Spills in the Arctic and the Arctic Guide for Emergency Prevention, Preparedness and Response. A special place in most of the above documents is given to Health, Safety and Illness Management Systems.

The most recent result of the work of the Arctic Council on the development of various assessment reports and guidance documents on sustainable environmental development in the Arctic in the context of the development of the oil and gas industry in the region: the report of the EPPR Working Group "The Arctic Offshore Oil and Gas Guidelines: Safety Management Systems and Safety Culture 2014", Guide for Oil Spill Response in Snow and Ice, as well as in the Arctic Council's Ikaluit Declaration, a decision to proceed with the implementation of the developed Framework Plan for Cooperation on Prevention of Oil Pollution from Petroleum and Maritime Activities in the Marine Areas of the Arctic. The development of these documents is also largely the result of the work of the Arctic Council Task Force on Arctic Marine Oil Pollution (TFOPP), co-chaired by Norway and Russia. Nevertheless, it should be noted that, from the point of view of both countries in chair, the developed Framework Plan for Cooperation on Prevention of Oil Pollution from Petroleum and Maritime Activities in the Marine Areas of the Arctic is not specific enough and requires further, more specific international agreements.

5 DISCUSSION

The accomplished part of the research concerned with comparative legal analysis reveals the regional specifics of international and national legal regulation, which influence the corporate compliance with ESG criteria.

However, another important aspect of the problem that is still under consideration by the author are the best corporative practices.

Describing the issue in general, there could be pointed several main groups of corporative best practices, as follows:

- information disclosure (for example, the annual reports of Rosneft on sustainable development are always published in Russian and English and contain detailed information on the company's Arctic projects),
- public consultations with indigenous and local communities (for example, as a result of consultations on the Point Thompson pipeline project in Alaska with the Kaktovik community, ExxonMobil redesigned a number of pipeline sections to prevent possible interference with the caribou deer migration routes),
- emergency damage response programs (for example, at the stage of elaboration of the project on drilling the Universitetskaya-1 well, Rosneft and ExxonMobil together with the responsible Russian government departments developed an Emergency Oil Spill Response Plan)
- environmental responsibility programs (for example, in 2014, Rosneft together with ExxonMobil, Eni and Statoil signed a Declaration on environmental protection and biodiversity conservation during exploration and development of the mineral resources of the Arctic continental shelf of the Russian Federation, in 2013 Rosneft and the World Wildlife Fund Russia signed a Protocol of intent for cooperation in the field of environmental protection, including the polar bear protection, etc.).

At the moment of publication, the author is under way with her semi-structured interviews with representatives of the corporate sector, aimed to elicit the most relevant information about the activities oil and gas companies undertake in the abovementioned areas.

At the same time, the scientific contribution of the research already produced and presented herein consists in the analysis of the ESG legal basis as applied to the Arctic, which shows it both requires for a special approach to the ESG assessment in the region, and can serve as a paragon in many ESG aspects.

6 CONCLUSION

The research is still ongoing and will be finished by May 2019. The midterm results show that the process of management approaches' harmonization with the modern ESG requirements is still under way, ESG criteria also being simultaneously developed and specified by international organizations and expert society. A major influence on the ESG management in the Arctic has the regional – both international and national – regulation. The final research results will be ready for presentation at the VI WPC Youth Forum.

ACKNOWLEDGEMENTS

This research was carried out under the aegis of the VI World Petroleum Council Youth Forum.

REFERENCES

Coast Guard Polar Icebreaker Modernization: Background and Issues for Congress. Congressional Report Service, viewed 05.03.2019, https://www.fas.org/sgp/crs/weapons/RL34391.pdf

Federal Law "On Ecological Expertise" [Federal'nyy zakon "Ob ekologicheskoy ekspertize"] of November 23, 1995 N 174-ФЗ

Federal Law "On Environmental Protection" [Federal'nyy zakon «Ob okhrane okruzhayushchey sredy»] of January 10, 2002, No. 7-FZ. Art. 64-69.

Federal Law "On Environmental Protection" [Federal'nyy zakon «Ob okhrane okruzhayushchey sredy»] of January 10, 2002, No. 7-FZ. Article 32

Federal target program "Liquidation of accumulated environmental damage" for 2014-2025 [Federal'naya tselevaya programma «Likvidatsiya nakoplennogo ekologicheskogo ushcherba» na 2014 – 2025 gody].

Regulations on the procedure for licensing the use of subsoil [Polozheniye o poryadke litsenzirovaniya pol'zovaniya nedrami] (approved by the Decree of the Armed Forces of the Russian Federation of July 15, 1992 N 3314-1). P. 19.4.

Regulations on the procedure for licensing the use of subsoil [Polozheniye o poryadke litsenzirovaniya pol'zovaniya nedrami] (approved by the Decree of the Armed Forces of the Russian Federation of July 15, 1992 N 3314-1). Pp 6.6., 14.3., 15.2.

Regulations on the Unified State System of Environmental Monitoring [Polozheniye o Yedinoy gosudarstvennoy sisteme ekologicheskogo monitoringa] (approved by the Order of the Ministry of Natural Resources of the Russian Federation of 09.02.1995 No. 49).

Russian International Affairs Council (RIAC) official website, viewed 05.03.2019, http://russiancouncil.ru/arcticoil

Port of Churchill official website, viewed 05.03.2019, http://www.portofchurchill.ca/

Arctic Council official website, viewed 05.03.2019, http://www.arctic-council.org/index.php/ru/about-us/permanent-participants/sc

Alaska Highway Pipeline Project official website, viewed 05.03.2019, http://www.emr.gov.yk.ca/oilandgas/ahpp.html

Anthology "*The Arctic Region: Problems of International Cooperation*" [Arkticheskyi region: problem megdunarodnogo sotrudnichestva], RIAC, 2013, viewed 05.03.2019, http://russiancouncil.ru/common/upload/Arctic%20Anthology%20Vol%202.pdf

Ashabov A.M., Burzev I.N., Kuznezov S.K. & Timonina N.N. "*The Arctic Vector of Geological Research: Oil, Gas and Mineral Resources*", 2014. Vestnik IG Komi SC UB RAS, September, 2014, № 9, P. 3-10.

Barber B.M., Morse A., Yasuda A. "*Impact Investing*", 2018, viewed 05.03.2019, http://www.ayakoyasuda.com/papers/2017/Paper15wp.pdf

Doyle Timothy M. "*Ratings that Don't Rate*", 2018, viewed 05.03.2019, https://corpgov.law.harvard.edu/2018/08/07/ratings-that-dont-rate-the-subjective-world-of-esg-ratings-agencies/

Gassiy V. "*Indigenous Communities in the Arctic Change in Socio-Economic and Environment Perspective*", 2018. Action Group on Indigenous Involvement, International Arctic Science Committee (IASC), Kuban State University, Russia, viewed 05.03.2019, https://www.intechopen.com/online-first/indigenous-communities-in-the-arctic-change-in-socio-economic-and-environmental-perspective

Hludeneva N.I. *Ecological law in the Arctic region*, 2012 [Pravovaja ohrana okruzaushej sredy Arctiki]. Materials of the International Scientific Symposium. Moscow, Spezkniga press, 2012. P. 23.

Kukushkina A.V., Shyshkin V.N. *The Arctic: ecological law for the oil and gas sector*, 2010 [Arctika: ohrana okruzhauschej sredy pri dobyche uglevodorodov. Mezhdunarodno-pravovye aspecty]. Moscow, Afrika press, 2010. P. 20.

Mogunova M., Westphal K. "Offshore Hydrocarbon Resources in the Arctic", 2016, viewed 05.03.2019, https://www.swp-berlin.org/en/publication/offshore-hydrocarbon-resources-in-the-arctic/

Piotrovskiy A.S., Yarygin G.A., Fedorenko A.V., Khvastina A.S. *Solving environmental problems during the development of the Shtokman gas condensate field*, 2008 [Resheniye ekologicheskikh problem pri osvoyenii Shtokmanovskogo gazokondensatnogo mestorozhdeniya], viewed 05.03.2019, https://helionltd.ru/reshenie-ekologicheskih-problem-pri-osvoenii-shtokmanovskogo-gazokondensatnogo-mestorozhdeniya/

Renneboog L., Ter Horst J.R., Zgang C. "Socially Responsible Investments", 2007, viewed 05.03.2019, https://pure.uvt.nl/ws/portalfiles/portal/840728/dp2007-31.pdf

Solovyanov A.A. *On the preservation of the ecology of the Arctic Zone of the Russian Federation*, 2011 [O sohranenii prirodnoy sredy Arkticheskoy zony Rossiiskoy Federazii], viewed 03.03.2019, http://www.ibrae.ac.ru/images/stories/ibrae/arktika_magazine/94-103_solovyanov.pdf

Tysiachniouk, M., Henry L. A., Lamers M., Tatenhove, J.P.M. van. "Oil and Indigenous People in sub-Arctic Russia", 2018, viewed 05.03.2019, http://vbn.aau.dk/files/272757728/Oil_and_indigeno_CC_BY_NC_ND_4.0_us_people_in_sub_Arctic_Russia_Rethinking_equity_and_governance_in_bene_t_sharing_agreements_.pdf

Youth Technical Sessions Proceedings – Litvinenko (ed)
© 2019 World Petroleum Council, London, UK, ISBN 978-0-367-34668-3

Augmented reality system and maintenance of electromechanical equipment in industrial production

N. Koteleva, K. Bekenev & V. Valnev
Saint-Petersburg Mining University, Saint-Petersburg, Russian Federation

ABSTRACT: The oil&gas enterprises have their own safety concerns list. Poor maintenance of equipment is one of the top safety concerns for many enterprises. There are a lot of methods of increasing of safety, but its shouldn't be used alone. A complex approach and addition the modern technologies in safety ensuring make possible not only increase the quality but increase the efficiently of production. There are a lot of kinds of quality improving of maintenance of electromechanical equipment, one of them are the augmented reality system. This is a new technology and there are no methods and effective estimation for developing and using this technology in industrial production. Developing the special unified modules of augmented reality for maintenance of electromechanical equipment is the actual task. The aim of this research - to identify possible uses, to estimate the effectiveness in ensuring the safety of production and to develop a way of integrating into typical process control systems of the augmented reality system for maintenance of electromechanical equipment.

1 INTRODUCTION

Access to energy and sustainable development of the fuel and energy sector is still a big problem. It is known that today one of the main strategic goals of companies is a course on energy saving and the introduction of information technologies that can enhance the interaction between the consumer and the power grid (Kislitsyn,2017; Zhukovskiy, 2017). Energy saving and digitalization are actual trends in various industries (Vasiliev, 2017; Safina, 2017). The last decades have been characterized by the rapid development of technology, the economy and society. There the cardinal changes are taking place (high technologies, population growth of the planet, global climate change, etc.) affecting the energy business, presenting it all new and new requirements (Abramovich, 2016; Kazanin, 2017).

Among the most significant changes in the development of society and the economy, including the energy sector, the researchers include the following (Annual statistics of world energy):

1. Deficit of sources of electrical energy.
2. Constantly growing requirements for reliability and quality of power supply from consumers.
3. The constant increase in the cost of electrical energy throughout the world.
4. Aging and growing shortage of qualified personnel in the energy industry.
5. Growth of requirements of stakeholders to the results of the activities of industrial companies.
6. The requirements of environmental and industrial safety of the operation of industrial and energy objects.
7. Reducing system-wide costs.

Thus, the main factors that determine the need for fundamental changes in the power industry under the influence of the prevailing conditions can be grouped as follows.

Technological progress factors:

- the emergence and development of new technologies, devices and materials potentially applicable in the field of electric power production, and in the first place the increasing pace and scale of development of digital and information technologies;
- intensive growth in the number of high-tech equipment that generates (primarily renewable) energy sources;
- general tendency to increase the level of process automation.

Factors increasing consumer requirements:

- increasing requirements for recruitment (range) and quality of services;
- the expectation of reducing the price parameters of services in industries;
- requirements for informational transparency of the system of relations between participants of electric power markets, primarily with consumers.

Reliability factors:

- increasing level of equipment depreciation;
- the need for massive investment in the renovation of fixed assets;
- reducing the overall level of reliability of energy supply systems;
- high level of losses in the conversion, transmission and distribution of electricity.

Factors increasing requirements in the field of energy efficiency and environmental safety:

- the need to reduce the impact on the environment;
- the need to improve energy efficiency and energy saving.

The energy sector of oil and gas enterprises is not the only one that penetrates new technologies, such as the Internet of things, big data and artificial intelligence, etc. (Kuznetsov, 2018, Saveliev, 2017) The changes relate to safety systems, reliability, geology, transportation, automation, mechanization, etc.

At the same time, the energy sector of oil and gas enterprises is not the only one that penetrates new technologies, such as the Internet of things, big data and artificial intelligence, etc. The changes relate to safety systems, reliability, geology, transportation, automation, mechanization, etc. In the struggle for competitive advantage, enterprises are looking for various ways to improve production efficiency. One of such ways may be reducing the cost of servicing electrical equipment and simultaneously increasing the reliability of its operation (Khokhlov, 2018; Vasilyeva, N., Koteleva, N. and Ivanov P. 2018). Today, there are various systems and technologies that simplify the maintenance of electromechanical equipment. Remote maintenance systems, technical condition prediction systems (which can be used for reducing the frequency of maintenance), adaptive equipment repair planning systems, and other systems. One of technologies to increase the efficiency of the maintenance of electrical equipment is the augmented reality system (Zhukovskiy Y, Malov D. 2018; Zhukovskiy Y, Koteleva N 2018).

2 MAINTENANCE OF ELECTOMECHANICAL EQUIPMENT IN INDUSTRIAL PRODUCTION

Electric motors are subject to depreciation. During their operation there are often malfunctions, breakdowns or work with parameters different from the nominal values (Novoseltseva,2016). Since electric energy is converted into mechanical energy in an electric motor, it is obvious that electrical motor malfunctions can be caused by both malfunctions in electrical and electromagnetic systems and defects in the mechanisms. These deviations can be caused by both technological errors in the manufacture of the engine, incorrect operating conditions, and depreciation during operation (International renewable energy agency, Vasilev, 2018).

Maintenance is a set of operations to maintain the operational state of the equipment and reduce its level of depreciation. In the oil and gas industry, conservative maintenance methods are used to detect equipment defects:

- By time (the turnaround cycle is set as a calendar period; weakly correlated with the actual state of electro mechanical equipment);
- By operating time (machine hours, production volume, mileage, etc.);
- By state (regular repair is necessary when certain parameters are reached).

Traditional services methods associated with stopping the process entail serious costs. Even greater damage is caused by emergencies associated with the depreciation of equipment; The percentage of equipment depreciation in the oil and gas industry today is 57%. This leads to an increase in the number of equipment failure. The average share of equipment with an excess of service life for the enterprises of the oil and gas complex reaches almost 30%. Such equipment is constantly in a "pre-refund" state, which leads to an increase in the share of costs for its maintenance in the cost of production, as well as a security risk.

Violations of the work of individual components of the machine, which were not detected in a timely manner, can lead to severe accidents at work, as well as unplanned downtime and losses. In difficult operating conditions of the electric drive, the assessment of the technical condition allows to implement methods based on the analysis of the currents consumed by the engine. Technical diagnostics of electrical equipment in the oil and gas industry has a special character; the complexity is provided by the specifics of the execution and operating modes, and in some cases the lack of direct access to the object being diagnosed (Nikolaichuk, 2016).

The presence of several diagnostic parameters of a different physical nature, allowing to classify a certain type of damage, makes it possible to carry out a more accurate analysis of the defect that has arisen and to eliminate the error of a false definition as much as possible, as well as discard the arising interference (Vasilyeva N.V., Fedorova E.R., 2018; Vasilyeva N.V., Fedorova E.R., Koteleva N.I., 2018). At the same time, some of these parameters can be obtained by modelling on diagnostic models (Boikov A V, Savelev R V and Payor VA, 2018).

One of the main trends in the development of technical production is the growth of controlled parameters. In these conditions, it is necessary to develop new information technologies for monitoring the technical condition, based on intellectual methods, because the use of old approaches in these conditions is ineffective and impractical. Timely and effective monitoring and diagnostics of electrical equipment based on intelligent state monitoring technologies reduces maintenance time (Zhukovskiy Y L, Starshaia V V, 2019; Zhukovskiy Y, Korolev N, 2018).

3 AUGMENTED REALITY SYSTEM

Augmented reality (AR) is a relatively new form of interaction between man and machine. The basis of technology is the imposition of real and virtual information and is a mixture of resources of the real and virtual worlds.

In the era of "digital transformation" and complication of the component composition of power and electromechanical equipment, the time spent on servicing a single device tends to decrease. At the same time, the number of pieces of equipment per industry per worker increases. Indeed, in the paradigm of the fourth industrial revolution, there is a tendency to reduce the number of employees per unit of output while increasing efficiency. As a result, the number of devices and equipment that need to be operated and maintained will only increase, as a result, the risks associated with personnel error increase.

AR has good potential in terms of improving the performance of labour tasks and equipment maintenance, as well as reducing the risk of accidents. In addition, researchers have identified other applications for AR performance measurement, diagnostics, training, and control operations.

Thus, an AP can be useful in many situations where service and additional information is required. In addition, with proper use and design, AR visualization capabilities can transform maintenance and repair processes and allow for a switch to a predictive strategy. However, AR research in maintenance applications is still at the prototype stage; there are no systems with full AR integration in the maintenance systems.

Augmented reality technology, an increase in real-world objects occurs by adding virtual information objects in the form of graphs, text information, 3D models, sounds and virtual controls.

From the point of view of implementation, there are several ways of comparing augmented reality and real objects: augmented reality using a marker; marker less type; augmented reality based on projections and augmented reality based on superposition.

The technology of augmented reality is being introduced rapidly at almost all spheres of human activity: education, industry, medicine, engineering, entertainment, education, etc. The use of augmented reality technology at power supply facilities is not so widespread, but is only beginning to take turns to implementation. Such a delay is due to a number of reasons, mainly one of the most important of which is the unreadiness of technical specialists to use and maintain systems based on augmented reality technologies.

The review of successful implementations of augmented reality technology covers a wide range of industries - from industrial applications to everyday use by ordinary users.

In most cases, regardless of those branches of human activity, where augmented reality technology is used, augmented reality technology is aimed at solving standard tasks, among which the main ones are assistance and expert support in carrying out any work.

In industrial enterprises, the technology of augmented reality can be implemented to solve various problems, with the main tasks being the following:

- assistance in manual assembly on a conveyor production line;
- assistance in the maintenance of equipment or systems;
- providing expert support at different levels;
- providing support for quality control;
- providing support in obtaining information about the object and its characteristic parameters in real time;
- assistance in organizing videoconferences between various specialists;
- assistance in practicing skills on specialized computer simulators.

2 THE RESEARCH OBJECT

To ensure the safety at the oil and gas enterprises, various devices operating according to the "smart grid" concept are put into operation (U.S. Energy information administration, International Energy Agency). One of these elements is the Smart Shield (Intelligent Low Voltage Bundle Device) by Schneider Electric. In this work, as an object of study, an intellectual Smart Shield is presented, established in the laboratory of the Mining University (Figure 1).

"Smart Shield" - allows for: distribution of electricity and load management; technical accounting of electricity; operational service; power quality monitoring. In the Schneider Electric laboratory of the Mining University there are 2 types of "Smart Shield", which have small functional differences and differences in the layout.

Figure 1. The research object

Smart Shield 1 is equipped with the following type of equipment: 3 Acti9 Smartlink digital tires; 20 Circuit Breakers with an additional latching contact; Automatic switch Compact NSX100B with electronic Micrologic control block; Microserver with Com'X200 Ethernet Gateway; Switch; Shield indicator with touch screen; 4 Multi-tariff electricity meters.

Smart Shield 2 is equipped with the following type of equipment: 9 remote controlled circuit breakers; 3 Acti9 Smartlink digital tires; Microserver with Com'X200 Ethernet Gateway; Switch; WiFi– router; Shield indicator with touch screen.

The main functional features of an intelligent "Smart Shield" are to analyse the electricity consumption in general and in groups, timely identify problems associated with energy consumption and implement management of various elements of the power supply system in order to optimize and improve the efficiency

5 THE AUGMENTED REALITY SYSTEM FOR MAINTENANCE OF ELECTOMECHANICAL EQUIPMENT IN INDUSTRIAL PRODUCTION

In order to increase the availability of on-line information in real time about the actual condition of the serviced electrical equipment and providing support in obtaining information about the serviced equipment for the Smart Shield object, a software module was developed based on augmented reality technology. The development environment for the augmented reality module was chosen by Unity (Unity Technologies) with the augmented reality platform installed and the Vuforia augmented reality software developer toolkit (Qualcomm).

Figure 2 shows a diagram of the interaction between an intelligent Smart Shield and an augmented reality module.

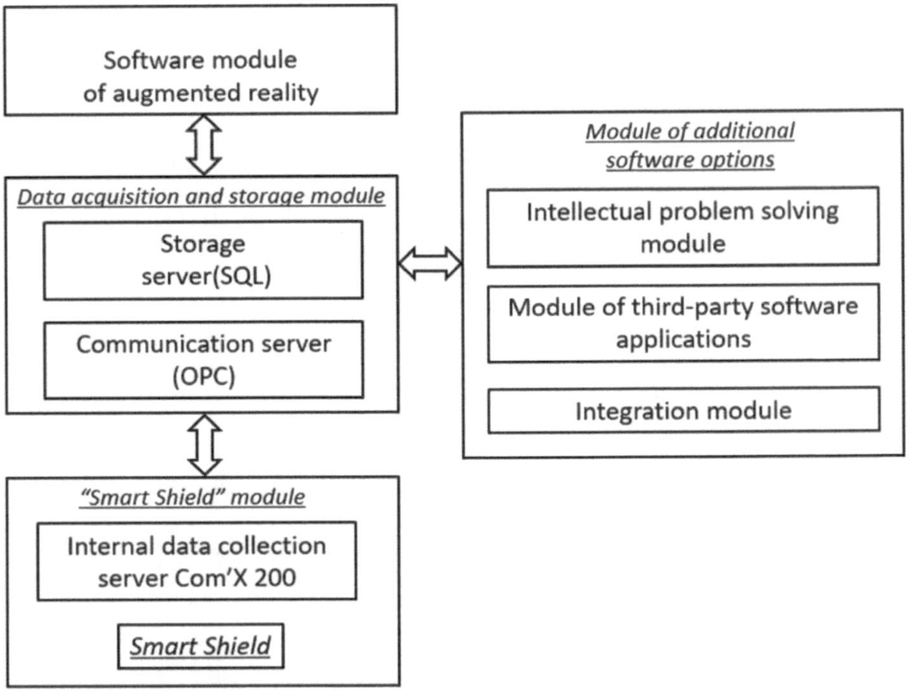

Figure 2. A diagram of the interaction between an intelligent Smart Shield and an augmented reality module

The interaction scheme shown in Figure 2, besides the interactions between the research object and the augmented reality module being developed, contains a module for implementing additional software options. The development of additional software modules is not presented in the work; however, this diagram shows the possibility of expanding the functions of the module being developed. A full-featured module can additionally provide the following functions: providing expert support at different levels; assistance in organizing video conferences between various specialists; assistance in practicing skills in specialized computer simulators, etc.

The user interface of the developed module has six modes (scenes) of information display.

Realized scenes:

- The "Menu" scene - the main and initial window, has a two-way connection with all other application scenes.
- Scene "Mode" contains information about the network operating parameters changing in real time: average and between phases voltage, average and phase currents, frequency, power and other parameters.
- Scene "Equipment" - contains information about what equipment is connected to the panel, the location of this equipment in the room. For example, when you hover on a switch, the system shows the specific equipment, which is the active load for the switch, etc.
- Scene "Resource" - contains information about the residual resource, the number of equipment inclusions in total and per day.
- Scene "Energy" - contains information about the energy consumed from switching on, per day and per month, as well as the duration of switching on.
- Scene "Control" - control buttons for automatic switches are displayed.
- Scene "Quality" - contains information about TDD, current and voltage harmonics.

Also in the "Menu" there is a "Search" button for direct access to the Internet from a browser installed on the device, and an "Exit" button.

In all scenes, the same-name button for returning to the "Menu" scene is implemented, as well as the "Next parameters" button - for ease of perception of information by the workers, the displayed parameters placed on each scene are separated by "sub-prices", and with this button you can choose to display that parameter, which interests the worker at the moment.

6 RESULTS

Figure 3 demonstrates the testing of the developed application. In the course of testing, the performance of all the developed scenes was proven, and the system's drawback was revealed - difficulty in positioning virtual reality. The development of a more contrast marker or the development of a marker less augmented reality module can correct this shortcoming.

With the present and forecasted in the future diversity of energy and electromechanical equipment, the assessment of how much and what equipment should be served by one person is a significant difficulty. Testing of the developed algorithms showed the physical possibility and feasibility of introducing software modules of augmented reality into real-life objects of the oil and gas industry. The amount of time required for service personnel to obtain information about the facility being serviced has been reduced to a few seconds. Prior to the introduction of software modules, information about the object was collected through communication with dispatch personnel. It took at least one minute to receive one parameter, while 2 specialists were involved. A rough estimate gives a reduction in electrical equipment maintenance time when monitoring 10 parameters for 9 minutes. In addition, the ability to consider changes in these parameters over time, the general monitoring of several parameters at once, and the recording of changes in the operation of electrical equipment directly in the system in real-time mode further reduces the maintenance time for one unit of electrical equipment.

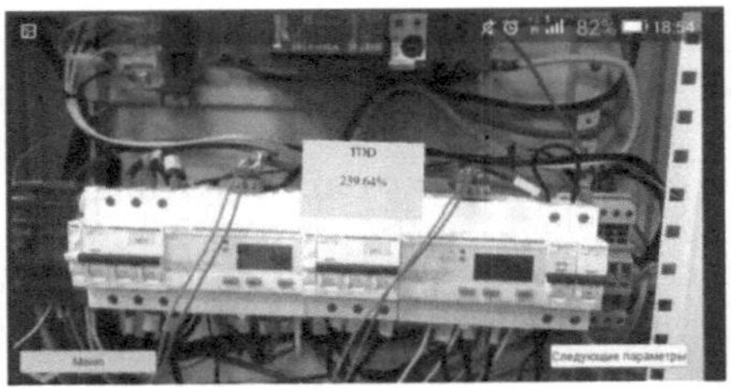

Scene "Quality". The TDD parameter

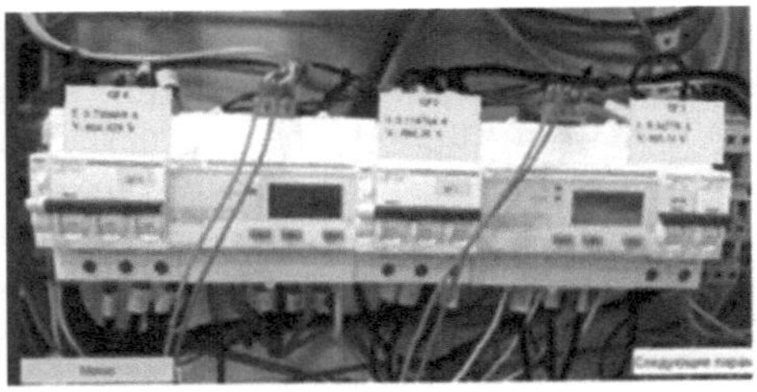

Scene "Mode". Parameters - average current and voltage

Main window

Scene "Resource". The number of starts

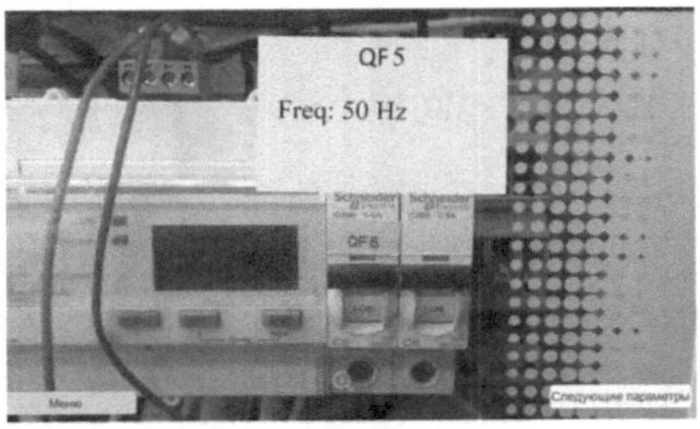

Scene "Mode". Parameter - frequency

Scene "Energy". The amount of energy per month

Figure 3. The testing of the developed application

7 CONCLUSION

Oil and gas enterprises operate a significant number of electrical equipment. The safety of conducting technological processes directly depends on the quality and timely maintenance of electrical equipment. Quality improving, reducing costs, reducing the time spent on servicing a piece of equipment during the its' maintenance affects the increase in efficiency. The introduction of modern technologies, such as an augmented reality technology, in maintenance processes will not only increase the efficiency of equipment maintenance, but also, as conceptually dictates, Industry 4.0 will help the quantity and quality of "collaborations" between processes, machines and humans.

REFERENCES

Abramovich B, Sychev Y. 2016. Problems of ensuring energy security for enterprises from the mineral resources sector. Zapiski Gornogo Instituta - *Journal of Mining Institute*, 217(1),132-139

Annual statistics of world energy URL: https://yearbook.enerdata.ru

Boikov A V, Savelev R V and Payor V A 2018 DEM Calibration Approach: design of experiment *Journal of Physics: Conference Series* 1015(3) 032017

Boikov A V, Savelev R V, Payor V A 2018 DEM Calibration Approach: Random Forest, *Journal of Physics Conference Series*. 1118 012009

International Energy Agency URL: https://www.iea.org

International renewable energy agency URL: http://www.irena.org
Kazanin O I, Sidorenko A A, Vinogradov E A 2017 Substantiation of the technological schemes of intensive development of gas-bearing coal beds *ARPN Journal of Engineering and Applied Sciences* 12 (7) 2259-2264
Khokhlov, S., Safina, E., Vasiliev, V. 2018 Risk-oriented approach implementation in departments ranking and teaching staff motivation International *Journal for Quality Research*, 8 (8) 39-47
Kislitsyn E.V., Panova M.V., Shishkov E.I. 2017. Problems of enterprises of the oil and gas complex of Russia: trends and solutions. *Internet magazine "SCIENCE"* Vol. 9, No. 3
Kuznetsov P. Solyonyj S. Sychev Y. Abramovich B. 2018. *MATEC Web of Conferences (St. Petersburg: EDP Sciences)* 161 2010
Nikolaichuk Liubov A., Pavel S. Tsvetkov, 2016 Prospects of Ecological Technologies Development in the Russian Oil Industry*International Journal of Applied Engineering Research* 11(7),5271-5276.
Novoseltseva M V, Masson M V, Pashkov E N, 2016 Investigation of Input Signal Curve Effect on Formed Pulse of Hydraulic-Powered Pulse Machine Novoseltseva *IOP Conference Series: Materials Science and Engineering* 127(1) 012007
Safina, E., Khokhlov, S. 2017 Paradox of alternative energy consumption: Lean or profligacy? *International Journal for Quality Research*, 11 (4), 903-916
Saveliev A, Malov D. Tamashakin M, & Budkov V. 2017 Service and multimedia data transmission in IoT networks using hybrid communication devices *MATEC Web of Conferences. – EDP Sciences* 113 02010
U.S. Energy information administration URL: https://www.eia.gov
Vasilev B U, Grigorev P S, Shulgenko V M 2018 Configuration and energy supply of promising types of underwater pumping complexes for transportation of hydrocarbons from the shelf. *Neftyanoe Khozyaystvo - Oil Industry* 3 77-81
Vasiliev B U 2017 Factors of environmental safety and environmentally efficient technologies transportation facilities gas transportation industry. *IOP Conference Series: Earth and Environmental Science* 50 (1) 012003
Vasilyeva N.V., Fedorova E.R. 2018. Statistical methods of evaluating quality of technological process control of trends of main parameters dependence. *Journal of Physics: Conference Series* № 1118, 012046
Vasilyeva N.V., Fedorova E.R., Koteleva N.I. 2018. Real-time control data wrangling for development of mathematical control models of technological processes *Journal of Physics: Conference Series*, № 1015, 32067
Vasilyeva, N., Koteleva, N. and Ivanov P. 2018. Quality analysis of technological process control. International *Journal for Quality Research* 12(1),111-128.
World energy council URL: https://www.worldenergy.org/
Yu L., Zhukovskiy, Y.L., Vasilev, B.U. & Koteleva, N.I. 2017, "Quality estimation of continuing professional education of technical specialists", *Proceedings of the 2017 International Conference "Quality Management, Transport and Information Security, Information Technologies", IT and QM and IS 2017* 704-707.
Zhukovskiy Y L, Starshaia V V, Batueva D E, Buldysko A D. 2019. Analysis of technological changes in integrated intelligent power supply systems. *Innovation-Based Development of the Mineral Resources Sector: Challenges and Prospects - 11th Conference of the Russian-German Raw Materials* 249-258
Zhukovskiy Y, Korolev N, Koteleva N 2018 About increasing informativity of diagnostic system of asynchronous electric motor by extracting additional information from values of consumed current parameter *Journal of Physics: Conference Series* 1015(3) 032158
Zhukovskiy Y, Koteleva N 2018 Development of augmented reality system for servicing electromechanical equipment. *Journal of Physics: Conference Series* 1015(4) 042068.
Zhukovskiy Y, Malov D. 2018. Concept of Smart Cyberspace for Smart Grid Implementation *IOP Conf. Series: Journal of Physics* 1015(4) 042067

Youth Technical Sessions Proceedings – Litvinenko (ed)
© 2019 World Petroleum Council, London, UK, ISBN 978-0-367-34668-3

Creating long term shared value while developing new energy projects

A.V. Illarionova
PhD in international economy, sustainability expert, WPC YPC representative (alternate), Moscow, Russian Federation

ABSTRACT: The aim of the research paper is to showcase how sustainability drives the energy industry transformation and how the creating shared value (CSV) approach, applied throughout the value chain and life cycle of the energy projects, can help maximize positive environmental and social impacts and catch up new opportunities for business and local stakeholders.

1 INTRODUCTION

Today energy business more and more understands that it operates in the interdependent environment, in an ecosystem, with different actors involved that impact the operational activiies. To "survive" in a long run and stay competitive on the market, energy companies start to address social challenges such as fuel poverty, access to water and infrastructure, quality education, health & well-being, climate action (that make part of 17 United Nations Sustainable Development Goals, 2015) with the same attention as business targets. Sustainability in its environmental, social and governance-related aspects is becoming a driver for the industry and requires redefining strategies and business processes, in order to take into account different stakeholders' needs and create sustainable value in a long run (IPIECA, 2017). This can be achieved through introducing positive approaches such as philanthropy or corporate social responsibility (CSR) but in many cases they have a more reactive and short term impact, becoming a "burden" and "additional spending" for the companies, in other words, being "on top of business" and more "reputation and communication oriented". The concept and methodology of creating shared value (CSV), proposed in this paper, are focused on the win-win solutions while the value is created for all stakeholders involved and is tightly linked to the projects life cycle: from business development through engineering and construction to operation and maintenance phases; from discovering fields to providing products to end consumers; from power generation and transmission to its distribution and providing access to energy and services to customers (Shared Value Initiative, 2012).

2 LITERATURE REVIEW

The research leverages on the creating shared value (CSV) business concept in which companies find business opportunities in tackling social problems, firstly introduced in 2011 (Porter M. E. Kramer M.R., 2011) and later enriched by the ecosystem and collective impact concept (Kramer M.R. Pfitzer M.W., 2016). The shared value journey is promoted by the Shared Value Initiative association that engages corporate leaders of various sectors in a platform through which they can learn, connect, and evolve the shared value practices. Therefore, in order to highlight some business practices from the energy sector where the CSV methodology was applied, the author studied specific case studies (Chevron, Shell, the Rockefeller Foundation Initiative on tackling the energy poverty, etc.). Energy trends and sustainability drivers are studied through BP Energy Outlook 2019 and IPIECA (International Petroleum Industry Environmental Conservation Association) publications.

3 METHODOLOGY

Interdisciplinary analysis (analysis of corporate practices and case studies, social and environmental impact assessment), system analysis, expert method.

4 ANALYSIS

Nowadays the energy industry is going through significant transformation processes. The whole energy landscape is changing, being driven by the growing energy demand, mostly from emerging economies, led by India and China (BP Energy Outlook, 2019) and at the same time driven by a great challenge of climate change risks mitigation. The industry is impacted by renewable energy sources growth (the price of kWh is already competitive to the traditional thermal generation in many countries), digitalization trend, innovative technologies boost (such as augmented and virtual reality, artificial intelligence, automation and robotics, etc.), increasing attention to sustainability aspects from the investor community, the need of the industry to gain return on investments in a much shorter period than it used to be and to develop new skills and qualifications for people who work in energy.

All these tendencies, sometimes controversial but very influential, requires from the energy business to act quickly, responsibly, be flexible and ready to listen to its stakeholder needs and be open for partnerships in order to guide and shape the energy transition. The concept of openness and stakeholder engagement is becoming more and more important for the energy business since the growth strategies could no longer be defined without taking into account global environment and strong interconnections. Moreover, no new energy services and products can be developed without tackling the global social challenges such as lack clean water, low quality education, poverty, hunger and pollution. The energy industry, thanks to its DNA, is stimulating social progress in a long run by providing affordable and reliable access to energy (ex. by providing 24/7 electricity supply to rural businesses, Rockefeller's Smart Power for Rural Development Initiative helped raise the income for thousands of rural Indians, Shared Value Initiative, 2015), therefore integrating different actors (governments, communities, academia, private companies) in an ecosystem where, in order to achieve the goals, they have to share a common agenda and co-design mutually reinforcing activities. This is how the creating shared value approach takes its real shape by turning the methodology into practical beneficial initiatives.

Talking about operative steps of the CSV approach application in the energy business, it is important to focus on the project development path starting from the business development phase (origination, preliminary development, feasibility, advanced development), to engineering and construction phases and operation and maintenance phases (IPIECA, 2018). In Tab. 1 there are presented some shared value analysis tools that allow to work out relevant actions and solutions for the whole project life cycle.

This finalized action plan, created within the business development phase of the energy project, previews also necessary actions in the engineering and construction phase (ex. local capacity building, job creation, environmental impacts mitigation, circular economy solutions on site, access to infrastructure and ICT services or communities and employees, safety trainings, etc.) and is monitored and updated within the operation and maintenance phase.

As a result of applying this methodology, the energy company is not only bringing the social progress to the territory where it operates but through engaging local stakeholders maintains its competitive edge by optimizing operational costs, catching up new business opportunities such as new services and products.

Among successful examples of applying CSV approach in energy projects, it is possible to mention the Dawson Creek Reclaimed Water Project, a joint venture between Shell and the City

Table 1. Application of the creating shared value approach through specific analysis tools

Proposed analysis tools	Description and impact
Local context analysis	Social, economic and environmental context of the area that include relevant information related to current and historic situations, main indicators, criticalities, opportunities and provides a general context where the energy project might be started
Mapping and management of relevant stakeholders involved/potentially involved in the project (local authorities, suppliers, employees, associations, local communities, universities, local businesses, etc.)	Identifying key actors, keeping transparent track of contacts and engagement actions to monitor their attitude and influence over the project life time
Environmental and social risk analysis	Identification and evaluation, prioritization of potential projects' impacts and risks on the social and environmental context (human rights, infrastructure, jobs, access to primary resources, waste management, etc.)
Materiality assessment of stakeholder and business needs through internal and external surveys, media coverage, direct stakeholders' requests	Analysis of social, economic and environmental issues, relevant both to stakeholders and to business, understanding the most important areas to select actions at local level
Action plan with shared value initiatives for a specific business energy project, supported by the financial impact evaluation and proposal of potential partners	Selected set of actions with a shared value effect, based on all previous analysis, that is implemented, monitored and amended during the whole life of the project

of Dawson Creek in British Columbia in Canada, where the energy company helped develop a water reclamation plant to treat waste water and make it useful for municipal and industrial activities, and guaranteed itself a reliable source of water in a long run (Adrian R., 2014).

To build prosperity in African regions and improve the operating environment, Chevron collaborated with an international partner to use a data-driven approach (by assessing trends and current expenditures in Chevron's supply chain) and identify new market opportunities and local solutions to unemployment, sharing those solutions with other partners (Hidalgo C., Peterson K., Smith D., Foley H., 2014).

5 DISCUSSION

The creating shared value (CSV) approach and its application through the business project value chain is currently discussed by different academic, business and institutional representatives. Among key discussion points, it is possible to highlight the following:

- The "non-originality" of the CSV idea as it is also represented by corporate social responsibility initiatives, social innovations or stakeholder management. It is possible to argue at this point that for the energy industry the idea of sharing value with the stakeholders, engaging them proactively from the beginning of business project, thus mitigating potential risks (related to environment, infrastructure, relocation, etc.) and opening up new ways of collaboration for a company, sometimes in an unusual and creative way (ex. job creation for small and medium enterprises that provide services for the energy company during the construction works, crowdsourcing ideas, circular economy solutions, cultural development and valorization of the territory) is really beneficial and original, since it helps prepare an action plan with a long term perspective that responds to the needs of the territory where the company operates (Hurst G.,,2016).
- The correlation between social and economic targets of business. Surely, in order to prepare an action plan (CSV plan) with relevant initiatives for local stakeholders, it is important to

take into account many factors and also make social investments in the territory. However, existing methodologies allow not only calculate positive impact of social and business returns but also prove that investments in local territory and its citizens at the early stage of the project bring significant saving in the future and even additional returns. The key point here is to plan correctly and precisely the required initiatives, leveraging on the risk assessment results and local context analysis and engaging possible partners in solving issues that the company cannot do alone.

- Creating shared value is not something that drives the business compliance. Talking about the energy sector where lots of operations are licensed and/or respond to the high international management standards, the idea of responsibility and accountability for people and environment is embedded in business by its nature. However, there is always space to do more in the fast-moving and complex economic environment in order to maintain a competitive edge in a long run. It can be done through introducing new sources of energy at affordable price (unconventional, renewable), offering new services to clients (es. flexible demand response, e-mobility, smart technologies for residential and commercial customers), by ensuring the inclusive growth of the local territories and fostering job creation. These business opportunities create real measurable impact and add a lot to business compliance.
- Creating shared value confirms that the usual role of corporations in the society "to solve the problems that need solving". Many energy projects are launched at the territories that require social support and the company quite often becomes a city forming entity, in other words, the expectations from the energy business are always high. That's why, analyzing and valorizing the opportunities of the territory and its stakeholders, finding relevant partners (universities, startups, NGOs, etc.) help reconsider the concept of a "social burden" and transform it into synergy, cost saving and additional benefits.

6 CONCLUSION

The co-creation is gradually becoming a vital word for the energy sector: business projects are defined and implemented together with the local stakeholders and customized to the peculiarities and needs of the local territory. The examples of shared value analysis tools confirm the simplicity, importance, and necessity of the creating shared value approach application. Being fully aware of the transformation processes and emerging industry trends, energy companies tend to reconsider their supply chain by ensuring its transparency, respect of human rights and diversity, creating new job opportunities, introducing circular economy and innovative solutions, and putting the long term sustainable value creation in the heart of their business strategies.

ACKNOWLEDGEMENTS

This paper focuses on sustainability in energy business through creating shared value approach. It makes part of the VI World Petroleum Council Youth Forum – Future Leaders Forum's agenda that indicates Sustainability is one of its key pillars, which will be discussed during the event, together with Leadership and Innovation.

REFERENCES

Adrian R. 2014. Case Study. Shared Value in the Extractives Sector: https://www.sharedvalue.org/groups/shared-value-extractives-sector

Hidalgo C., Peterson K., Smith D., Foley H. 2014. Case Study. Chevron's Partnership Initiatives in the Niger Delta: https://www.sharedvalue.org/groups/chevrons-partnership-initiatives-niger-delta

Hurst G., The Shared Value Initiative. 2016. Case Study. Enel: redefining the value chain: https://www.sharedvalue.org/sites/default/files/resource-files/Enel%20Case%20Study_December%202016.pdf

Kramer M., Pfitzer M.W. 2016. The Ecosystem of Shared Value. *Harvard Business Review*: https://hbr.org/2016/10/the-ecosystem-of-shared-value

Porter M.E., Kramer M. 2011. Creating Shared Value. *Harvard Business Review*: https://hbr.org/2011/01/the-big-idea-creating-shared-value

Shared Value Initiative. 2012. About Shared Value: https://www.sharedvalue.org/about-shared-value

Making the business case for Corporate Social Responsibility (CSR). Social performance tools. 2018. IPIECA. 28 p.

Improving social and environmental performance. 2017. IPIECA: http://www.ipieca.org/resources/good-practice/improving-social-and-environmental-performance/

Shared Value Initiative. 2015. The Rockefeller Foundation: Smart Power for Rural Development. Creating a Sustainable Market Solution to Energy Poverty: https://www.sharedvalue.org/sites/default/files/resource-files/Smart%20Power%20Case_Final.pdf

BP Energy Outlook. 2019: https://www.bp.com/content/dam/bp/business-sites/en/global/corporate/pdfs/energy-economics/energy-outlook/bp-energy-outlook-2019.pdf

United Nations Sustainable Development Goals. 2015: https://www.un.org/sustainabledevelopment/sustainable-development-goals/

New Approaches in Project Management

Youth Technical Sessions Proceedings – Litvinenko (ed)
© 2019 World Petroleum Council, London, UK, ISBN 978-0-367-34668-3

Calibration of pre-FEED estimation tools for oil & gas upstream projects

M. Tkachenko
Ingenix Group, Moscow, Russian Federation

ABSTRACT: Development of complex projects in oil and gas industry requires a profound search for possible methodical approaches to an integrated economic evaluation. One of the most influential challenges is to determine appropriate cost assumptions and to obtain reliable data including capital and operational expenditures for well-prepared decisions. Integrated method uniting modular analogue objects database and cost models to perform early phase cost estimation of oil and gas projects can serve as a solution to increase accuracy and transparency of estimates in current conditions. Results of actual data vs. estimates deviation analysis are presented in the case study from Russian oil and gas industry showing benefits one can derive from calibration and combined usage of two advanced cost estimation tools.

1 INTRODUCTION

Sustainability inevitably requires selection of promising projects to build well-balanced and profitable portfolio, ensuring company's growth and development. Accurate early phase estimate is one of the key challenges for project teams, especially in capital-intensive oil and gas industry where wrong decision usually have a very high price. Evaluation of capital expenditures on a pre-FEED phase is usually performed when input data is limited and even insufficient; often it is not possible to involve multi-disciplinary team. At the same time decisions taken on early phases have the highest impact on the total project cost curve whilst early phase cost estimations are characterized by low accuracy as a result of high uncertainty of input parameters. Current industry trend can be easily traced in many international oil and gas companies which is aimed at higher demand of investors and shareholders in capital expenditures (CAPEX) estimation accuracy and validity.

Widely used aggregative per unit metrics depending on one parameter (e.g. maximum capacity for site facilities) cannot anymore meet expectations on demanding level of accuracy. New normality demands new approaches to cost estimation as an important element of cost management ensuring successful execution of the project. Application of modular analogue objects cost database and parametrical cost models in Russian oil and gas industry are considered as advanced solutions in this research.

2 LITERATURE REVIEW

International cost estimation software (IHS Que$tor, CostOS, Cleopatra Enterprise etc.) applications are not looking good for Russia mainly because of difference in cost and deliverables structures; this results in unavailability to analyze deviations and adjust estimates to industrial specifics. Besides there are no recognized approaches to methodology and expected accuracy range at each project stage (the best example of such kind classification system, probably, is the AACE International Recommended Practice "87R-14: Cost Estimate Classification System - As Applied for the Petroleum Exploration and Production Industry").

Nevertheless Russian oil and gas companies understand the importance of accurate early phase estimation of capital expenditures and make efforts on development of corporate cost estimation methods.

PJSC LUKOIL has decided to apply existing Russian software, specially developed for the purpose of early-phase estimation (Bozieva, I.A. & Zinnatullin, D.F., 2016). In-built database is enlarged with corporate data; both types of data are stored securely and protected against data modification. Beforehand corporate data are being reconciled and unified according to software database structure as historical data are presented in different formats and are stored in unrelated sources, and this fact currently makes their effective use complicated.

PJSC Gazpromneft focuses efforts on development of multi-factor cost models based on accumulated historical data in regions where Company operates (Khasanov, M.M. et al., 2014). These models are intended to provide estimates for specific objects (roads, wells and etc.). To ensure comparability the one cost model is used through the project stages: with the project development users are able to specify parameters of the cost model, in case if required information is lacked data are obtained from cumulative historical database. For example, key parameters of a road as a linear objects ('Region', 'Category of the road', 'Conditions') are assigned at the stage "Assess", later at the stage "Select" technical parameters can be specified with transportation range, depth of fill, travel width etc.

PJSC ANK Bashneft has chosen to model typical objects and their modules followed by calculation of estimated costs (Atnagulov, A.R. et al., 2015) because of the lack of actual data from previous periods. Principle of decomposition was applied: objects were divided into smaller blocks to provide user a possibility to choose the most appropriate combination. Natural environment and climatic conditions were taken onto account through development of typical projects for 3 macroregions (the Far North, Siberia and Central Russia). According to article, cost deviation is about 3% between costs of analogue and object modeled from typical modules in the developed database. The result is satisfactory, but further work is required to verify consistency and accuracy comparing with larger number of objects and projects.

One can see that these three companies have chosen different approaches, reflected in articles and presentations at the conferences.

Two independent surveys (Beck & Dhole, 2016; Andreev & Tkachenko, 2016) on capital projects cost estimation processes point out that current satisfaction of corporate experts with estimating activities is about 50%. Those surveys demonstrate similar results despite different levels and number of respondents: the one (Beck & Dhole, 2016) was conducted globally while the other was focused on Russian oil and gas industry (Andreev & Tkachenko, 2016).

Taking into account that companies are experimenting with various methods as well as surveys' findings that industry experts are not satisfied with available tools and processes together with current industry trends, it is obvious that the search for solution to increase accuracy and transparency of early phase estimates in oil and gas industry is a pressing issue.

3 METHODOLOGY

Analysis of current situation of early phase estimation approaches within a company was performed to define common factors. Based on it advanced tools were selected for further study.

To calibrate available modular analogue objects database and multi-factor cost models initial technical inputs on project A were provided by one Russian oil producing company. Based on those inputs estimates were calculated using cost database and parametrical cost models. Afterwards estimates were compared with actual data on project A received from the company. Analysis of deviation was carried out; after receiving additional data, specifying features of project A, database and models were calibrated to increase level of accuracy for the future estimations.

Special cost estimation software Ingenix Cost Manager to approach the task was used in the study. This software provides access to comprehensive database and cost models of oil and gas objects.

4 ANALYSIS

Recently executed upstream projects are a valuable source of up-to-date information: historical data are commonly used as a cost estimation basis for next projects. However in real-world contexts it is often mistakenly believed that collected data (especially from accounting systems) is enough by itself to provide accurate estimates, though they answer the question how much did it cost to us, but not how much it will cost in the future for the next projects. For that reason data collection is only one stage together with data management process planning, data processing and analysis. Main aim of historical cost database is not to repeat but to improve (Stephenson, H. Lance, ed., 2015). Companies in Russia usually develop their internal corporate databases to have a possibility to estimate promising projects and improve effectiveness of existing projects. Further the analysis is focused on the first (namely estimation) task.

Ideally historical database (Figure 1) should include all subsequent estimates related to one project such as feasibility study, summary cost estimate, detailed cost estimates and actual costs to track changes and their grounds to fix lessons learned. In practice estimation departments enjoy often a limited access to data related to one or two stages maximum.

Figure 2 demonstrates typical process of early phase estimation approaches development within the company: as a general rule firstly the estimation activities are based on per unit metrics; when the accuracy level is not sufficient or existing limitations of this tool became critical, next steps such as development of comprehensive cost database and cost models could be taken. The combination of last two advanced tools (analogue objects from database + cost models) can be described as an integrated method.

Aggregative per unit cost metrics depending on one parameter (e.g. maximum capacity for site facilities) are widely used in Russia to get early phase estimates. As a general rule these metrics are represented by averaged values and are based on actual costs received from the accounting system without deep analysis of technical inputs. This method is simple, fast and understandable demonstrating acceptable results in case if key technical parameters of the current object and base object/set of base objects are similar, but also has its own limitations. For example, the length of linear object is an important cost driver because shorter pipelines or roads cost more per unit (per km usually). For this reason to make accurate estimates it is important to know the initial lengths of constructed pipelines included in set composing the basis for per unit metric, and that in most cases remains unknown for the end user. As a result estimate accuracy is highly dependent on chosen analogue. Another limitation is a lack of broad scale variation of key parameters.

Structured and analyzed data open wide possibilities for their application in cost estimation processes. For that reason key success factor making the application of the integrated method possible is a database containing data on costs and other parameters collected from previously executed and/or designed projects. Each upstream project is unique, for that reason additional technical parameters are required to make relevant estimates. Sufficient breakdown and standardized classifiers make database flexible and suitable for cost modeling of various oil and gas objects (Chizhikov, S.V. et al., 2013).

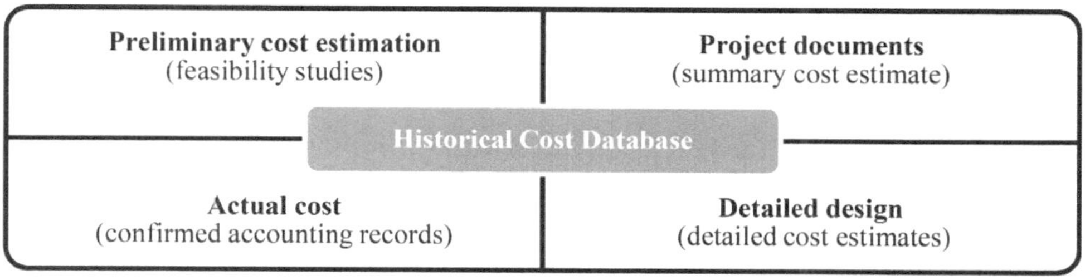

Figure 1. Historical cost database and results gained at subsequent cost estimation stages as its possible components

	Per unit metrics	Comprehensive cost database	Cost models
Short description	Cost metrics depending on one key parameter	Structured multilevel data on various objects	Calculative interrelations among cost and technical parameters to model cost of particular object / module
Advantages	- Widely used; - Simple and fast calculations; - Based on actual costs from accounting systems.	- Structured and unified data; - Access to various levels of objects; - Secure storage of project information.	- Transparent and easy-to-use tool; - Wide possibilities to perform analysis and adjust if necessary.
Disadvantages	- Data are usually averaged without deep analysis; - Difficult to adjust to project specifics; - No access to details on objects included in calculation basis.	- Require initial amount of historical data; - Cost and time demanding development; - Developed structure should in advance meet requirements of all potential user groups.	- Require initial structured data on objects; - Time and high-quality analytical skills demanding development.

Figure 2. Typical development of early phase estimation approaches within a company

Figure 3 shows the comparison among first estimate based completely on analogue object from database, second estimate after receiving more detailed information and actual costs received from oil and gas company. This case proved once again that estimation accuracy critically depends on technical inputs. In case of minimum technical inputs estimation based on analogue objects from database resulted in 25% of deviation from actual costs. When inputs were specified (after receiving site plan) and the initial analogue object was edited by in-built means of the database, deviation was reduced to 9%. Key ground of deviations between first

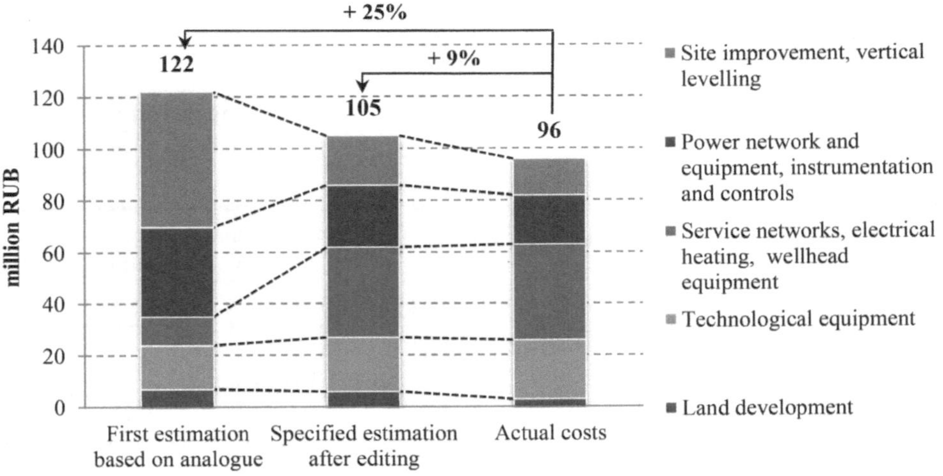

Figure 3. Analogue objects database: estimates vs. actual data for well pad

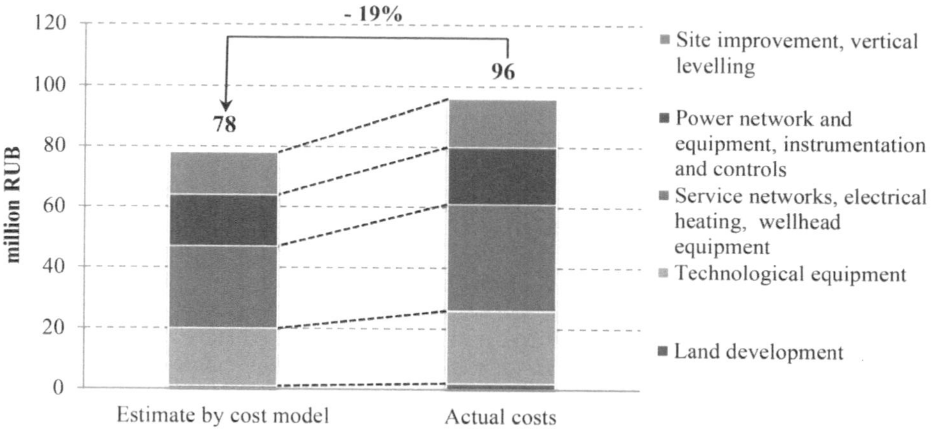

Figure 4. Cost model: estimate vs. actual data for well pad

estimation based on analogue and specified estimation after more detailed information was received is largely due to smaller well spacing of actual well pad. Key grounds of deviations between specified estimation and actual costs were:

- Difference in cost structure: analogue object from database included remediation in land development costs whereas oil company according to internal standards includes it in other budget item;
- Difference in equipment list: analogue object included electrical submersible pump (ESP) control station cost whereas actual costs did not.

Gained from performed analysis information on deviation causes was used to calibrate modular analogue objects database and cost models mainly related to cost items reallocation and technological equipment cost update.

Developed database provides a perfect basis and powerful capabilities to built cost models with identified cost drivers and their interrelations. It is a transparent and easy-to-use tool with wide possibilities to perform analysis and adjust calculation to individual specifics of each project if necessary (Chizhikov, S.V. et al., 2017). Development of calculative interrelations among cost and technical parameters to model cost of particular object or technological module require initial structured data on objects and is time and high-quality analytical skills demanding.

Figure 4 demonstrates the comparison between estimate done by means of multi-factor cost model and actual data on well pad.

After analysis of deviation promising areas for calibration were defined: to increase accuracy it is necessary to update technological equipment costs, service networks construction costs to include water supply line and power network, instrumentations and control costs to include custom well spacing.

5 DISCUSSION

This paper contributes to an analysis of three possible cost estimation tools related to continuous advancement of estimation process within a company and sharing experience on real case how cost models and objects from database are able to benefit cost estimation processes in oil and gas industry. This work highlights that high quality work in the early phases including preliminary cost estimations is very important for capital intensive oil and gas projects because it is a basis for conceptual decisions which are difficult and expensive to change afterwards. The significance of early phase estimation tools development from per unit metrics till specialized database and cost models is shown. After calibrating tools for well pads cost estimation further work is required to revise tools for other types of oil and gas objects.

6 CONCLUSION

New normality demands advanced approaches to project management including cost estimation to ensure successful project execution. Oil and gas industry in general and in Russia in particular is currently looking for solutions to increase efficiency of collected historical data use for estimation purposes. Existing international out-of-the-box solutions fail to meet the requirements due to difference in cost and deliverables structures resulting in unavailability to analyze deviations and adjust estimates to local specifics. Russian companies are forced to develop their own corporate solutions; some companies are already at the stage of their comprehensive databases development and update whilst only few are working upon cost multi-factor models. Designed and executed projects provide valuable up-to-date information enlarging corporate historic database. Calibration of advanced cost estimation tools (cost databases and multi-factor models) is of key importance to increase accuracy of estimates for feasibility studies of oil & gas upstream projects minimizing the gap between estimates and actual costs. Based on performed estimates vs. actual costs analysis promising areas (e.g. technological equipment, well spacing, structure rearrangement) to make objects in database and cost models more flexible to provide broader variety of engineering solutions, including specific for particular company, were identified.

REFERENCES

AACE International Recommended Practice 87R-14: Cost Estimate Classification System - As Applied for the Petroleum Exploration and Production Industry. 2015, viewed 19.03.2019, https://web.aacei.org/docs/default-source/toc/toc_87r-14.pdf

Andreev, A.F. & Tkachenko, M.A. 2016. Cost estimating of upstream projects in early phases of development [Otsenka zatrat na rannikh stadiyah razvitiya neftegazovykh proektov]. *Oil, gas and business [Neft, gaz i biznes]* 7: 29-34.

Atnagulov, A.R. et al. 2015. Developing CAPEX Database for Oil Field Surface Facilities Construction at Bashneft PJSOC. *Oil Industry [Neftyanoe Khozyastvo]* 8: 98-101.

Beck, R. & Dhole, V. 2016. How Does Your Organization Stack Up Against Your Peers in Bidding, Estimating and Project Performance?, viewed 28.02.2019, https://www.aspentech.com/en/resources/white-papers/benchmarking-organizations-during-lean-times.

Bozieva, I.A. & Zinnatullin, D.F. 2016. Aspects of Corporate Information System Development to Generate the Costs of Construction Facilities and Oil and Gas Fields Infrastructure Development. *Oil Industry [Neftyanoe Khozyastvo]* 2: 114-117.

Chizhikov, S.V., Dubovitskaya, E.A. & Pashchenko, A.D. 2013. Problems and proposed solutions for oil and gas projects cost estimation in Russia. *Oil Industry [Neftyanoe Khozyastvo]* 9: 92-95.

Chizhikov, S.V., Dubovitskaya, E.A. & Tkachenko, M.A. 2017. Costs modeling: support point in a changing world. *Oil Industry [Neftyanoe Khozyastvo]* 10: 64-68.

Information on software in domestic software register [Svedeniya o programmnom obespechenii v reestre otechestvennogo programmnogo obespecheniya], viewed 19.03.2019, https://reestr.minsvyaz.ru/reestr/141575/

Khasanov, M.M. et al. 2014. Improvement of CAPEX Estimation Accuracy During Early Project Stages. *Oil Industry [Neftyanoe Khozyastvo]*12: 22-27.

Stephenson, H. Lance, ed. 2015. Total Cost Management Framework: An Integrated Approach to Portfolio, Program, and Project Management: a Continuing Project of the AACE Technical Board. AACE international.

Youth Technical Sessions Proceedings – Litvinenko (ed)
© 2019 World Petroleum Council, London, UK, ISBN 978-0-367-34668-3

The energy company's competitive leadership: Enviromental vector and development of renewable energy sector

A.A. Konareva, D.V. Senchenkova & N.V. Trifonova
Saint-Petersburg State University of Economics, Saint-Petersburg, Russian Federation

ABSTRACT: The question of the formation of a rational and effective energy balance is strategically important for each country. Today, renewables and electrical networks are the most popular segments of the energy sector for investment. Oil and gas production is an energy-intensive process, so the electric power industry has always been a related industry for oil and gas. The Russian Market of Electrical Energy/Power has a great potential for the future development, it can be attractive for oil and gas companies all over the world. In this article we considered the world trends of improving and increasing efficiency of the electricity industry, the activity of the world's leading oil and gas companies in renewables and made the estimate of investment attractiveness of The Russian Market of Electrical Energy/Power for the potential investors.

1 INTRODUCTION

The defining role in shaping the potential of the economy of any country belongs to energy. It is one the most systematically important and life-supporting industry. In the context of globalization of the world economy, oil and gas companies are faced with the task of forming a sustainable energy system that combines three important criteria: high-energy, high-economic and high-ecological efficiency. Given the growing influence of environmental factors such as global warming, renewable energy is actively considered by oil and gas companies as new opportunities. Today, the use of renewables contributes to improving competitiveness and provides leadership to energy companies.

The aim of the scientific work is to address renewable energy from the prospective of a tool for ensuring a sustainable, flexible and safe energy supply in a changing climatic, economic and political conditions and as a result, to estimate the investment attractiveness of The Russian Market of Electrical Energy/Power for the world's leading oil and gas companies.

The following tasks were set: to define the global trends of improving, increasing competitiveness and ensuring leadership of the energy company; to consider and define the behavior of the leaders of the oil and gas industry as the world's leading hydrocarbons import and export countries of as dynamically diversifying their investment portfolios by integrating into the renewable energy sector and innovative technological sectors associated with the development of the latter; to estimate the investment attractiveness of the Russian Market of Electrical Energy/Power for the world's leading oil and gas companies from the point of view of the necessity to stimulate the development and expansion renewables technologies in Russia.

2 LITERATURE REVIEW

Despite the fact that the first papers about renewables appeared in the 1970s, there are four large-scale papers were published for the last two years in the next fields: geopolitics of renewables (Scholten D. 2018), environmental aspects of using renewables (Singh A. 2018), work optimization of renewable systems (Erdinc O. 2017). Also, the digest of articles, which represents the most comprehensive work about renewables, was republished in 2018 (Sorensen B. 2018).

Renewables in Russia don't have such widespread use and distribution as the countries of the European Union. However, a large number of Russian papers are devoted to the issues of introduction and using renewables in the Russian energy practice. SKOLKOVO Energy Centre considers renewables from the point of view of the implementing it in the Russian energy practice (Hohlov A. 2017.; Grushevenko E. & Hohlov A. 2017) and the estimate of the potential of the distributed energy in Russia (Hohlov A., Melnikov Y., Veselov F., Kholkin D. & Datsko K. 2018). The analysis of the green energy in the EU countries was conducted by scientists of the Higher School of Economics (Ermolenko G.V., Tolmacheva I.S., Ryapyn I. Y., Fetisova Y.A., Matsura A.A. & Reutova A.B. 2016). The Government of the Russian Federation is also interested in renewables in the fields of distributed energy (Grigoriev L. 2017a) and the digitalization of the electrical networks (Grigoriev L. 2017b). Also, the review of the global trends of the world's leading oil and gas companies was presented by scientists from the Higher School of Economics.

3 METHODOLOGY

The estimate of the investment attractiveness of The Russian Market of Electrical Energy/Power was based on the next it's parts: The Federal Wholesale Market of Electrical Energy/Power and The Retail Electricity Market. The calculation of the growth rates was made on the basis of the growth in gross value added in the produced GDP. The degree of the market concentration was calculated as the ratio of the sales volume of the first four companies on the market to the total sales volume of all companies on the market. The qualitative analysis of the competitiveness of the Russian power industry was made by the Porter five forces analysis. The quantitively analysis of the investment attractiveness was made by the GE/McKinsey Matrix. There are two groups of estimates were used in calculations: competitiveness and industry attractiveness. Each of the factors has a specific gravity in the groups and estimates from 1 to 10 depending on its severity. The final score determines as the amount of multiplies of specific gravity and estimate of each factor in the groups. Also, the following methods was used: benchmarking activities of the leading world oil and gas companies, statistical methods, information analysis-synthesis method, marketing analysis, risk identification and analysis.

4 ANALYSIS OF GLOBAL TRENDS OF IMPROVING AND INCREASING THE ENERGU EFFICIENCY OF THE ELCTRICITY INDUSTRY

The liberalization of the electricity and capacity markets all over the world from the 1970s to the 2000s led to the market pricing of resources and increasing of the household's electricity prices compared to the non-household's electricity prices. The electricity prices for the household and non-household consumers of the countries of the world for the 2017 are presented in Figure 1.

Households' electricity prices are higher than non-household's electricity prices in Europe and the USA, because households connecting to electricity networks is more difficult than non-households connecting. It is economically justified. Since the electricity and capacity markets in Europe and the USA are liberalized, the growth of electricity prices for the households has to be controlled by non-regulatory methods. The current state of the energy of the EU is characterized by the following features: high dependence on energy imports in conditions of a limited number of large energy suppliers, high volatility of energy prices, increasing risk of security of energy supply and environmental problems. All of this contributes to the development and implementation of technologies based on renewables.

In the electric power industry, the spread of renewables takes places in the distributed energy. The growth of the spread of the microgeneration based on renewables among households is facilitated by high electricity prices, simplified process of electricity networks connecting, government support and guaranteed sale of electricity by long-term contracts. The measures used by the countries of Europe to stimulate investment in renewables and its's active use and distribution are presented in Table 1.

Figure 1. The electricity prices for the household and non-household consumers of the countries of the world for the 2017, euro per kWh

Table 1. The measures of stimulating investments in renewables and it's active use and distribution in Europe.

Measure	Description	Countries
Financial subsidies, grants and discounts	Financial subsidies are one-time payments to developers or purchasers of renewables equipment to offset initial investments. Grants is a one-time payment for the implementation of a renewable project. Discounts are granted to owners of an improved renewable equipment	Austria, Belgium, Great Britain, Hungary, Germany, Greece, Denmark, Spain, Italy, Cyprus, Luxemburg, Malta, Netherlands, Portugal, Slovenia, Finland, France, Czech Republic, Sweden
Tax incentives	To generate electricity produced by renewables	Austria, Belgium, Germany, Greece, Denmark, Spain, Italy, Netherlands, Portugal, Slovakia, Finland, Czech Republic, Sweden
Investment tax incentives	For private investors in renewables for income tax reimbursement	
Exemption from VAT	Household who purchase a renewable microgeneration equipment are exempted from payment	Belgium, Great Britain, Hungary, Germany, Greece, Denmark, Italy, Latvia, Malta, Netherlands, Poland, Portugal, Slovakia, Slovenia, Finland, France, Czech Republic, Sweden
Accelerated depreciation	To optimize the taxation of a renewable power plant	
Microcredits	For households and small companies for the purchase of renewable equipment	Austria, Bulgaria, Great Britain, Hungary, Germany, Greece, Denmark, Italy, Lithuania, Netherlands, Poland, Portugal, Romania, Slovenia, France, Czech Republic, Sweden, Estonia.
Interest-free loans		
Loan guarantees	From the European Investment Bank in relation to the exchange rate or loan to decrease investment risks	
Open auction/ tendering	Open competitive auction for fixed volume of power	Belgium, Denmark, Ireland, Italy, Cyprus, Latvia, Poland, Portugal, Slovenia, Finland

Thus, the range of the proposed instruments for attracting investment is quite wide, includes direct and indirect instruments of the investment regulation and focuses on the different groups of potential investors.

Renewables technologies have contributed to the development of the distributed energy. The connection of separate generating equipment requires special technologies to ensure efficiency and reliability of the electricity transmission and distribution processes. The digitalization of the electricity networks is the most common technology in the EU. It consists of three parts: infrastructure modernization, infrastructure digitalization and business processes of managing smart grids transformation. Russia is also considering the issue of modernization and digitalization of the electricity network infrastructure. In 2017 Russia approved «The

Digital Economy of the Russian Federation» program. The National Technology Initiative Energy Net was created in the frameworks of its program and it's aimed to create competitive technology-based solutions by 2035. As a part of this Initiative pilot projects were developed and implemented in Russian regions.

The development of the energy based on renewables is mainly realized to solve two problems: environmental issues and increasing energy self-sufficiency of countries. In Russia these factors are not so important as in the EU, because the country fully provides itself with the necessary energy recourses at a relatively low cost. The possible direction of the renewable energy development in Russia is household microgeneration in settlements that are not connected to the electricity networks, or enterprises that have a connection but it needs to increase the reliability of energy supply. The creation of domestic renewable technologies can be a stimulating factor for the development of a new sector of the Russian Economy. The most attractive areas for electricity power sector investments in Russia are infrastructure modernization, digitalization modernization and using of storage devices. The experience of the EU is presents as an example of the efficient use of a variety incentive investment instruments.

For the period 2012-2017 the most invested energy sectors are renewables and electricity networks, the least – nuclear energy and hydrocarbons. The dynamics of the share of investments in the various energy industries are presented in Figure 2 (*World Energy Investment*. 2018). The obviously trend of growth of investment in renewables and electricity networks amounted 40,2% and 39,8% in 2017, respectively. The trend of investment in nuclear energy and hydrocarbons has declined and amounted 17,6% and 2,3% in 2017, respectively.

The increasing investments in the electricity energy sector from the oil and gas companies is observed in the world (Ermolenko G.V. 2017). The summary comparative analysis of investments in the electricity energy sector and renewables from the world and Russian oil and gas companies is presented in Table 2.

Figure 2. The dynamics of the share of investments in the various energy industries for the period 2012-2017, %

Table 2. The summary comparative analysis of investments in the electricity energy sector and renewables from the world and Russian oil and gas companies

№	Name of company	Country	Electricity producing	Renewables strategy	Renewables subsidiary	Type of project	Renewables purchase	Sale of oil and gas assets
1	Lukoil	Russia	+	+	+	Sun, wind, water	+	-
2	Rosneft		-	-	-	-	-	-
3	Gazprom		+	+	-	Water	-	-
4	NOVATEK		-	-	-	-	-	-
5	Surgutneftegas		+	-	-	-	-	-
6	Tatneft		+	+	-	-	-	-
7	Bashneft		-	-	-	-	-	-
8	Saudi Aramco	Saudi Arabia	+	+	+	Sun, R&D	+	+
9	Shell	The Netherlands	+	+	+	Sun, wind, biofuel, hydrogen, R&D	No data	+
10	Total	France	+	+	+	Sun, biofuel, R&D	+	+
11	Eni	Italy	+	+	+	Sun, wind, biofuel, R&D	+	+
12	Equinor (ранее Statoil)	Norway	+	+	+	Wind, R&D	+	+

13	SINOPEC	China	+	+	+	Biofuel, geothermal, R&D	+	+
14	Engie	France	+	+	+	Sun, wind, biofuel, water, tidal, R&D	+	+
15	REPSOL	Spain	+	+	+	Sun, wind, biofuel, R&D	+	+
16	PetroChina	China	+	+	+	Geothermal	-	+
17	Reliance Industries	India	+	+	+	Sun, R&D	-	+
18	Idemitsu Kosan	Japan	+	+	+	Sun, wind, biofuel, geothermal	-	+
19	Kuwait Petroleum Corporation	Kuwait	+	+	+	Sun, R&D	-	+
20	Hindustan Petroleum Corporation	India	+	+	+	Sun, wind, biofuel	-	+
21	Nippon Oil & Energy Corporation	Japan	+	+	+	Sun, hydrogen, R&D	-	+
22	ExxonMobil	USA	+	+	+	Sun, R&D	+	+

There is a weak activity and ambiguous attitude of Russian oil and gas companies to the electricity and renewable energy sectors compared with the world's leading oil and gas companies. Unlike the Russian oil and gas companies (only four of the seven considered companies), all of the world's leading oil and gas companies presented in Table 2 produce electricity. Three of the seven considered Russian companies have renewables strategy and only one company has renewable subsidiary. None of the Russian companies use the practice of selling oil and gas assets, which is a popul ar practice among world players. Despite the active sale of oil and gas assets and renewables technology and R&D development, not all of the world's leading companies of oil and gas sector invest in the purchase of renewables assets. Basically, these are the European companies. It can be concluded that the Russian oil and gas companies have an interest in investment in renewable energy sector, but they are wary of it and the implementation of renewables in their practice is situated at the starting stage of the development.

5 THE ESTIMATE OF INVESTMENT ATTRACTIVENESS OF RUSSIAN ELECTRIC POWER INDUSTRY FOR WORLD LEADING OIL AND GAS COMPANIES

The investment attractiveness of the industry is estimated by defining such indicators as the cyclical development phase of the industry, the industry average profitability, the level of the industry prospects, the level of intra-industry competition and the level of government support.

The cyclical development phase of the industry is defined as the combination of two factors: the industry lifetime and the industry growth rate. The average growth rate of the industry for the period 2012-2017 has amounted 1,07%. The electricity as a branch emerged in the 1920s in the USSR. Based on the obtained results, the Russian power industry is at the maturity age.

There are 118 power producers compete in The Federal Wholesale Market of Electrical Energy/Power. According to the calculation of the degree of the market concentration of The Federal Wholesale Market of Electrical Energy/Power, it's 0,61, which is an indicator of the average level of monopolization. The industry average profitability amounted 17,47% from the period 20212-2017 (*Obzor elektroenergeticheskoy otrasli Rossii.* 2017). The results show the high level of profitability of the Russian power industry.

5.1 *Porter's five forces analysis*

Applying the Porter five forces analysis of the Russian power industry next conclusions were made.

Industry Rivalry (high level) - there are 118 power producers compete in The Federal Wholesale Market of Electrical Energy/Power, who produces similar product. The level of monopolization of the industry is average.

Bargaining power of Suppliers (average level) – the transmission and distribution sector is a natural monopoly. So, in the case of an emergency situation the probability of a delivery failure is high.

Threat of New Entrants (low level) – the risk of the emergence of new entrants is small due to the high level of required investments and initial costs to enter the industry.

Threat of Substitutes (average level) – electricity is an irreplaceable product, but consumers may not to connect to the integrated energy system, the can buy their own generator, which may be the motivation to efficiently distribute their money.

Bargaining Power of Buyers (average level) – there are 28 large buyers in The Federal Wholesale Market of Electrical Energy/Power, who can influence the final electricity/power price. In the Retail Electricity Market consumers have an opportunity to buy their own generation.

5.1 *The GE/McKinsey Matrix*

The quantitively analysis of the investment attractiveness was made by the GE/McKinsey Matrix. There are two groups of estimates were used in calculations: competitiveness and industry attractiveness. Each of the factors has a specific gravity in the groups and estimates from 1 to 10 depending on its severity. The final score determines as the amount of multiplies of specific gravity and estimate of each factor in the groups.

In order to introduce the digital technologies and distributed energy based on renewables, the Russian power industry needs to upgrade its electricity networks due to their high level of depreciation. Therefore, the estimate of the investment attractiveness of the Russian power industry was made, because renewables as an industry doesn't exist in the country.

The Russian power industry attractiveness and competitiveness was made on the basis of an expert survey conducted by the North-West Center for Strategic Research, during which 78 people were interviewed. These people are representatives of all sectors of the Russian power industry, organized consumer groups, scientific and analytical organizations, government organization of market management. Among them: Veselov F.V., the Head of the Department for the Development and Reform of the Electric Power Industry at the Energy Research Institute of the Russian Academy of Sciences, Melentyeva S.B., Corresponding Member of RAS; Dod E.V., the Chairman of the Board - General Director of JSC «RusHydro»; Drel V. Y., Director of the Directorate for Pricing and Economics in the Electric Power Industry of the Institute of System Monitoring; Yemelyanov S.M., General Director of LLC «Transneftenergo»; Penkin V.T., Professor of the Department of Electric Power, Electromechanical and Biotechnical Systems of the Moscow Aviation Institute, the State Technical University; Uprety A., partner and managing director of BCG;. Khakhanov Y.M., Project Manager, Energy Efficient Technologies Cluster, «Skolkovo» Foundation; Chuvaev A.A., General Director, OJSC «Fortum», Executive Vice President of the Fortum Group; Yasin E.G., Scientific Director of the National Research University Higher School of Economics.

The compiled assessment allowed the authors to determine the level of competitiveness and attractiveness of the industry. The final result is presented in Table 3.

Table 3. The quantitively analysis of the investment attractiveness of the Russian power industry

	Factor's specific gravity	Factor's severity estimate	Multiply
Industry attractiveness criteria			
Product uniqueness	5%	10	0,5
Needs satisfaction	8%	10	0,8
Brand power	5%	1	0,05
Resources sufficiency	19%	6	1,14
Flexibility and adaptation	20%	8	1,6
Low level of competition	27%	3	0,81
Slow reaction of competitors	16%	3	0,48
Final score	100%		5,38
Industry competitiveness criteria			
High level of sales	30%	10	3,0
High growth	15%	3	0,45
Few players	10%	3	0,3
No advertising investment in the segment	5%	10	0,5
There are opportunities to expand the range	3%	1	0,03
Low level of the culture of use the product	5%	2	0,1
Low competing brand power	10%	10	1
Presence of hidden and unmet needs	15%	9	1,35
Long-term growth is projected	8%	2	0,16
Risk of influence of external factors	6%	4	0,24
Final score	100%		7,13

Based on the made estimate presented in Table 3, the Russian power industry has a high level of the investment attractiveness and the average level of competitiveness due similarity, indispensability and uniqueness of the product.

The Russian electric power industry shows the high potential of development, but there are some barriers of industry entrance exists. They are economic barriers – the high level of required investments and initial costs to enter the industry; technological – underdeveloped intersystem and backbone communications in the power system; organizing – the high level of the government participation and industry regulation; regulatory – the presence of legal acts establishing strict requirements for regulating The Russian Market of Electricity/Power; informational – the low level of consumer awareness about certain types of electricity generation; social – the possibility of non-acceptance of the object by the society.

The main factors of slowing development of the Russian electricity power industry are the high level of depreciation of fixed assets and the presence of cross-subsidies in tariff policy. The average level of the depreciation of fixed assets in the sector of providing energy, gas and steam, air conditioning is 46,33% for the period 2012-2017 (*Stepen iznosa osnovnyih fondov na konets goda po vidam ekonomicheskoy deyatelnosti*, n.d).

6 CONCLUSION

The behavior of the leaders of the oil and gas industry as net import and net export countries of hydrocarbons has been defined. They dynamically diversify their investment portfolios by integrating into the renewable energy sector and innovative technological sectors associated with the development of the latter. At the current stage of the world energy sector development, a high dependence on energy imports can be clearly seen as well as a high volatility in energy prices. The threat of climate change is increasing and the risk of energy supply security is growing due to the insufficient level of links between the countries' energy systems.

Renewables contributes to reducing the cost of fuel imports, improves trade, economic and energy balance. The development leads to the prevention of CO2 emissions. The development of technologies using renewable energy is accompanied by a reduction in the cost of construction and operation of energy facilities. The creation of technological cycles of industrial production contributes to increasing competition and efficient use of renewable energy at the global level.

To eliminate the identified risks, and as a result, to reduce financial costs, the development of renewable energy is formed specifically to create certainty in the income and expenditure of these projects.

REFERENCES

Average Price of Electricity to Ultimate Customers by End-Use Sector, viewed 9.03.2019 https://www.eia.gov/electricity/monthly/epm_table_grapher.php?t=epmt_5_6_a

«Bashneft» Annual report 2017 [Godovoy otchet PAO «Bashneft» za 2017 god], viewed 11.03.2019 http://www.bashneft.ru/files/iblock/edc/Годовой_отчет_Общества_ГОСА_2017.pdf

Electricity prices for household consumers – bi-annual data (from 2007 onwards), viewed 9.03.2019 http://appsso.eurostat.ec.europa.eu/nui/show.do?dataset=nrg_pc_204&lang=en

Electricity prices for non-household consumers – bi-annual data (from 2007 onwards), viewed 9.03.2019 http://appsso.eurostat.ec.europa.eu/nui/show.do?dataset=nrg_pc_205&lang=en

Electricity storage and renewables: Costs and markets to 2030, viewed 9.03.2019 https://www.irena.org/publications/2017/Oct/Electricity-storage-and-renewables-costs-and-markets

Energy Strategy 2020, viewed 9.03.2019 https://ec.europa.eu/energy/en/topics/energy-strategy-and-energy-union/2020-energy-strategy
Erdinc O. 2017. *Optimization in Renewable Energy Systems: Recent Perspectives*. Istanbul: Butterworth-Heinemann
Ermolenko G.V. 2017. *Analiz deyatelnosti veduschih neftegazovyih kompaniy v oblasti vozobnovlyaemoy energetiki* [Analysis of the activities of the world's leading oil and gas companies in the renewables field], viewed 18.03.2019 https://energy.hse.ru/data/2017/10/25/1157689079/Анализ%20деятельности%20ведущих%20нефтегазовых.ласти%20возобновляемой%20энергетики.pdf
Ermolenko G.V., Tolmacheva I.S., Ryapyn I. Y., Fetisova Y.A., Matsura A.A. & Reutova A.B. 2016. *Spravochnik po vozobnovlyaemoy energetike Evrosoyuza* [The green energy in the EU], viewed 14.03.2019 https://energy.hse.ru/data/2017/10/04/1159483435/Справочник%20ВИЭ%20в%20ЕС.pdf
Hohlov A. 2017. *Vozobnovlyaemyie istochniki energii: novaya revolyutsiya ili ocherednoy puzyir?*[Renewables: a new revolution or another bubble?], viewed 10.03.2019 https://www.forbes.ru/biznes/343591-vozobnovlyaemye-istochniki-energii-novaya-revolyuciya-ili-ocherednoy-puzyr
Hohlov A., Melnikov Y., Veselov F., Kholkin D. & Datsko K. 2018. *Raspredelennaya energetika v Rossii: potentsial razvitiya* [Distributed energy in Russia: development potential], viewed 10.03.2019 https://energy.skolkovo.ru/downloads/documents/SEneC/Research/SKOLKOVO_EneC_DER-3.0_2018.02.01.pdf
Itogi reformyi elektroenergetiki v Rossiyskoy Federatsii (Po rezultatam espertnogo oprosa. 2014. [Results of the reform of the Russian electric power industry (According to the results of the expert survey], viewd 11. 03.2019 http://www.csr-nw.ru/files/publications/reform_ee_2014.pdf
«Gazprom» Annual report 2017 [Godovoy otchet PAO «Gazprom» za 2017 god], viewed 11.03.2019 http://www.gazprom.ru/f/posts/57/287721/gazprom_annual_report_2017_rus.pdf
Grigoriev L. 2017a. *Mikrogeneratsiya na osnove VIE* [Renewables microgeneration], viewed 10.03.2019 http://ac.gov.ru/files/publication/a/13570.pdf
Valovaya dobavlennaya stoimost po otraslyam ekonomiki, n.d. [Gross value added by industry], viewed 18.03.2019 http://www.gks.ru/free_doc/new_site/vvp/vvp-god/tab10-3.xls
Grigoriev L. 2017b. *Tsifrovyie tehnologii v setevom komplekse* [The digital technologies in the electricity networks], viewed 15.03.2019: http://ac.gov.ru/files/publication/a/14737.pdf
Grushevenko E. & Hohlov A. 2017. *Rossiya na globalnom ryinke vie* [Russia on the global renewables market], viewed 10.03.2019 http://russiancouncil.ru/analytics-and-comments/analytics/rossiya-na-glo balnom-rynke-vie/
«Lukoil» Annual report 2017 [Godovoy otchet PAO «Lukoil» za 2017 god], viewed 11.03.2019 http://www.lukoil.ru/FileSystem/9/289721.pdf
Mashnina V.I. The results of the electric power industry reform in The Russian Federation (Accordind to the results of the expert survey) [Itogi reformy electroenergetili v Rossiiskoi Federacii (Po resultatam expertnogo oprosa], viewed 15.02.2019 http://www.csr-nw.ru/files/publications/reform_ ee_2014.pdf
«NOVATEK» Annual report 2017 [Godovoy otchet PAO «NOVATEK» za 2017 god], viewed 11.03.2019 http://www.novatek.ru/common/tool/stat.php?doc=/common/upload/doc/NOVATEK_AR_2017_RUS.pdf
Obzor elektroenergeticheskoy otrasli Rossii. 2017. [The Russian Electricity Power Industry overview], viewed 14.03.2019 https://www.ey.com/Publication/vwLUAssets/EY-power-market-russia-2018/$File/EY-power-market-russia-2018.pdf
«Rosneft» Annual report 2017 [Godovoy otchet PAO «Rosneft» za 2017 god], viewed 11.03.2019 https://www.rosneft.ru/upload/site1/document_file/a_report_2017.pdf
Scholten D. 2018. *The Geopolitics of Renewables*. Delft: Springer.
Singh A. 2018. *Talking Renewables: A Renewable Energy Primer for Everyone*. San Rafael: Morgan & Claypool Publishers
Sorensen B. 2018. *Renewable Energy: Four Volume Set*. Routledge: Earthscan

Stepen iznosa osnovnyih fondov na konets goda po vidam ekonomicheskoy deyatelnosti, n.d. [The level of the depreciation of fixed assets at the end of the year by the type of economic activity], viewed 12.02.2019 http://www.gks.ru/free_doc/new_site/business/osnfond/STIZN_ved.htm
«Surgutneftegas» Annual report 2017 [Godovoy otchet PAO «Surgutneftegaz» za 2017 god], viewed 11.03.2019 https://www.surgutneftegas.ru/download.php?id=5333
«Tathneft» Annual report 2017 [Godovoy otchet PAO «Tatneft» za 2017 god], viewed 11.03.2019 http://www.tatneft.ru/storage/block_editor/files/cbcecc546778f2e5c6bcf0f60bfd692aa13d4afe.pdf
World Energy Investment. 2018 [Mirovie investicii v energetiku], viewed 9.03.2019 https://www.iea.org/wei2018/

Youth Technical Sessions Proceedings – Litvinenko (ed)
© 2019 World Petroleum Council, London, UK, ISBN 978-0-367-34668-3

Intellectualization tools that increase the efficiency of oil and gas condensate field development

A.S. Novikova & O.V. Eremenko
National University of Oil and Gas «Gubkin University», Orenburg, Russian Federation

ABSTRACT: Modern problems of exploitation of oil and gas condensate fields necessitate technological breakthrough and reorientation of the economy on the innovation-oriented path of development. This is especially important in the conditions of growing geopolitical loneliness of Russia; local military conflicts on the territories of oil and gas producing countries; strengthening of sectoral sanctions; lack of access to investment resources for the implementation of major projects, climate change. The most effective way to solve these problems are the tools of intellectualization at all levels of the oil and gas production chain. Despite the fact that the issues of technical efficiency of intellectualization have already been well studied, in this article we studied the impact of digitalization on improving the efficiency of operations and reducing costs by rationalizing energy management, reliable means of communication in the field of repair of fixed assets, improving the efficiency and representativeness of the flow of information in various segments of the regional (Orenburg) oil and gas complex. The authors, taking into account fifty years of experience in the operation of the Orenburg field, identified factors affecting the effectiveness of production at the stage of falling production; developed the necessary tools of intellectualization and evaluated the effectiveness of their implementation.

1 INTRODUCTION

Orenburg field is among the ten largest in Russia in terms of reserves. The composition of the extracted raw materials is unique. In addition to methane and liquid hydrocarbons, it contains sulfur, helium, mercaptans, which made it possible to create a full-fledged gas chemical complex that combines the extraction, processing, transportation and distribution of hydrocarbon raw materials (Eremenko 2017). In the first years of development of the field not only in the USSR, but also worldwide there was no experience of production of hydrogen sulfide-containing gas that became the reason of lack of technical solutions of problems, to the main of which it is necessary to carry: excessive aggressiveness of the extracted raw materials causing high rates of destruction of the equipment, wear of pipelines' walls; inefficient system of inhibitory protection; the complexity of the geological structure of the field, assuming heterogeneity development on the horizons; lack of technologies of tail gas post-processing and utilization of associated low pressured gases, etc. (Eremenko 2012). The geological model (field development Project) formed at that time could not take into account all these features. In the future, systematic violations of technologies exacerbated the situation and led to significant deviations from the Project, decline of extraction, increased costs and reduced profitability (Novikova 2017). Therefore, the urgency of solving these problems is obvious. At the same time, the means of intellectualization are considered by us as the most affordable tools for increasing the economic-technological potential of oil and gas production. It is digital technologies that increase the speed and quality of decisions, create the possibility of processing and interpretation of huge flows of information, allow to implement scenario modeling tools, remote control of operation of production facilities in the «on-line» mode, significantly reducing costs and increasing the degree of safety of operations. Therefore, in this paper an attempt is made to form means of intellectualization based on the analysis of data on the functioning of the objects in Orenburg complex. To this end, the production and economic indicators of the field development, energy consumption and technical condition of fixed assets were studied, which allowed to establish priority areas and methods of implementation of digital technologies. Some of them can be considered universal and possible to use by other subsoil users.

2 LITERATURE REVIEW

Many works of foreign authors (Al-Dhubaib 2011, Begg 2001, Howard 1998, Walls 1996, etc.) are devoted to the concept, methods and tools of digitalization. This problem has long been studied in relation to the development of oil and gas fields (Alberto 2008, Coopersmith. 2007, Cramer 2012, Jamal 2013, Khan 2015, Karakaya 2012, Stefano 2014, etc.). However, for the Russian science and practice the digitalization and intellectualization are considered new fields of research, therefore, publication on those issues minor (Adishev 2010, Berezina 2014, Demarchuk 2014, Eremenko 2012, Kaligin 2016, Milovidov 2008, Novikova 2017, 2018 etc.). At the same time, they fragmentary reflect certain aspects of digitalization (in particular, in subsoil use). In this regard, there is a need for a deeper understanding of the methodological and practical aspects of the introduction of digital technologies, the creation of digital culture and platforms, taking into account the characteristics of the sectors of oil and gas companies and the specifics of the fields. Therefore, the article reflects the results of factor analysis of the activity and efficiency of intellectualization in various parts of oil and gas production; the effectiveness of the introduction of digital technologies in the regional oil and gas complex, including the de-velopment of oil and gas condensate field. At the end, the means of intellectualization, the most relevant to modern problems in the development of the Orenburg complex, are presented.

3 METHODOLOGY

The methodological and theoretical basis of the article is the work of foreign and regional researchers in the field of innovation management and digital technologies; general scientific methods of collection, analysis, interpretation and evaluation of information. In particular, the methods of comparative analysis, classification and ranking, expert assessments and traditional methods of evaluating the effectiveness of innovative solutions are applied. The information base of the study was made up of open data of PSC «Gazprom» (2014-2018), Russian statistical Agency, materials of industry journals. In addition, the paper uses data on the state of development of the field; physical and cost consumption of electricity by objects of the Orenburg oil and gas complex, including the results of daily hourly consumption in January-December 2018, obtained by continuous observation. On the basis of these data the factor analysis of deviations of actual indicators in comparison with the project of development of the Orenburg oil and gas condensate field is carried out.

4 ANALYSIS

Today, the Orenburg field is characterized by falling production as a result of significant depletion of reserves, systematic violation of the methods of production and preparation of hydrocarbons, the inability to solve many technical problems, which in turn leads to significant deviations from the development Project (Novikova 2018). So, the level of water cut of the collector in 2018 by 28.4% above the design value; the fraction of dry wells in the operating stock 31.3% below; the rate of humidity of the raw material is 99.7% instead of 60.1% on the Project. As for liquid hydrocarbons, the content of free and emulsion water, chloride salts, resins, asphaltenes has been significantly increased in their composition. The rate of reduction of reservoir pressure from the design value was 74.5%. Such deviations from the main parameters of the Project (especially in terms of reducing the volume of construction of new wells and geological-technical measures) caused an increase in the number of technological complications; acceleration of the formation of salt deposits and falling out of retrograde condensate; reduction of oil and gas recovery rates, the growth of negative environmental impact, etc. (Figure 1).

As can be seen from the Figure 1, the greatest negative impact of the deterioration of the technological parameters of the field development was manifested in an increase in the cost and the occurrence of a lose from the production and processing of raw materials of Orenburg oil and gas condensate field. The most significant impact on the increase in the cost of gas production was the factor of increasing electricity tariffs (by 7.3% with a physical increase in consumption by 4,5%) (Tab. 1).

Due to the fact that the reduction in hydrocarbon production was faster than physical consumption electricity energy, the specific consumption per 1000 m^3 of gas increased by 6.8%. Electricity consumption in 2018 increased by 10 million kWh, that was caused by the need of growth of compression ratio of booster compressor stations at decrease in reservoir pressure in productive deposits of the field. For this reason the number of hours of operation of electric gas pumping units and the additional inclusion of backup EGPU has increased. This is reflected in the increase in the share of EGPU in the total energy consumption (by an average of 1.8% per year). It should be noted that the increase in the cost of fuel and energy resources is due not only to the increase in tariffs in accordance with the resolutions of the Government of the Russian Federation, but also to deviations of the actual consumption of resources from

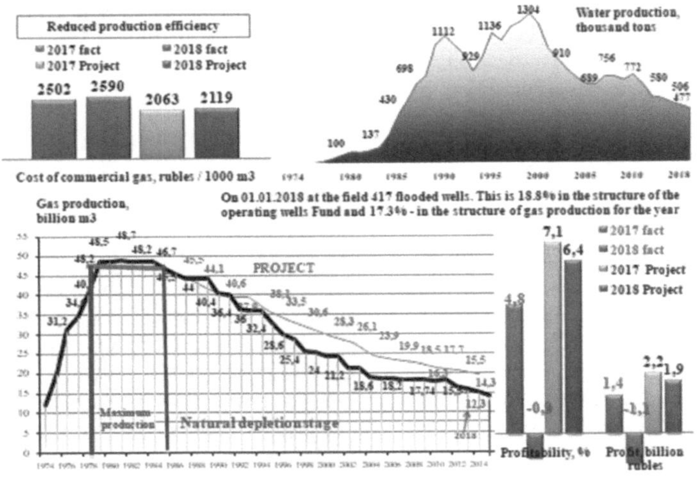

Figure 1. Performance indicators of the Orenburg oil and gas condensate field (OOGCF) development and deviation from the design level of gas production

Table 1. Electricity consumption in the segment of Orenburg oil and gas condensate field

Indicators	2014	2015	2016	2017	2018
Electricity consumption in hydrocarbon production, million kWh., total, including:	205	235	248	270	274
- electric gas pumping units	160	189	203	224	234
Share in consumption structure, %	78,1	80,4	81,9	82,9	85,4
Rate of change in total consumption,% («+»- growth, «-» - decrease)	-	+2,3	+1,5	+1,0	+2,5
Weighted average price, rub./kWh.	2,775	2,902	2,871	2,907	3,119
Rate of change in weighted average price,% («+» - growth, « - » - decrease)	-	+4,6	-1,1	+1,3	+7,3

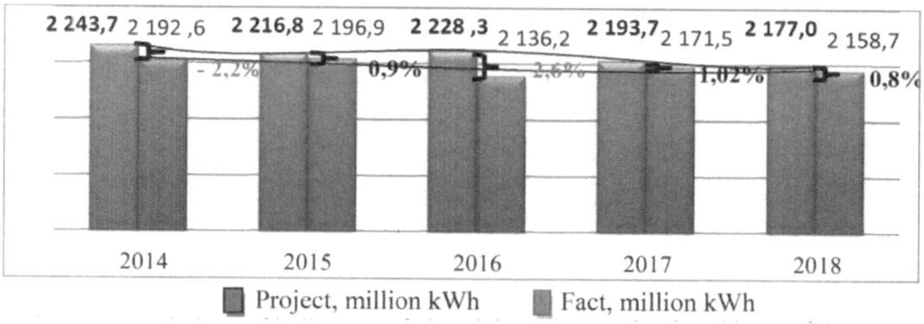

Figure 2. Deviations of indicators of electricity consumption by objects of the OOGCF

those stated in contracts with energy supply organizations. They arise in the absence of data on changes in the operating modes of equipment, tools for reliable planning of the need for fuel and energy resources for the future. Therefore, the priority need of the enterprise is digital innovation aimed on optimizing the planning of daily power consumption, taking into account the availability and possibility of processing a large array of data on the operation of field and gas processing facilities [12].

For these purposes, the most efficient tool is a digital platform for hourly planning of energy consumption by the actual operating time of power consuming equipment and its management in real time. The platform will allow oil and gas company to purchase electricity at the wholesale market price without the use of a progressive scale of increasing coefficients (if the consumption deviates by more than 2% (+,-) of the application, energy is paid at a double tariff). Although, as shown in figure 2, over the past five years, such deviations were allowed twice – in 2014 and 2016, nevertheless, the improvement of the energy consumption planning system is relevant. Only in the designated periods, the increase in electricity costs amounted to 141886,131 thousand rubles and 262512,756 thousand rubles, respectively.

The energy savings indicated in the schedule is not only due to improper planning of needs, but also energy-saving measures that are carried out in the units of the company. So, the installation of intelligent sensors in lighting network of gas production facilities of OOGCF resulted in savings over five years 3392,95 thousand kWh. Futhermore, within the framework of the digital platform, the oil and gas enterprise is will be presented as a single object on the basis of the General schedule of energy consumption of its structural units. Last circumstance is very important, since the contract provides for the supply of electricity to the enterprise as a whole, while energy consumption schedules are planned separately by sector. This fact makes it impossible to quickly recalculate the volume of supply when changing the mode of operation of the equipment in any segment of the company, and then include an increase in the cost of buying electricity in the cost of products (services) of this division. Payment at a double tariff is made by the company, which complicates the calculation of performance indicators of individual business processes. As example, the graphs of hourly electricity consumption of individual units on Orenburg oil and gas complex – gas processing (1) and helium plants (2) (Figure 3).

Deviations on the gas processing plant are less than 2%. At the helium plant: in the first shift, the need was exceeded due to the unscheduled commissioning of additional technological

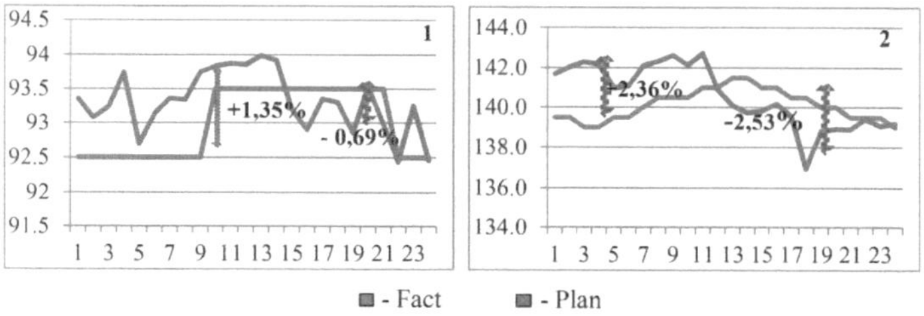

Figure 3. Hourly daily schedule of electricity consumption at plants, 28.12.2018

equipment, in the third shift - due to the unscheduled output in the repair of the technological installation. And such cases in the work of plants arise often, but, according to the existing contract with the energy supply organization, currently, there is no possibility to write off the growing costs on a specific facility of production. In such circumstances, the digital hourly planning platform using a significant number of intelligent devices will reduce the cost of electricity consumption, as in real time will provide an account of the actual time of operation of energy-intensive equipment and will be able to adjust the course of the production process.

Ways to solve such problems are extremely relevant, especially for Mature oil and gas fields. The fact is that the Orenburg complex is no exception to Russian practice. As the analysis of Russian indicators shows, the digitalization of oil and gas companies is on average higher than in other sectors of the economy (*Indicators of the digital economy 2018*) (Tab.2).

As shown in Table 2, the leader in intellectualization is oil and gas processing, and the outsider is the gas distribution segment. In general, there are five factors that affect to the activity of the oil and gas business in the implementation of digital technologies: x_1 - the share of costs for digitalization in the total volume of shipped products, %.; x_2 - the share of costs for the formation of the digital platform in the total volume of shipped products,%.; x_3 - the share of the cost of intellectualization in the total investment in fixed capital, %; x_4 - share of costs for training of digital personnel in the salary of employees, %; x_5-share of oil and gas company divisions that have implemented digital technologies in the total number of divisions, % (Tab.3).

The results of the correlation analysis show that the factor x_2 is the determining. It's greatest influence is manifested in oil and gas chemistry, and the least – in oil and gas processing. There is a negative correlation between the level of digital activity and factor x_5, which confirms the fact that the important thing is not the number of units, and the amount of their funding.

The second unsolved problem at OOGCF facilities is significant depreciation of fixed assets, which increases the risk of accidents. This makes it necessary to prevent and improve the

Table 2. Digital tools used in the activities of Russian companies (the share of organizations using devices in the total number of organizations in the economy, %), 01.01.2017

Digital technology	Average in Russia	Oil and gas production	Oil and gas processing	Gas distribution
Broadband Internet	80,5	88,8	91,3	80,1
«Cloud» services	20,5	17,7	23,3	16,2
RFID - technology	5,8	10,3	8,7	4,4
ERP-system	17,3	24,5	22,1	13,3
EDM-system	62,3	65,5	68,9	69,6
Devices for calculations with large data sets	55,4	61,9	70,0	59,4
Automated process control systems	20,9	40,1	40,5	25,7
Digital design tools	17,1	32,3	30,9	24,6
Training program	15,3	29,6	16,5	19,9
Digital devices for solving organizational, technical and economic problems	54,0	64,0	64,8	56,5

Table 3. Value of pair correlation coefficients

	Result parameter Y			
Variables	Oil and gas company as a whole	Oil and gas production	Oil and gas processing	Oil and gas chemistry
x_1	0,263	0,635	-0,472	-0,179
x_2	0,602	0,559	0,277	0,684
x_3	0,338	0,789	-0,349	-0,118
x_4	0,141	0,777	-0,181	0,442
x_5	-0,121	-0,333	-0,528	-0,202

Table 4. Composition and age structure of pipelines in Orenburg region

Type of pipeline	Length, km	Service life, years									
		>1		1-10		11-20		21-33		<33	
		km	%	km	%	km	%	km	%	km	%
Pipelines of clean gas	1301	-	-	36	2	75	6	410	32	780	60
Product pipelines for condensate, wide fraction of light hydrocarbons, ethane, methanol	2590	-	-	1300	50	274	11	645	25	371	14
Oil pipeline	69	-	-	-	-	-	-	31	45	38	55
Total	3960	-	-	1336	34	349	9	1086	27	1189	30

responsiveness to emergency situations, since most of the facilities are classified as particularly dangerous. In particular, this applies to the system of pipelines operated by the enterprise. In 2018 their length on the territory of the Orenburg region amounted to 3960,147 km (Tab. 4).

Taking into account the average service life of 22 years, it can be noted that only 43% of pipelines form safe category of facilities. The remaining 30% must be written off and replaced, 27% are on the verge of liquidation and their use is high risk. It is impossible to solve the situation in the absence of sufficient sources of financing for the construction of new transport networks, so it is necessary to create a digital platform for the organization of measures to reduce the risk of accidents and stops. Today, these problems are solved by linear workers' during the inspection of tracks. However, it is associated with a number of problems: high laboriousness of the inspection route; lack of uniformity in the name of route maps and tablets; the complexity of measurement laboriousness when updating the routes; difficult quality control of technical inspections; lack of structuring and systematization of data accumulated during the works on technical operation of pipelines; inability to quickly identify areas of disconnect in the localization of accidents.

5 MEANS OF INTELLECTUALIZATION FOR THE SOLUTION OF THE REVEALED PROBLEMS OF THE ORENBURG OIL AND GAS COMPLEX

The problems that the digital energy management platform should additionally solve include: elimination of differences between the functional goals (often conflicting) of the company's divisions and the peculiarities of their activities; lack of predictive tools in the practice of energy management, ways to optimally combine the interests of business processes and the efficiency of coordination of their energy consumption. Development of the platform includes the following steps: 1) development of the program and its blocks in General for the Orenburg complex, as well as in the context of its structural units; 2) development and installation of technical devices (probes, control sensors, intelligent electronic devices, routers, remote controls, etc.) to detect and record changes in the operation of energy-intensive equipment, which will send information to the computer processors; 3) data transfer to all users of the digital platform; 4) data integration and processing to obtain representative information about the energy consumption process in the context of key business processes; 5) the use of intelligent technical means (cloud technologies, servers, information protocols, streaming Analytics systems, cognitive technologies, machine learning, etc.) to automate the process of planning the power supply of objects, adjusting the operation of equipment in real time, predicting optimal energy consumption in the enterprise. The formation of a full-fledged platform involves, among other things, the use of 3D models that provide rapid prototyping of energy consumption; digital copying of assets of oil and gas company units; the creation of digital twins and a single digital chain (Adishev 2010). According to preliminary estimates, the use of the digital platform only for the first year (taking into account the deviations of previous years) should save 202.2 million rubles for electricity.

The created tool of intellectualization of works on maintenance of pipelines implies development of system of internal accounting of networks of gas distribution and constructions on

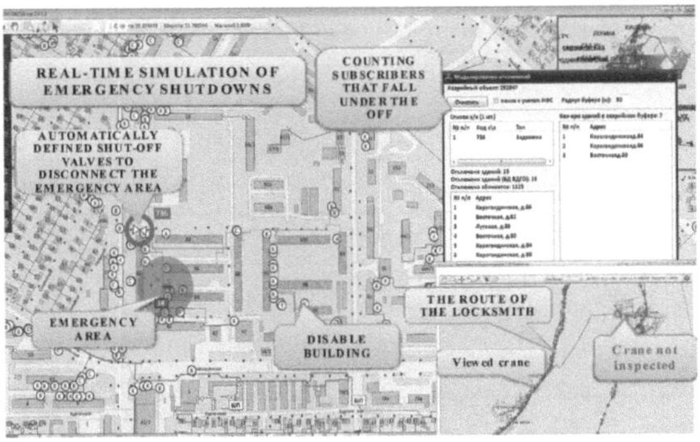

Figure 4. A reflection of the situations of the inspection and shutdown of the pipeline on the monitor of the manager

them. The basis of the system is an electronic archive of executive and technical documentation, which combines the coordinates of pipelines, their characteristics, vector substrate settlements, public map of the «Russian registry», telemetry and telemechanics data. According to these data, electronic passport of the pipeline is forming, in which it is necessary to specify its method of laying (ground surface, subterranean), length, material, date of construction, purpose and location. After entering the pipelines into the electronic map, routes and schedules of technical inspection, its frequency and complexity are built. Each locksmith is given a tracker to reflect the indicators obtained during the inspection, and the ability to monitor the implementation of the necessary operations (Milovidov 2008). Sections of the pipeline are equipped with sensors reflecting the technical condition (wear of the wall, the possibility of an accident, the presence of stress-strain state, etc.). They continuously monitor complications in real time (Figure 4).

When using the proposed tools of digitalization, the efficiency of the Orenburg oil and gas condencate field will increase due to the growth of productivity and safety of business processes, as well as reducing operating costs, creating stability with changes in internal factors of production. It should be borne in mind that to solve these problems will require a huge flow of information to capture and process which can only intellectual technical devices, analytical systems for collecting and processing large amounts of data. In other words, the introduction of digital technologies should be accompanied by the development of information maturity of the oil and gas company.

6 CONCLUSION

Spreading the practice of technological innovations, digitalization, activisation of efforts to introduce technologies without human participation reveal limitless opportunities for the Russian oil and gas business, stimulate the intellectualization of the assets used. In this regard, the role of emerging fundamental systems providing a high level of efficiency and security of key business processes and digital culture is increasing. This is especially relevant for fields at the stage of falling production, such as the Orenburg oil and gas condensate field. After studying the current problems of its operation and establishing prospects for development, two platforms were developed as priority tools of the digital economy. The first is a system of energy consumption management based on hourly planning of needs and adjustment of equipment operation in real time. The second platform solves the problem of organizing measures to improve the efficiency and reliability of control over the technical condition of pipelines, prevention and elimination of accidents. As a result of the application of these technologies, among other things, the operating costs of the oil and gas company will be reduced. In the first case - by optimizing the cost of buying electricity; in the second – by reducing the cost of repair of pipelines and elimination of the consequences of accidents. Already today the company has started digitalization, but it is only a small part of information technologies. Therefore, it is safe to say that the oil and gas business is on the threshold of a grandiose intellectualization of assets and business processes that involve improving the efficiency of subsoil use.

REFERENCES

Al-Dhubaib, T.A. 2011. Intelligent Fields: Industry's Frontier and Opportunities. Society of Petroleum Engineers - SPE Middle East Oil and Gas Show Conference, Manama, Bahrain: P.12

Adishev, E. 2010. The Digital field. Trends in the CIS and the world: a report of 39th conference «Modern it in the oil and gas industry».

Alberto, F. 2008. Metrics for economic evaluation of innovative technologies in the oil industry. Society of Petroleum Engineers - SPE Saudi Aramco:111996.

Berezina, A.A. 2015. The feasibility of the transition to the smart field concept in the modern problems of the oil and gas industry. *Journal of problems of economics and management of oil and gas complex*. №2:42-44.

Demarchuk, V.V. 2014. Perspectives for the implementation of «intelligent» oil and gas. - Conference of young scientist, Moscow 19: 284-289.

Eremenko, O.V. 2012. Efficiency of development of resource potential of oil and gas production enterprise by innovative method of intelligent well systems. *Journal: Intelligence. Innovations. Investment*. № 2: 24-27.

Eremenko, O.V. 2017. Innovations aimed at increasing the life cycle and efficiency of development of oil and gas condensate fields in the stage of falling production. *Scientific and practical journal «Economics and management: problems and solutions»*. 10(vol.1): 65-71.

Kaligin, A. 2016. Digital mining. *Business magazine about it, communications and broadcasting in Russia and the world «STANDARD»*. №7/8 (162/163): 28-31.

Milovidov, K.N. 2008. Innovative technologies in exploration and production of oil: governance and performance: study guide: 272. Moscow: MAKS Press.

Begg, S.H. 2002. The value of flexibility in managing uncertainty in oil and gas investments. Society of Petroleum Engineers – SPE Annual Technical Conference, San Antonio 77586:10.

Coopersmith, E.M. 2007. A practical approach to evaluating the value of information and real option decisions in the upstream petroleum industry. - SPE paper 77582.

Cramer, R. 2012. A measure of the digital oil field status - is it the end of the beginning? Society of Petroleum Engineers - Intelligent Energy International held in Utrecht. Netherlands: 149957.

Howard, R.A. 1998. Decision Analysis: Practice and Promise. *Journal of Management Science*, Vol. 34 (6): 679-695.

Jamal, M. 2013. Effective well management in Sabriyah intelligent digital Oilfield. Society of Petroleum Engineers - SPE Kuwait Oil and Gas Show and Conference held in Mishref, Kuwait 167273: 12.

Khan, F. 2015. Back to the Future: Applying Lessons learnt to a new Generation of Digital Oil Field Applications. Society of Petroleum Engineers:173443.

Karakaya, Erkan. 2012. Perceived value and popularity of decision analysis tools in upstream asset valuations. Society of Petroleum Engineers: 162675.

Novikova, A.S. & Eremenko, O.V. 2017. Innovative methods of well processing as a basis for extending the profitable operation of fields with falling production. – «Fundamental and applied research: problems and results». materials of the I International scientific-practical conference. In 2 volumes (vol.1). Grozny: 619-623.

Novikova, A.S. 2018. Innovations that ensure the growth of profitability of the Orenburg oil and gas condensate field. X All-Russian scientific-practical conference with international participation "Young Russia". Kemerovo http://science.kuzstu.ru/wp-content/Events/Conference/RM/2018/RM18/pages/Articles/64105-.pdf

Russell, Louis. 2016. Petroleum enterprise intelligence in the digital oil field. Society of Petroleum Engineers: 177812.

Saputelli, L.A. 2013. Best practices and lessons learned after 10 years of digital oilfield (DOF) implementations Society of Petroleum Engineers:- SPE Kuwait oil and gas show and conference, Kuwait City 167269:1.

Steve, R.2012. Choices, changes, and challenges for the digital oil field. *JPT*, 5: 20-22.

Stefano, M. 2014. Digitizing oil and gas production.

Walls, M.R.1996. Developing an exploration decision support system (EDSS): A strategy for combining information and analytics. *Nonrenewable Resources*, v.5 (3): 181-192.

Youth Technical Sessions Proceedings – Litvinenko (ed)
© 2019 World Petroleum Council, London, UK, ISBN 978-0-367-34668-3

Abandonment fund: Financing schemes, default security mechanisms and decommissioning of oil and gas wells

I. Khalidov
Gubkin Russian State University of Oil and Gas (National Research University), Moscow, Russian Federation

ABSTRACT: Currently, no standard and regulation govern the establishment of abandonment fund and oil and gas well decommissioning in Russia. That leads to uncertainty in taxation and correct accounting for these expenses in the financial statements becomes an issue to be addressed. Aim of the article – evaluation of the financing schemes and default security mechanisms within the framework of the establishment of abandonment fund, as well as the synthesis and identification of key issues in oil and gas well decommissioning.

1 INTRODUCTION

According to estimates of Rystad Energy (2019) and Wood Mackenzie (2018) in 2018 a record indicator was reached on decommissioning and restoration obligations (D&R), the costs of which for the global oil and gas industry amounted to 11.7 billion US dollars. In 2021 around 32-36 billion US dollars will be spent on these commitments worldwide. However, laws and regulations on D&R in many countries remain incomplete or untested in practice.

In 2013-2017 only on the continental shelf of the UK the exploitation of 16% of 472 fields was stopped. The main reason was the fall in oil prices and the breakeven point offset.

According to Wood Mackenzie (2018) in the next 10 years on the continental shelf of the UK the D&R costs will be about 30 billion US dollars considering that almost all oil and gas production is in the offshore zone.

Approximately 700 exploited fields around the world including only about 150 in the UK starting in 2020 can stop production in the next 5 years depending on oil prices. D&R is the most mature in the US Gulf of Mexico, where on average more than 100 platforms per year have been taken out of operation since 1985.

About 9000 wells worldwide are located in fields that are currently struggling to stay profitable at oil prices of 60 US dollars per barrel. This is a relatively high breakeven price, which became the Achilles' heel for many fields in 2018. In 2013-2014, when oil prices were high a very few companies implemented plans for D&R.

In 2013-2017 the most active market for D&R was Europe mainly due to the UK, which accounted for more than 50% of all global expenditures according to Rystad Energy (2019), Oil & Gas UK's (2018) and Wood Mackenzie (2016). However, D&R activities will grow significantly in other parts of the world, especially in the near term in Asia, Latin America and North America.

2 DECOMMISSIONING OPERATIONS ANALYSIS

D&R operations are comparable in complexity to new field development projects. They include the following main stages (BCG, 2017) (Figure 1):

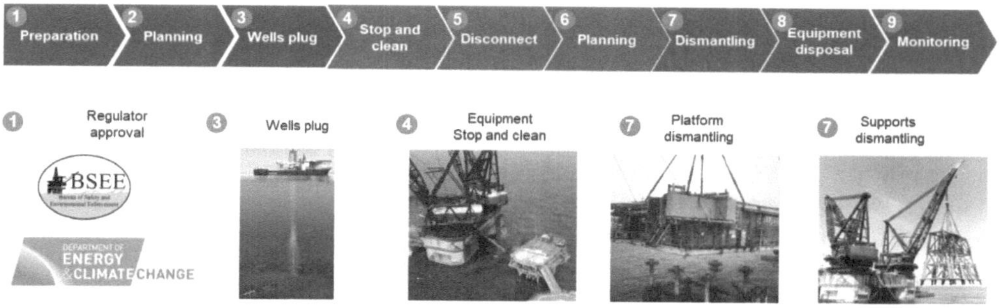

Figure 1. Decommissioning main stages

At the same time the risks are high, and the sector is less attractive for contractors and operators. An example of field development considering the D&R costs is shown in Figure 2:

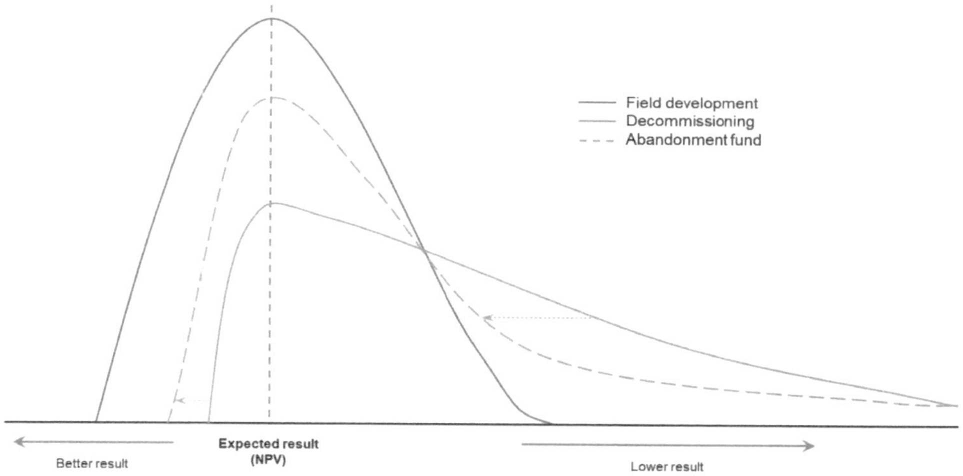

Figure 2. Field development example including abandonment fund

This problem is complicated by the fact that there are no uniform rules and standards in the world governing the procedure for D&R, the establishment of abandonment fund, financing schemes and default security mechanisms.

Petroleum Resource Management System (PRMS) of the Society of Petroleum Engineers (SPE) offers only some guidelines for D&R and establishment of abandonment fund.

To address these issues, it is necessary to identify key steps in state regulation, standardization and innovation that promote the development of cooperation and sharing of best practices at the international level and systematic improvement in the quality of implementation through involvement of all interested parties.

Oil and gas companies should be more transparent and ready to cooperate. They should to openly share their forecast of timing and abandonment expenditures (Abex), along with the major influences on those estimates for the run up to cessation of production (COP), as well as asset-specific information.

Without the established initial parameters, the industry cannot set realistic market expectations decommissioning performance.

3 ABANDONMENT FUND FINANCING SCHEMES

Both government agencies and oil and gas companies should plan the size of abandonment fund costs considering D&R, since the volume of these works is substantial, and previous forecasts of liquidation costs may turn out to be unreliable and understated, which as a rule happens in practice. There are various financing schemes for liquidation expenses, the most common of which are (Khalidov I.A., 2018):

1. Write-down according to the unit-of-production method, where estimated future Abex is recovered gradually by operators via a predetermined portion of production and deposited in a fund, either throughout life of field or form a trigger point onwards, when 50-80% of reserves have been produced. Governments are wary of the trigger mechanism as operators can provide misleading estimates. The system allows operators to recover initial costs faster and to delay the negative impact on cash flow.
2. Depreciation charges over field life, where an estimate of Abex is deposited into a fund and 'recovered' pro-rata over some years and being classified as operational expenditures.
3. Loss carry-back, such as the system variant in the UK, where Abex is carried back for the maximum period possible and offset against taxable income (via a tax rebate) for the years during which D&R occurs or for the years immediately preceding D&R. The system is claimed to have the least negative impact on cash flow throughout the project's lifetime.
4. State grants, such as the refund system proposed in Norway, which operates outside the country's oil taxation system, and apportions Abex between the government and operators using a ratio that considers the amount of tax paid by the latter. The Norwegian government currently provides a direct grant of between 70-80% towards Abex. In 2005, Norway introduced new legislation based on "realization principle" to govern the timing of taxation, where a benefit is taken as income in the year it accrues to the taxpayer, and costs are deducted in the year they arise. Hence, where D&R is concerned, the costs incurred for wells plugged and abandoned and installation removal are deductible when the work is carried out. There is a risk that an operator may not have taxable income to shelter such Abex, particularly if it does not have other producing assets. However, Abex will be allowed to be carried forward, and the refund system upon (COP) will ensure that the State picks up the tax value of the losses, for as long the refund system is in place.

Regardless of which scheme is chosen, there is always the risk that D&R may lead to a default by the oil producing company. Therefore, there are mechanisms that can be used to protect against such potential threats.

4 SECURITY DEFAULT MECHANISMS

Protective mechanisms are important for government agencies seeking to ensure the availability of funding for Abex and to protect any of the parties from a possible default. There are several types of default security mechanisms, the most common of which are (Khalidov I.A., 2018):

1. Letters of credit and facility arrangements with third-party financiers: usually issued by a bank, but which still has a risk that funds may not be provided at a future date. Host governments usually insist the guarantor has a minimum credit rating that is reviewed periodically and, if the guarantor falls below a stipulated rating, the operator must provide alternative security or be fined. Third party guarantees, such as letters of credit, are often used by smaller companies, who prefer to avoid upfront payment schemes for cash flow reasons. However, the bank will need some form of collateral, and the extent of this could mean that a single asset company may be unable to get cover.
2. Bond issuance by a bank or institution having a certain minimum credit rating: where operators insure against the consequences of unforeseen expenditure and earlier than anticipated (COP) dates, but premiums are generally high.

3. participants contribution into designated trust agreements: of which a variant is used in the UK, are regarded as the safest security mechanism, requiring the payment of cash into a fund with the requisite amount being established from a ratio of present production to anticipated future production. Problems may arise when trying t reconcile the operator's desire for high returns with the host government's need for security.

5 RUSSIAN PRACTICE

The contributions to the state-owned abandonment fund in Russia were initially set at 10% of the recurrent payments for the minerals extraction tax and 10% of deductions for the mineral reserves replacement tax (MMRT). After the MMRT deductions were abolished in 2002 and the new tax code was introduced in 2007, this source of financing was lost (Mazurina, E.V. 2010).

Currently, the abandonment fund is envisaged only for production fields developed under the terms of Production Sharing Agreements (Russian Federation Law, 1996). The existing legislation does not contain a provision on the establishment and employment of abandonment funds for the oil and gas extraction companies, although in 2015 the question was raised about the preparation of mechanisms for the formation of abandonment fund (*The creation of liquidation funds of oil companies postponed until 2019, 2015*). Abex, as a rule, are shown only at the time of planned abandonment, i.e. when wells are decommissioned and, later, when the entire field is abandoned. These costs are counted as expenses for the liquidation of assets, which are taken out of operation, and, in accordance with paragraph 8 of Art. 265 of the Tax Code, refer to non-sale expenses that reduce the corporate income tax base during the liquidation period.

The requirement to abandon wells and to liquidate oilfield facilities, including production and injection wells, as well as the requirement to reclaim the land after the field development is completed is established in the RF Law "About subsoil", which is the main legislative act regulating the relationship between the state and a subsoil user (Kryukov, V.A. & Anashkin, O.S. 2012). According to the procedure adopted in this Law (Section II; Art. 22 "The basic rights and obligations of a subsoil user"), a subsoil user is obliged to ensure,

> *"... the abandonment, in the duly established manner, of mine workings and boreholes that are not subject to further exploitation" and "the re-cultivation of lands and other natural objects disturbed by the subsoil operations into a state suitable for their further use".*

Currently, the Russian practice offers a fairly simple solution to the problem of how the basic production assets (BPA) can be liquidated that is to assign the Abex to the expenses incurred at other facilities as well as at other, more profitable, fields (Kryukov, V.A. & Anashkin, O.S. 2012). Regarding the establishment of an abandonment fund, in the Russian practice it is appropriate to start with a "hybrid form", which implies the establishment of this fund through contributions from depreciation accounts and tax loss carry-back.

Field abandonment implies the development of a new field that is required to compensate for the loss of hydrocarbon reserves, which means that when some wells are abandoned, new wells are brought into exploitation. In this case, they can be seen as newly commissioned BPA, which, in its turn, is consistent with para. 14 of the Accounting Policy Regulations "Asset accounting" PBU 6/01 (version of the Order of the Ministry of Finance of the Russian Federation dated 11/27/2006, No. 156н//Rossiyskaya Gazeta – 12/31/2006, No. 297),

> *"... changes in the initial value of fixed assets, in which they are accepted for accounting, are allowed in cases of completion, additional equipment, reconstruction, modernization, partial liquidation and revaluation of fixed assets".*

This approach allows to deduct Abex from the net profit and to add them to the production costs in the form of depreciation over the entire field life, during which the estimated Abex are accumulated in a special fund and "remitted" in proportional shares in several years while classified as operating expenses (Simlogic, 2019).

To determine the amount of Abex, both for the purpose of estimating the size of the required abandonment fund and for taking into account the direct financing during the investment planning process, the standard practice today is the following: 1) use specific indicators of the abandonment operation costs per well, taking into account the number of wells planned for decommissioning and 2) apply the standard norms of abandonment costs relative to the costs of construction and installation of the field facilities (Mazurina, E.V. 2010). Regarding the second value, the experience of PAO LUKOIL and PAO Gazprom shows that when a feasibility study on the development of a hydrocarbon field is carried out, a factor equal to 10% of the BPA value is used for all groups of field facilities (except for wells) (Isachenko, V.M. 2004).

6 ABANDONMENT FUND IMPACT ON REPORTED RESERVES OF PRODUCING FIELDS APPROACHING ABANDONMENT

Sections of IAS (16 "Property, Plant and Equipment" and 37 "Provisions, Contingent Li-abilities and Contingent Assets") offer only some guidelines for field decommissioning and establishment of an abandonment fund. Thus, correct accounting of these expenses in financial statements needs to be ascertained, especially because such costs influence the classification of hydrocarbon reserves as recoverable at the final stage of exploitation of mature fields. These costs are always significant, so it is important for oil and gas companies to calculate the value of reserves and to review it annually. The rules for determining the economic limit test (ELT), which do not take into account any costs associated with the termination of project development, are described in the PRMS as the date when the net operating cash flow (NOCF) becomes negative. Its value determines the amount of reserves until the end of commercial field life.

In foreign and Russian practice, the final stage of field development is usually determined by the date when the exploration and development license expires or this period is calculated based on the estimated oil and gas reserves and the planned annual extraction rate (Medvedeva, N.V. 2008). Consequently, the exploitation of some fields continues when NOCF is negative in order to delay the costs associated with the termination of project development (Abex) and to maximize extraction. Therefore, the field decommissioning costs must be included in subsequent calculations of the net present value (NPV) of the project under development; however, this inclusion moves the NPV value in the negative territory and a discrepancy may appear despite the positive ELT forecast. To classify recoverable hydrocarbons as reserves, its amount must be commercially significant. Thus, the situation when mature fields continue to be exploited becomes increasingly common, although the reserves present there may need to be reclassified as contingent (Vaughan, A. 2017). Oil and gas companies find much more difficult to get bank loans or to attract investments when their resource base decreases.

Consider the ELT and NPV estimates of a developed project to find out why they may be inconsistent. As noted earlier, to illustrate the magnitude of the values under consideration, we will use a simple discounted cash flow (DCF) model of an average by size satellite field in the Nenets Autonomous Okrug, commissioned in the early 2000s and connected with two export oil pipelines several kilometers long to central infrastructure. Oil comes from several oil wells, pressure support for which is provided with a few water injection wells.

ELT is the moment when maintenance costs exceed extraction revenues. By this moment, the oil and gas company must formulate, write and submit for approval a field decommissioning plan with an approximate schedule of operations and associated Abex (Thornton, W. 2016).

Table 1 shows an example of the ELT calculated for a project under development until 2023 (the year when the NOCF becomes negative) that has 7.9 million tons of oil reserves and a positive NPV forecasted in the amount of 62.8 million Russian rubles.

Table 2 shows that undiscounted Abex, calculated for the development, amount to 177.8 million Russian rubles, while the potential tax reduction in relation to the previously paid profit tax and additional fees is 23.0 million Russian rubles, which gives a total forecasted value of field decommissioning liabilities in the amount of 154.8 Russian million rubles.

However, if Abex are included in this example, the estimated NPV value becomes -0.2 million Russian rubles, which will lead to further loss of 63.0 million Russian rubles. This means

Table 1. DCF model to compute ELT ignoring decommissioning

Year	Oil Production, thousand tons	Income, million rubles	OPEX, million rubles	CAPEX, million rubles	Income Tax, million rubles	Cash Flow, million rubles	DCF 15%, million rubles
2019	2424	69.3	-26.9	0.0	-8.5	33.9	29.5
2020	1781	50.9	-25.6	0.0	-5.1	20.2	15.3
2021	1816	51.9	-25.0	0.0	-5.4	21.5	14.1
2022	1172	33.5	-23.8	0.0	-1.9	7.8	4.4
2023	752	21.5	-23.0	0.0	0.3	-1.2	-0.6
Total	7944	227.1	-124.3	0.0	-20.6	82.2	62.8

Table 2. DCF model to compute point forward NPV to end of project life including decommissioning costs and tax refunds

Year	Oil Production, thousand tons	Income, million rubles	OPEX, million rubles	CAPEX, million rubles	Income Tax, million rubles	Cash Flow, million rubles	DCF 15%, million rubles
2019	2424	69.3	-26.9	0.0	-8.5	33.9	29.5
2020	1781	50.9	-25.6	0.0	-5.1	20.2	15.3
2021	1816	51.9	-25.0	0.0	-5.4	21.5	14.1
2022	1172	33.5	-23.8	0.0	-1.9	7.8	4.4
2023	752	21.5	-23.0	0.0	0.3	-1.2	-0.6
2024	-	-	-	-98.7	14.0	-84.7	-36.6
2025	-	-	-	-79.1	9.0	-70.1	-26.4
Total	7944	227.1	-124.3	-177.8	2.4	-72.6	-0.2

that the residual hydrocarbon deposits may be counted as sub-commercial. It is expected that the extracted 7.9 million tons of oil are more appropriate to classify as contingent resources.

Table 3 shows an approximate DCF model, which was modified to analyze the NPV break-even point in order to estimate the field decommissioning contributions due to a trust fund that cover the forecasted field decommissioning liabilities.

Since the DCF model describes a theoretical situation. rather than a fully modeled financial state of a company. it suggests that the remaining revenue received at the field presented as an example cannot support NOCF until the well is abandoned in 2022. Thus. if the appropriate coverage was not previously provided for the field decommissioning and it must start from

Table 3. Breakeven NPV model to estimate the annual payments into abandonment fund, in period 2019-2023, to cover forecast Abex in 2024-2025

Year	Oil Production, thousand tons	Income, million rubles	OPEX, million rubles	CAPEX, million rubles	Income Tax, million rubles	Cash Flow, million rubles	DCF 15%, million rubles
2019	2424	-	18.8	-	-	18.8	16.3
2020	1781	-	18.8	-	-	18.8	14.2
2021	1816	-	18.8	-	-	18.8	12.4
2022	1 172	-	18.8	-	-	18.8	10.7
2023	752	-	18.8	-	-	18.8	9.3
2024	-	-	-	-98.7	14.0	-84.7	-36.6
2025	-	-	-	-79.1	9.0	-70.1	-26.4
Total	7 944	-	94.0	-177.8	2.4	-60.8	0.0

2019. then the modeled ELT can be moved to another moment. thus reducing the amount of estimated reserves. If an oil and gas company that develops a mature field does not accumulate sufficient reserve funds as it approaches its abandonment and does not get access to credit lines to cover its field decommissioning liabilities. it could go bankrupt. The decision to continue field exploitation or to terminate project development may depend on the total financial flows generated by the asset. and must not result in the opposite decision when taking into analysis all relevant costs. including those related to the abandonment expenses. As soon as the field NOCF becomes negative and regardless of the origin of the costs. the oil company (that does not expect a significant increase in the hydrocarbon prices) must decide to terminate project development.

7 CONCLUSION

The establishment of abandonment fund and the development of standards governing the oil and gas well decommissioning in Russian projects that involve the formation of financing schemes and default security mechanisms can be considered. at the initial stage. as a combination of successful international practices (employed. for example. in the US. UK and Norway) and proposals submitted by the relevant Russian ministries and oil and gas companies. which take into account the interests of both the state and the local businesses.

REFERENCES

BCG. 2017. Decommissioning of oil and gas wells market analysis.

Isachenko. V.M. 2004. Estimation of project capital intensity of oil field development. *Dissertation for Degree of Candidate of Economic Sciences.* Federal State Budget Educational Institution of Higher Education "Industrial University of Tyumen".

Khalidov. I.A. 2018. Handling decommissioning and restoration liabilities. *Microeconomics* (6): 74-81.

Kryukov. V.A. & Anashkin. O.S. 2012. Oil funds – a sterilization tool or a modernization tool? XIII International Scientific Conference on Economic and Social Development. *Moscow*: National Research University Higher School of Economics.

Mazurina. E.V. 2010. On the formation of liquidation funds of hydrocarbon deposits. *Petroleum Geology – Theoretical and Applied Studies* (T.5).

Medvedeva. N.V. 2008. Fixed assets liquidation fund and fields restoration. Corporate financial reporting. International standards. viewed 08.12.2018. https://www.cfin.ru/appraisal/other/o&g.shtml

Oil & Gas UK's. 2018. Decommissioning Insight. Annual survey and report. viewed 08.12.2018. https://oilandgasuk.cld.bz/Decommissioning-Insight-2018/38/

Rystad Energy. 2019. Global decommissioning set to hit record 36 billion US doll. over the next three years. viewed 13.03.2019. https://www.rystadenergy.com/newsevents/news/press-releases/Global-decommissioning-set-to-hit-record-$36-billion-over-the-next-3-years/

Simlogic. 2019. How to account for decommissioning provision under IFRS. viewed 12.02.2019. https://www.ifrsbox.com/decommissioning-provision-ifrs/

The creation of liquidation funds of oil companies postponed until 2019. 2015 [Sozdaniye likvidatsionnykh fondov neftyanykh kompaniy otlozhili do 2019 goda]. viewed 09.12.2018. https://www.interfax.ru/business/468899

Thornton. W. 2016. Decommissioning and abandonment. *Technology focus article. JPT.* January 2016.

Vaughan. A. 2017. Shell begins huge task of decommissioning Brent oil rigs. Article by the Energy correspondent of The Guardian newspaper. 6th February 2017.

Wood Mackenzie. 2016. UKCS decommissioning: challenges in the current price environment. viewed 08.12.2018. https://www.woodmac.com/reports/upstream-oil-and-gas-ukcs-decommissioning-challenges-in-the-current-price-environment-37636260

Wood Mackenzie. 2018. 32 billion US doll. to be spent on decommissioning worldwide in 5 years. viewed 08.12.2018. https://www.woodmac.com/reports/upstream-oil-and-gas-us32-billion-of-decommissioning-worldwide-over-the-next-five-years-is-the-industry-ready-9599

Youth Technical Sessions Proceedings – Litvinenko (ed)
© 2019 World Petroleum Council, London, UK, ISBN 978-0-367-34668-3

Digital model of project information management

R. Khakimov & V. Kapustin
Gubkin State University of Oil and Gas (National Research University), Moscow, Russia

ABSTRACT: Worldwide practice of major companies proves the importance of digital models as the basis of engineering information management systems and the pattern of complex projects, data integrated both with entities involved in the project life cycle and a set of information systems applied (Eastman, 2011). The paper suggests the draft of a typical design "A Digital Model of Project Information Management", which is a complex of architectural-and-technical, organizational and regulatory decisions defining the project MIS structure and principles of its operation.

1 INTRODUCTION

Application of digital models for project information management is the basic tendency in improving generated business-processes for implementation of complex engineering projects, enhancement of performance efficiency at every stage of the industrial project life, as well as optimization of costs and time for design/construction of facilities (Weygant, 2011).

The first to turn to this subject were developed countries with a strongly pronounced and closely integrated interactive system of industrial structures (engineering, producing organizations), academic community, regulatory and industrial agencies - UK, USA and EU nations. Currently, a substantial base of industry and civil standards, guidelines on the use of components and information modeling technologies has been built up, finding its application in such industries as oil and gas, power generation (for the most part, nuclear and alternative), shipbuilding, metallurgy and mining.

In the Russian Federation digital models of project information management are in the making, international standards are under adaptation, pilot projects are under implementation using means of automation and production management, engineering and technical data acquisition, knowledge and workflow management at all life-cycle stages (Digital McKinsey, 2017). Oil-and-gas holding companies, metallurgical and mining enterprises, civil engineering developers are in the initial testing of these technologies.

Market presence of software solutions and technologies ensuring integrated use of digital models allows design and producing companies to decide in favor of priority goals of business processes optimization, without unnecessary adaptation to the architecture of software platforms, but rather relying solely on guidance methodologies (McKinsey&Company, 2016).

The use of CAD tools for the management of workflow, requirements and knowledge, visualization of an integrated model, systems of scheduling-and-network planning and project management, etc. is the basis for digital model applications, yet not being a limiting factor in execution of works. Processes and problems solved within the life cycle are abstract from specific software and can be implemented using various types and combinations of software products.

Studies of international consulting companies and developers of standards first evaluate the importance of shaping and rebuilding of existing business-processes while software application for solving applied problems is just an extra advantage (PAS 1192-2:2013).

Thus, a digital model of project information management is actively used at all life-cycle stages of an entity. All models may differ in data completeness provided by information slices, and integration with different outer information systems (ISO 15926). Yet, models are based on the object-oriented unified element base that ensures data consistency and their exchange among users.

2 METHODOLOGY

The studies were based on a typical business segment project "Oil Refining". An assessment methodology for technological maturity of project information management has been developed with the aim to determine the current technological maturity level of information management processes and application practices of IT tools in the priority projects organizations. The objective of the Methodology is to develop an approach to assessing the current technological maturity of business processes and the level of automation of the priority projects organization activities. The assessment tool is Technological Maturity Model. The Model is a description of the evolution/development stages (maturity levels) that organizations go through as they (organizations) define, implement, measure, control and improve information management processes and IT tools application practices (ISO 16739).

On the basis of international methods, practices, standards, as well as existing IT tools, there have been studied all business processes related to project information management, divided into separate project areas (function units), as a convenience: follow-up and schedule control, management and estimation of project costs, contract and procurement management, stakeholder management and communications handling, organizational project capacity management, risk management, workflow management, engineering data management, knowledge base (Project Management Institute Commends, 2012).

3 ASSESSMENT OF TECHNOLOGICAL MATURITY

Any project information management system is an integrated information management solution in the selected business segment, in this case in the Oil Refining business segment. As a project information management system includes a complex of corporate information systems within which the project information management is carried out, the approach to building up the system will differ for each particular organization and the elaboration rate of the system will strongly depend on the technological maturity level of the company as a whole. Therefore, in each case, it is important to assess technological maturity basing on the initial investigation results of all company's business processes by certain criteria (Sliger, 2008).

The article will describe the general methodology for the technological maturity assessment of project information management developed in the process of the studies.

3.1 *Description of the approach*

In accordance with the approach technological maturity is defined as a formalized level of development, allowing to assess current information management processes and application practices of IT tools.

The basis for development of the general methodology to asses technological maturity was the model for technological maturity assessment of project management processes PMMM based [Project Management Maturity Model] - a tool developed on the best practices presented at the Project Management Institute (PMI) and used to assess the project management maturity of companies (PMI Fact File, 2018). PMMM is used to develop a progressive strategic plan for company's project management improvement. Two models were also used: the Model of Knowledge Management Technological Maturity, developed on the KMMM basis [Knowledge Management Maturity Model], characterizing the maturity of the company's knowledge management process and the Model of Technological Maturity developed on the CMMI basis [Capability Maturity Model Integrated] - characterizing a scale of five levels based on how consistent the company is in following information management processes (CMMI-survey, 2007). The later has been developed by the Research Center SEI.

The Technological Maturity Assessment is performed on a 5-point scale of the maturity level in which the level of maturity is the company's development stage in accordance with standardized assessment models (Kumar, 2004). A brief description of the technological maturity levels is presented in Table 1.

A conceptual approach to application of maturity assessment models includes identification of an initial maturity level, decisions on criteria and steps to an increased maturity level and development of a step-by-step increased maturity plan (roadmap).

For an object under study the quality of each criterion compliance to a technological maturity level is identified based on gradation: full compliance - the maturity level corresponding to the criterion is assigned; partial compliance - the maturity level is assigned based on a combination of fully and partially corresponding criteria; discrepancy - the maturity level corresponding to the criterion is considered unachieved (Aachen, 2005).

3.2 *Evaluation criteria*

The technological maturity assessment of automated functions performed by organizations is carried out in accordance with the described tools of the Technological Maturity Model.

For each project area of the project information management system the processes for information management implementation were elaborated as a preliminary to the steps of the processes according to the applied Technological Maturity Model. Next, for each step of a process, we have compiled the evaluation criteria list, in which each criterion corresponded to one or two levels of technological maturity (Karsten, 2002).

Table 1. Technological maturity levels (Kochikar, 2000)

Stage 1 - Initial	• scattered information storage • poor information retrieval • predominant paper documents • no information management tools • no procedures and rules
Stage 2 - Controlled	• availability of separate information management tools, no integration • requirements development for electronic document transfer • manual verification and validation of project data • electronic document management
Stage 3 - Specific	• formalization of processes and procedures for managing data/documents • application of data-centric tools • integration of MIS basic components • a united center of responsibility for project information quality and distribution
Stage 4 - Measurable	• standardized information management processes and tools KPI and analytics integrated • centralized keeping of registers • integration with asset operation systems • electronic operations of information management • integrated MIS with possible automated quality control of data and documents
Stage 5 - Optimizable	• end-to-end integration with existing corporate systems • mobile device management for information remote handling • dominance of data application instead of documentation • compliance with PAS55, ISO-55000, ISO-15926 and other industry standards • integrated data system (IDS) of information management throughout the life cycle • use of augmented reality resources, etc.

For *follow-up and schedule control* the following criteria were chosen: decisions on the scope of work and workflow, project schedule working out, project schedule monitoring, project schedule integration.

For *management and estimation of project costs* the following criteria were chosen: costs estimation, project budget preparation, project performance analysis, project cost control.

For *organizational project capacity management* the following criteria were chosen: organizational planning, staff recruiting, personnel development.

For *contract and procurement management* the following criteria were chosen: procurement planning, preparation of procurement documentation, choice of suppliers, contract management and contract closing.

For *stakeholder management and communications handling* the following criteria were chosen: communication planning, information dissemination, project progress report preparation, identification and problem management

For *risk management* the following criteria were chosen: risk identification, risk assessment, risk response, risk control, risk documentation.

For *engineering data management* the following criteria were selected: engineering data and digital asset data management, engineering data management in designing, engineering data management in construction planning, engineering data management in precommissioning control, engineering data management at the facility commissioning stage.

For *workflow management* the following criteria were chosen: receipt and initial processing of project design documentation, analysis and examination of project design documentation, storage and version control, archiving and retrieval.

For *knowledge base* the following criteria were selected: corporate culture of priority projects knowledge, information, technological infrastructure of knowledge management, dedicated applications of knowledge management.

4 CONCEPT OF BUILDING A DIGITAL MODEL

Application of digital models for project information management suggests a whole range of software tools in the areas of automation of all the above function units, but the present concept and business requirements elaborated are abstract from software selection. Selection of software platforms, detailed elaboration of business requirements and their comparison is the target of the next development stage and is beyond the scope of this concept.

More generally, a digital model of project information management is to include generated business-processes of project participants relationship, a set of regulations to approve classification of model elements, principles of modeling and its application, security policy, as well as integration with related information systems and a software toolchain for successful problem solving.

When developing a certain methodology and processes of project participants relationship, ultimately a digital model of project information management is to allow undocumented information transfer scheme.

Digital model business requirements are developed in view of organizational and functional scope of its application. A digital model both as independent and integrated with related information systems is to provide solutions to a number of problems, yet not replacing (introducing) related system functionality, but only complete the package of automation tools and provide users with additional data visualization and analysis tools.

Below is a brief description of digital model business requirements to manage project information.

A digital model is to compile a list of activities and major events under the project, shape network diagrams, provide feedback from project results, appoint responsible persons, define labor content according to local estimates, record operation time and associated risks, develop schedules by various procedures, design resource and target project plans, provide information import on the project content, develop and document rough project costs based on the imported information, maintain cost estimating basic methods, generate versions for basic and specialized project budget, scheduling for accounting of indirect taxes, predict project costs, generate a post-project predicted budget on the execution phase report, generate a bill of work, generate a materials' plan, compile/import a sheet of assembly and technology facilities, compile a job cost records of assembly and technology facilities, develop requirements for active jobs, pack procurement documentation, record all changes, compile a full list of stakeholders, keep the calendar of meetings, ensure separation of access rights to information depending on the role model of project participants, generate protocols and lists of assignments, provide remote access to design and technical documentation, make up records of comments on documentation, register various versions of documents, develop engineering data objects, ensure engineering data import, track running state of design execution, compile a list of identified collisions, store up a historical knowledge base, store up patterns and forms of typical solutions under the project (Yeona, 2010).

5 REPORTING

A digital model of project information management is to be the foundation for possible scheduled reporting under approved forms and set patterns.

A digital model is to include an adjustable word processor of reporting forms with a feature to present information in text and graphical forms (similar in functionality to MS Word). The word processor is to design layouts of coded reporting documents containing description of database elements and various functions with them for further automatic reporting.

The system is to allow creation and storage of an arbitrary number of report layouts displayed for visualization, printing, export in PDF/Word/Excel on user's demand.

6 INTEGRATION OF RELATED SYSTEMS

While developing a digital information management model for a project, it is important to consider integration solutions with such systems as Design Automation Systems (CAD), including engineering calculations and cost estimates, Archival Document Management System (MDM), Reference Data System (RDS), Enterprise Resource Planning System (ERP), Electronic Document Management System (EDMS), Calendar and Network Planning System (CSP), Enterprise Asset Management (EAM) and Process Control System (SCADA).

7 CONCLUSION

Implementation of the Project Digital Information Management Model aims to improve quality and reduce total project dates, over-budgeting risks and design changes, improve inter-participants communication, increase costs control and forecast precision.

Enhancement of performance efficiency is achieved by data centralization and management of works in uniform informedia, unification and development of business-processes types, provision of opportunities for engineering workflow management, job design, optimization of internal operation, thus finally resulting in reduction of expenses, time and cost of works under the project.

REFERENCES

BIM Handbook: A Guide to Building Information Modeling for Owners, Managers, Designers, Engineers and Contractors, 2nd Edition. Chuck Eastman, Paul Teicholz, Rafael Sacks, Kathleen Liston. ISBN: 978-0-470-54137-1. 2011.

BIM Content Development: Standards, Strategies, and Best Practices. Robert S. Weygant. ISBN: 978-0-470-58357-9. 2011.

Digital Russia. New reality. July 2017. Digital McKinsey. Tsifrovaya Rossiya. Novaya real'nost'. Iyul' 2017.
Imagining construction's digital future. Capital Projects and Infrastructure June 2016. McKinsey&Company.
ISO 15926. Industrial automation systems and integration – Integration of life-cycle data for process plants including oil and gas production facilities.
PAS 1192-2:2013. Specification for information management for the capital/delivery phase of construction projects using building information modelling.
PAS 1192-3:2014. Specification for information management for the operational phase of assets using building information modelling.
ISO 16739. Industry Foundation Classes (IFC) for data sharing in the construction and facility management industries.
Michele Sliger and Stacia Broderick (2008). The Software Project Manager's Bridge to Agility. Addison-Wesley. ISBN 0321502752 p.26: The five people, who founded the Project Management Institute were James Snyder, Gordon Davis, Eric Jennett, A.E. Engman, and Susan C. Gallagher.
"PMI Fact File". PMI Today. Project Management Institute: 4. September 2018. Retrieved 15 Sep 2018.
"Project Management Institute Commends ISO 21500 Standard for Alignment with PMBOK Guide". Pmi.org. September 6, 2012. Retrieved2014-06-05.
mini CMMI-survey. SQI Hungarian Software Quality Consulting Isntitute Ltd. (2007).
Denis M. Aachen, Aaron Clause, Richard Turner CMMI: An Integrated Approach to Process Improvement. Practical introduction to the model.— M: "IFC", 2005, 300 p. ISBN 5-85389-082-4.
Ehms, Karsten, and Manfred Langen. "Holistic Development of Knowledge Management with KMMM". Siemens AG (2002).
Kochikar, V. P. "The Knowledge Management Maturity Model - A Staged Framework forLeveraging Knowledge". Infosys Technologies Limited. 2000.
Kumar, Santosh, and Manis Chand. "5iKM3 Knowledge Management Maturity Model". Tata Consultancy Services (2004).
Jang, Yeona, Dr. Managing Knowledge with IT, McGill University (2010).

Youth Technical Sessions Proceedings – Litvinenko (ed)
© 2019 World Petroleum Council, London, UK, ISBN 978-0-367-34668-3

Development of an economical approach for investment optimization in heavy oil industry

J. Chacón Solar
Novosibirsk State University, Novosibirsk, Russian Federation

ABSTRACT: The aim of this investigation is to propose an economical approach to take the most optimal investment decision in heavy oil fields, taking into account key variables such as exchange rate, oil benchmarks spread, technology, discount rate, capital and operating costs, taxes and environmental expenses. The method for this research is based on economical, mathematical and statistical methodologies, as well as sensitivity analyzes. As any investment analysis consists in forecasting costs and revenues within an intertemporal framework, this document's subject assess to develop an empirical methodology which could develop a better system for increasing companies' revenues when producing heavy oil restricted to the product and market conditions. Analysis showed that exchange rate and oil benchmarks fluctuations are not a considerable threat for the model, due to the fact that there is low volatility in the spread among WTI and Brent oil benchmarks, as well as Canadian and US Dollars. On the other hand, increases in technology ratios applied to production highly affect total income and revenues, meaning that the future of economic investments in non-conventional oil is highly related to technological advances.

1 INTRODUCTION

Heavy oil analysis and modeling is vital for today's world society. As the global economy continues to depend on oil and oil derivatives, and taken into consideration the enormous volatility of oil markets, developing methods to economically optimize investments in the industry of heavy oil -which has not even reached its potential- appears to be a strategic matter for companies and nations willing to take advantage on international energy markets.

Conventional oil resources are naturally declining after decades of massive consumption, and traditionally oil producers such as Iraq, Syria, Iran or Libya are facing colossal security difficulties, which affect international global market and cause even more volatility than that already existed. Long run perspectives for oil markets are far away from being clear and prices equilibrium tend to change constantly, affected not just by internal oil industry behavior but by exogenous variables. On the other side, reserves of unconventional oil are enormous worldwide and interest on them increases on a daily basis. Despite the existence of serious challenges in the sector, problems in heavy oil production are partially solved with the constantly enhance of new sophisticated recovery methods; however, new technology is still needed to reduce risks and costs.

Researchers have produced an enormous quantity of investigations on heavy oil, fundamentally from the perspective of engineering, geology, chemistry or physics; however, the economic perspective is still to grow and this document expects to contribute to the increase of academic literature concerning the economical aspect of the sector. The idea of this paper is topical because heavy oil resources are essential to satisfy global energy consumption for next decades; moreover, an appropriate economic valuation of them will encourage large-scale investments which will stimulate economic growth, create jobs, provide needed energy and generate significant revenues for companies, countries and the society itself, fulfilling climate change policies at the same time.

The aim of this investigation is to propose an economical approach to take the most optimal investment decision in heavy oil fields, taking into account key variables such as exchange rate, oil benchmarks spread, technology, discount rate, capital and operating costs, taxes and environmental expenses.

2 LITERATURE REVIEW

Several authors have considered the topic of analyzing investments in energy sectors, specifically in oil industry. However, to this day there is not still a total agreement about which model is the most accurate. One of the most important problems these investment models face relies on the incapacity of properly forecasting and then managing the risk and the uncertainty of this sector, especially when considering the stochastic dynamic of oil prices and costs.

The most important question of investment in oil and gas projects is which method to use. Lin and Ji (2007) perform a very interesting comparison between real option methods and financial options, and determined that the first is more suitable for oil/gas project, then they built up a new model extended from the Black-Scholes and tested it within a computational case. Moreover, they identify risk managing as a key factor to take investment decisions. Some years before, Cortazar, Schwarts and Casassus (2001) also used a real options model under price and geological-technical uncertainty which was successfully applied to a copper exploration prospect. Their findings were useful because they highlight the option of stopping the exploration investment schedule. Under a similar objective of deciding the best method for optimizing investments, Midthun, Fodstad and Hellemo (2015) mentioned that Combinatorial optimization models might help decision-makers to take better decisions on which elements to invest in, at what time and with what capacity, they also presented a optimization model for analysis of system development for natural gas fields, processing and transport infrastructure. Qin, Wang and Xue (2015) built a multi-factor analysis with a real options model for investment in deep-water oil and gas exploration projects, addressing the importance of oil price, geology and engineering uncertainties. As for the evaluation of deep-water oil and gas projects, they used a numerical analytical method and found that the evaluation results show that a single-factor model is not as accurate as a multi-factor real options model. Zhang and Wang (2011) preferred to use system dynamics to analyze and forecast investment scale and structure in upstream sector for oil companies and found that upstream investment of the oil company is significantly affected by expected oil and gas reserves and expected oil and gas yield and international oil price. On another aspect, Ettehadtavakkol, Jablonowski and Lake (2016) compared Monte Carlo (MC) simulation and stochastic programming for development optimization and uncertainty analysis; they concluded that while MC method generates more useful information, stochastic programming method is more computationally efficient in determining the optimal solution.

Gaspar, Barreto and Schiozer (2016), presented and optimization design for oil exploitation strategy, in which they use the Producer Well Economic Indicator (PWEI) as a valid indicator for scheduling a well opening -it was first developed by Ravagnani et al. (2011) -. Branka, Zdravka and Tea (2014) presented a brief review of Binomial and Trinomial models for Valuation Options and of Finite Difference Method in Option Pricing. Although their work is focused on a computational framework, it is important to consider it in the sense that it gives tools to understand better oil prices behavior, which are based on derivatives traded using both: American and European methods. Zhong and Zhao (2016) showed a model for optimal oilfield development investment, which included only two types of investment: exploration and development. In this paper, the author considers one more type: investment in transportation.

Probably one of the most important papers written on this topic, was done by Brennan and Schwarts (1985), showing that techniques of continuous time arbitrage and stochastic control theory may be used to value natural resource investments, as long as optimal policies for developing, managing and abandoning them. They argue that traditional standard techniques of evaluation take expected cash flows from an investment model and discount them at a rate which takes into consideration the risk; later, this result is compared with the cost of the project. When output prices vary above a specific rate –due to inevitable market dynamic-, the model stops being accurate and fail in prevailing risk and uncertainty. As a result, authors built a model where output prices are assuming to be stochastic and where managerial decisions of reducing, increasing or even shutting down the production might be taken. In addition, according to authors Smith and McCardle (Smith and McCardle, 1999), almost every capital expenditure firm evaluates its investments options using sensitive analysis to identify risks and uncertainties, before developing models to calculate expected Net Presented Values (NPV) and distributions on NPVs. Another approach is presented by Constance Helfat (Helfat, 1989). She develops a model where firms take into consideration not a particular investment return but the covariance of returns of several investments done by the firm. A portfolio model is used to describe how an average petroleum firm reduces its risk minimizing the variance of the forecast present value per dollar over all projects (*ex-ante* discounted cash flows divided by initial fixed costs), subject to a required expected rate of return constraint on the portfolio.

Several documents were taken into consideration to specifically analyze heavy oil shields. Canadian information was taken mainly from documents published by Canadian Energy Research Institute, Alberta Oil Sands Industry –Quarterly Update, published by the Alberta Government and Suncor Energy INC. annual report 2015. Venezuelan information was taken mainly from the official website of PDVSA and Center on Global Energy Policy from the University of Columbia. The Oxford Institute for Energy Studies offered pertinent information about Russia's shale oil, as well as Russian companies such as Lukhoil and Tatneft. Furthermore, Congressional Research Service of the United States offered an excellent description of the advances done by that country in non-conventional oil production. World Energy Outlooks from International Energy Agency (IEA) offered remarkable information about global energy markets and trends. The objective of its projections is not to exactly forecast what will happen, but to propose what might happen given specific assumptions and methodologies. In the Reference case scenario, for example, a business-as-usual projection was done with the aim of estimating trends of future global energy consumption and production. For this, an energy model was built and demographical, economic and technological variables, among others, were included.

3 METHODOLOGY

A simplified Net Present Value (NPV) investment model was built. A NPV model was chosen over a Real Options approach considering that in this paper neither any managerial intuitive decisions nor non-economical strategic thinking is allowed to be considered. Real Options has the advantage of allowing managerial flexibility when taking investment decisions, as well as adjusting the model to new variables that are not considering in NPV (such as the possibility of obtaining market power, entering into a new market, or weaken competition, even if these give short-term negative revenues as a result). Although this approach seems to be more feasible for oil investments, it is not consider in this document and therefore opens opportunities for further research.

4 MODEL

To the NPV model was added a variable describing the effect of technology on the increase of production, which therefore affects the revenues. This variable (A), starts at 1 and grows at an exponential rate "α=1%" assumed by the author.

$$NPV = \sum_{t=0}^{20} \frac{-I_t}{(1+\delta)^t} + \sum_{t=0}^{20} \frac{A_t^{\alpha} P_t Q_t}{(1+\delta)^t} \tag{1}$$

Although NPV is frequently used when taking oil investing decisions, it is not accurate because it is not a dynamic model able to properly manage the fluctuation and volatility of the variables (oil price, quantity of oil sold, quantity of oil produced, etc.). In order to solve this problem, as already explained, two important variables (oil price and quantity produced) were divided under three categories with different minimum and maximum values, rather than forecasting their exact daily, weekly or monthly value for next 20 years.

This model doesn't take specifically into consideration one important variable: cost. Therefore, a new model was built in order to offer more accurate results:

$$NPV = \sum_{T=0}^{20} \frac{A_T^{\alpha} P_T^{\beta} Q_T^{\gamma} - I_T^{\mathrm{T}} - \theta_T - \lambda_T - ROY_T - \tau_T - \pi_T - \rho_T - T_T}{(1+\delta)^T} \tag{2}$$

Where: $A_T^{\alpha} P_T^{\beta} Q_T^{\gamma}$, Represents the total amount of oil produced and sold; I_T^{T}. Represents the total amount of investment (Total Operating Costs and Total Capital Costs); θ, Represents Operating Working Capital; λ_T, Represents Fuel (Natural Gas); ROY_T, Represents Royalties; τ_T, Represents Income Taxes; π_T, Represents Emissions Compliance Costs; ρ_T, Represents Abandonment Costs; T_T, Represents Transportation Costs; δ, Sigma, represents discount rate.

It is important to mention that these new costs are not represented in the variable for Investments (I), since it only considers capital and operational costs.

Assuming Technology increases at an annual rate of 1%, the following charts show how much yearly income and yearly revenues change through time. It is worth seen the influence of technology in the variables' behavior in the long run, creating a spread with the variables without technology for both, income and revenues. Taken data from Canadian Energy Research Institute (CERI) and author's own calculations, under the described circumstances the revenues are negative until 2023, when they start to growth with an exponential form. With the aim of determining the behavior of revenues for different values of technology, the following analysis is done. If technology annual growth changes from 1% to 2%, then total value of NPV revenues (meaning, the sum for all the 20 years) will increase 73,1%. If it changes from 2% to 3% (assuming a 5% discount rate), then total value of NPV revenues will increase 48,3%. Total revenue is affected positively by an increase in technology, as expected. Higher technology increases production growth which increases income and revenues. Theory of decreasing marginal revenues explains why Total NPV increases every time in a smaller proportion as Technology gets higher.

An increase in discount rate affects negatively Total revenue. Considering discount rate is dividing the revenues, a higher value of sigma increases the denominator and reduces the value of NPV. Theory of increasing marginal revenues explains why NPV decreases more with higher values of sigma. For example, if Sigma changes from 1% to 2%, Total NPV Revenues will decrease in 21,9% (assuming there is no technological growth).Despite discount rate also affects Investment (which affects negatively NPV as well), the value for Income is higher than the value of investment, therefore the relation among sigma and NPV is inversely proportional.

The proposed investment model considers eight different kinds of expenses and only one kind of income: oil sales revenues (gas revenues are not taking into consideration). Reviewing the annual report about oil sand costs, it was created a table of costs per barrel for oil sands in the Alberta region under SAGD extraction method (from 2011 to 2015, excepting 2012 because it was not presented in the reports).

The analysis shows how a 1% change in every cost affects total revenues for the entire period. First, as it is expected, an increase of 1% of any cost affects in the same way the three

Table 1. Costs per barrel for oil sands in the Alberta region under SAGD extraction method

SAGD/bbl	Average (USD)	Proportion
Fixed Capital (Initial & Sustaining)	16,86	24%
Operating Working Capital	0,38	1%
Fuel (Natural Gas)	4,23	6%
Other Operating Costs (incl. Elec.)	9,85	14%
Royalties	7,46	11%
Income Taxes	2,64	4%
Emissions Compliance Costs	0,30	0%
Abandonment Costs	0,30	0%
TOTAL PRODUCTION COSTS	**41,75**	**59%**
Transportation Costs	17,46	25%
TOTAL COSTS	**59,21**	**83%**

Source: CERI, 2017. Study No. 163 and author's own calculations.

scenarios. For example, if Fixed Capital rises in 1%, it will increase the total costs in 0,2848% for each of the three scenarios: high, reference and low. Second, transportation and fixed capitals are the variables that affect the most total revenues, 0,2848% and 0,2948% respectively. Third, an extra one percent expense in royalties and taxes reduce total revenues in almost 0,17%. Fourth, emission compliance costs and operating working capital basically don't affect total NPV in any proportion and finally, other operating costs (including electricity) change in 0,17% total revenues.

5 RESULTS

In brief, calculations found that investments in heavy oil are fundamentally affected by improvements in technology. Addition of new techniques and methodologies to explore, extract and transport oil is core for the project's revenues. Fluctuation in the exchange rate between USD and CAD and between WTI and Brent benchmarks don't have big effects on revenues. Moreover, findings show that increases in discount rate (due to increases in inflation, for example) negatively affects revenues, for which an increase in technology rate is proposed in order to maintain the goal of profits. Under certain assumptions, a table of values for technology and discount rate is proposed. It is important to mention that technology has a cost, which was not considered in this document and might be a topic for future researches.

6 CONCLUSIONS

The research's topic is highlighted when considering outlooks from powerful oil companies, such as British Petroleum, Total or Exxon Mobil, which consider non-conventional oil fields as key for the future energy production and consumption. The importance increases when taking into account the dimension of global investment in energy during the next two decades.

Despite the vast amount of academic literature related to energy investments, the field of specifically economic literature focused on investment optimization for heavy oil sector has not reached its peak. Traditional investment models applied to this industry lack of dynamism and are built in a rigid way, as a consequence their utility is limited when including external shock in oil prices, supply and demand. Although previous researches have addressed to the importance of improving investment models, the novelty of this research relies on its intention to develop a more appropriate economical approach which could guide to the development of a new investment optimization model that can be successfully applied to non-conventional oil, using oil sands in Alberta, Canada. The key point in this paper was to examine the influence of technology in investments' revenues and dynamics.

Analysis showed that exchange rate and oil benchmarks fluctuations are not a considerable threat for the model, due to the fact that there is low volatility in the spread among WTI and Brent oil benchmarks, as well as Canadian and US Dollars. On the other hand, increases in technology ratios applied to production highly affect total income and revenues, meaning that the future of economic investments in non-conventional oil is highly related to technological advances.

The consequences of increment in costs were also analyzed with a sensitivity analysis, finding that those related with production and transportation are more important than other ones, such as royalties, taxes or emission compliance. This means that technology improvement might reduce more investment's expenses than legal or political costs.

Regarding the discount rate, which is one of the most important factors when building investment models, it was found that technology can neutralize the harmful effects of increases in discount rate. With this is mind, a simulation was done with the aim of obtaining the equilibrium among discount rate and technology. Therefore, analysts can know how much technology must change when opportunity costs vary in order to maintain the equilibrium.

One of the most important challenges that affected this investigation was confidentiality: too much data about costs and strategies are kept in secret by companies and governments. Besides, companies do not specifically present in their financial reports detailed data for their activities in heavy-oil fields. In addition, although the main framework of this research is economical, the complexity of the topic makes necessary to consider and analyze data and researches from politics, geopolitics, management, geology, physics, chemistry and engineering.

In a more specific economic aspect, there is not a generally accepted methodology for forecasting oil prices and predicting oil shocks. Therefore, there is a lot of uncertainty when developing models for income and revenue in the industry, in particular if it's a long run forecast. Equilibrium prices tend to vary due to external and company-internal reasons, making impossible to identify one general oil equilibrium price or one discount rate for the entire industry. On the other hand, the quantity of oil produced is not affected just by financial or economic reasons, but also for qualitative-managerial, legal or environmental reasons; mining the accuracy of any economic approach developed.

REFERENCES

Ana Teresa F.S. Gaspar, Carlos Eduardo A.G. Barreto, Denis J. Schiozer, 2016. *Assisted process for design optimization of oil exploitation strategy*, Journal of Petroleum Science and Engineering, Volume 146, October 2016, Pages 473-488, ISSN 0920-4105, http://dx.doi.org/10.1016/j.petrol.2016.05.042.

Bernanke, B., Gertler, M., & Watson, M. (2004). *Reply: Oil Shocks and Aggregate Macroeconomic Behavior: The Role of Monetary Policy. Journal of Money, Credit and Banking, 36*(2), 287-291. Retrieved from http://www.jstor.org/stable/3839021

Bernanke, Ben S., Mark Gertler, and Mark Watson (1997). "*Systematic Monetary Policy and the Effects of Oil Price Shocks*". Brookings Papers on Economic Activity1, 91-142

Branka, M., Zdravka, A., Tea, P. (2014). '*Numerical Methods versus Bjerksund and Stensland Approximations for American Options Pricing*'. World Academy of Science, Engineering and Technology, International Science Index 88, International Journal of Social, Behavioral, Educational, Economic, Business and Industrial Engineering, 8(4), 1023-1031.

Brennan, M., & Schwartz, E. (1985). *Evaluating Natural Resource Investments. The Journal of Business, 58*(2), 135-157. Retrieved from http://www.jstor.org/stable/2352967

Cortazar, G., Schwartz, E. S. and Casassus, J. (2001), *Optimal exploration investments under price and geological-technical uncertainty: a real options model. R&D Management*, 31: 181–189. doi:10.1111/1467-9310.00208

Elder J., & Serletis, A. (2010). *Oil Price Uncertainty. Journal of Money, Credit and Banking, 42*(6), 1137-1159. Retrieved from http://www.jstor.org/stable/40784879

Ettehadtavakkol, A., Jablonowski, C. & Lake, L. Nat Resour Res (2017) 26: 177. doi:10.1007/s11053-016-9308-1

Ferderer, J. Peter. (1996) "*Oil Price Volatility and the Macroeconomy*". Journal of Macroeco-nomics, 18, 1-26.

Gibson, Rajna, and Eduardo S. Schwartz. (1990) "*Stochastic Convenience Yield and the Pricing of Oil Contingent Claims.*" Journal of Finance, 45, 959-976.

Graça, Ermida, *Strategic decisions of international oil companies: Arctic versus other regions*, Energy Strategy Reviews, Volume 2, Issues 3–4, February 2014, Pages 265-272, ISSN 2211-467X, https://doi.org/10.1016/j.esr.2013.11.004.

Hamilton, J., & Herrera, A. (2004). *Comment: Oil Shocks and Aggregate Macroeconomic Behavior: The Role of Monetary Policy. Journal of Money, Credit and Banking, 36*(2), 265-286. Retrieved from http://www.jstor.org/stable/3839020

Hamilton, James D. (1988) "*A Neoclassical Model of Unemployment and the Business Cycle*". Journal of Political Economy, 96, 593-617.

Hamilton, James D. (2003) "*What Is an Oil Shock?*" Journal of Econometrics, 113, 363-398.

Hamilton, James D. (2009) "*Causes and Consequences of the Oil Shock of 2007-08*". Brook-ings Papers on Economic Activity, 215-259.

Helfat, C. E. (1989). *Investment in Offshore Oil by Diversified Petroleum Companies. The Journal of Industrial Economics, 38*(1), 79–93. http://doi.org/10.2307/2098400

Henderson, James. *Tigh Oil Developments in Russia.* WPM 52. October, 2013. The Oxford Institute for Energy Studies. Page 1.

Hooker, Mark A. (1996) "*What Happened to the Oil Price-Macroeconomy Relationship?*" Journal of Monetary Economics, 38, 195-213.

Kilian, Lutz. (2009a) "*Not All Oil Price Shocks Are Alike: Disentangling Demand and Supply Shocks in the Crude Oil Market*". American Economics Review, 99, 1053-1069

Kilian, Lutz. (2009b) "*Comment on "Causes and Consequences of the Oil Shock of 2007-08.*" Brookings Papers on Economic Activity, 1, 267 -IS.

Kjetil Trovik Midthun, Marte Fodstad, Lars Hellemo, *Optimization Model to Analyse Optimal Development of Natural Gas Fields and Infrastructure*, Energy Procedia, Volume 64, 2015, Pages 111-119, ISSN 1876-6102, http://dx.doi.org/10.1016/j.egypro.2015.01.014

Lee, Kiseok, Shawn Ni, and Ronald A. Ratti. (1995) "*Oil Shocks and the Macroeconomy: The Role of Price Variability*". The Energy Journal, 16, 39-56.

Lin Z., Ji J. (2007) *The Portfolio Selection Model of Oil/Gas Projects Based on Real Option Theory.* In: Shi Y., van Albada G.D., Dongarra J., Sloot P.M.A. (eds) Computational Science – ICCS 2007. ICCS 2007. Lecture Notes in Computer Science, vol 4489. Springer, Berlin, Heidelberg

Manescu, Cristiana and Van Robays, Ine, *Forecasting the Brent Oil Price: Addressing Time-Variation in Forecast Performance* (December 14, 2016). CESifo Working Paper Series No. 6242. Available at SSRN: https://ssrn.com/abstract=2906230

Marion, N. P., & Lars E. O. Svensson. (1984). Adjustment to Expected and Unexpected Oil Price Changes. *The Canadian Journal of Economics/Revue Canadienne D'economique, 17*(1), 15–31. http://doi.org/10.2307/134934

Monaldi, Francisco. "*The impact of the decline in oil prices on the economics, politics and oil industry of Venezuela*". September, 2015. Center on Global Energy Policy. University of Columbia.

Qiu, XH., Wang, Z. & Xue, Q. Pet. Sci. (2015) 12: 525. doi:10.1007/s12182-015-0039-4

Ratner, Michael and Tiemann, Mary "*An overview of Unconventional Oil and Natural Gas: Resources and Federal Actions*". Congressional Research Center. April 22, 2015.

Ravagnani, A.T.F.S.G., Muñoz Mazo, E.I., Schiozer, D.J., 2011. *A case study of the structure of the process for production strategy selection.* Int. J. Model. Simul. Pet. Indu. 4-5, 9-15.

Ruoran Chen, Tianhu Deng, Simin Huang, Ruwen Qin, *Optimal crude oil procurement under fluctuating price in an oil refinery, European Journal of Operational Research*, Volume 245, Issue 2, 1 September 2015, Pages 438-445, ISSN 0377-2217, http://dx.doi.org/10.1016/j.ejor.2015.03.002

Smith, J. E., & McCardle, K. F. (1999). *Options in the Real World: Lessons Learned in Evaluating Oil and Gas Investments. Operations Research, 47*(1), 1–15. Retrieved from http://www.jstor.org/stable/222889

Suncor Energy INC. annual report 2015.

Yihua Zhong, Jiao Zhao, *The optimal model of oilfield development investment based on Data Envelopment Analysis*, Petroleum, Volume 2, Issue 3, September 2016, Pages 307-312, ISSN 2405-6561, http://dx.doi.org/10.1016/j.petlm.2016.04.004.

Zhang, B. & Wang, Q. Pet. Sci. (2011) 8: 120. doi:10.1007/s12182-011-0124-2

Oil and Gas Field Development, Production and Processing

Youth Technical Sessions Proceedings – Litvinenko (ed)
© 2019 World Petroleum Council, London, UK, ISBN 978-0-367-34668-3

Effects of top and bottom sides of fracture blockage in gas assisted gravity drainage

M. Hasanzadeh, R. Azin & R. Fatehi
Persian Gulf University, Bushehr, Iran

S. Zendehboudi
Memorial University, Newfoundland, Canada

ABSTRACT: Gas injection under gravity force is the significant oil recovery method known as Gas Assisted Gravity Drainage (GAGD). In this investigation, effect of three different configurations of fracture sides on recovery factor were examined. In the first configuration, the top and bottom parts of fractures were unblocked and open to injection and production. In the second configuration, the lower parts of fractures were blocked and direct production were just from the matrix. In the third configuration, the top and bottom parts of fractures were closed and only matrix directly contributed in injection and production. In this configuration, fractures contributed in injection and production only by perforated surface between matrix and fracture. Results showed that in the open fracture model, gas invaded into fractures and early gas breakthrough occurred and this was the reason of lower recovery factor at model with unblocked fracture sides. Also, higher final recovery achieved in the close fracture model. Therefore, in gas assisted gravity drainage process recovery improved if gas could not invade directly into fractures and oil could not produce from fractures.

1 INTRODUCTION

A large proportion of world's oil reserves are in naturally fractured rocks (Saidi, 1983). This significant amount of oil could be recovered efficiency by gas injection methods (Rezaveisi et al., 2010). Moreover, gravity drainage is one of the most important recovery mechanisms in fractured reservoirs. Accordingly, gas injection under gravity force is the significant oil recovery method known as Gas Assisted Gravity Drainage (GAGD) (Carlson, 1988; Fassihi & Gillham, 1993; JR, 1988; Kantzas, 1988). One type of GAGD is Free Fall Gravity Drainage (FFGD). In FFGD displacement, gravity is the only driving force and there is no another external force. In this process oil can be produce if only gravitational force be more than capillary force. During FFGD, gas spontaneously permeates into the block and oil produces from bottom of the block (Zendehboudi et al., 2011). Another type of GAGD is Forced Gravity Drainage (FGD). In this process gas injects into the top of the model and oil displacement by gas depends on relative magnitude of viscose and gravity forces by considering capillary force (Zendehboudi et al., 2011). Cardwell and Parsons described the concept of free fall gravity drainage theoretically for the first time and presented a theory for estimating the gravity drainage rate of a liquid out of a sand column (Cardwell Jr & Parsons, 1949). Then Briks described forced gravity drainage for special cases (Saidi, 1987). The objective of this study is to investigate the effect of top and bottom sides of fracture blockage on wet phase recovery factor during FFGD and FGD. To this aim, some experiments were designed with a fractured sand pack in three different fracture configurations sides. Recovery factor and residual liquid saturation in matrix at different fracture configurations sides are the main results of these experiments.

2 METHODOLOGY

In this paper, a novel experimental setup is used for study of gas assisted gravity drainage. The setup consists of two parts; the inner rectangular part that filled by sorted sea sands and utilized as the porous media and two rectangular spaces in both sides of the inner rectangular section simulating vertical fractures. In order to join matrix and fractures, using laser technology to perforate the surface between them. Figure 1 showed the schematic diagram of the experimental setup. The properties of the employed sand packed model are given in Table 1. In this investigation, distillated water was used as wet phase, air during FFGD and nitrogen during FGD were used as non-wet phases. Properties of the fluids are given in Table 2. The range of operational condition was listed in table 3. During this exprimental work, the following procedure is utilized at room condition:

a. The sea sands were packed in the inner rectangular part of the setup (matrix) carefully.
b. The air packets in the matrix and fractures were flushed out by using vacuum pump.
c. In the next step, model saturated by distillated water from the bottom at very law rates to ensure complete saturation of the matrix block.
d. Finally, during FGD process gas injected from the top of the model at constant rate. Recovery factor and residual liquid saturation at different fracture sides blockage were the main aims of this investigation. To this goal, nitrogen in a constant rate (20 cc/min) was

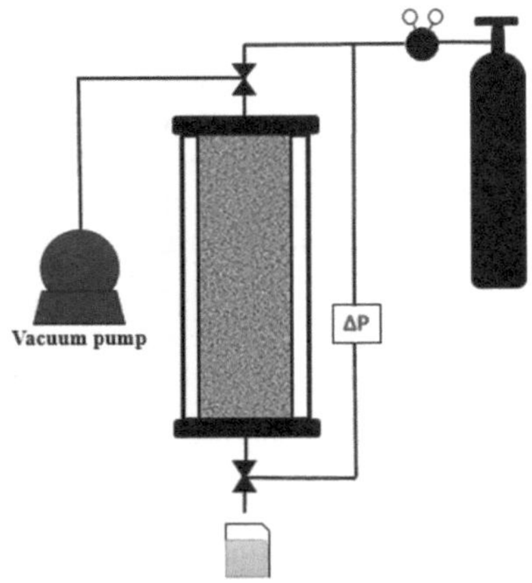

Figure 1. Schematic diagram of the experimental setup.

Table 1. Physical properties of setup

Sand size (micron)	Matrix porosity (%)	Matrix permeability (mD)	Model height (cm)	Fracture aperture (mm)
150-250	35.2	1440	100	1

Table 2. Physical properties of test fluids

Fluids	Molecular weight (g/gmol)	Viscosity (cp)	Density (gr/cm^3)
Distilled water	18.00	0.89	1.0
N_2	14.00	0.017	0.00114
Air	-	0.018	0.0012

Table 3. Range of operational condition of the experiment

Parameter	Range of parameter
Pressure	4 -18 (Psi)
Temperature	25 °C (constant)
Gas injection rate	20 (cc/min)

injected from the top of the model and distilled water as the wetting phase was produced from the lowest part of the model. In FFGD experiments, all steps prior to step d were the same as FGD tests. But during FFGD tests in the final step, the top and bottom valves of the model were opened to atmosphere suddenly.

In this investigation three different configurations of fracture sides were examined. In the first configuration, the top and bottom parts of fractures were unblocked and open to injection and production. The injected gas in FGD and air in FFGD could enter the fractures and liquid could produce from matrix and fractures straightly. In the second configuration, the lower parts of fractures were blocked and production were directly only from matrix. In the third configuration, the top and bottom parts of fractures were closed and only matrix directly contribute in injection and production. In this configuration, fractures contribute in injection and production only by perforated surface between matrix and fracture (Figure 2). Recovery factor and residual liquid saturation were recorded during the experiments.

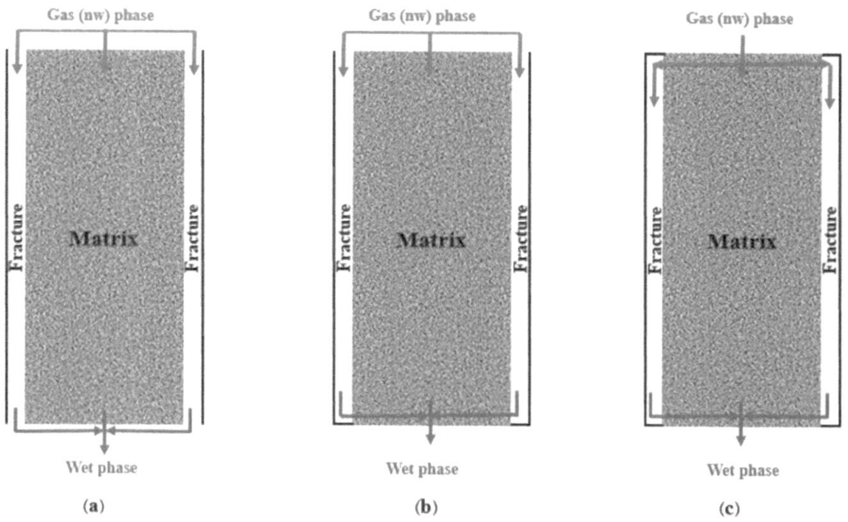

Figure 2. Schematic diagram of different fracture configurations sides. a) fractures with open top and bottom sides. b) fractures with open top and blocked bottom. c) fractures with blocked top and bottom sides.

3 DISCUSSION

In this paper, the effect of top and bottom sides of fracture blockage on recovery factor was investigated. At the beginning of each test everything was almost the same, except the fracture top and bottom sides blockage. During FGD, nitrogen was injected into the top of the model and liquid phase was produced from bottom of the model. When both sides of fractures were open, first gas progresses through fractures and surround the matrix, the density difference between the gas in fractures and water in the matrix, makes a driving force for water to drain. By invading gas into the matrix, water saturation is changed and replaced by injected gas. The gas quickly invaded into fractures and early gas breakthrough occurred and this is the reason of lower recovery factor in model with open top and bottom sides of fractures. Figure 3 shows that the model with closed both sides fractures has maximum final recovery. In this model, fractures contribute in injection and production only by perforated surface between matrix and fracture. Final recovery factor in this model was 23.5% more than model with open sides fractures. The observations showed that residual water was completely within matrix, and in fractures nothing is remained except gas. Figure 4 shows residual liquid saturation in matrix during FGD. It is evident that, minimum residual liquid saturation is in model with closed both sides fractures because of its more final recovery. Figure 5 and 6 show the recovery factor and residual liquid saturation in different top and bottom sides of fracture blockage during FFGD. Results showed that during FFGD, when both sides of fractures were open, ultimate recovery factor was 48.14%. Also, it was 53.26% when bottoms of fractures were closed. Furthermore, when both sides (upper and lowest parts) of fractures were blocked, the final recovery reached 53.72%. Results showed that during FFGD, in model with blocked fracture sides, the final recovery was around 6% higher than the model with unblocked fracture sides.

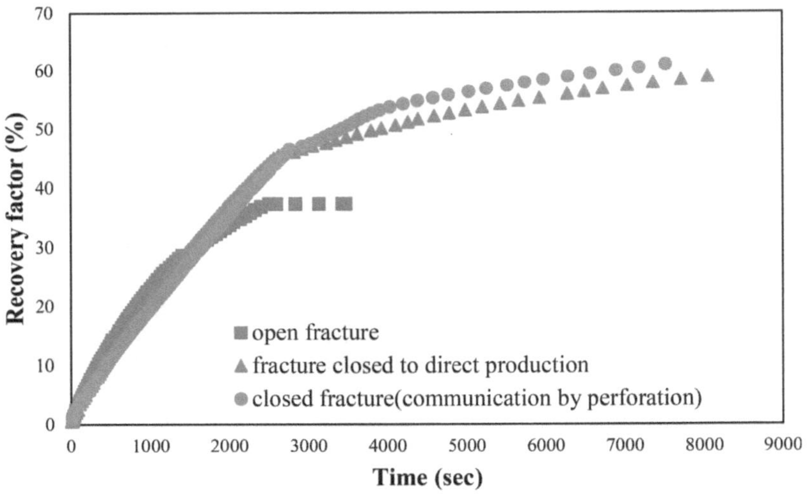

Figure 3. Effects of top and bottom sides of fracture blockage on recovery factor during FGD.

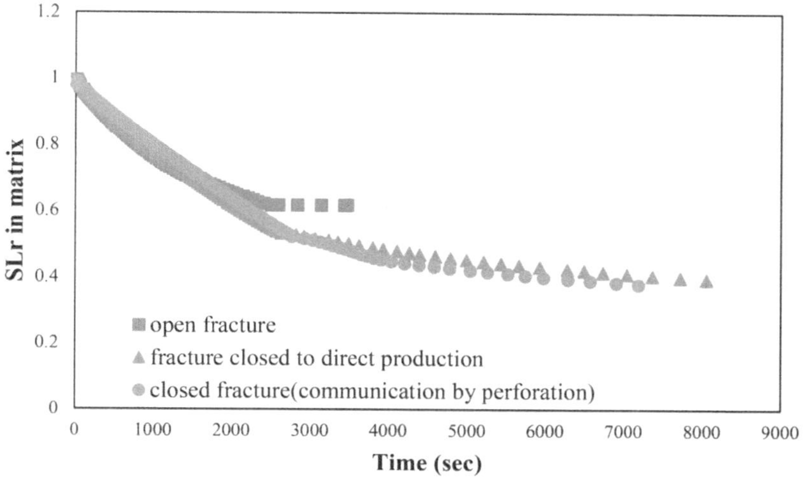

Figure 4. Effects of top and bottom sides of fracture blockage on residual liquid saturation in matrix during FGD.

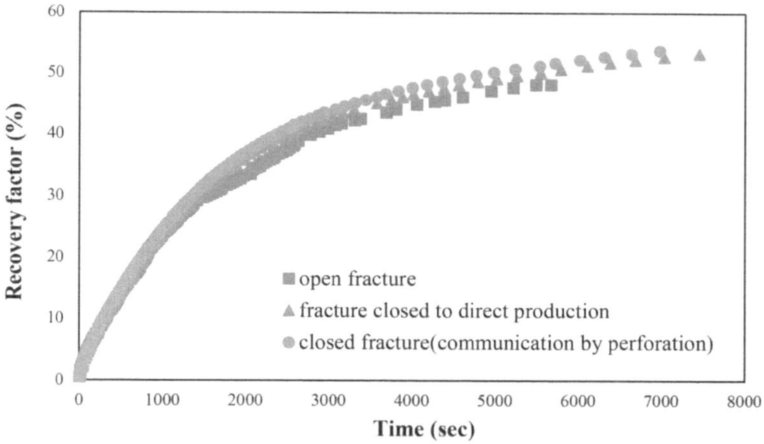

Figure 5. Effects of top and bottom sides of fracture blockage on recovery factor during FFGD.

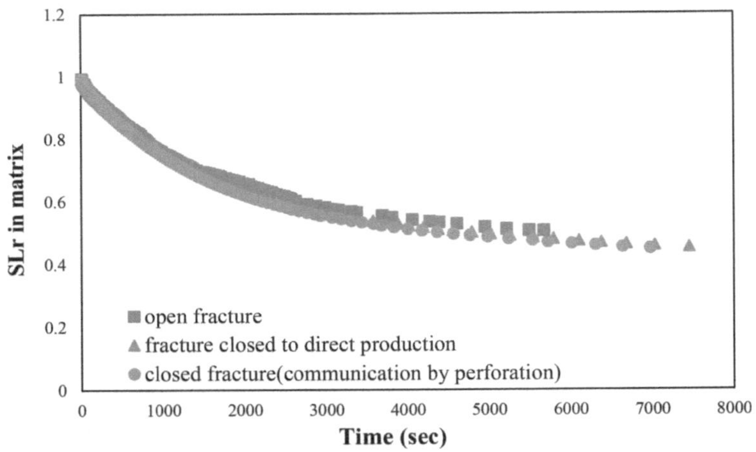

Figure 6. Effects of top and bottom sides of fracture blockage on residual liquid saturation in matrix during FFGD.

4 CONCLUSION

In this study, experimental results were obtained in visual laboratory model, for a single matrix block with two vertical fractures. Experiments have been conducted in order to investigate the effect of fracture top and bottom blockage on recovery factor. Results showed that, during both FFGD and FGD, in the open fracture model, gas invades into fractures and early gas breakthrough occurred and this is the reason of lower recovery factor at model with unblocked fracture sides. Therefore, in gas assisted gravity drainage process, recovery improved if gas could not invade into fractures and oil could not produce from fractures directly.

REFERENCES

Cardwell Jr, W., & Parsons, R. (1949). Gravity drainage theory. *Transactions of the AIME, 179*(01), 199-215.

Carlson, L. O. (1988). *Performance of hawkins field unit under gas drive-pressure maintenance operations and development of an enhanced oil recovery project.* Paper presented at the SPE Enhanced Oil Recovery Symposium.

Fassihi, M., & Gillham, T. (1993). *The use of air injection to improve the double displacement processes.* Paper presented at the SPE Annual Technical Conference and Exhibition.

JR, J. (1988). Weeks Island gravity stable CO2 pilot. *paper (SPE/DOI 17351) presented at the SPE/DOI symposium on enhanced oil recovery, Tulsa.*

Kantzas, A., Chatzis I, Dullien FAL. (1988). Mechanisms of capillary displacement of residual oil by gravity-assisted inert gas injection. *paper (SPE 17506) presented at the SPE rockey mountain regional meeting, Casper, Wyoming.*

Rezaveisi, M., Rostami, B., Kharrat, R., Ayatollahi, S., & Ghotbi, C. (2010). Experimental investigation of tertiary oil gravity drainage in fractured porous media. *Special Topics & Reviews in Porous Media: An International Journal, 1*(2).

Saidi, A. (1983). *Simulation of naturally fractured reservoirs.* Paper presented at the SPE Reservoir Simulation Symposium.

Saidi, A. (1987). Reservoir engineering of fractured reservoir- Fundamental and practical aspects book, Totl edition press, Paris.
Zendehboudi, S., Chatzis, I., Mohsenipour, A. A., & Elkamel, A. (2011). Dimensional analysis and scale-up of immiscible two-phase flow displacement in fractured porous media under controlled gravity drainage. *Energy & Fuels*, *25*(4), 1731-1750.

Youth Technical Sessions Proceedings – Litvinenko (ed)
© 2019 World Petroleum Council, London, UK, ISBN 978-0-367-34668-3

Hydrodynamic simulation of the displacement processes in WAG and SWAG enhanced oil recovery techniques: A Case study on Pannonian Basin oil field, Serbia

A. Avramovic
Department of Petroleum Engineering, Gubkin University, Moscow, Russian Federation

ABSTRACT: The aim of this study is to investigate the perspectives for application different gas EOR methods in the "Sm" oil reservoirof the "X" oil-gas field in the southeastern part of the Pannonian Basin, Republic of Serbia. Special attention is given to the Water Alternating Gas (WAG) and Simultaneous Water Alternating Gas (SWAG) EOR techniques. Studied reservoir is a massive one, with a light oil in place. Reservoir rocks are the sandstones, alevrolites and the cap rocks are the clays and marls. Study is based on the hydrodynamic modeling of the oil displacement in case of water, gas (CO2, nitrogen, saturated gas) and water-gas mixtures injection. It includes the creation, initialization and calculation of the 3D compositional hydrodynamic model using the tNavigator 18.2 software. The model is created as a segment of the initial geological model of the reservoir. It is initialized with the PVT model of reservoir fluids and the petrophysical model of the reservoir rocks. Based on the simulation results and evaluation of the main development parameters (cumulative oil production and oil recovery factor on the first place), it is shown, that the "Sm" reservoir is perspective for the application of the gas EOR methodsin general. The best EOR cases with the maximum oil recovery (up to 46%, comparing with 25% without injection) are continuous gas injection, SWAG and WAG techniques with the use of are CO2 and saturated petroleum gas.

1 INTRODUCTION

Gas enhanced oil recovery (EOR) methods stay widely used for increasing oil production in the petroleum industry worldwide. Many field development project are successfully implemented in the practice (Christensen, et al., 1998). However, the useofthe continuous gas injection (CGI) techniques showed some disadvantages, mainly connected with the fast gas breakthrough and uneven oil displacement profile due to low viscosity of injected gases (CO2, nitrogen, petroleum gas etc.) (Amin, et al., 2013; Wu, et al., 2004). Different techniques of water alternating gas (WAG) EOR method are recommended to achieve better control of the oil displacement processes. The WAGtechnique combines some advantages of pure gas displacement of the oil on the microscopic scale and, in the same time, the water displacement on the macroscopic scale (Nangacovie, 2012). These two features give significant improvement in displacement efficiency and finally – in increasing the oil recovery by WAG EOR methods (Amin, et al., 2013). Displacement of the oil with the water and gas portions provides more even displacement profile, which is the main reason for increase of oil displacement ratio (Fatemi & Sohrabi, 2013).

In the most cases, hydrodynamic modelling (HDM) of the oil displacement in the gas EOR project comes to the first place, when there is a need to evaluate some field project perspectives and its efficiency. By using the results of the HDM, it is possible to get the principal laws and quantitative parameters of gas-related EOR project. Based on this kind of results, the general conclusion on the efficiency of the gas EOR project could be made. In addition, the HDM is a base for planning and setting the further research and detail experiments.

Regarding that kind of approach, the case study is set with the aim to evaluate the perspectives and technological efficiency of the application the (S)WAG EOR techniques in comparation with the other gas EOR techniques. The study is conducted on the oil reservoir "Sm" from the "X" oil-gas field in the south-eastern part of the Pannonian basin, Republic of Serbia. In the study, we used hydrodynamic modelling of the oil displacement in case of water and gas injection.

2 RESEARCH OBJECT – GAS-OIL FIELD "X", PANNONIAN BASIN, SERBIA

"Sm[1]" oil reservoir from the "X" gas-oil field represents the main research object of this study. The "field is located in the southeastern part of the Pannonian basin withinin the Banat Region, Republic of Serbia.The field was discovered in the 1960 by the regional geophysical prospecting. The reservoir rock are connected to the Paleozoic magmatic-metamorphic complex and Neogene (Sarmatian, Panonnian, and Pontian) age sedimentary rocks (STC "NIS-Naftagas" Ltd., 2018). Very active and multiphase tectonic is present on the field (Marovic, 2001).

In the 1963, the first prospect well on the field was drilled. It was positive to oil and gas. In the period from 1963 to 2018, around 220 exploration and productive wells were drilledon the field (STC "NIS-Naftagas" Ltd., 2018). By the data, provided byZivkov, et al. (2012), the oil-gas filed "X" contains 15 oil, gas and oil-gas reservoirs. Summary geological reserves of the hydrocarbons in the "X" field are: (1) 25 000 thousand cubic meters of oil, (2) 3 million cubic meters of dissolved petroleum gas and (3) 334 million cubic meters of gas (STC "NIS-Naftagas" Ltd., 2018).

The great interest for increasing oil production of the "X" field has its biggest oil reservoir – "Sm+Pz". Ercegovac (2004) classifies this oil reservoir as a massive one with the bottom formation water (Figure 1.b). The reservoir is divided into two parts – (1) Sarmatian part, formed with alevrolites, sandstones and conglomerates and (2) Paleozoic part, formed with magmatic and metamorphic rocks (granite, greenschist etc.). The cap rocks of this reservoir are Pannonian and Lower Pontian marls and clays (Zivkov, et al., 2012) (Figure 1.b).Regarding the low information on the geology and the reservoir characteristics in the Paleozoic part of the reservoir, and also the fracture type of the reservoir rock, we chose the Sarmatian part (Sm) of the "Sm+Pz" oil reservoir as a main research object for this case study.

Reservoir rocks in the Sarmatian part of the reservoir are mainly terrigenous sedimentary rocks. Sandstones, alevrolites and conglomerates represent the most part of reservoir. Average petrophysical parameters for the Sarmatian (Sm) reservoir are showed in Table 1. These rocks are very good reservoir and its great heterogeneityis consequence of stratigraphic and lithological changes (Zivkov, et al., 2012). Summary geological reserves of oil in the "Sm+Pz" reservoir are 18, 2 million cubic meters (STC "NIS-Naftagas" Ltd., 2018).

Table 1. Petrophysical parameters of Sarmatian age reservoir rocks (Zivkov, et al., 2012).

Petrophysical property, unit of measure	Interval of Values	Average Value
Porosity, %	8 – 39	19
Absolute permeability, mD	0.5 – 869	174
Water saturation, %	-	0.27
Oil saturation, %	-	0.73
Effective thickness of reservoir, m	2 – 41	7.4

1. All titles and geographical location of the research object and it's part (field, reservoir etc.), location and registration codes of wells, and all depth references in this study are replaced with conditional titles due to need to preserve confident geological and business information of the data owner – "Scientific and Technological Center "NIS-Naftagas" Ltd." (Novi Sad, Republic of Serbia).

With the system of hydrodynamic isolated faults, reservoir is divided into three blocks – A, B and C. Water-oil contacts in the blocks are on the different depths: -1907 m (block A), -1921 m (block B) and -1906 m (block C) (Figure 1.a). Average value of the reservoir pressure is 19.5 to 20.5 MPa. Reservoir temperature is 111.5°C (STC "NIS-Naftagas" Ltd., 2016).

The "Sm" reservoir is saturated with the light, low viscous and paraffinic oil (density – 830.6 kg/m3, viscosity 0.566 cP, 13% of paraffin's). Oil contains about 126 m3/m3 of dissolved wet gas (82% methane). The reservoir water is neutral (pH 7, 04) and high mineralized (around 14 mg/l, mainly Cl, K, Na ions) (Zivkov, et al., 2012). By the results of oil and gas PVT analysis from the depth sample (STC "NIS-Naftagas" Ltd., 1971), the bubble-point pressure of oil is determined at 22.6 MPa. Oil formation volume factor is 1.47.

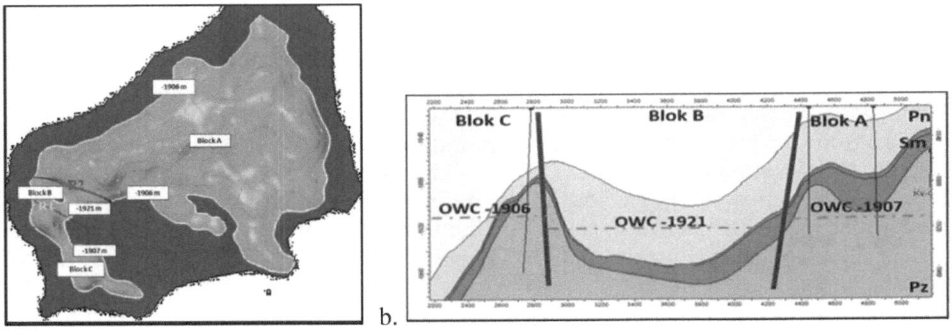

Figure 1. a. Structural map of the "Sm+Pz" reservoir, mapped on the top of the Sarmatian sediments. b. Schematic geological profile of the "Sm+Pz" reservoir, showing reservoir heterogeneity and block structure (STC "NIS-Naftagas" Ltd., 2018).

The "Sm+Pz" oil reservoir is in the production since 1965 (Zivkov, et al., 2012). Cumulative oil production from the reservoir is 5 million tons (as for 2018), which represents 70% of the "X" field total oil production. Oil recovery is around 33% (STC "NIS-Naftagas" Ltd., 2018). There is a perspective for the continuing field development, by using new approaches and EOR methods.

3 CASE STUDY METHODOLOGY AND TECHNIQUES

The methodological workflow of this study is set to provide good quality evaluation of the perspectives for application different gas EOR methods at the "Sm" oil reservoir. The base for this study represents hydrodynamic modeling of the oil displacement with the injection of gas, water and its mixtures. General workflow of this study is showed at Figure 2.

As an essential start of the study, the literature review on gas EOR methods, especially WAG techniques is conducted. Great attention is put on the physical mechanism of oil displacement in gas EOR, methodological approaches for projecting EOR and the techniques for conducting hydrodynamic simulation of researched processes.

The initial part of this study is represented by the screening for the application of the gas EOR (especially WAG) on the Serbian part of the Pannonian basin oil fields. This is done with the aim to narrow the research process to the specific object and to select the best appropriate reservoir/filed for this research. The screening criteria are determined based on the publications (Nazarova, 2011) and the efficiency of already conducted commercial projects (Christensen, et al., 1998). Screening included simple statistical comparation of the reservoir average parameters with the formed criteria base. Based on the screening results, the main research object, with the most perspectives for the application of gas EOR methods is selected. It is necessary to acquire and analyze all reservoir data that is connected with the planned

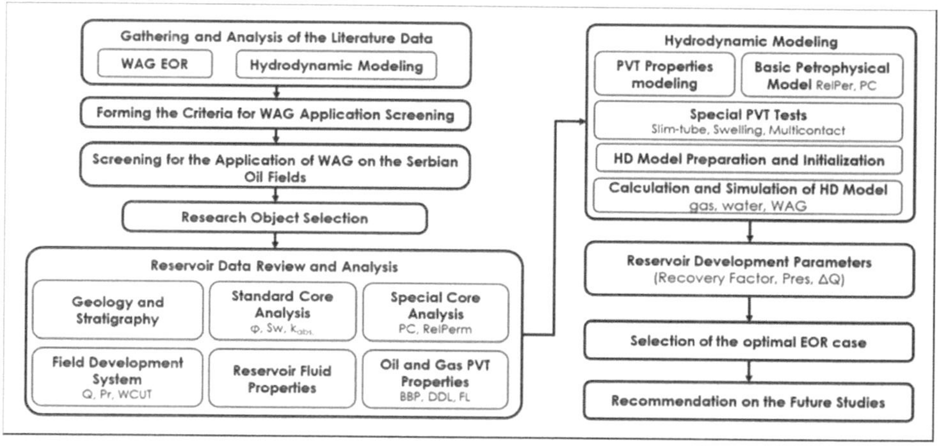

Figure 2. Methodological workflow of the study.

research. That data complex includes reservoir geology, petrophysics, properties of reservoir fluids, development and production history etc.

The main part of the research is representedbythe preparation, initialization, simulation and evaluation of the hydrodynamic model of oil displacement in the studied "Sm" reservoir due to injection of water, gas and its mixtures. Hydrodynamic modeling (HDM) requires several important methodological phases, that are providing the maximum available similarity of the created model and the real research object (Kanevskaya, 2002). At the all phases of the HDM in this project, we put an effort to follow industry standards and methodological recommendations on analysis and preparation the complex of entry data for the model (A.P. Krylov All-Russian Oil and Gas Research Institute j.s.c., et al., 2000). In addition, we respected the standards for creating and calculation of the model, and the rational evaluation of the technological efficiency of the studied technologies based on obtained HDM results (VNIIOENG j.s.c., 2003).

Updated geological model of the "Sm" reservoir (STC "NIS-Naftagas" Ltd., 2018) is used as a base for creating the hydrodynamic model in this study. The initial geological model is upscaled and modified in Petrel software (Schlumberger, 2017) to fit the required model size and calculation time. For this study, we decided to use HDM on the closed rectangular reservoir segment, due to large size of the initial geological model and not enough data to do the hystory matching and simualate present saturation for the whole reservoir. The reservoir segment is chosen based on the ammount of the residual oil reserves, effective thickness of the reservoir rocks and the dynamics of reservoir pressures, by the methodic showed in Zahoor et al. (2011).

One of the most important phase of the research, when is up to preparation of hydrodynamic model, is the modeling of the PVT properties of oil and gas, based on the results of laboratory experiments. The PVT modeling included interpretation of the main properties of reservoir fluids (oil viscosity and density, gas solution factor, oil and gas formation volume factor, gas viscosity etc.) as a function of the thermobaric conditions in the reservoir. This procedure is done using mathematical correlations in software PVTi 2010.2 (Schlumberger, 2011) with the standard approach. The compositional model of oil and gas is selected to ensure the best evaluation of the reservoir fluids behavior during the displacement processes.

Beside the standard PVT model, we conducted the special injection-based PVT tests by the industry approved methodic (Schlumberger, 2013). The aim of these tests is to provide the better understanding of the oil displacement and injection processes, its physical specifics and to improve the basic PVT model (STC "NIS-Naftagas" Ltd., 2016). Several special PVT test are conducted, using computer simulators, including: slim-tube, swelling and multi-contact mixing test.

With the aim to ensure better understanding of the reservoir rock properties, the basic petrophysical modeling is created. It includes the porosity and permeability models, and the initial and residual fluid saturation models. Important part of this phase represents the creation of relative phase permeability model (Zolotukin & Ursin, 2000)based on the laboratory data from core experiments (Zivkov, et al., 2012). Last, but not the least, is the creation of capillary pressure model (Brooks & Corey, 1966) to ensure realistic evaluation of the reservoir saturation profile.

The 3D hydrodynamic model is initialized using correlated reservoir fluids PVT model and obtained petrophysical models. The initialization, calculation and evaluation of the HDM is done using hydrodynamic simulator tNavigator 18.2 (Rock Flow Dynamics, 2017). The compositional type of HD model with 16 components is chosen, to ensure better simulation of the gas injection processes. On this kind of model, the different cases of gas EOR processes are simulated and calculated: (1) continuous water flooding, (2) continuous gas injection (CO2, nitrogen, and saturated petroleum gas), (3) water alternating gas and (4) simultaneous water alternating gas. All cases are created with the comparable injection pressures and flow rates, to follow the similarity criteria of the model.

Evaluation of the EOR efficiency is done by comparing the main reservoir development parameters (cumulative oil production, oil recovery, reservoir pressure, water cut etc.) by the methodic of Darvishney had, et al. (2010).According to that parameters, the comparation between different simulated EOR cases is done to ensure the selection of the most optimal case. The evaluated reservoir development period in all simulated EOR cases is 20 years, regarding methodologic instructions (Pyatibratov, 2015). Regarding the fact that the reservoir segment model is used for evaluation of the reservoir development parameters, the history matching procedure is not performed. The present saturation is modeled using the fluid saturation coefficients for different regions. This approach is based on the practical techniques, showed in STC "NIS-Naftagas" Ltd., (2016).

4 RESULTS AND DISCUSSION

As a result of the conducted screening for the application of the gas EOR methods (including (S)WAG techniques) on the 40 Serbian oil-gas fields (data provided by STC "NIS-Naftagas" Ltd., 2016) by the addopted criteria and methodology, the "Sm" oil reservoir from the "X" oil-gas field is selected as a perspective for application of gas EOR methods.

By correlating the PVT laboratory analysis results (flash test, differential liberation test) with the analytical models of the fluid behavior, the general PVT model of studied oil and gas is composed. The model consists of ten components and one pseudo component (heavy oil fraction). Component composition of oil is obtained from the laboratory results of the gas chromatography, using computer simulation of the oil recombination. The four main parameters, significant for the further hydrodynamic modeling of oil displacement is correlated using regression technique with the laboratory PVT experiments: oil viscosity and density, gas and oil formation volume factor and oil-gas ratio. The obtained PVT model is a good quality one and it is used as input parameter for the initialization of the hydrodynamic model. The results of special PVT tests (oil swelling factor, first contact and multiple contact minimal mixing pressure etc.) are used to update obtained PVT model for the water and gas injection processes.

The petrophysical modeling of reservoir properties gave the porosity, permeability and saturation models as input parameters for the HD model initialization. In addition, the relative phase permabilities and capillary properties of reservoir rock are obtained from petrophysical modeling. Reservoir rocks are very lithological heterogeneous, which makes interpretation of petrophysical properties more difficult.

The 3D compositional hydrodynamic model of the reservoir sector is created from the geological model of the "Sm" oil reservoir. It represents the compositional model with

16 components. The model is discretized by 45x35x22 grid blocks of 20 m size with 27557 active cells. In total, the model contains 5 wells – 1 injection (Inj-1) and 4 production (Prod-1,2,3,4) (Figure 3.a).

The results of 3D hydrodynamic modeling of the oil displacement processes in case of the injection of the gas, water and its mixtures shows, that "Sm" reservoir is suitable for the application of the gas EOR processes. Based on simulated development parameters for the project period of 20 years (cumulative oil production and oil recovery factor), all studied gas EOR techniques give significant increase in the oil recovery, when compared with the base case (25% oil recovered). The best EOR cases are: (1) CO2 and saturated gas continuous injection (42.2 % and 46.1% of oil recovered, respectfully), saturated gas and CO2 SWAG (45.2 % and 43.0%) and CO2 and saturated gas WAG (34.2 % and 33.9%).

Obtained results has more qualitative nature than the pure quantitative, due to complexity in simulation the pore scale processes during the water and gas injection. There is need for more detail modeling of the physical mechanisms of the oil displacement. For the better understanding of the oil displacement and more accurate prognosis of reservoir development it is necessary to: (1) conduct more detail hydrodynamic modeling, (2) to create different model types (e.g. pseudo-compositional model of Todd-Longstaf), (3) to apply several methodological approaches for the HDM, (4) to test few PVT model interpretation techniques and especially - (5) to conduct the laboratory filtration experiments on cores, that will provide the physics of the studied processes.

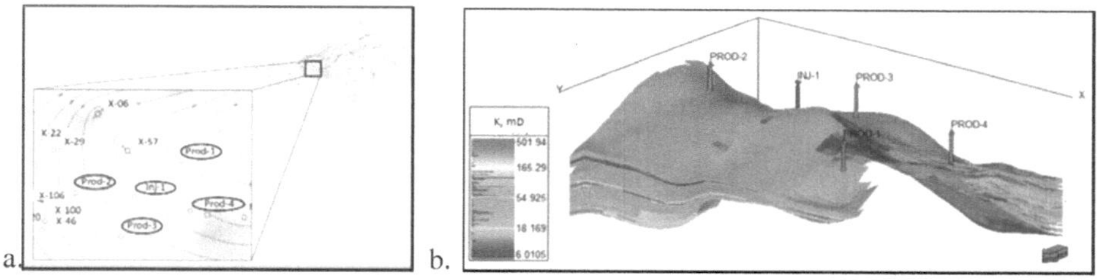

Figure 3. Composed 3D hydrodynamic model of the closed rectangle "Sm" reservoir sector, based on the updated geological model. a. The reservoir segment scheme on the effective thicknesses and reservoir pressures map. b. 3D model of permeability.

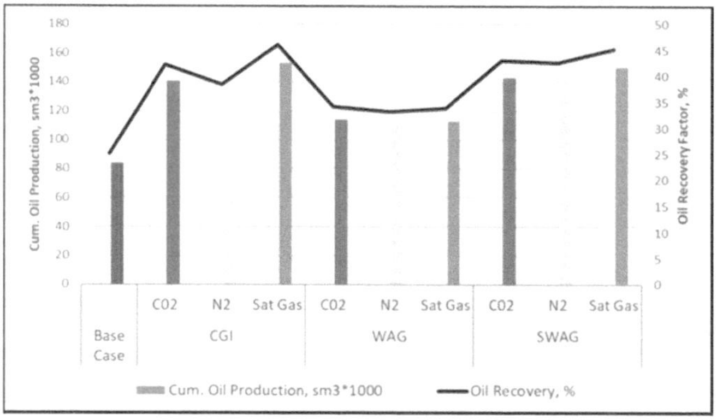

Figure 4. Comparation of the cumulative oil production and recovery factors for studied gas EOR cases for the development project period of 20 years.

5 CONCLUSION

This case study is carried out with the aim to evaluate the possibilities for application different gas EOR methods (including WAG and SWAG techniques) at the "Sm" oil reservoir from the "X" oil-gas field in the Pannonian basin, Serbia. The main part of the study represents the creation, initialization and simulation of the oil displacement due to injection of water, gas and its mixtures, using 3D hydrodynamic compositional model of the reservoir segment. By the evaluation of the main development parameters (cumulative oil production and oil recovery factor on the first place) it is shown, that "Sm" reservoir is perspective for implementing the gas EOR projects in general. The best EOR cases, which gives maximum oil recovery from the reservoir, are CO2 and saturated gas injection, with CGI, SWAG and WAG techniques. These techniques provide up to 46% of oil recovery from the studied reservoir, comparing to the 25% without injection. It is recommended to do further research on the improvement of hydrodynamic modelling techniques. In addition, it will be useful to conduct the filtration experiments on cores, to provide better understanding of the physical mechanisms on the pore scale, occurring during the oil displacement.

ACKNOWLEDGEMENTS

The material, showed in this paper, is a result of research-oriented internship during Summer Internship Programme at Division for Geology and Field Development of Scientific and Technological Centre NIS-Naftagas in July and August 2018. The author would like to express his gratitude to Mr. Andrey Antropov, MSc, expert-coordinator for hydrodynamic modelling for his supervision and generous help during research. Results of this research are included in the author's B.Sc. thesis at the Department of Petroleum Engineering of the Gubkin University (Moscow, Russia) under the mentorship of the Assoc. Prof.Irena Yazyinina, PhD.

REFERENCES

A.P. Krylov All-Russian Oil and Gas Research Institute j.s.c., Central Geophysical Expedition j.s.c, Gubkin Russian State University of Oil and Gas & Institute for the Problems of Oil and Gas Industry, 2000. *Regulations on the creation of permanent geological and technological models of oil- gas and oil fields (RD 153-39.0-047-00)*, Moscow: Ministry of Energetic of the Russian Federation.

Amin, M. E., Zekri, A. Y., Almehaideb, R. & Al-Attar, H., 2013. Optimization of CO2 WAG Processes in a Selected Carbonate Reservoir: an Experimental Approach. *International Journal of Engineering Research and Science & Technology*, 2(2), pp. 19-31.

Brooks, R. H. & Corey, A. T., 1966. Proporties of Porous Media Affecting Fluid Flow. *Journal Irrigation and Drainage Division (ASCE)*, Volume 92, pp. 61-68.

Christensen, J. P., Stenby, E. H. & Skauge, A., 1998. *Review of WAG Field Experience*. Villahermosa, Mexico, Society of Petroleum Engineers, pp. 357-371.

Darvishneyhad, M. J. et al., 2010. *Study of Various Injection Patterns in Water, Gas and WAG Injection Methods in an Iranian Fractured Reservoir*. Teheran, 14th Intl. Oil, Gas and Petrochemical Congress.

Ercegovac, M., 2004. *Petroleum Geology*. Belgrade, Novi Sad: Faculty of Mining and Geology, University of Belgrade; Society of Engineers and Technicians "NIS-Naftagas".

Fatemi, M. S. & Sohrabi, M., 2013. Experimental Investigation of Near-Miscible Water-Alternating-Gas Injection Performance in Water-Wet and Mixed-Wet Systems. *SPE Journal*, Issue 2, pp. 114-123.

Kanevskaya, R. D., 2002. *Mathematical Modeling of the Hydrodynamic Processes in the Hydrocarbon Field Development*. Moscow-Izhevsk: Institute for Computer Research.

Marovic, M., 2001. *Geology of Yugoslavia*. Belgrade: Faculty of Mining and Geology, University of Belgrade.

Nangacovie, H. L. M., 2012. *Application of WAG and SWAG Injection Techniques in Norne E-Segment*, Trondheim: Norwegian University of Science and Technology.

Nazarova, L. N., 2011. *Development of the Oil Reservoirs with Hard-To-Recover Reserves*. Moscow: Publishing Center, Gubkin Russian State University of Oil and Gas.

Pyatibratov, P. V., 2015. *Hydrodynamic Modeling in Oil Field Development*. Moscow: Publishing Center, Gubkin Russian State University of Oil and Gas.

Schlumberger, 2013. *Miscible Gas Injection Studies*, Calgary: Schlumberger.

STC "NIS-Naftagas" Ltd., 1971. *Report on PVT analysis of the fluid sample from well XX-22*, Novi Sad: Laboratory Upstream, Scientific and Technological Center "NIS-Naftagas" Ltd.

STC "NIS-Naftagas" Ltd., 2016. *CO2 EOR*, Novi Sad: Department of Geology and Reservoir Engineering, Scientific and Technological Center "NIS-Naftagas" Ltd.

STC "NIS-Naftagas" Ltd., 2018. *Deep Dive Geological Model of the Sm-Pz Reservoir, "X" Field (Phase 2)*, Novi Sad: Department of Geology and Reservoir Engineering, Scientific and Technological Center "NIS-Naftagas" Ltd.

VNIIOENG j.s.c., 2003. *Guidelines for creating a permanent geological and technological models of oil-gas and oil fields (Part 2. Filtration models)*. Moscow: VNIIOENG j.s.c.

Wu, X., Ogbe, D. O. & Khataniar, S., 2004. *Critical Design Factors and Evaluation of Recovery Performance of Miscible Displacement and WAG Process*. Calgary, Alberta, Petroleum Society of Canada.

Zahoor, M. K., Derahman, M. N. & Yunan, M. H., 2011. WAG Process Design - an Updated Review. *Brazilian Journal of Petroleum and Gas*, 5(2), pp. 109-121.

Zivkov, J., Bejatovic, M. & Trifunovic, P., 2012. *The Study on the Oil, Gas and Gas Condesate Reserves of the "X" Field*. Novi Sad: Scientific and Technological Center "NIS-Naftagas" Ltd.

Zolotukin, A. B. & Ursin, J.-R., 2000. *Fundamentals of Petroleum Reservoir Engineering*. Kristiansand S: Høyskoleforlaget AS – Norwegian Academic Press.

Youth Technical Sessions Proceedings – Litvinenko (ed)
© 2019 World Petroleum Council, London, UK, ISBN 978-0-367-34668-3

Development of oil recovery stimulation technology for carbonate reservoirs using chlorine-free acid systems

D.V. Nuriev
PJSC TATNEFT, TatNIPIneft Institute, Russia

ABSTRACT: In oilfield practice, acid treatments with hydrochloric acid are the most widely used. However, hydrochloric acid contributes to the formation of organochlorine compounds both in reservoir and in oil processing. The presence of organochlorine compounds in the produced fluids leads to hydrochloric corrosion of equipment, which is very dangerous for oil refining processes. Furthermore, a large number of iron ions enter the underground reservoirs with hydrochloric acid. Iron ions form secondary sediments, which damage formation and decrease the efficiency of acid treatments. From what has been said above it is clear that, it is very important to develop hydrochloric-free acid systems for matrix acidizing. Aim of this study was development of new oil recovery stimulation technology for carbonate reservoirs using chlorine-free acid system. The components of chlorine-free acid systems were sulfamic acid, ammonium acetate, surfactant, acetic acid, corrosion inhibitor. The complex of laboratory studies included researching of interfacial tension, rate of steel corrosion, secondary sedimentation after neutralization of acid, rate of dissolution of marble, compatibility of new chlorine-free acid systems with reservoir fluids. The compatibility of new chlorine-free acid systems with reservoir fluids was investigated by using brine water and crude oils selected from carbonate reservoirs of various fields of PJSC TATNEFT. The developed acid systems reduced the interfacial tension at the boundary with crude oil to less than 0.001 N/m. Corrosion activity of the new acid systems was at a low level, which allowed injecting large volumes of acid without significant damage of oilfield equipment. Secondary sedimentation of salts after neutralization of acid was absent. The developed chlorine-free acid systems dissolved the marble much slower compared to other acids. The new acid systems are fully compatible with formation fluids. New well stimulation technology has been tested since 2017. The absolute average increase in oil production from treated wells was 0.7 tons/day (35 %). The absolute average increase of the fluid flow rate was 1.0 tons/day (41.6 %). By this method more than 3,000 tons of additional oil was produced with the continuing benefit from treatment.

1 INTRODUCTION

One of the main ways to increase production of hydrocarbons from carbonate reservoirs is well stimulation through increase of permeability of the bottomhole formation zone by acid treatment (Michael J. Economides et al., 1994).

In oilfield practice, acid treatments with hydrochloric acid are the most widely used (Glushchenko et al., 1992, Tokunov et al., 2004, Glushchenko et al., 2010). However, hydrochloric acid contributes to the formation of organochlorine compounds both in reservoir and in oil processing. The presence of organochlorine compounds in the produced fluids leads to hydrochloric corrosion of equipment, which is very dangerous for oil refining processes (Tatianina et al., 2017).

Furthermore, a large number of iron ions enter the underground reservoirs with hydrochloric acid, because of corrosion of oilfield equipment during injection of acid and corrosion of acid containers during long-term acid storage (Glushchenko et al., 2010, Folomeev et al., 2013).

Iron ions form solid hydrates of iron hydroxide when the acid neutralized. Iron ions also initiate the precipitation of asphaltenes. These sediments damage formation and, ultimately, decrease the efficiency of acid treatments (Galimov et al., 2010, Davletshina et al., 2016, Sharov et al., 1983).

From what has been said above it is clear that, it is very important to develop hydrochloric-free acid systems for matrix acidizing.

2 CHLORINE-FREE ACID SYSTEMS

Sulfamic acid (HSO_3NH_2) is a good alternative to a hydrochloric acid especially for low-temperature reservoirs, typical for the fields of PJSC TATNEFT (an average of 30 °C). Sulfamic acid effectively dissolves calcium and magnesium carbonates to form water-soluble salts (reaction 1 and 2) (Amiyan et al., 1970, Kazakova et al., 2005).

$$2HSO_3NH_2 + CaCO_3 = (NH_2SO_3)_2Ca + H_2O + CO_2 \tag{1}$$

$$2HSO_3NH_2 + MgCO_3 = (NH_2SO_3)_2Mg + H_2O + CO_2 \tag{2}$$

Sulfamic acid does not contain chlorine, so the possibility of the formation of organochlorine compounds in oil is completely excluded. Sulfamic acid is less active than hydrochloric acid. It has lower corrosion activity and lower speed of reaction with carbonates. In addition, the sulfamic acid is a powder, so the acid can be prepared immediately before injection. So, the use of sulfamic acid significantly reduces the amount of iron ions getting into the reservoir.

To increase the efficiency of matrix acidizing, TatNIPIneft specialists have developed chlorine-free acid systems. Russian patents protect new acid systems available in two modifications, CFAS-1 and CFAS-2. The main components of the chlorine-free acid compositions are presented in Table 1 below. CFAS-1 system is used for new wells, which have never been acidized before. CFAS-2 system has a higher dissolving capacity, thus it can be used for repeated acidizing.

Table 1. Chlorine-free acid systems

CFAS-1		CFAS-2	
Component	Action	Component	Action
Sulfamic acid	Solvent of carbonate rock	Sulfamic acid	Solvent of carbonate rock
Ammonium acetate	pH-regulator, retarder, corrosion inhibitor, iron control additive	Acetic acid	Increasing of dissolving capacity, iron control additive
Surfactant	Reducer of interfacial tension, wetting agent, corrosion inhibitor	Ammonium acetate	pH-regulator, retarder, corrosion inhibitor, iron control additive
		Surfactant	Reducer of interfacial tension, wetting agent, corrosion inhibitor

3 LABORATORY STUDIES

The complex of laboratory studies confirms the efficiency of the developed acid systems. Physical and chemical parameters of CFAS-1 and CFAS-2 are presented in Table 2.

The developed acid systems reduce the interfacial tension at the boundary with crude oil to less than 0.001 N/m. This effect is achieved due to use of nonionic surfactants, which are stable in a wide range of pH and salinity. After acid neutralization, the acid systems retain a low interfacial tension, which ensures better extraction of the spent acid after reaction. Corrosion activity of the new acid systems is at a low level, which allows injecting large volumes of acid without significant damage of oilfield equipment. Secondary sedimentation of salts after neutralization of compositions is absent.

Table 2. Physical and chemical parameters of CFAS-1 and CFAS-2

Parameters	Value	
	CFAS-1	CFAS-2
Outward appearance	Colorless transparent liquid	Colorless transparent liquid
Interfacial tension at the boundary with formation oil		
• before reaction	< 0.001 N/m	< 0.001 N/m
• after neutralization of acid	< 0.001 N/m	< 0.001 N/m
Rate of steel corrosion, g/(m2·h)	0.1-0.2	0.1-0.2
Secondary sedimentation after neutralization of acid	Not found	Not found

One of the main factors that determines the efficiency of the acid treatment is the nature of interaction between the acid and the carbonate rock. In this work, the interaction acid with the rock was estimated by determining the rate of dissolution of marble (calcium carbonate content of 99.7 %) and the dissolving capacity of calcium carbonate for the following acid systems: inhibited hydrochloric acid (HCl (in.)), nitrogen acid (HNO_3), sulfamic acid without additives (NH_2SO_3H), CFAS-1, CFAS-2. The results are presented in Figure 1 and Figure 2.

The results show that both developed chlorine-free acid systems dissolve the marble much slower compared to other acids (Figure 1). At the same time, the dissolving capacities of CFAS-1 and CFAS-2 have a sufficiently high level (Figure 2). Thus, the use of CFAS-1 and CFAS-2 provides a deeper penetration of acid into the formation while maintaining reactivity. These effects increase the efficiency of acid treatments.

The compatibility of new chlorine-free acid systems with reservoir fluids was investigated by using brine water and crude oils selected from carbonate reservoirs of various fields of PJSC TATNEFT. The properties of these fluids are presented in Table 3 and Table 4.

All presented samples of oil from carbonate reservoirs have high content of asphalt-resinous substances, which are the main stabilizers of water-oil emulsions. According to the density value, three of the four samples of oil belong to the bituminous type, and the sample of oil from the Ersubaikinskoye field belongs to the heavy oil type. All oil samples are particularly high-sulfur. Brine waters from all fields belong to chloride-calcium type.

The compatibility test was carried out in the following sequence: mixing of acidic compounds with formation fluids, exposing these mixtures to a reservoir temperature and visually

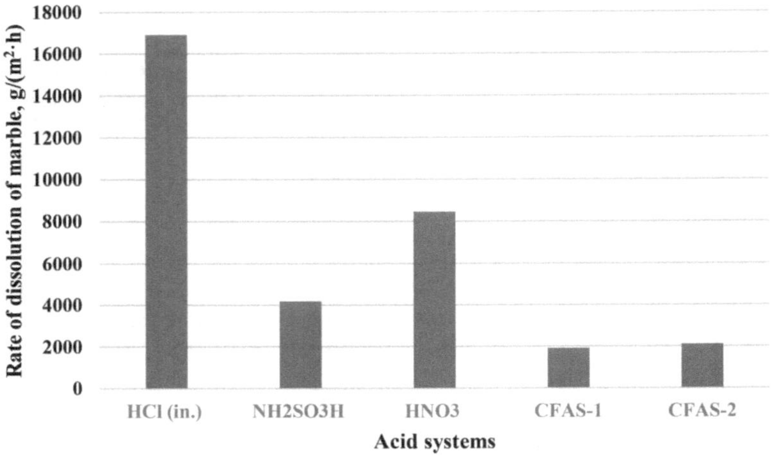

Figure 1. Rate of dissolution of marble for different acids

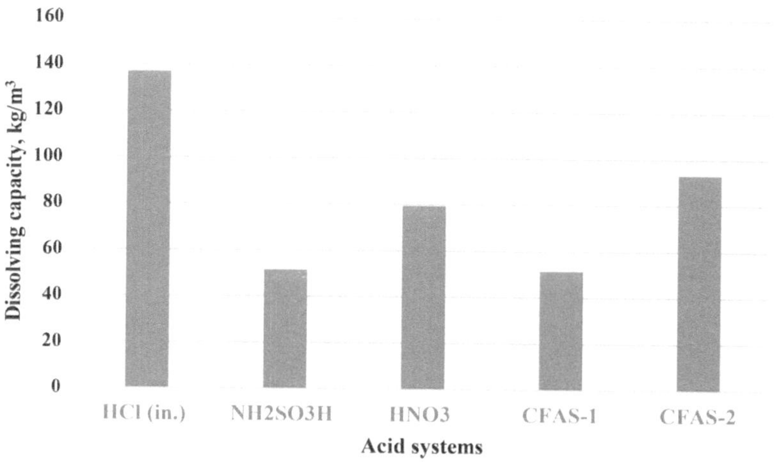

Figure 2. Dissolving capacity for different acids

Table 3. Composition, physical and chemical properties of crude oils

Oil field	Concentration, wt % Sulfur	Asphaltenes	Resins	Dynamic viscosity at 20 °C, Pa·s	Density at 20 °C, g/cm^3
Arhangelskoe	3.98	6.38	13.88	0.1252	0.916
Yamashinskoe	4.22	6.72	16.04	0.0856	0.916
Ersubaikinskoe	3.58	6.96	16.34	0.0359	0.880
Shegurchinskoe	3.56	5.51	20.55	0.0654	0.907

Table 4. Composition, physical and chemical properties of of brine water

Oil field	Concentration of ions, g/dm^3						Type of water	Density at 20 °C, g/cm^3	The ionic strength of the solution, mol/dm^3
	Cl^-	SO_4^{2-}	HCO_3^-	Ca^{2+}	Mg^{2+}	Na^+, K^+			
Arhan-gelskoe	136.460	0.513	0.122	10.020	3.040	70.614	calcium chloride	1.144	4.073
Yam-ashinskoe	143.939	0.733	0.366	10.421	4.134	77.319	calcium chloride	1.156	4.520
Ersubai-kinskoe	101.396	0.634	0.238	9.118	2.554	53.082	calcium chloride	1.119	3.216
Shegur-chinskoe	136.140	0.247	0.110	10.521	3.526	72.757	calcium chloride	1.152	4.257

estimating the separation of the mixtures. Emulsifying properties of the compositions were determined by estimating the stability of the emulsions prepared from oil and acid composition versus the emulsion prepared from oil and brine water. The results are presented in Table 5.

The obtained results (Table 5) show that CFAS-1 and CFAS-2 are fully compatible with formation fluids. No emulsions and secondary sediments are formed.

Table 5. Results of compatibility test

Acid system	Sediments after mixing with crude oil	Sediments after mixing with brine water	Stabilization of high-viscosity emulsions
Arhangelskoe			
CFAS-1	Not found	Not found	Not found
CFAS-2	Not found	Not found	Not found
Yamashinskoe			
CFAS-1	Not found	Not found	Not found
CFAS-2	Not found	Not found	Not found
Ersubaikinskoe			
CFAS-1	Not found	Not found	Not found
CFAS-2	Not found	Not found	Not found
Shegurchinskoe			
CFAS-1	Not found	Not found	Not found
CFAS-2	Not found	Not found	Not found

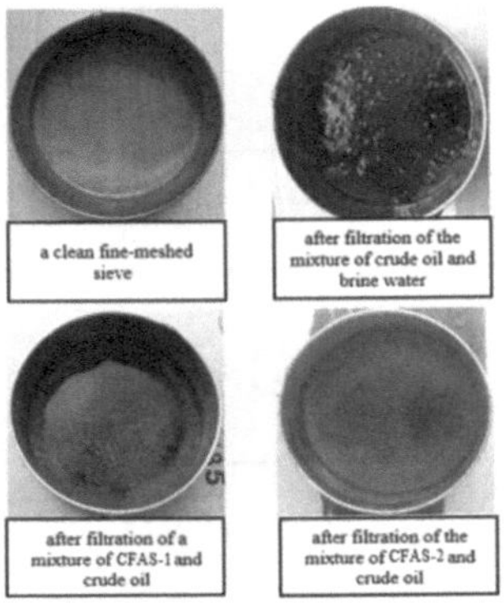

Figure 3. Photos of fine-meshed sieve

The possibility of the precipitation of asphaltenes, resins and paraffins after contact of oil with acid systems was investigated by filtering the mixture acid and crude oil from the Ersubaikinskoye field through a fine-meshed iron sieve (Figure 3). It can be seen from Figure 3, that the use of CFAS-1 and CFAS-2 prevents the precipitation of asphaltenes, resins and paraffins.

According to the results of laboratory studies, the new effective acid systems were obtained.

4 PILOT TEST

New well stimulation technology has been tested since 2017. Table 6 shows the results of ten treatments of producing wells (as of Feb 1, 2018).

The volume of injected acid composition varied in the range from 15 m3 to 35 m3. The absolute average increase in oil production from treated wells is 0.7 tons/day (35 %). The absolute average increase of the fluid flow rate is 1.0 tons/day (41.6 %).

By this method more than 3,500 tons of additional oil was produced with the continuing benefit from treatment. Average cost of one treatment is RUB 600,000.

The expected economic effect is presented in Table 7.

Table 6. Results of pilot test of new well stimulation technology (as of Oct 1, 2018)

Well number	Oil flow rate, tons/day Before CFAS injection	After CFAS injection	Increment	Fluid flow rate, tons/day before	after	diff.	Volume of acid system, m^3	Benefit from treatment (month)*
28431	1.2	1.9	+0.7	1.7	2.3	+0.6	30	15
3926	2.2	3.3	+1.1	2.5	3.5	+1.0	20	12
8039	2.8	2.9	+0.1	2.9	3.0	+0.1	15	15
1228	2.8	3.5	+0.7	3.0	4.2	+1.2	20	15
4653	2.3	3.4	+1.1	2.6	4.9	+2.3	20	13
3860	2.4	2.8	+0.4	2.6	3.1	+0.5	20	12
2019	3.1	3.1	0	4.5	4.2	-0.3	35	15
1099	0.6	1.4	+0.8	0.6	1.4	+0.8	32	15
1923	1.8	2.3	+0.5	2.4	4.1	+1.7	15	10
1518	0.9	2.3	+1.4	1.2	3.3	+2.1	15	15
av.vol.	**2.0**	**2.7**	**+0.7**	**2.4**	**3.4**	**+1.0**	**22.2**	**13.7**

* **Note: technological effect continues.**

Table 7. Expected economic effect from commercial introduction of new acid chlorine-free systems

Item	Indicators	Value
1	Additional oil, tons	600.0
2	Oil sales price, RUB/tons	13,800.7
3	Mineral extraction tax, RUB/tons	6,697.9
4	Net proceeds, RUB	8,280,420
5	OPEX (oil production), RUB	465,000
6	Costs for introduction of technology, RUB	600,000
7	Mineral extraction tax, RUB	4,018,700
8	TOTAL OPEX, RUB	5,083,700
9	Production costs of additional oil, RUB/tons	8,472.9
10	Profit from sales, RUB	3,196,700
11	Income tax, RUB	639,300
12	Net profit (actual effect), RUB	2,557,300
13	Cash flow, RUB	8,280,400
14	Discounted expenses, RUB	5,723,100
15	Profitability index	1.45

5 CONCLUSION

- New stimulation enhanced oil recovery technology for carbonate reservoirs using chlorine-free acid compounds was developed.
- Laboratory studies confirmed the efficiency of the new acid compositions.
- The preliminary results of the pilot test and analysis of expected economic efficiency of new well stimulation technology show high efficiency.

REFERENCES

Michael J. Economides, A. Daniel Hill, Christine Ehlig-Economides Petroleum Production Systems, PTR Prentice Hall, 1994 – 611 p.

Glushchenko V.N., Pozdeyev O.V. On Effectiveness Increase of Acid Solutions for Well Treatment. Moscow, All-Russian Scientific Research Institute for Organization of the Management and Economics of the Oil and Gas Industry, 1992, 52 p. (In Russian)

Tokunov V.I., Saushin A.Z. Process Liquids and Mixtures to Increase Production from Oil and Gas Wells. Moscow, Nedra-Business Centre LLC, 2004, 711 p. (In Russian).

Glushchenko V.N., Silin M.A. Oil Field Chemistry. In 5 vol. Vol. 4. Acid Treatment of Wells. Moscow, Interkontact Nauka, 2010, 703 p. (In Russian)

Tatianina O.S., Abdrahmanova L.M., Sudikin S.N., Djilina E.V. Evaluation of the influence of hydrochloric acid on the formation of organochlorine compounds of oil. Collection of scientific papers TatNIPIneft, 2017, V. 85, P. 363-369 (In Russian)

Folomeev A.E., Vakhrushev A.S., Mikhaylov A.G. On the Optimization of Acid Compositions for Geotechnical Conditions of Oilfields of Joint-Stock Oil Company "Bashneft". Neftyanoe khozyaistvo = Oil Industry, 2013, No. 11, P. 108–112. (In Russian)

A multifunctional Acid Solution (MAS): patent 2451054 Russian Federation, IPC C09K 8/74.Author – I. M. Galimov; applicant and patent holder – Research and Production Enterprise "NefteServisKomplekt" CJSC; No. 2010152459/03; applied December 22, 2010; published May 20, 2012; Bulletin No. 14, 8 p. (In Russian)

Davletshina L.F., Tolstykh L.I., Mikhailova P.S. About Reliance on Analysis of Hydrocarbon's Behavior for Improvement of the Acidizing Effectiveness. Territorija "NEFTEGAS" = Oil and Gas Terrotiry, 2016, No. 4, P. 90–96. (In Russian)

Sharov V.N., Gusev V.I. An Operator in Chemical Treatment of Wells. Moscow, Nedra, 1983, 145 p. (In Russian)

Amiyan V.A., Ugolev V.S. Physical and Chemical Methods for Increased Oil Production. Moscow, Nedra, 1970, 280 p. (In Russian)

Solid Solution Basis for Bottomhole Acidizing: patent 2257467 Russian Federation, IPC E21B 43/27. Authors – L.V. Kazakova, A.I. Mikov, T.V. Chabina, et.al.; applicant and patent holder – POLYEX CJSC; No. 2004105615/03; applied February 24, 2004; published July 27, 2005; Bulletin No. 21, 9 p. (In Russian)

Youth Technical Sessions Proceedings – Litvinenko (ed)
© 2019 World Petroleum Council, London, UK, ISBN 978-0-367-34668-3

Increasing of corrosion resistance of welded radiant and convection coiled-pipes in tubular furnaces at kinef crude oil refinery

B. Issa, V.Y. Bazhin, N.M. Telyakov & A.N. Telyakov
Saint-Petersburg Mining University, Saint-Petersburg, Russian Federation

ABSTRACT: The presence of heavy metals, metallic compounds and other impurities in crude oil directly and adversely affects the corrosion resistance of equipments and installations of the crude oil refineries. Improving the corrosion resistance of heating tubular furnaces — especially in a cheap way by using a special corrosion inhibitor — plays a very important and vital role in oil industry, since failure of coiled-pipes of tubular furnaces due to corrosion will be very expensive in terms of safety, environmental damage and losses. Moreover, all this directly leads to an improvement in the productivity and efficiency of tubular furnaces, a reduction in losses and an increase in the safety level of the operation of tubular furnaces, obtaining additional profits and reducing environmental disasters. The results of the experiments showed that the primary and secondary types of local corrosion and the studied inhibitor is very effective for increasing the corrosion resistance of the coiled-pipes of the tubular furnace up to 97.94%.

1 INTRODUCTION

Kinef crude oil refinery in the city of Kirishi - Russia has faced many accidents along its history which led to huge losses and stop the production at different periods vary depending on the damage caused to the units, installations and equipment of the refinery. In this article has been studied in detail one of the big problems that has faced kinef crude oil refinery which lead to the longest period of closure of the facility and the most expensive in terms of reforms.

This article is devoted to the study of welded radiant and convective coiled-pipes for tubular furnaces of crude oil refining and petrochemical, chemical and gas industries, working at a coiled wall temperature up to 760°C and under pressure up to 16 MPa.

One of the main problems encountered us at the refinery is the process of corrosion in the heating tubular furnaces, which in turn can lead to a catastrophic disaster. The high-temperature operating system of tubular furnaces significantly accelerates the processes of damage and destruction of metallic structures, including coiled-pipes alloys of tubular furnaces.

This article discusses in detail the coiled-pipes of heating tubular furnaces. Based on scientific results of our experiments, very important conclusions are shown, which can lead to a noticeable increase in the degree of protection of the coiled-pipes of heating tubular furnaces.

The purpose of this article is to study the role of chloride anions in process of damaging of coiled-pipe alloys of tubular furnaces, as well as studying the ability of a particular inhibitor to reduce the corrosion rate and therefore to minimize rate of accidents in tubular furnaces.

2 METHODOLOGY

During the study, the following methods were used:

- Chemical analysis for determination the content of metals and metallic compounds in West Siberian crude oil by using atomic absorption spectroscopy — VARIAN-SPECTRA AA 220 FS;

- Chemical and microstructural analysis for determination the content of metals and metallic compounds in coiled-pipes of tubular furnaces by using analytical scanning electron microscope — VEGA3 TESCAN;
- Microspectral X-ray analysis for the determination of the phases existed in coiled-pipes alloys by using oxford INCA energy EDS (energy dispersive X-ray spectroscopy);
- Analysis of the corrosion ability of the coiled-pipes by the method of weight loss (corrosion measurements).

3 THE RESULTS OF THE EXPERIMENTAL STUDIES AND THEIR DISCUSSION

3.1 *Object of study*

Heating tubular furnaces at crude oil refineries as part of complex metallurgical research, in order to reduce fire risk, were investigated failed coiled-pipes of tubular furnaces of crude oil refineries (Kirishinefteorgsintez - (KINEF crude oil Refinery), KIRISHI, Leningrad Region).

3.2 *Welded radiant and convection coiled-pipes of tubular furnaces*

The alloy brand of welded radiant and convection coiled-pipe steel of tubular furnaces is "15X5M" — martensitic grade steel — low-alloy and heat-resistant steel, the recommended temperature is up to 650°C approximately. The chemical composition of steel "15X5M" is shown in table 1.

In this article, the coiled-pipes took from Kirishinefteorgsintez — Kinef oil refinery in the town of Kirishi, Leningrad region, which are studied after an accident had happened (cause of the accident: burnout the wall of the coiled-pipes). This company receives raw materials through a pipeline from distribution centers in the Yaroslavl region. The refinery receives West Siberian and Volga-Ural crude oil.

Studies have been carried out on the section of the tubular furnace in the installation ELOU - AVT-6. ELOU-AVT — electrical desalting unit, installed before the atmospheric column, which in turn stands before the vacuum one.

To study the causes of damage and destruction of the coiled-pipe in tubular furnaces, it was necessary to take coiled-pipe fragments at various sites in the tubular furnace to analyze them, and therefore to identify the composition of the coiled-pipe, as well as the quantitative transition of the main elements through the corrosion process to the coiled-pipe material.

3.3 *Microstructure and composition of coiled-pipe alloy*

The structure of the coiled-pipe is shown under a scanning electron microscope in spectra, which are studied carefully to compare the microstructure, corrosion performance and heat-affected zones during combustion, as shown in figure 1 and the composition of the spectra shown in table 2.

- For non-affected areas, without any signs of corrosion: The composition of nine spectra for the sample of the coiled-pipe, shown here on the table 2 together with figure 1 of their location:
- For affected areas, corroded areas, spectra 14 and 22 are shown in figure 2 and 3 as an example of several spectra have studied with a table of their composition in table 3 and 4.

Table 1. Chemical composition of steel "15X5M"

C, %	Si, %	Mn, %	Ni, %	S, %	P, %	Cr, %	Mo, %	W, %	V, %	Ti, %	Cu, %	Others
to 0.15	to 0.5	to 0.5	to 0.6	to 0.025	to 0.03	4.5-6	0.45-6	to 0.3	to 0.05	to 0.03	to 0.2	Fe – Rest

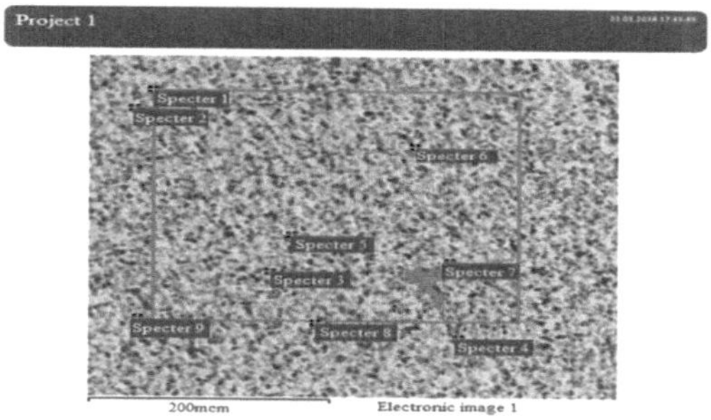

Figure 1. Electronic image - position of nine spectra of the coiled-pipe in non-affected areas — scale 1: 200

Table 2. Analytical composition of nine spectra of coiled-pipe

# Spectrum	O,%	Si,%	Cl,%	Cr,%	Fe,%	Total
Spectrum 1	—	10.20	—	5.47	84.33	100
Spectrum 2	—	37.90	0.62	3.84	57.63	100
Spectrum 3	—	6.42	—	4.56	89.02	100
Spectrum 4	—	22.20	—	4.17	73.63	100
Spectrum 5	—	2.87	—	4.68	92.45	100
Spectrum 6	—	12.74	—	5.41	81.85	100
Spectrum 7	—	4.24	—	5.34	90.41	100
Spectrum 8	9.98	19.64	—	4.27	66.12	100
Spectrum 9	—	4.04	—	7.55	88.41	100
Min.	—	2.87	—	3.84	57.63	
Max.	9.98	37.90	0.62	7.55	92.45	

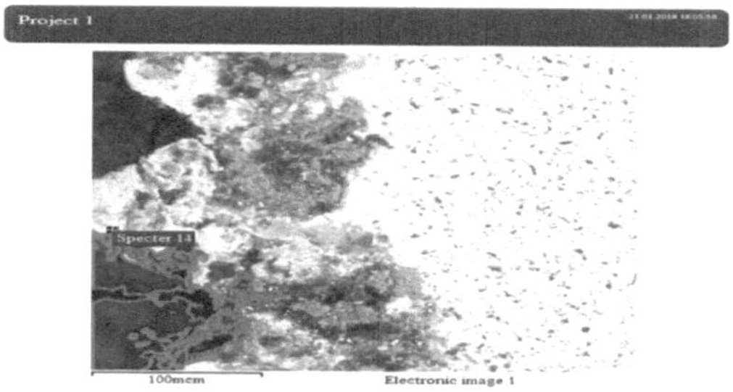

Figure 2. Electronic image - spectrum 14 of coiled-pipe in the affected-area — scale 1: 100

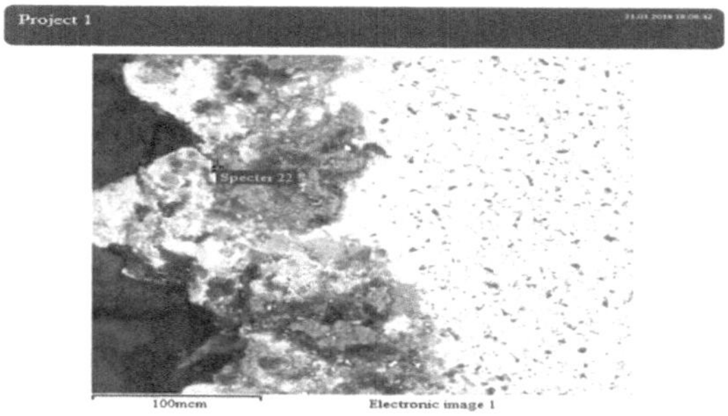

Figure 3. Electronic image - spectrum 22 of coiled-pipe in the affected-area — scale 1: 100

Table 3. Analytical composition of the spectrum 14 of coiled-pipe

Element	Weight %	Atomic %
OK	39.38	66.42
AlK	0.66	0.66
SiK	1.91	1.83
SK	0.99	0.83
ClK	8.01	6.1
CrK	14.88	7.72
FeK	32.89	15.89
CuK	1.28	0.54
Total	100.00	

Table 4. Analytical composition of the spectrum 22 of coiled-pipe

Element	Weight %	Atomic %
OK	33.74	60.45
AlK	2.71	2.88
SiK	5.47	5.58
SK	1.67	1.49
ClK	4.44	3.59
KK	1.04	0.76
CaK	2.46	1.76
CrK	3.24	1.79
FeK	21.14	10.85
CuK	22.35	10.08
ZnK	1.75	0.77
Total	100.00	

It should be noted that in the spectrum of 14 there is a high percentage of oxygen up to 39% and a high percentage of chloride up to 8%, which play a significant role in the corrosion process.

It should be noted that in the spectrum 22 there is a high percentage of oxygen up to 34%, as well as a high percentage of copper up to 22%, in addition to a large proportion of chloride up to 4.5%, which play an important role in the corrosion process.

In figure 2 and 3 the right side is the material of the coiled-pipe, and the left side is the scale, formed as a result of a long stay of the coiled-pipe under the burners of the tubular furnace. And between them there is a transition zone of corrosion.

Analyzing figure 2 or 3 showed that in the coiled-pipes of the tubular furnace, there are many signs of corrosion. We can indicate the types of corrosion as follows in figure 4:

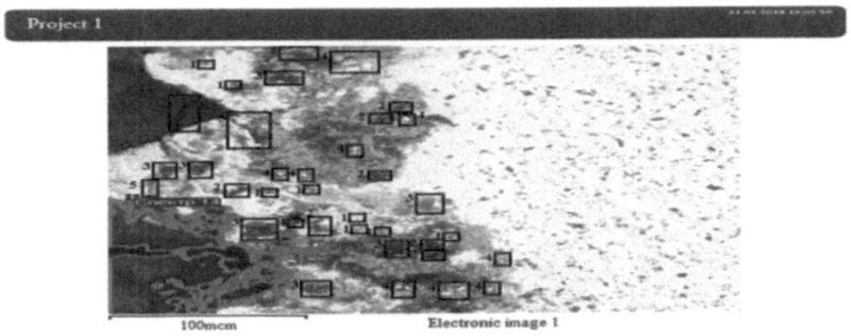

Figure 4. Types of corrosion in the transition zone

1 - Pitting corrosion; 2 - corrosion by ulcers, 3 - corrosion by spots, 4 - intergranular corrosion, 5 - corrosion by cracks.

By analyzing the electronic images that are shown in this article and others are not shown here, our results can be explained as follows:

- The main type of local corrosion is pitting corrosion, and the secondary type of local corrosion is corrosion by microelectrochemical cells.
- Since the coiled-pipe alloy contains a relatively low chromium content in its composition, it is likely that this alloy is less resistant to pitting corrosion; therefore, the main type of corrosion in the coiled-pipe alloy is pitting corrosion.
- High local concentration of HCl in the pits led to the dissolving of corrosion products, such as Fe_3O_4 and the most important the following compound Cr_2O_3 (which is responsible for protection the coiled-pipe alloy through a protective film), both of them dissolve well in the pitting solution, which leads to a deterioration of the protective layer inside the pits, which leads to a sharp acceleration of corrosion deep inside the metal, which is the most likely cause of destruction or damage to the coiled-pipe material in an accident.

3.4 *Metals in West Siberian crude oil*

The content of heavy metals in West Siberian crude oil was determined and shown in table 5, which shows the highest concentrations of metals found in West Siberian oil.

Table 5. The content of heavy metals in West Siberian crude oil

Metal	Fe	V	Al	Ni	As	Zn	Mn	Co	Cr
C_M, *ppm*	24.9	21.68	18.4	12.75	2.9	2.88	1.1	0.15	0.09

It is clearly noted that vanadium, aluminum and nickel, as it shown here, are high in this crude oil.

3.5 *Corrosion of welded radiant and convection coiled-pipes in the electrolyte and the effectiveness of the inhibitor SNPKS-6030-b*

The experiments were carried out at two different temperatures of 100°C (using a water bath type heater) and 150°C (using an oil bath type heater) with a special salt electrolyte having the composition shown in table 7. First, without using the inhibitor, and second, with the inhibitor in three different concentrations of inhibitor 25 ppm, 50 ppm, 200 ppm. Coupons are made of a parallel-shaped with the following dimensions, as shown in table 6.

The results of our experiments table 8 and figure 5 showed that the inhibitor SNPKS-6030-b reduces the weight loss by about half.

From figure 5 we can notice that by increasing temperature, weight loss increases because of corrosion.

The results of our experiments table 9 and figure 6 showed that the used inhibitor significantly reduces the rate of corrosion, so it is recommended to use it as a corrosion inhibitor in crude oil refinery installations.

From figure 6, we can notice that:

- The best results were achieved using the inhibitor SNPKS-6030-b under the following conditions: t = 48 h, C = 200 ppm, T = 100°, where the effectiveness of the inhibitor was reached 97.94%.
- The worst results were achieved using the inhibitor SNPKS-6030-b under the following conditions: t = 3 hours, C = 25 ppm, T = 100, where the effectiveness of the inhibitor was 81.82%.

And, therefore, the inhibitor SNPKS-6030-b is successfully used in oil refining processes on the ELOU-AVT-6 unit, since its efficiency can be achieved up to 98%.

Table 6. Dimensions of coupon of the coiled-pipe

Coupon dimensions, *cm*	Length	Width	Thickness
	5.2	2.7	0.13
Total surface area of coupon, cm^2	30.134		

Table 7. Composition of the salt electrolyte by ions

Ions	Na^+	Ca^{2+}	Mg^{2+}	SO_4^{2-}	Cl^-
C, ppm	115	12	0.972	3.8	198.8

Table 8. Weight Loss of coupon

		Δm, mg (T=100°C)		Δm, mg (T=150°C)	
t,hr	C,ppm	без ингибитора	с ингибитором	без ингибитора	с ингибитором
3	25	30.54	16.8	46.27	25.4
24	50	15.85	8.3	32.85	17.5
48	200	3.76	1.9	20.59	10.7

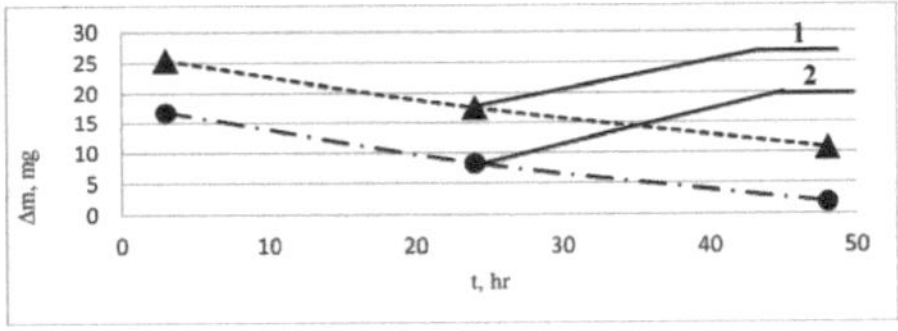

Figure 5. Weight loss with SNPKS-6030-b — time exposure at two different temperatures: 1-150°C; 2—100°

Table 9. Effectiveness of the inhibitor SNPKS-6030-b and the corrosion rate of this inhibitor

t, hr	C, ppm	%-100°C	%-150°C	mdd-100°C	mdd-150°C	mpy-100°C	mpy-150°C
3	25	81,82	82,16	121824	181440	$4.2 * 10^{-5}$	$6.4 * 10^{-5}$
24	50	91,02	87,71	7516,8	15552	$2.7 * 10^{-6}$	$5.5 * 10^{-6}$
48	200	97,94	92,48	864	4848,336	$3.05 * 10^{-7}$	$1.72 * 10^{-6}$

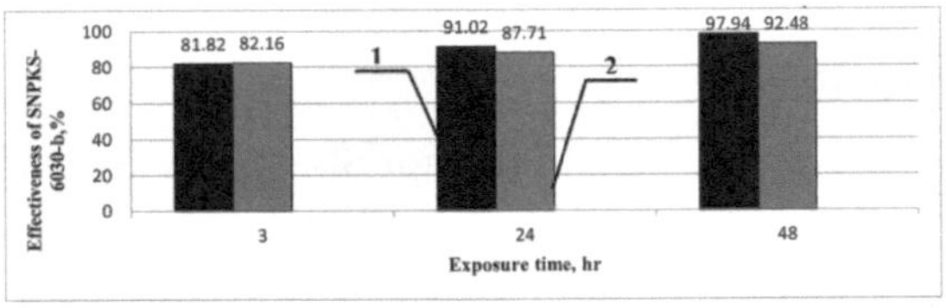

Figure 6. Effectiveness of the inhibitor SNPKS-6030-b —time exposure at two different temperatures: 1-100°C; 2-150°C

4 CONCLUSION

- The most likely type of local corrosion in coiled-pipes is pitting corrosion, and the secondary type of local corrosion is corrosion by microelectrochemical cells.
- Since the coiled-pipe alloy contains a relatively low chromium content in its composition, it is likely that this alloy is less resistant to pitting corrosion; therefore, the main type of corrosion in the coil alloy is pitting corrosion.
- High local concentration of HCl in the pits led to the dissolving of corrosion products, such as Fe_3O_4 and the most important compound Cr_2O_3 (which is responsible for protecting the coiled-pipe alloy through a protective film), both of them dissolve well in the pitting solution, which leads to a deterioration of the protective layer inside the pits, which leads to a sharp acceleration of corrosion deep inside the metal, which is the most likely cause of destruction or damage to the coiled-pipe material in an accident.
- No adverse effects of the inhibitor SNPKS-6030-b were observed to use it in refining conditions (except for its toxicity).
- The inhibitor SNPKS-6030-b is successfully used in oil refining processes at the unit ELOU-AVT-6, since its efficiency can be achieved up to 98%.
- The best results were achieved with the use of the inhibitor SNPKS-6030-b under the following conditions: t = 48 h, C = 200 ppm, T = 100, where the effectiveness of the inhibitor was reached 97.94%.
- The worst results were achieved using the inhibitor SNPKS-6030-b under the following conditions: t = 3 hours, C = 25 ppm, T = 100, where the effectiveness of the inhibitor was 81.82%.
- The ability of the inhibitor to protect weakened after prolonged exposure to preliminary corrosion in these conditions. Moreover, the corrosion rate increases with increasing temperature.
- The inhibitor has a weak ability to protect against pitting corrosion and, therefore, this weak ability to inhibit leads to the formation of deep spherical cavities.
- In the case of existing, old facilities in refineries, inhibiting corrosion by chemicals is the only and the cheapest solution.
- The mechanism of corrosion has a nature of an irreversible process, which ultimately leads to deterioration of the properties of the metal.

REFERENCES

Todt F. Corrosion and corrosion protection. - L.: Chemistry, 1967. - 709p.

Evans Yu.R. Corrosion and oxidation of metals. - M.: Mashinostroitelnoy lit., 1962. - 857p.

Zhuk N.P. Course of the theory of corrosion and protection of metals. M.: Metallurgy. - 1976. - 472p.

Alekseev M.V., Smirnov V.M. Fire prevention in the technological processes connected with the circulation of flammable and flammable liquids. M.: Ministry of Communal Services of the RSFSR, 1955. P. 22-52.

Alekseev M.V. Basics of fire prevention in technological processes of production. M., Higher School of the Ministry of Internal Affairs of the USSR, NIIRIO, 1972. - P. 111-112, 151-167.

Baratov A.N., Korolchenko A.Ya., Kravchuk G.N. and others. Fire and explosion hazard of substances and materials and means of extinguishing them; - M., Chemistry, 1990. p. 34-48.

Fire safety rules in the oil industry (PPBO-85). M.: Nedra, 1987.– P. 66-75.

Ming Shen, Alla Furman, Rita Kharshan, and Tim Whited, Development of Corrosion Inhibitors for Prevention of Top of the Line Corrosion (TLC), Paper no. 2509, NACE CORROSION 2013 Conference, NACE International, Houston, TX, USA, 2013, 12 p

B.M. Miksic, A.Y. Furman, M.A. Kharshan, Effectiveness of the Corrosion Inhibitors for the Petroleum Industry Under Various Flow Conditions, Paper no. 09573, CORROSION 2009, NACE International, Houston, TX, USA, 2009, 9 p.

Youth Technical Sessions Proceedings – Litvinenko (ed)
© 2019 World Petroleum Council, London, UK, ISBN 978-0-367-34668-3

Enhanced oil recovery technology for low-permeable polymict reservoirs

A. Kuznetsova & M. Rogatchev
Saint-Petersburg Mining University, Saint-Petersburg, Russian Federation

ABSTRACT: About 25% of the oil reserves of the West Siberian oil and gas basin are contained in deposits that are characterized by complex structure, high clay content and low reservoir properties. The interaction of clay minerals with water causes an increase in their volume and, accordingly, blockage of the pore channels. It can stop filtration in low-permeability collectors. We have developed and patented the composition of surface-active substances, designed for practical use in the artificial flooding of low-permeability polymict reservoirs. The composition allows to reduce the interfacial tension at the interface "oil-water", diffuse from the aqueous solution into the oil and have a dispersing effect on asphaltenes and inhibitory effect on the hydration of clay minerals in the composition of the reservoir rock. Based on the composition, a technology has been developed for the physicochemical effect on the oil reservoir, based on the injection of the surfactant composition developed through the injection wells of the RPM system. The results of the hydrodynamic simulation of the waterflooding with this composition showed an increase of 4.2% in the oil recovery rate from the reservoir site model.

1 INTRODUCTION

The Jurassic reservoirs of Western Siberia consist of the Vasyugan and Tyumen oil and gas complexes and comprise more than 25% of the total oil reserves of the basin. However, the reservoir rocks of these rocks are characterized by high clay content and low filtration characteristics, which prevents the extraction of significant volumes of oil from them.

Collectors of this type are characterized by high water holding capacity, specific surface and adsorption, and, therefore, the process of extracting oil from such reservoirs has a number of difficulties.

First of all, it concerns the process of water flooding. The interaction of clay minerals with water causes an increase in their volume and, accordingly, clogging of the pore channels. In the case of low-permeable polymictic reservoirs, in which most of the pores are represented by capillaries and subcapillaries, even a slight increase in the volume of clay minerals leads to a significant decrease in oil permeability. Therefore, in such collectors it is necessary to pay high attention to the quality of water treatment, as well as to take into account the possibility of using chemical compositions to inhibit the hydration of clay minerals.

Increasing the efficiency of waterflooding for low-permeable polymictic reservoirs is achieved by adding specially selected multifunctional surfactants to the water for injection, which can not only improve the oil-displacing properties of water and the rheological properties of oil but also reduce the intensity of clay swelling in reservoir rocks.

In modern realities, it is necessary to use more effective and cheap surfactants, which will allow a complex effect on the oil-water-rock system. And surfactants for use in low-density polymict reservoirs should simultaneously affect the clay component of reservoir rocks (to inhibit clay swelling) and the process of oil displacement from them (in order to reduce interfacial tension at the interface between the oil-water-rock system).

2 LITERATURE REVIEW

Surfactants have the ability to reduce the interfacial tension at the oil-water interface, as well as change the properties of the surface layer of the rock. Their addition to the injected water of the RPM system allows improving the water flooding method, improving the properties of water. Academician PA Rebinder is the founder of the use of surfactants in the Russian oil industry. He created the theoretical basis for the use of surfactants for many technological processes.

Surfactants have the ability to reduce the interfacial tension at the oil-water interface, as well as change the properties of the surface layer of the rock. Their addition to the injected water of the RPM system allows improving the water flooding method, improving the properties of water. Academician PA Rebinder is the founder of the use of surfactants in the Russian oil industry. He created the theoretical basis for the use of surfactants for many technological processes.

The first works on the use of surfactants belong to the 50s of the XX century. They analyze the results of laboratory studies of the effect of surfactants on the ability of water to wash oil, on how its properties change depending on the component composition (Tumasyan A.B. 1973).

In the early 60s of the 20th century, the idea of using mixtures of various surfactants appeared because different surfactants that can be included in a composition have different properties and perform specific functions (Gimatudinov Sh.K., 1962). Consequently, sharing each surfactant with others is more effective than their single use, which means a synergistic effect is achieved.

The first pilot works on the use of surfactants as an additive to the water injected into the reservoir were carried out on the Nagayevsky dome of the Arlanskoye field in 1964. The aqueous solution of the non-ionic surfactant OP-10 was used as a displacement agent. As a result, it was possible to increase oil recovery due to an increase in the anhydrous period of oil production. The result was considered positive and the method was recommended for use on an industrial scale in the fields of Tatarstan, Bashkiria, Azerbaijan, Western Siberia and Ukraine (Babalyan G.A, 1976 , *1983*).

Despite the predominantly successful results of industrial experiments, currently, the use of surfactants for water flooding in the Russian practice of enhanced oil recovery is not widespread. This is due to factors complicating their use, such as significant adsorption on the rock surface, chemical instability under reservoir conditions, insufficient reduction of surface tension at the fluid boundary, as well as their high cost. Therefore, in recent years, preference has been given to combined types of flooding, simultaneously affecting several components of oil recovery factor, such as surfactant-polymer, surfactant-alkaline, and alkali-surfactant-polymer flooding. The latter is known abroad as ASP (alkaline/surfactant/polymer) technology, a pilot project of which is being implemented in the experimental area of the West Salym field (Panikovsky E.V., 2006).

3 METHODOLOGY

To determine the interfacial tension used device Easydrop (State Standard 29232-91 "Determination of the critical micelle concentration." - M.: Publishing house of standards, 1992. - 7 p.). The size of the clay particles was determined using a laser graininess meter. And to determine their size under reservoir conditions, the gas sorption analyzer The NOVA 1000e (Quantachrome) was used (State Standard R 8.777-2011 "Dispersed composition of aerosols and suspensions. Determination of particle size by laser diffraction". - M.: Standardinform, 2012. – 12p.).

Filtration experiments were carried out on natural samples of the core of the Jurassic deposits of Western Siberia on the Autoflood-700 equipment (Vinci technologies) using high-pressure pumps, Hasler core holder (State Standard 26450.0-85 "Mountain Species. General requirements for the selection and preparation of samples for the determination of reservoir properties ". - M.: Publishing house of standards, 1985. - 6 p. Industry Standard 39-195-86

"Oil. Method for determining the coefficient of oil displacement of water in the laboratory." - M.: Printing house of the Ministry of Oil Industry, 1986. - 20 p.).

The geological reservoir was modeling in the Roxar software package, and then re-scaled was transferred to the tNavigator software package, in which hydrodynamic modeling of the process of oil displacement with surfactant solutions was performed.

4 RESULTS

The selection of surfactants for the waterflood system (RPM system) of a particular field is not an easy task. There is still no uniform system of criteria by which it is possible to draw unambiguous conclusions about the effectiveness and scope of a particular reagent. According to researchers (Devlicamov V.V., 1976, Petrov N.A., 2004), an effective surfactant should meet the following requirements:

- Dissolve in formation water;
- To reduce the interfacial tension at the interface of the "water-oil" phases;
- Have minimal adsorption;
- Not to completely lose its activity upon contact with fluids, surfaces of downhole equipment;
- Prevent swelling of clay rocks;
- To be available;
- Be environmentally friendly.

Traditionally, nonionic surfactants are used for flooding, because they most fully meet all these requirements. But nonionic surfactants, like anionic surfactants, in distinction from Cationic ones, do not affect the surface quality of clay minerals. then eating does not spin them swelling. But cationic, in contrast to non-ionic surfactants, reduce the interfacial tension at the oil-water interface much worse.

Thus, the optimal solution was the composition of two nitrogen-containing surfactants, the use of which allowed both to reduce the interfacial tension and suppress clay swelling (Kuznetsova, Sukhih, Rogachev, 2018).

The research results showed that, compared with the known surfactants, the developed composition of surfactants has less adsorption on the pore surface of collection rocks, can more effectively reduce the interfacial tension at the oil-water interface and suppress the hydration of clay minerals, as well as reduce the discharge pressure in the reservoir model.

To determine the effectiveness of the developed surfactant composition under conditions of a low-permeable polymict collector, as well as to select the optimal conditions for its injection, we conducted a hydrodynamic simulation of the oil displacement process. In the RoxarRMS software package, a lithofacies model of an element of the reverse nine-point waterflood was built. The re-scaled distribution of filtration-capacitive properties was used to create a hydrodynamic model. The constructed model made it possible to take into account all the most important effects of surfactants on the fluid and rock properties, including such factors as miscibility, change in wettability, and magnitude of adsorption. Also, the model took into account the effects of changes in pore size as a result of clay swelling during the flooding of clay reservoirs with fresh water, which led to a change in reservoir permeability.

The concentration of surfactant in injected water was 0.1% by mass for the Cationic surfactant, and 0.05% by mass for non-ionic.

We have considered 5 options for the development of deposits.

1. Freshwater injection (basic option).
2. For the first 10 years fresh water was used as a flooding agent, and then a surfactant solution.
3. Injection of formation water.

4. For the first 10 years, formation water was used as a waterflood agent, and then a surfactant solution.
5. From the very beginning to the end, the surfactant solution was pumped.

In the first variant, the clay is hydrated, therefore, the pore size decreases, and the interfacial tension at the oil-water interface remains rather high. During the injection of formation water (option #3), only a high value of interfacial tension was taken into account, but the pore size did not change. For option # 2, it was taken into account that with the addition of surfactants, the wettability parameters changed, and also the hydrated clay minerals partially returned to their original sizes. For option # 4, after 10 years, the wettability parameters and, accordingly, the oil recovery factor changed when it was displaced with an aqueous surfactant solution compared to water.

The simulation results are presented in Figure 1. According to figure 1, the minimum amount of oil can be produced using fresh water injection (option #1). But at the same time, the injection of a reservoir (saline) water, allows extracting almost 3000 tons of oil more. Due to the fact that the developed surfactant composition is capable of not only inhibiting clay swelling, but also makes this process reversible, when applying the composition even 10 years after the development began, the difference between options 1 and 2 is about 2000 tons of oil. The maximum effect is achieved when using the injection of a surfactant solution from the beginning of development (option # 5). The difference between this option and the base (option # 1) is almost 8,000 tons of oil.

Figure 2 presents the values of the final oil recovery rate for all flooding options. From the graphs, it can be seen that the difference between the base case (freshwater) and option # 5 for the constructed model of the reservoir section is more than 4%. Accordingly, it is possible to recommend the developed composition of surfactants for the waterflooding of low-permeable polymictic collectors.

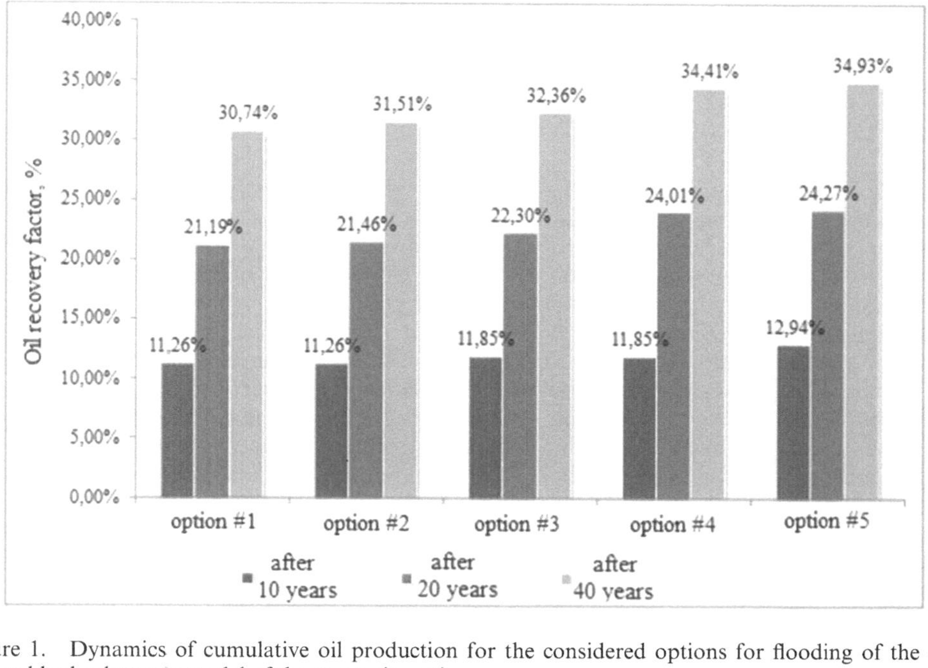

Figure 1. Dynamics of cumulative oil production for the considered options for flooding of the constructed hydrodynamic model of the reservoir section

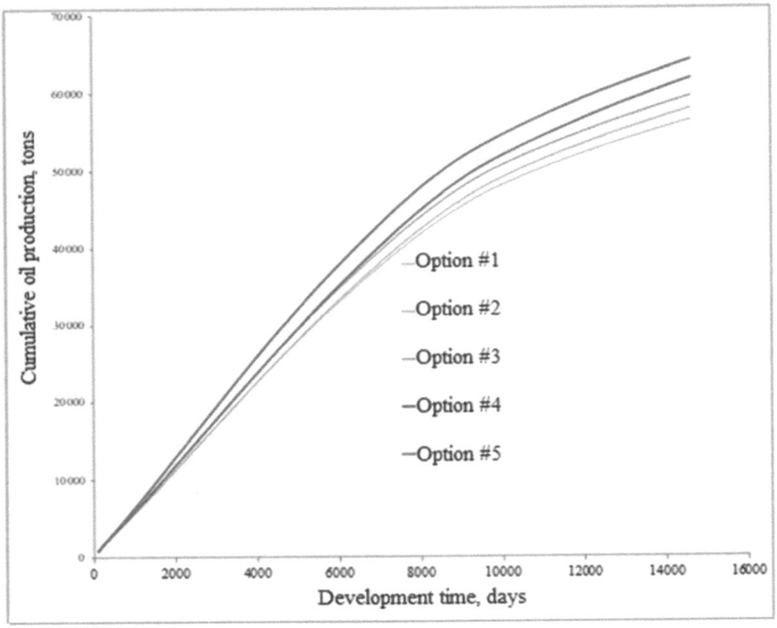

Figure 2. The magnitude of the oil recovery factor for the considered options for flooding of the constructed hydrodynamic model of the reservoir section

The results of hydrodynamic modeling of the low-permeable polymict reservoir flooding process showed an increase of 4.2% in the oil recovery rate from the reservoir model when using the developed surfactant composition compared to the fresh and saline waters traditionally used in the RPM system, as well as allow recommending the developed surfactant composition for industrial use for artificial flooding of collectors of this type.

5 CONCLUSION

Increasing the efficiency of artificial flooding of low-permeable polymictic reservoirs is possible by injecting aqueous solutions of specially selected multifunctional surfactants into the formation, which, in addition to improving the oil-displacing properties of water, can reduce the intensity of clay swelling in reservoir rocks. For this purpose, a surfactant composition has been developed and patented for injection into a low-permeable clay layer.

The results of hydrodynamic modeling of the waterflooding of a low-permeable polymict reservoir using the developed surfactant composition showed a decrease in residual oil saturation compared to standard water flooding and an increase in oil recovery by 4.2%.

The complex technology of flooding of low-permeable polymict reservoirs was developed and proposed for industrial implementation, which includes the technology of injection of a low-concentrated solution of the developed surfactant composition into the reservoir through a system of injection wells of the RPM system.

REFERENCES

About a state and use of mineral raw material resources of the Russian Federation in 2015. State report. Ministry of Natural Resources and Environment of the Russian Federation, 2017 [O sostoyanii i ispol'zovanii mineral'no-syr'evyh resursov Rossijskoj Federacii v 2015 godu. Gosudarstvennyj doklad. Ministerstvo prirodnyh resursov i ekologii Rossijskoj Federacii], viewed 15.02.2018, http://www.mnr.gov.ru/

docs/o_sostoyanii_i_ispolzovanii_mineralno_syrevykh_resursov_rossiyskoy_federatii/gosudarstvennyy_doklad_o_sostoyanii_i_ispolzovanii_mineralno_syrevykh_resursov_rossiyskoy_federatsii/

Tumasyan A.B. Effect of surfactants on the phase permeability of a porous medium for oil and water/A. B. Tumasyan, V.G. Panteleev//*Oil industry*. - 1973. - №10. - pp. 37-39.

Babalyan G.A. Development of oil fields using surfactants/G.A. Babalyan, A.B. Tumasyan, B.I. Levi, E. M. Khalimov. - M.: Nedra, 1983. - 216 p.

Gimatudinov Sh.K. On the mechanism of the washing action of waters of different composition in the displacement of oil from a porous medium/Sh.K. Gimatudinov//*Oil industry*. - 1962. - №10. - P. 43-48.

Babalyan G.A. The use of surfactants for enhanced oil recovery/G.A. Babalyan, A.B. Tumasyan, B.I. Levi, E.M. Khalimov//*Oil industry*. - 1976. - №7. - P. 7-16.

Panikarovsky, E.V. Prospects for the use of physico-chemical methods to increase the productivity of wells/E.V. Panikarovsky, V.V. Panikarovsky, I.I. Kleschenko//*Oilfield business*, 2006 - №3 - C.20-25.

Kuznetsova A. N., Sukhih A. S., Rogatchev M. K., Surfactant solutions for low-permeable polimictic reservoir flooding/IOP *Conf. Series: Earth and Environmental Science, № 194*, 2018. C 1-5.

Devlikamov, V.V. Effect of surfactants on the rheological properties of oil/V.V. Devlikamov, M.K. Rogachev//*Oil industry*. - 1976. - №6. - P. 29-31.

Petrov, N.A. Cationic surfactants - effective inhibitors in technological processes of the oil and gas industry/N.A. Petrov, B.S. Izmukhambetov, F.A. Agzamov, N.A. Nogayev, ed. F. Agzamov. - SPb.: Nedra, 2004. - 408 p.

Youth Technical Sessions Proceedings – Litvinenko (ed)
© 2019 World Petroleum Council, London, UK, ISBN 978-0-367-34668-3

Substantiation of a well killing technology for fractured carbonate reservoirs

Sh.R. Islamov, A.V. Bondarenko & D.V. Mardashov
Saint Petersburg Mining University, St. Petersburg, Russia

ABSTRACT: The article presents an analysis of the efficiency of production well killing processes in an oil and gas condensate fields of the Volga-Ural province for 2017-2018. In order to prevent fluid loss and gas breakthrough during well service, a specific selection of compositions and technologies is required, based on the field data analysis and laboratory tests. The results of laboratory studies on the development of an invert-emulsion composition with a multi-fractional microcalcite, which can prevent the absorption of the well killing fluid with a fractured reservoir and increased repression, are presented. The technology of well killing is presented using the developed composition in the conditions of producing an oil and gas condensate fields of the Volga-Ural province.

1 INTRODUCTION

The process of the well killing operations in the oil and gas condensate field in Volga-Ural province is complicated due to the following conditions: fractured carbonate reservoir; abnormally low reservoir pressure; high gas-oil ratio; high hydrogen sulfide content; acid fracturing and hydrochloric acid treatments requirement.

To prevent these complications in the oil and gas industry, two basic physical principles of reducing the absorption of a well killing fluid in the bottomhole formation zone are widely used: increasing the viscosity of the blocking fluid for well killing and blocking of pores and channels of filtration by solid particles (Fig. 1). These principles are implemented in blocking systems, which are invert-emulsion, lime-bitumen and polymer compositions, which can additionally contain solid clogging agents (Petrov et al. 2008; Tokunov & Saushin 2004; Jouenne et al. 2016).

Increased viscosity provides reduced filtration of blocking systems into the reservoir in the absence of natural and artificial fractures and super collectors (permeability not more than 1 μm^2) and with slight repression (not more than 5 MPa) (Zhelonin et al. 2015).

Suspension blocking systems in a solid phase can prevent the absorption of plugging solution in wells with hydraulic fracturing, natural fractures (super collectors) in the formation section and reservoir pressures from 0.3 to 1.8 hydrostatic (Zhelonin et al. 2015).

Currently, to increase the well killing efficiency in the oil and gas condensate field, blocking compositions with increased viscosity are used: invert-emulsion composition and biopolymer composition. The following components are used for emulsion preparation: an emulsifier, commercial oil and an aqueous solution of calcium chloride (or sodium) and the following reagents are used to prepare the biopolymer composition: a water-soluble organic polymer, an inorganic crosslinking agent, and a crosslinking regulator.

At the same time, the well killing technology consists of injecting the indicated blocking compositions from the condition of overlapping the perforation interval with a margin of 100-200 m according to (Akhankin et al. 2007; Nikolaev & Leusheva 2016; Ryabokon 2009).

In order to increase the cost efficiency, the remaining part of the wellbore is filled with an aqueous solution of calcium chloride salts and, less commonly, sodium chloride. However, the

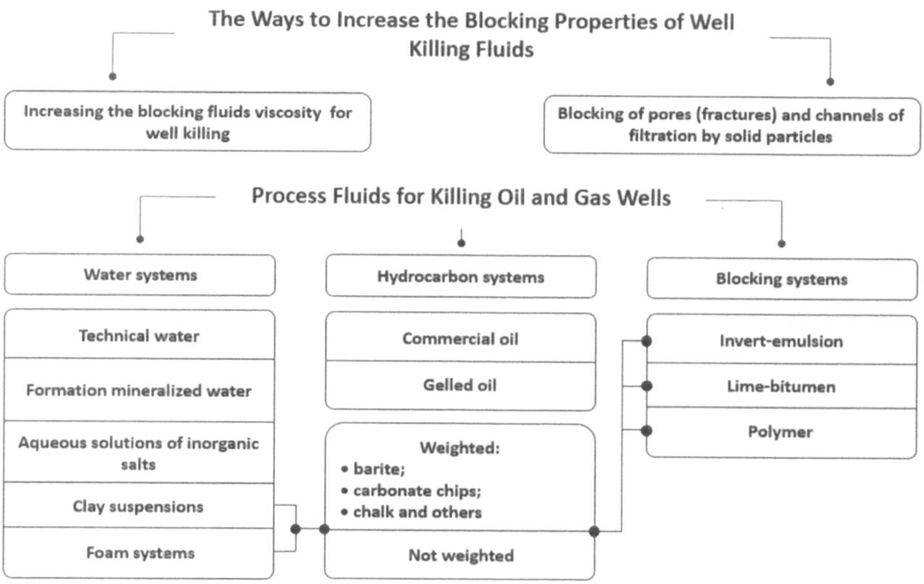

Figure 1. The main types of well killing fluids.

use of such systems, as will be shown below, does not always lead to reliable blocking of fractures in the bottomhole formation zone during the well killing.

Analysis of the efficiency of well killing in the oil and gas condensate field for 2017-2018 showed that the use of these blocking compositions is not effective enough due to repeated well killing cycles (up to 5) (Fig. 2); increased consumptions of blocking compositions forged into the bottomhole formation zone (up to 100 m^3 per well); long periods for the well to start operating (up to 4 days).

Reducing the number of well killing cycles, in addition to preserving the filtration properties of the bottomhole formation zone, will lead to a reduction in the volume of the composition

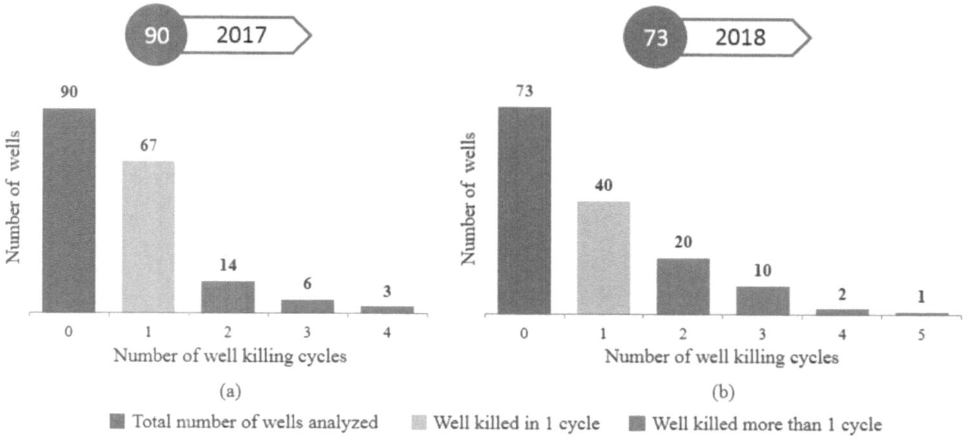

Figure 2. Well distribution by the number of well killing cycles.

forged into the bottomhole formation zone, a reduction in the time required for the well to start operating and a reduction in the cost of workover.

According to the analysis of the reasons of repeated well killings in the oil and gas condensate field, it was found that more than half of all repeated well killing (65-69%) are related to geological reasons: unreliable (overestimated value) reservoir pressure used to calculate the density of the well killing fluid; low reservoir pressure (formation pressure anomaly ratio up to 0.55), requiring the use of well killing fluids with a density of less than 1 g/cm^3; catastrophic absorption due to the lack of information about the volume of caverns and the size of fractures in the open and extended horizontal wellbore, as well as from previous operations to stimulation of production. At the same time, about 31-35% of repeated well killings are due to technical factors: emergency underground equipment; reached the upper gas reservoir without cutting off the lower one with less pressure; gas bubble when the packer breaks down; lack of reliable information on the position of the circulation valve.

Thus, on the basis of a retrospective analysis of patent materials and periodicals, as well as the field data studied, a well killing technology was proposed using an invert-emulsion composition with a solid phase that can prevent the absorption of a well killing fluid with a fractured reservoir and increased repression. Calcium carbonate multi-fractional (microcalcite and marble crumb), which can be removed with further acid treatment, was recommended as a solid phase. The role of a multi-fractional colmatant is to create a filter cake on the surface of the collector (on fructures). As a result, a quick and effective blockage of the mouth of fractures occurs and a strong layer of dispersed particles is formed, which prevents deeper penetration not only of the solid phase but also of the blocking filtrate (Fig. 3) (Orlov et al. 1991; Sharath et al. 2017; Elkatatny 2016).

The aim of the research is to increase the well killing efficiency during well service operations in the conditions of fractured carbonate reservoirs, high gas-oil ratio, and abnormally low reservoir pressure.

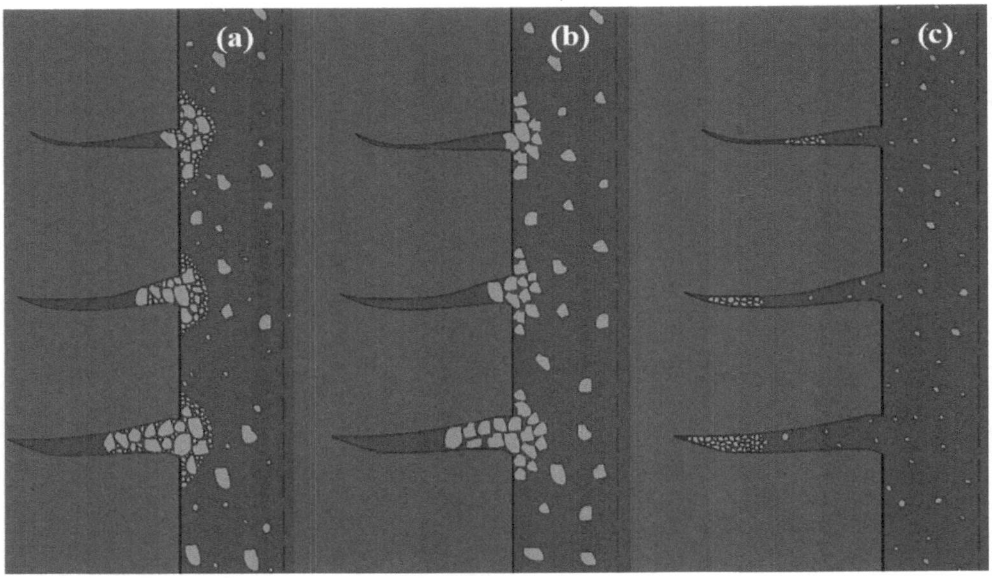

Figure 3. The scheme of formation of the blocking screen when used in the invert-emulsion composition of multi-fractional colmatant: (a) multi-fractional colmatant; (b) large fractional colmatant; (c) small fractional colmatant.

2 MATERIALS AND METHODS

Laboratory studies of blocking properties of well killing fluids under fractured reservoir conditions were carried out using a modernized high-pressure filter press from FANN, designed to test the penetrating ability of cement materials, as well as measuring static filtration of process fluids into the bottomhole formation zone (Fig. 4-5).

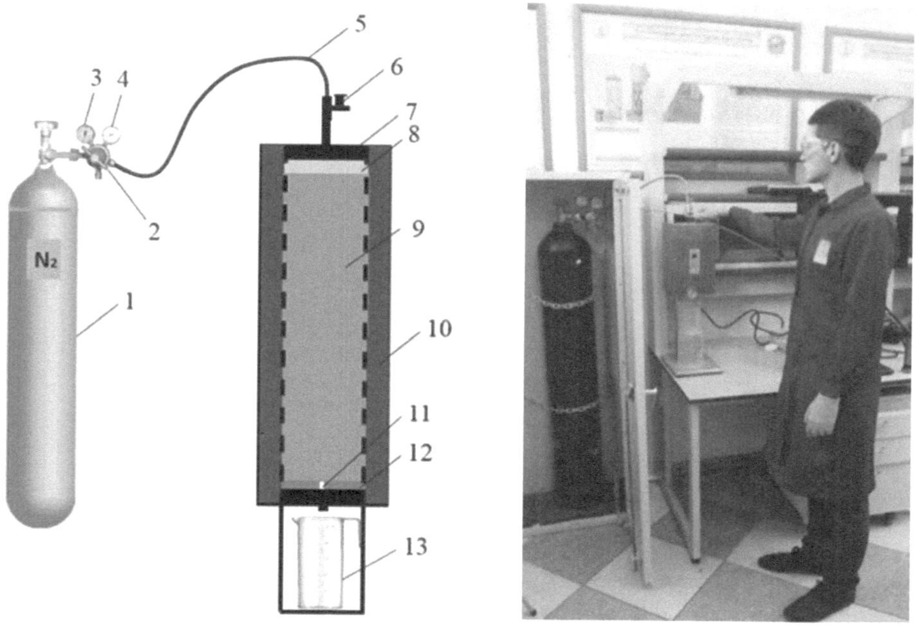

Figure 4. Process functional diagram of the filter press with modelling fracture: 1 – gas cylinder; 2 – pressure reduction valve; 3 – pressure indicator (storage pressure); 4 – pressure indicator (flow line pressure); 5 – gas line (gas supply); 6 – overhead valve (open/close); 7 – overhead cap, 8 – gas; 9 – blocking composition; 10 – thermo-shirt, 11 – fracture; 12 – fractured metal disc; 13 – measuring cylinder.

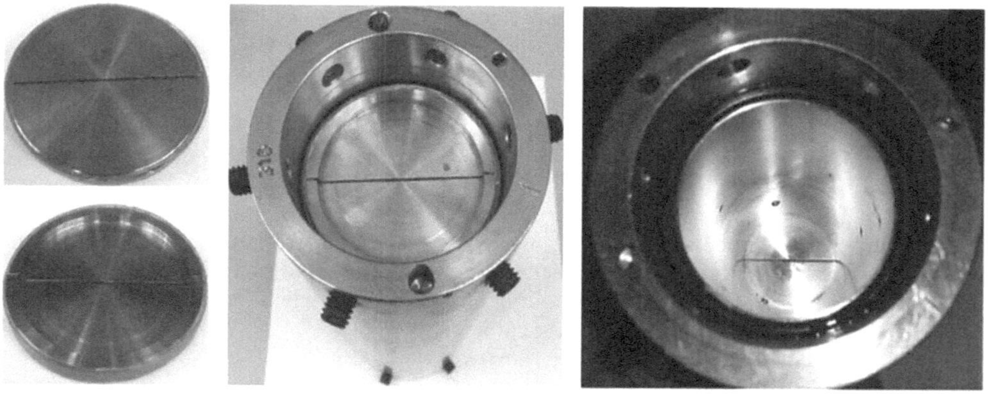

(a) (b) (c)

Figure 5. The appearance of the cell filter press: (a) fractured metal disk; (b) the lower part of the cell filter press (place exit blocking composition of the fracture); (c) the upper part of the cell (the location of the blocking composition above the fracture).

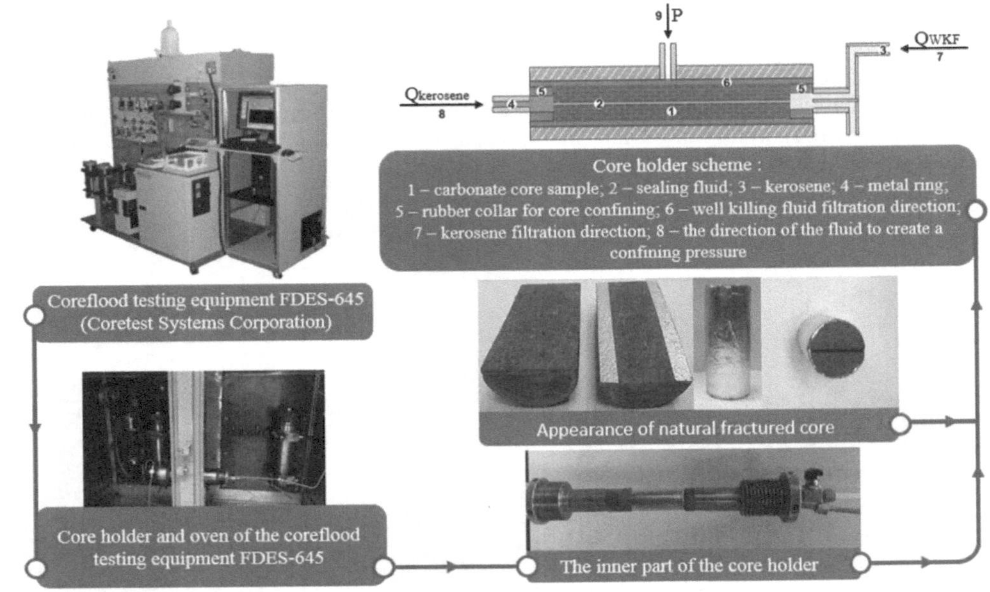

Figure 6. Coreflooding tests of compositions in modelling fractured reservoir.

In studies, a minimum fracture opening of 0.1 mm was used, corresponding to the minimum size of natural microfractures according to (Golf-Rakht 1986; Glushchenko & Silin 2010). Literature review revealed that microfractures are widespread in carbonate reservoirs of the Volga-Ural oil and gas province, whose openings are less than 0.5 microns. Among them are fractures developed in styllolites, the width of which does not exceed 1-2 mm (Kirkinskaya & Smekhov 1981; Lirio Quintero et al. 2017; Vikrant Wagle et al. 2018).

The efficiency of the blocking fluids was evaluated on the basis of the results of coreflooding tests on their effect on the change in permeability of reservoir models (cores) with fractures of varying degrees of the opening (0.1-1.5 mm). Coreflooding tests were conducted using the FDES-645 formation damage assessment unit (Coretest Systems Corporation) in the reservoir conditions. In this case, a modernized design of the core holder was used (Fig. 6).

Coreflooding tests were carried out in the conditions of the oil and gas condensate field when modelling the processes of "well killing" and "development" of production wells. Repression during the modelling of well killing was 1 MPa (5% of the reservoir pressure).

3 RESULTS AND DISCUSSION

The results of laboratory studies assessing the blocking properties of well killing fluids showed the ability of the developed composition, in contrast to the invert-emulsion and biopolymer compositions used in the oil and gas condensate field, to withstand high pressure drops (more than 8 MPa), as well as to restore coreflooding characteristics, when modelling of a fractured reservoir (fractures opening up to 1.5 mm) (Table 1).

These results can be explained by the presence in the composition of a multi-fractional microcalcite, due to which the formation of a durable and impermeable screen is ensured.

Figure 7 shows the appearance of a metal disk inside a filter press cell after conducting an experiment with an invert-emulsion composition mixed with microcalcite. Visually, one can see that a fine filter cake with a height of about 3-5 mm has formed over the fracture, blocking the fracture and preventing filtration of the emulsion composition. At the same time, this filter cake is easily destroyed when the blocking composition is drained from the cell.

Table 1. The results of filtration tests for evaluating the blocking properties of the compositions on the filter press.

Fracture opening, mm	The name of the blocking composition	Breakthrough pressure, MPa	Filtered volume composition, cm^3	Filtration time
0.1	Biopolymer	1	200	4 seconds
	Invert-emulsion	1	450	23 seconds
	Invert-emulsion + small fractional microcalcite	1	450	16 seconds
	Invert-emulsion + multi-fractional microcalcite	> 8	28	100 seconds
	Invert-emulsion + large fractional microcalcite	1	320	22 seconds
0.5	Biopolymer	1	150	2 seconds
	Invert-emulsion	1	450	8 seconds
	Invert-emulsion + multi-fractional microcalcite	> 8	6	40 minutes
1.5	Invert-emulsion + multi-fractional microcalcite	> 8	100	40 minutes

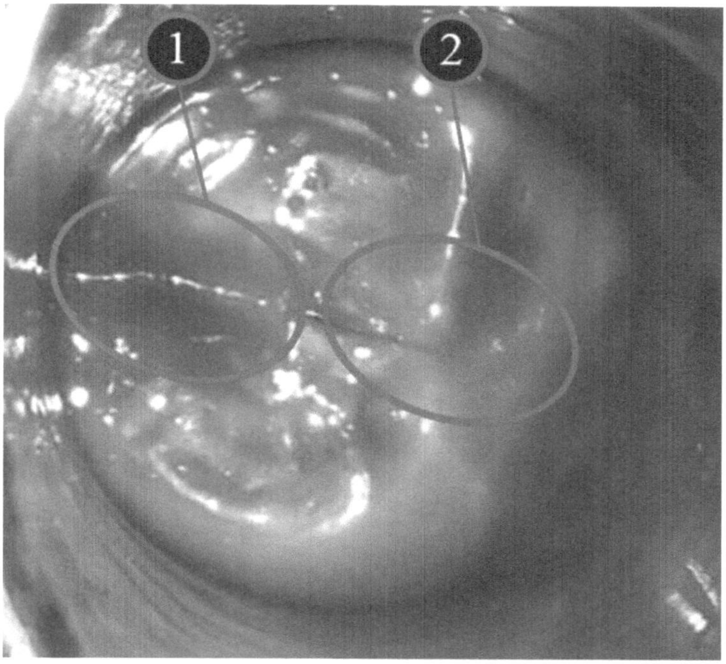

Figure 7. The appearance of the filter cake over the fracture after filtration of the invert-emulsion composition with microcalcite (fracture opening 0.5 mm): 1 – whole part of the filter cake; 2 – the destroyed part of the filter cake after draining the blocking composition from the cell.

The results of the coreflooding tests indicate that the addition of a multi-fractional microcalcite to the invert-emulsion composition makes it possible to increase the core permeability recovery coefficient with fracrures with an opening of 0.1 and 0.5 mm on average by 2 times.

This is due to the formation of a fine jumper from microcalcite at the entrance to the core, which prevents further emulsion composition from filtering into the fracture (Table 2).

To implement this technology, a standard arrangement of equipment is used for well killing operations. The equipment is also standard, only additional technological reservoir is used (Fig. 8). The technology of killing oil wells before underground repair using an invert-emulsion composition with microcalcite is to close the perforation interval with it without forcing it into the bottomhole formation zone. This effect is ensured by the use of a multi-fractional microcalcite as part of a well killing fluid (Fig. 9).

The economic efficiency of the application of the developed composition for well killing during their underground repair is to reduce the time required for the well to start operating,

Table 2. The results of coreflooding tests of blocking compositions.

Fracture opening, mm	The name of the blocking composition	Coefficient of relative change in core permeability, %	Maximum pressure gradient of the beginning of kerosene filtration after injection of a blocking composition, atm/m	The amount of penetrating blocking composition in the fracture, cm^3
0.1	Biopolymer	-79	493	0.8
	Invert-emulsion	-32	54.9	0.3
	Invert-emulsion + multi-fractional microcalcite	-15	50.7	< 0.2
0.5	Invert-emulsion	-93	209	14
	Invert-emulsion+ multi-fractional microcalcite	-43	22.2	< 0.2
1.5	Invert-emulsion	0	1.3	300
	Invert-emulsion + multi-fractional microcalcite	0	1.3	< 0.2

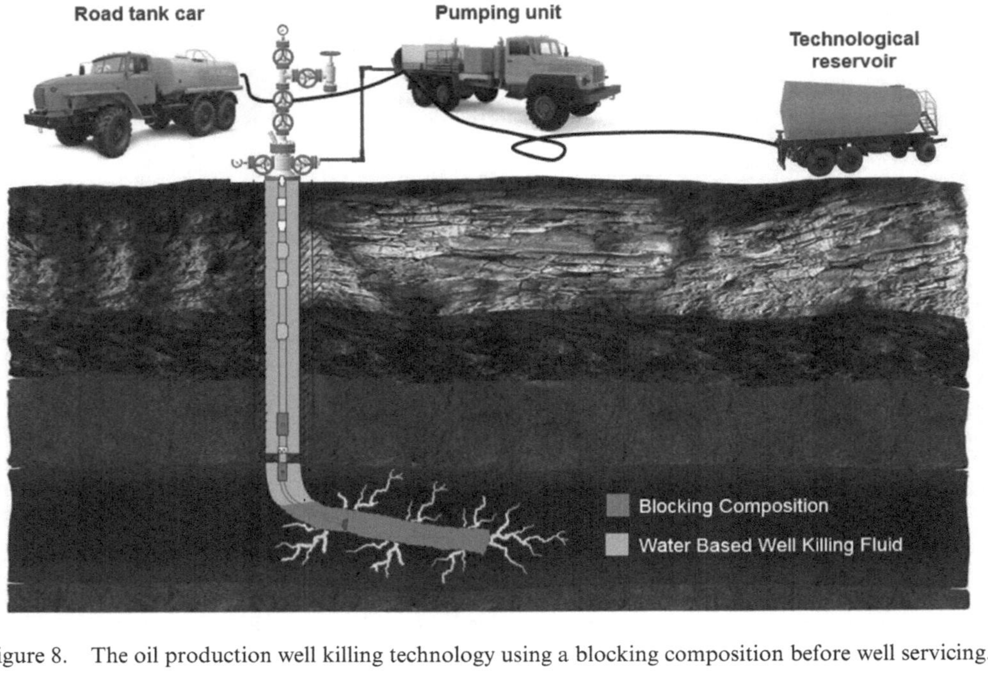

Figure 8. The oil production well killing technology using a blocking composition before well servicing.

Figure 9. The implementation of well killing technology using the developed composition on the example of one of the oil and gas condensate field in the Volga-Ural province in the summer and winter periods.

Table 3. Calculation of economic indicators in comparing the two compositions.

Indicators	Unit of measurement	New method	Basic method
Average duration of well killing	hours	12	24
Average response time	days	1	6
Number of well killing cycles	unit fraction	1	2
General costs	thousand rubles	821	1437
Cost differences	thousand rubles	616	
Incremental oil production as a result of saving time for well killing and response time	tons	65	
Profit due to improved well killing efficiency	thousand rubles	246	

by preserving reservoir properties of formations after carrying out repairs and reducing the costs of organizing and carrying out well killing operations.

The calculation of economic indicators in comparing the two compositions are presented in Table 3.

4 CONCLUSION

In order to prevent fluid loss and gas breakthrough during well service, a specific selection of compositions and technologies is required, based on the field data analysis and laboratory tests.

A well killing technology using a blocking emulsion with a multi-fractional microcalcite has been developed and recommended for implementation during well service operations at the oil and gas condensate field in the Volga-Ural province.

The developed technology will increase the efficiency of well killing operations in the oil and gas condensate field in the Volga-Ural province by:

- reducing the volume of the blocking composition by 1.5-3 times;
- reducing the number of repeated well kills by 10-20% of the total number of well kills;
- reducing the time required for the well to start operating up to 1-3 days.

The calculation of economic efficiency showed that the profit at the expense of increasing the efficiency of well killing amounted to 246.0 thousand rubles (3,729 thousand dollars), while the costs were reduced to 616.0 thousand rubles (9,338 thousand dollars), which made it possible to achieve the final economic effect for one well in the amount of 862.0 thousand rubles (13,067 thousand dollars). The developed well killing technology is 2 times cheaper than the previously used.

REFERENCES

Akhankin, O.B., Gusakov, V.N., Zdolnik, S.E., Latypov, A.R., Litvinenko, V.A., Telin, A.G. & Handriko, A.N. 2007. Well killing with controlled absorption in the conditions of intensified development of terrigenous reservoirs. *Oil Industry Journal* 830(11): 62-65.

Elkatatny, S.M. 2016. Determination the Rheological Properties of Invert Emulsion Based Mud on Real Time Using Artificial Neural Network. *SPE Kingdom of Saudi Arabia Annual Technical Symposium and Exhibition, doi: 10.2118/182801-MS, 25-28 April.* Saudi Arabia: Dammam.

Glushchenko, V.N. & Silin, M.A. 2010. *Oilfield Chemistry.* Moscow: Intercontact Science.

Golf-Rakht, T.D. 1986. *Fundamentals of Oilfield Geology and the Development of Fractured Reservoirs.* Moscow: Nedra.

Jouenne, S., Klimenko, A. & Levitt, D. 2016. Tradeoffs Between Emulsion and Powder Polymers for EOR. *SPE Improved Oil Recovery Conference, doi: 10.2118/179631-MS, 11-13 April.* USA: Tulsa.

Kirkinskaya, V.N. & Smekhov, E.M. 1981. *Carbonate Rocks of Oil and Gas.* Leningrad: Nedra.

Lirio Quintero, Ramakrishna Ponnapati & Mary Jane Felipe. 2017. Cleanup of Organic and Inorganic Wellbore Deposits Using Microemulsion Formulations: Laboratory Development and Field Applications. *Offshore Technology Conference, doi: 10.4043/27653-MS, 1-4 May.* USA: Houston.

Nikolaev, N.I. & Leusheva, E.L. 2016. Development of drilling fluids composition for efficiency increase of hard rocks drilling. *Journal of Mining Institute* 219(3): 412-420.

Orlov, G.A., Kendis, M.Sh. & Glushchenko, V.N. 1991. *Application of Invert Emulsions in Oil Production.* Moscow: Nedra.

Petrov, N.A., Soloviev, A.Ya., Sultanov, V.G., Krotov, S.A. & Davydova, I.N. 2008. *Emulsion Solutions in Oil and Gas Processes.* Moscow: Chemistry.

Ryabokon, S.A. 2009. *Technological Liquids for Completion and Servicing of Wells.* Krasnodar: Prosveshcheniye-Yug.

Sharath, S., Donald, W. & Jonathan, W. 2017. Acid-Soluble Lost Circulation Material for Use in Large, Naturally Fractured Formations and Reservoirs. *SPE Middle East Oil & Gas Show and Conference, doi: 10.2118/183808-MS, 6-9 March.* Kingdom of Bahrain: Manama.

Tokunov, V. I. & Saushin, A.Z. 2004. *Technological Liquids and Compositions for Increasing the Productivity of Oil and Gas Wells.* Moscow: Nedra.

Vikrant Wagle, Abdullah S. Al-Yami & Ali AlSafran. 2018. Designing Invert Emulsion Drilling Fluids for HTHP Conditions. *SPE Kingdom of Saudi Arabia Annual Technical Symposium and Exhibition, doi: 10.2118/192192-MS, 23-26 April.* Saudi Arabia: Dammam.

Zhelonin, P.V., Mukhametshin, D.M., Archikov, A.B., Zvonarev, A.N., Krayevsky, N.N. & Gusakov, V.N. 2015. Justification of the algorithm for selection of well killing technologies. *Scientific and Technical Bulletin of OJSC "Rosneft"* 39(2): 76-81.

Youth Technical Sessions Proceedings – Litvinenko (ed)
© 2019 World Petroleum Council, London, UK, ISBN 978-0-367-34668-3

Development and application of a metal seal soluble fracturing tool in tight oil horizontal wells

Z. Liu, J. Gui, Z. Wang, S. Guo, X. Xue, G. Ren & X. Meng
Changqing Oilfield Company of CNPC, Xi'an, Shanxi, China

ABSTRACT: The Ordos Basin is rich in tight oil and gas resources, which is an important resource base for 50 million tons of stable production in Changqing Oilfield. However, it is difficult for conventional technical means to develop them effectively. Relying on the national science and technology project "Tight Oil Development Demonstration Project in Ordos Basin", researchers have conducted the development of a new volume-fracturing tool, aiming at developing a soluble fracturing tool with the core of "metal seal, fast dissolution and drill-free". Through more than 20 times of formula adjustment and evaluation test, the technical problem of developing soluble metal with high ductility was finally broken through, developing a highly ductile soluble alloy, which broke the conventional rubber sealing method and realized metal sealing. The new fracturing tool solution has obvious advantages and broad application prospects compared with conventional tools.

1 INTRODUCTION

The Ordos Basin is rich in tight oil and gas resources, which is an important resource base for 50 million tons of stable production in Changqing Oilfield. The main stimulation technique for such reservoirs is close cut volume-fracturing, that is, "multi-cluster perforation + large volume + large fluid volume + low viscosity fluid" volume-fracturing mode, which can achieve full coverage of fractures to reservoirs and greatly increase single well production(Qi 2014, Ling 2014, Weng 2011, Wu 2011, Chen 2010, Wang 2010). For the horizontal wells in Changqing oilfield, soluble plug and fast drilling plug are mainly used in the early period. With the increase of the length of horizontal sections, the following problems arise after fracturing. The rubber of soluble plug dissolves slowly and incompletely, and bonds with solid particles in the wellbore such as proppant to cause secondary blockage of the wellbore. As to the fast drilling plug, the drilling efficiency is low because the drilling weight is difficult to apply in long horizontal sections.

Relying on the national science and technology project "Tight Oil Development Demonstration Project in Ordos Basin", researchers have conducted the development of a new volume-fracturing tool named MSF, aiming at developing a soluble fracturing tool with the core of "metal seal, fast dissolution and drill-free", and eventually forming an efficient fracturing method meeting the requirements of "staged seal, casing fracturing, multi-cluster fracturing and drill-free after fracturing".

2 LITERATURE REVIEW

Researches on solubility and related properties of metal materials have been discussed in the literature(Xu 2011, Zhang 2012, Mazyar O A 2013, Salinas B J 2012). In the field of fracturing tools, there are few reports on the use of soluble materials to achieve metal sealing. In the early stage, researchers have conducted researches on the use of soluble metals to manufacture fracturing balls(Li 2013), the use of insoluble metals to make ball seats (Ren 2017) and the study of solubility and strength of soluble metals (Shao 2018). Generally, soluble materials have perfect performance in pressure resistance, but poor shear resistance, tensile strength and low ductility. In order to achieve reliable metal seal of MSF, it is necessary to start with materials to improve the comprehensive performance while optimizing its structure and reducing the requirements for material properties.

3 METHODOLOGY

Metal sealing material and structure design are two key factors for the successful development of MSF. In terms of metal sealing material, the research team has developed material formulation research on the basis of mature metal soluble ball(Mg-Al alloy).Through more than 20 times of formula adjustment and evaluation test, the problem of developing highly ductile soluble alloy was finally broken through, which broke the conventional rubber sealing method and realized metal sealing. In terms of structure design, aiming at simplicity and reliability, the research team developed innovative self-locking structure design. Researchers carried out new materials research, scheme demonstration, 3D structure design, numerical simulation, prototype trial production, indoor test and ground simulation evaluation. Finally, a new fracturing tool named MSF with complete intellectual property was successfully developed.

3.1 *Technical idea*

In order to ensure the adaptability of MSF to the existing fracturing mode and keep the plug-perforation combined work unchanged, the main change is to increase MSF receivers among horizontal casings and develop fully soluble MSF. Separation between horizontal stages is achieved by the metal seal between MSF, receiver and soluble ball. The fracturing technology is schematically shown in Figure 1.

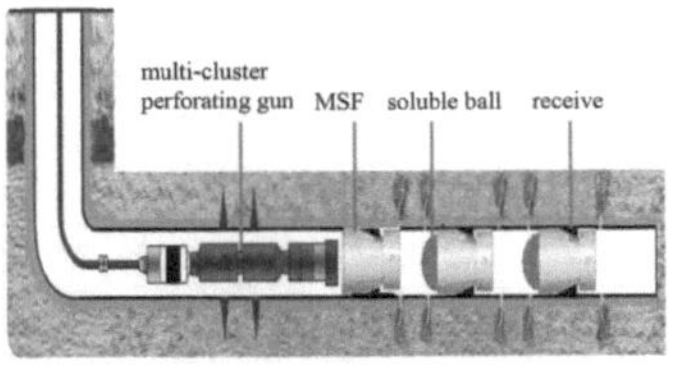

Figure 1. Schematic diagram of MSF fracturing technology

3.2 *Technology process*

(1) The MSF receivers are connected between casings, cemented and completed after entering the well.
(2) For the first stage, the tubing conveyed perforating is used to open the formation and smooth casing fracturing is conducted.
(3) The tool string including MSF and the MSF-perforation combined tools are delivered above a MSF receiver by hydraulic pumping. Fire to set and release the MSF, lift the string to finish multi-cluster perforation.
(4) Pull out the tool string, drop in a soluble ball, pump and push the ball and MSF to the receiver, sealing the upper stage and fracturing the next stage.
(5) Repeat step (3) and (4) to complete the whole fracturing in turn.
(6) The soluble ball and MSF are all dissolved, and the wellbore restores to full diameter and is put into production.

4 DEVELOPMENT, APPLICATION AND ANALYSIS OF MSF

4.1 *Development of key tools*

Key tools including fully soluble MSF, delivery tool, MSF receiver and large-size soluble ball.

4.1.1 *Fully soluble MSF*

The material of MSF is a fully soluble alloy and MSF consists of two upper lobes, two lower lobes and a base. The sides of the upper and lower lobes are designed with dovetail guide rails. The upper and lower lobes are connected by the guide rails, and the guide rails are at an certain angle with the axis. Therefore, when the upper and lower lobes move along the guide rails, the diameter can be expanded. After expansion, the outer diameter of MSF reaches 122mm, and the expansion rate exceeds 8%. The required closing stroke of MSF is 90mm. The common gunpowder setting tool can meet the requirements of the closing stroke. When the upper and lower lobes are completely closed, they are locked with the base. The states before and after setting are shown in Figure 2.

The difficulties in developing MSF are: (1) Because of the requirement of high pressure resistance, it is necessary to increase the strength of soluble material. (2) In order to ensure the metal sealing effect, the soluble material need to have a relatively high ductility. However, it is contradictory to increase the strength and ductility of soluble material at the same time. In order to solve this contradiction, researchers carry out work in two aspects. Firstly, optimize the sealing mechanism to mainly withstand positive pressure so as to avoid excessive shear force, and thus reduce the requirements for material strength. Secondly, adjust the formulation of soluble material and heat treatment parameters, so that the strength and ductility can reach a balance point(Shao Yuan 2018).

4.1.2 *Delivery tool*

The delivery tool is used to achieve setting and releasing of MSF. The core rod of the delivery tool is connected to the lower lobes by shear nails, and the cone is connected to the upper lobes. The relative movement of the upper and lower lobes can be achieved by the relative movement of core rod and cone, finishing setting of MSF. After that, shear nails are cut and the MSF is released. In order to prevent the premature setting during delivery, a start shear nail is designed between the cone and core rod. The structure of delivery tool is shown in Figure 3.

4.1.3 *MSF receiver*

The outer diameter of the receiver is 153.6mm, which is consistent with the $5^1/_2$" casing coupling to ensure the safety of well completion. The internal design includes a tapered step sealing surface, which can support the expanding MSF. Metal sealing is formed between the MSF receiver, MSF and a soluble ball to achieve segmental isolation during fracturing. After fracturing, the MSF and ball are completely dissolved, forming a large drainage and production passage, which does not affect the later stimulation operation. The structure and sealing principle of the MSF receiver are shown in Figure 4.

4.1.4 *Large-size soluble ball*

MSF is matched with 115mm soluble ball. But the segregation phenomenon of the large-sized soluble ball in casting process, that is, the uneven distribution of alloy elements, leads to

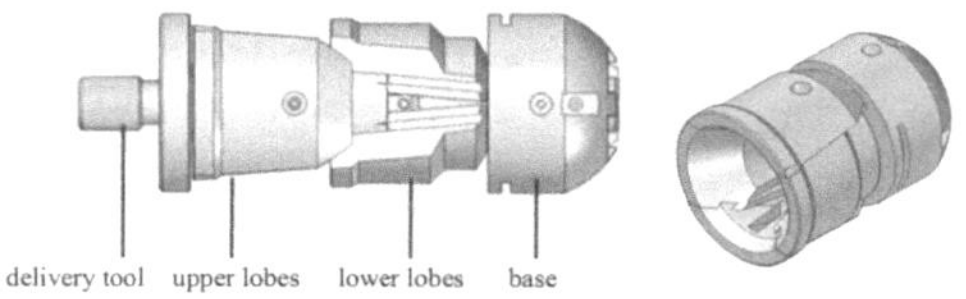

Figure 2. The states of MSF before and after setting

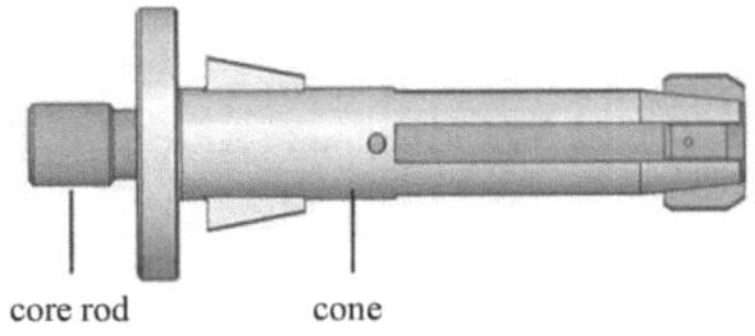

Figure 3. The structure of delivery tool

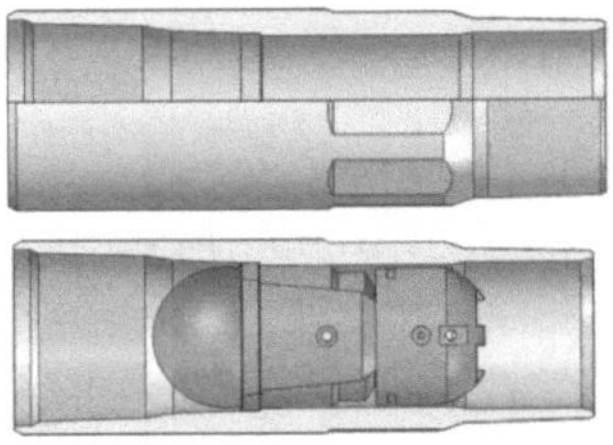

Figure 4. Structure and sealing principle of receiver

incomplete dissolution. Researchers propose a two-layer casting process, solving the problem of segregation, and achieve complete dissolution after fracturing.

4.1.5 *Key technological indexes of MSF*
Key technological indexes of MSF is shown in Table 1.

Table 1. Key technological indexes of MSF

Key tools	Outer diameter, mm	Inside diameter, mm	Pressure resistance, MPa	Dissolution temperature, °C	Dissolution time, h
Soluble ball	115	-	70	50	90
MSF	122	-	70	50	288
MSF receiver	153.6	117	70	-	-

4.2 Indoor test and evaluation

4.2.1 *Performance test of setting and releasing*
Due to the fact that gunpowder tools are strictly limited by relevant regulations and it is inconvenient to use for indoor test, a special hydraulic setting tool is designed to test the setting and releasing performance of MSF. The test results show that guide rails are flexible, starting the setting with a force of 2 tons and releasing at 4.5 tons, which meet design requirements. The test tool is shown in Figure 5.

4.2.2 *Test of pressure resistance*
Assemble MSF and soluble ball into a MSF receiver sequentially, and connect the upper end of the receiver to a hydraulic joint for hydraulic pressure. The test result shows that the MSF can bear a pressure of 70 MPa, as shown in Figure 6.

4.2.3 *Solubility test*
In a simulated flow-back environment, a constant temperature dissolving device was designed. A setting MSF was placed in 1% KCL solution at 50 °C. After 12 days, the MSF was completely dissolved, which meet the volume-fracturing requirements.

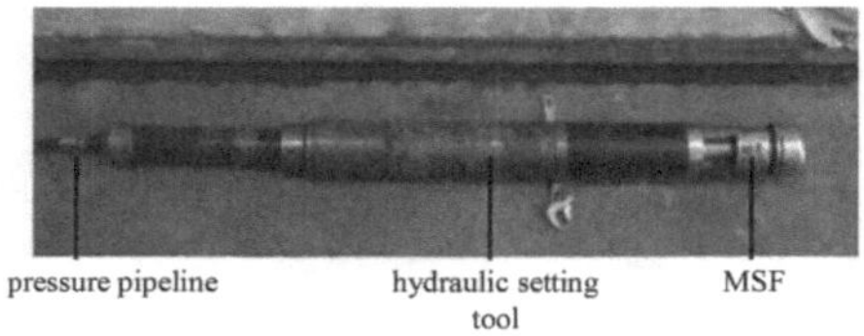

Figure 5. The test tool

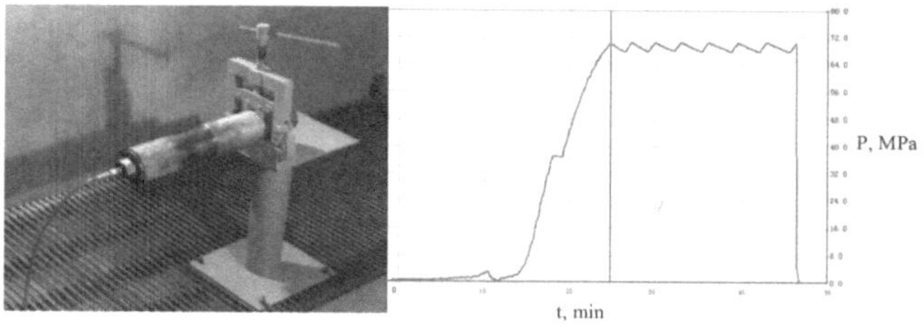

Figure 6. Test and result of pressure resistance

4.3 *Field test*

After the indoor test and evaluation, field test were carried out on 24 wells in 2 batches. In the first batch, two wells were selected for pilot test to find problems and solve them. In the second batch, 22 wells were selected and the longest horizontal section reached 1560 meters, with a maximum stage of 17. The performance was stable.

4.3.1 *Field operations*

The basic information of some fracturing wells is shown in Table 2.

(1) Structure of MSF-perforation combined string

From the bottom up, the structure is: MSF + Delivery Tool + Gunpowder Setting Tool + Cannon + Perforating Gun + Ignition Cannon Head + Fast-joint + Magnetic Positioning Section + Horse Head + Torpedo + Cable.

(2) Pumping parameters

In order to ensure the smooth delivery of MSF during pumping, the pumping parameters are shown in Table 3.

4.3.2 *Problems and improvement measures*

In the pilot test, the main problem encountered was that the pressure peak was not obvious after dropping in soluble ball and the separation between horizontal stages could not be accurately determined. The reason analysis and improvement measures are shown in Table 4.

Table 2. The basic information of some fracturing wells

Batch	Well	Horizontal section length, m	Application stages	Liquid volume, m^3	Sand, m^3	Volume, m^3/min
First batch	M135-4	935	1	1100	120	8
	F6-2	994	1	1300	160	14
Second batch	F6-5	1550	7	7730	1091	10-12
	F5-6	1560	11	14773	2374	10-14
	F5-7	1163	12	16494	2690	10-14
	F2-9	1487	17	18980	2704	10-12

Table 3. The pumping parameters

Well angle, °	Pumping volume, m^3/min	Cable entry speed, m/h
30	0.4	2000-2500
45	0.8	
60	1.2	
75	1.6	
90	1.8	

Table 4. The reason and improvement measures

Reason analysis	Improvement measures
There exists sand between MSF and the receiver, and the metal seal is not tight.	Double wellbore volume displacement, displacement volume> $10m^3$/min.
There exists a small amount of sand in guide rails of MSF. The resistance is large and it is not completely closed.	Increase the diameter of shear nails, increase the releasing force from 4.5 tons to 5.6 tons, overcome the frictional resistance, and completely close MSF.

4.3.3 *Application effect*

The fracturing curves of improved MSF show obvious pressure peak after dropping in soluble ball, and the sealing is reliable during fracturing. Compared with the soluble plug used in adjacent wells, pressure curves have same characteristics, as shown in Figure 7 and Figure 8. After fracturing, sand-washing pipe string was tested and passed smoothly, proving that MSF were completely dissolved, which solved the problem that the soluble plug was not completely dissolved and required to be drilled after fracturing.

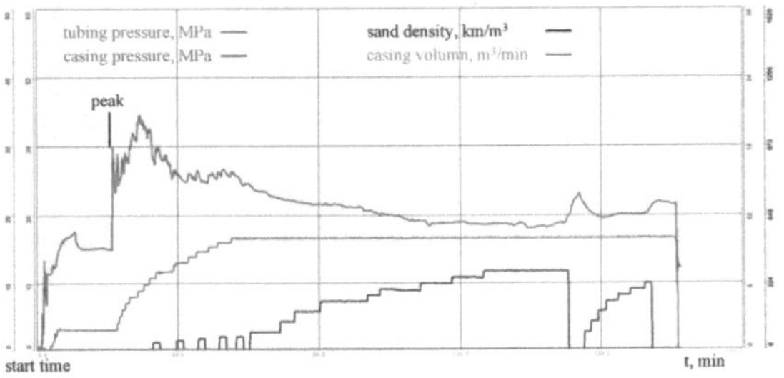

Figure 7. Fracturing curve of F6-5 using MSF

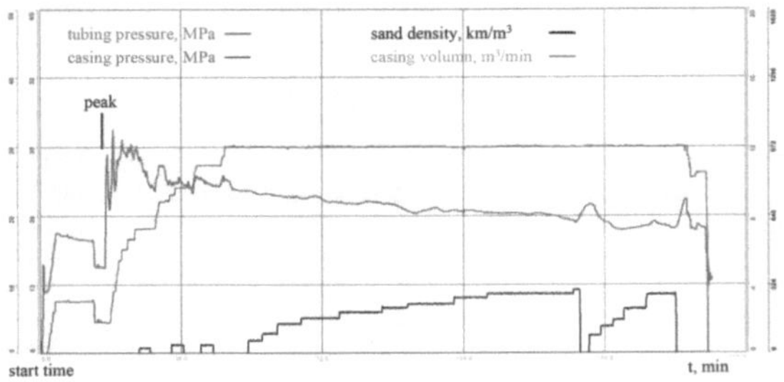

Figure 8. Fracturing curve of F5-3 using soluble plug

5 LONG-TERM EFFECT PREDICTION

By December 2018, MSF have been applied in 373 stages of 24 wells, with a success rate of 98%, a maximum length of 2704 m in horizontal section, a maximum record of 211 clusters in 43 stages in single well, a maximum volume of 14 m^3/min, a maximum injected fluid of 56,112 m^3 and a maximum sand of 10,066 tons in single well. After fracturing, wells can be put into production only by sand-washing, which avoids drilling and grinding operations. The post-fracturing processing time is shortened by 50%, the comprehensive cost is reduced by 30%, and the daily production at initial stage is 8 times that of adjacent vertical wells.

6 DISCUSSION

The size of the new tool has been reduced by 30% compared to conventional plugs, and the dissolve product is greatly reduced. The soluble fracturing tool can bear pressure of 70 MPa and temperature of 120°C and start initial dissolution at 40°C. It loses sealing in 50°C, 1% KCL solution for 2 days, loses anchoring in 7 days and completely dissolves in 12 days. Since the long horizontal section of wellbore is prone to sand settling, it may affect the smooth pumping and metal sealing. It is recommended to use double wellbore capacity and large volume to replace the wellbore to ensure the wellbore is clean.

7 CONCLUSION

MSF can meet the requirements of compact cutting volume-fracturing in the long horizontal sections, and dissolves completely, realizing drill-free after fracturing. Compared with soluble plug, the new tool adopts the soluble metal seal instead of the soluble rubber seal, which solves the problem that the current soluble rubber dissolves slowly and incompletely, and bonds with solid particles in the wellbore such as proppant to cause secondary blockage of the wellbore. The overall dissolution rate has been doubled, the tool cost is reduced by 40%, and the drilling and grinding cost is saved. It has significant advantages such as rapid dissolution and drilling-free production.

ACKNOWLEDGEMENTS

The research was funded by CNPC, the project "Special Project for Sustainable Development of Oil and Gas Reservoir Stimulation Technology", and it is carried out with the help of Xi'an Jiaotong University and Shanxi University of Science and Technology.

REFERENCES

Qi Yin. et al. 2014. Fracturing optimization and application of horizontal wells in ultra-low permeability reservoir. *FAULT-BLOCK OIL & GAS FIELD* 21(4):483-491.

Ling Yun. et al. 2014. New progress in fracturing technologies for tight sandstone gas reservoirs in the Sulige Gas Field, Ordos Basin. *NATURAL GAS INDUSTRY* 34(11):66-72.

Weng Dingwei. et al. 2011. Network fracturing techniques and its application in the field. *ACTA PETROLEUM SINICA* 32(2):280-284.

Wu Qi. & Xu Yun. 2011. The current situation of stimulated reservoir volume for shale in U.S. and its inspiration to China. *OIL DRILLING &PRODUCTION TECHNOLOGY* 33(2):1-5.

Chen Zuo. & Xue Chengjin. 2010. Proposals for the application of fracturing by stimulated reservoir volume(SRV)in shale gas wells in China. *NATURAL GAS INDUSTRY* 30(10):30-33.

Chen Zuo. & Wang Zhenduo. 2010. Hydraulic sand blasting segregated fracturing optimization and operation technology. *OIL DRILLING & PRODUCTION TECHNOLOGY* 32(3):72-75.

Xu Zhiyue. & Agrawal G. 2011. Smart nanostructured materials deliver high reliability completion tools for gas shale fracturing. *SPE 146586*.

Zhang Zhihui. et al. 2012. High strength nanostructured materials and their oil field applications. *SPE 157092*.

Mazyar O. A. & Johnson M. 2013. Method of controlling the corrosion rate of alloy particles, alloy particle with controlled corrosion rate, and articles comprising the particle: US, US2013/0029886 A1. 2013- 01-31.

Salinas B. J. et al. 2012. Controlled electrolytic metallics: An interventionless nanostructured platform. *SPE 153428*.

Li Jinbu. et al. 2013. Volume fracturing and its practices in Sulige tight sandstone gas reservoirs, Ordos Basin. *NATURAL GAS INDUSTRY* 33(9):65-69.

Ren Guofu. et al. 2017. Development and trial of casing ball seat fracturing tool. *DRILLING & PRODUCTION TECHNOLOGY* 40(5):76-80.

Guo Siwen. & Shao Yuan. 2018. Influence of Zn Content on the Degradation Rates of Degradable Aluminum Alloys. *Materials Review* 32(3):947-950.

Youth Technical Sessions Proceedings – Litvinenko (ed)
© 2019 World Petroleum Council, London, UK, ISBN 978-0-367-34668-3

Permeability evolution and the inner mechanism during hydraulic fracturing

S. Li, X. Wang, B. Cai, C. He & Y. Li
Hydraulic Fracturing Center, Research Institute of Petroleum Exploration and Development-PetroChina, Beijing, China

ABSTRACT: To evaluate the permeability changes and the underline mechanism during hydraulic fracturing, well shut-ins and flow back process, we perform a series experimental work by using tight core samples. Firstly, we build a physical model to simulate the whole hydraulic fracturing process by injecting fluids from different ends of core samples. According to the injection pressure and inject fluids flow rate, we calculate permeability according Darcy's law. What's more, we use nuclear magnetic resonance (NMR) method to detect aqueous phase migration to explain this phenomenon. Core flooding results showed that under no water invasion condition, flow-back pressure rise linearly at the beginning and reaches a plateau of 3.1 MPa. While under water invasion condition, flow-back pressure curves present a pressure peak and drop area then followed by a plateau. Then after different length of well shut-ins, the original pressure peak and plateau decreased which indicating rock permeability regained. Results demonstrated that permeability reduced 32.43% after water invasion while regained 11.0%, 24.7% and 32.9% after 2h, 4h and 7h's shut-ins. NMR results showed that aqueous phase redistribute and migrate to smaller pores spaces and deeper distance during well shut-ins. Via aqueous phase migration, near fracture water blocks can be weaken or removed and this is the reason for rock permeability enhancement. Study of permeability enhancement and aqueous phase migration can provide a reference to explain hydrocarbon production increase after hydraulic fracturing shut-ins.

1 INTRODUCTION

During the hydraulic fracturing of shale gas and tight oil reservoirs, scholars have fund that the average fracturing fluids flow back rate is only about 10%~15%, which is very low, and this phenomenon has caught scholars attention.

Interpretation has been proposed, for example, due to the high capillary pressure of the formation matrix, the fluid trapped within the fracture can be imbibed into formation via imbibition mechanism. However, these explanations are primarily hypothetical or based on experience and do not present convincing pore scales to justify their underlying mechanisms.

In this paper, we studied the permeability evolution after fluid intrusion and after well shut-ins using a core flooding system. We also studied the aqueous phase migration during well shut down using NMR methods. Via this, we try to explain permeability evolution and why well can increase oil and gas production.

2 LITERATURE REVIEW

Referring to the low recovery of fracturing fluids, Economides (2012) concluded that injected water would trapped in rock matrix near fracture face and would remains trapped due to the high capillary pressure in the matrix. Sharma (2013) found that fluids would also be trapped within fracture itself which will impact the cleanup and productivity. While some other scholar put the importance on imbibition process and many scholars have explained this phenomenon with mechanism of wettability alteration, relative permeability, capillary pressure and osmotic pressure using experimental or numerical methodology. Holditch (1979) described that during hydraulic fracturing, fluids will reduce gas relative permeability in invaded region because of clay swelling, solid precipitation and fines migration. While after hydraulic fracturing, imbibition will begin to alter the distribution of fluids. After that when production begins, fluids flow back and gas relative permeability will increase. He also demonstrated that complete water block of gas will not occur unless the rock permeability is serious damaged. Cheng (2012) reported oil output of a well in Marcellus shale reservoir improved significantly while its water output decrease sharply because of months of shut-in after hydraulic fracturing. Noe (2013) evaluated 80 Marcellus wells and other 120 wells and found that the shui-in event is not always harmful. The use of interfacial tension modifier fluid and high conductivity proppants can improve well performance.

3 METHODOLOGY

3.1 *Rock samples*

The unconventional rock samples used in our experiments came from the Chang7 formation of Changqing Oilfield, China. Six cylindrical water wet rock samples were used in our work. The first four samples were used for permeability evaluation experiments and the last two were used in NMR test. Table 1 gives detailed rock physics characteristics and comparative experimental schemes.

3.2 *Fluids*

To exclude or minimize the effects of clay swelling and fines migration, we used synthetic brine with a total salinity of 38.0 g/L as the injection fluid to match the formation brine. In addition, we chose fluorinated oil as a simulated oil. The reason for the use of fluorinated oil is that no NMR signal can be detected in the fluorinated oil.

3.3 *Core flooding procedure*

In the hydraulic fracturing field work, a part of the fluid penetrates into the formation matrix with a high pump pressure. Later on, when the bottom hole pressure drops below the reservoir pressure, hydrocarbon will flow back into the fracture and the wellbore. Under laboratory conditions, we simulated the process by injecting different fluids from different ends of the rock sample. As shown in Figure 1, the experimental process can be divided into four steps: initial flooding, fluid invasion, well shut-ins, and the second flooding.

In initial flooding, a dried and vacuumed core sample is attached to the core holder with a sealing pressure of 10 MPa, and inject simulated oil at a constant flow rate of 0.1 mL/min. Once the rock sample is fully saturated with simulated oil, record the flow rate and pressure

Table 1. Physical properties for the rock samples

Rock sample	Length /cm	Dimeter /cm	Porosity /%	Permeability /mD
cq001-1	5.01	2.51	10.2	0.21
cq001-2	5.06	2.53	10.6	0.25
cq001-3	5.03	2.49	11.2	0.27
cq001-4	4.96	2.52	9.7	0.19
cq002-1	5.51	2.51	9.3	0.11
cq002-2	5.50	2.48	9.6	0.16

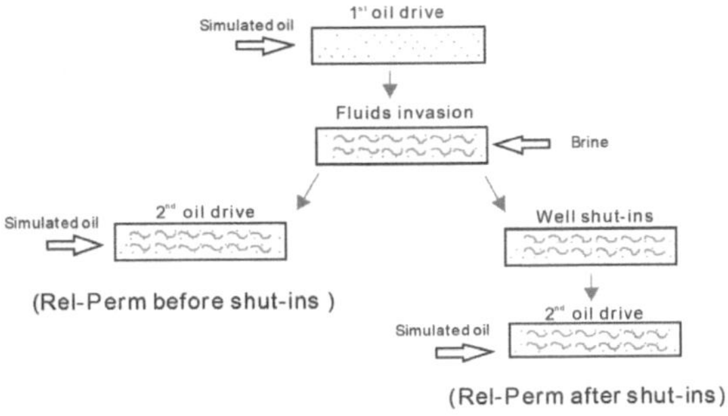

Figure 1. Core flooding procedure.

drop. The oil absolute permeability of the core is measured during this process. In order to simulate the invasion of fluid during hydraulic crushing, 1 PV synthesis brine is injected from the right end of the rock sample at 0.1 mL/min. Then inject the fluorinated oil from the left end at the same speed and calculate the relative permeability of the simulated oil before well shut-ins. Then, remove all pressure to confirm that there is no further pressure gradient throughout the rock sample and rest rock samples for various times to simulate well shut-in. In second flooding process, brine water was injected in the opposite direction at a constant flow rate of 0.1 mL/min. The sample reached the plateau, and only the simulated oil leaked out. Collective flow back to the measuring cylinder and calculate the relative permeability of the simulated oil at the residual water saturation.

3.4 *NMR detection principle*

Nuclear magnetic resonance (NMR) technology has been widely used in reservoir pore structure, fluid saturation distribution and oil/gas seepage experiments in recent years. This technique establishes nuclear magnetic variables and rocks based on the relaxation and diffusion principles of fluid spin nuclear magnetic moments. The calibration relationship of the fluid information enables online quantitative detection of the pore size and fluid distribution in the core, and has the advantages of fast, repeatable and high precision.

T_2 relaxation is transverse relaxation, also known as spin-spin relaxation, and the transverse component of the macroscopic magnetization vector gradually decays to zero during relaxation. Microscopically, the transverse relaxation is mainly the process of the phase of the spin system from focusing to divergence. The total energy of the system does not change, but the energy exchange between the spin hydrogen nuclei within the system. After the end of the pulse, the transverse component of the macromagnetization vector is attenuated in the form of an exponent. The characteristic constant T_2, called the T_2 relaxation time, is usually measured by the CPMG sequence.

Simulated crude oil and simulated formation water are used to simulate the oil-water migration process in the underground. At normal temperature, the diffusion coefficient of simulated crude oil is much smaller than the diffusion coefficient of water ($2\times10^{-9}m^2/s$), and can be utilized in two-dimensional spectrum. The difference is distinguished by a large difference in the diffusion coefficient. The middle value of the oil-water diffusion coefficient is selected as the boundary line between the oil and water on the two-dimensional spectrum. The signals with the diffusion coefficient greater than the value are all water signals, and the signals with the diffusion coefficient smaller than the value are all regarded as oil signals.

The transverse relaxation time T_2 of the fluid in the rock is related to the nature of the fluid itself, the interaction between the fluid and the pore surface of the rock. For the fluid in the rock sample, due to the different forces acting on the surface of the rock solid, the transverse relaxation time T_2 is caused. The greater the interaction force between the liquid and solid, the shorter the transverse relaxation time T_2, and vice versa. Therefore, for the low permeability and compact rock samples, the transverse relaxation time T_2 will be shortened. For fluids close to the pore surface of the rock, the transverse relaxation time T_2 is reduced to a few milliseconds, or even less than one millisecond. According to previous studies, the NMR transverse relaxation time can be expressed as

$$\frac{1}{T_2} = \rho_2\left(\frac{S}{V}\right) \tag{1}$$

$$T_2 = C \times r \tag{2}$$

where, C (ms/μm)is a constant conversion coefficient (ms/μm). Hence, if the relationship between pore radius and T_2 can be found, NMR T_2 spectrum can be converted to pore radius distribution curve and the inside pore fluids migration can be evaluated by NMR data.

3.5 *NMR detection procedure*

The core samples cq002-1 and cq002-2 will be used for NMR detection. (1) First, wash the rock sample without destroying the pore structure in order to remove all organic substances. In order to ensure that the rock samples are cleaned, high-pressure injection- of fluorinated oil is used to clean the rock samples. The rock sample was then dried in a 110 ° C oven for 3 days. (2) Test the permeability of the rock sample, and weigh the dry rock sample, then use the vacuum pump to vacuum the rock sample, saturate the brine water, and let it stand for 0h, 20h, 40h and 90h to ensure that the brine water saturates the rock sample. The rock samples were weighed to obtain parameters such as porosity and pore volume, and nuclear magnetic resonance experiments were carried out on the rock samples.

4 RESULTS AND DISCUSSION

4.1 *Relative permeability enhancement after shut-ins*

In Figure 2, the red column shows the relative permeability of the simulated oil before the well shut-ins, and the blue column shows the permeability after that. We can see that the rock sample had a relative permeability of 0.16 to 0.23 after brine water intrusion, and the permeability showed a different degree of enhancement from 0.19 to 0.27 after the well shut-ins. Permeability reduced 32.43% after water invasion while regained 11.0%, 24.7% and 32.9% after 2h, 4h and 7h's shut-ins. In the olive dotted line, we plotted the relative permeability regained, and we can see that the permeability increases rapidly at the beginning, and this growth rate gradually slows down during the long shutdown.

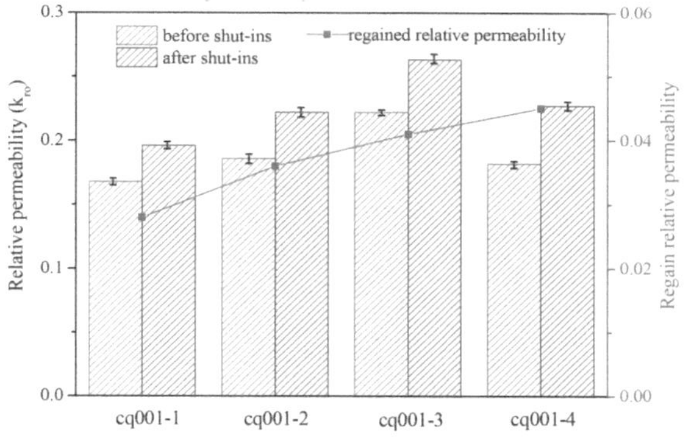

Figure 2. Relative permeability of simulated oil before and after well shut-ins for rock samples. cq001-1: 6h shut-in time; cq001-2: 12h shut-in time; cq001-3: 18h shut-in time; cq001-4: 24h shut-in time.

4.2 *T2 spectrum migration during shut-in period*

After comparing the core sample mercury injection curve with the NMR T_2 spectrum, scolaars gave an improved method for transferring the T_2 spectral distribution to the pore radius distribution and calculated the transfer parameter C in equation (2). For their previous work, we estimated the relationship between T_2 relaxation time, pore radius and pore type, as shown in Table 2, which we will use in the discussion that follows.

Figure 3 shows the T_2 spectral distribution of core sample cq002-1 and core sample cq002-2 at different shut-ins times after saline injection. The green line shows the T_2 spectrum at the end of the fluid injection. The other blue, magenta and red lines are T_2 spectra after 20 hours, 40 hours and 90 hours. In Figure 3a, the T_2 spectrum shows two amplitude peaks, indicating a smaller peak at 0.1 to 1 ms intervals and a larger distance of 1 to 100 ms. It should be noted that these two peaks are not well connected, indicating the separation of the micropores and mesopores. On the other hand, in Figure 3b, the T_2 spectrum has a long-span amplitude peak only at intervals of 0.1 ms to 100 ms. In both figures, the same curve moves as indicated by the red arrow. After blocking different lengths, one phenomenon is the T $_2$ spectrum moving to the left and the other is the T_2 amplitude when increasing the aperture radius. For example, in Figure 3a, after a 90-hour break, the transition from the right end of the 138.1 ms T_2 spectrum to 94.2 ms has increased from 28.9 to 430.2 at 22.1 ms.

As described above, the aperture radius and the T_2 relaxation time are interdependent and have a relationship with the large aperture comprehensive T_2 relaxation time. The extraordinary T_2 amplitude means a high fluid content. Injury to a small number of cases, the degree of local injection of water injection into the flooding start time is distributed in the middle of the large inner hole. During the period of residence, the existing synthetic water from the first 90 to 140 ms is shifted by 10 to 50 ms for interspersed and interlaced. The water transfer leads to an increase in the total water content, an increase in the water content, and an increase in the incremental T_2 magnitude.

Table 2. Relationship between T2 relaxation time and pore characteristics.

T2 relaxation time/ms	Pore radius/μm	Pore type
T2 relaxation time≤1	pore radius≤2	micropore
1<T2 relaxation time≤10	2<pore radius≤10	small mesopore
10<T2 relaxation time≤50	10<pore radius≤20	mesopore
50<T2 relaxation time	20<pore radius	macropore

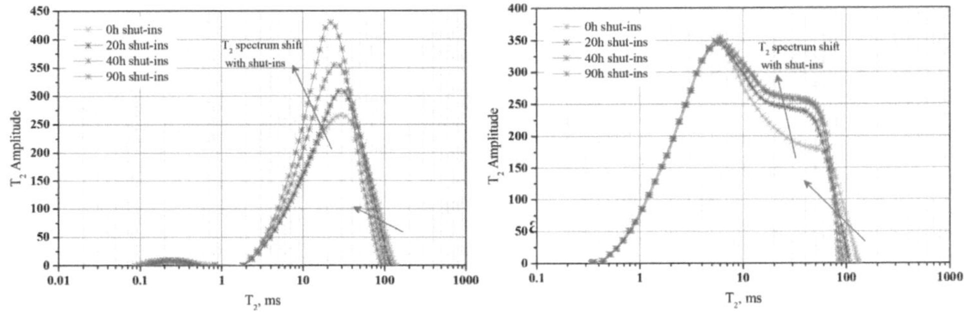

Figure 3. T2 spectrum shift during shut-ins period, drawn in semi-logarithmic coordinate system. (a): core sample cq002-1; (b): core sample cq002-2.

4.3 *Why shut-ins can increase permeability*

Our experiments show that the permeability increases after shut-in, which is mainly due to the redistribution and migration of the water phase during shut-in. Redistribution of the aqueous phase and migration to smaller pore spaces and deeper distances contribute to increased matrix permeability and subsequent oil and gas production because 1) the aqueous phase removes water blocks near the fracture. During shut-in, the aqueous phase migrates from the larger pore space to the smaller pore space, replacing the hydrocarbon phase by capillary-dominated countercurrent expansion. The redistribution and migration of the aqueous phase can release large pore spaces, weaken or even remove water resistance. 2) Reduce capillary discontinuity. Well shut-ins and water redistribution cause these water phases to no longer be distributed as a continuous whole, some of which enter smaller pore spaces, some enter deeper distances, thereby reducing the permeability of the hydrocarbon phase. 3) Increasing the relative permeability of the hydrocarbon phase. The swelling mechanism can increase the amount of oil gas phase, reduce the amount of water phase in the pores, and increase the relative permeability of the oil gas phase by changing the water saturation.

The effectiveness of the imbibition process depends on several parameters, including the size of the matrix block, the pore permeability of the rock, the viscosity of the fluid, the interfacial tension and the wettability of the rock, as well as the contact area of the wetting fluid and the matrix. Therefore, tight reservoirs rich in micro-cracks are advantageous for osmotic oil recovery because natural cracks will greatly increase the contact area of the fluid with the matrix. At this point, the tight reservoir is considered to be divided into a collection of innumerable matrix blocks by microcracks rather than a monolithic matrix block. During hydraulic fracturing and flow-back, this procedure may be underlined: (1) The high-pressure fracturing fulids is filtered from the cracks to the matrix, invading the large oleophilic pores that are connected to each other; (2) Water intrusion causes high water and low permeability areas near the fracture wall; (3) The water trapped near the crack obstructs the oil and gas flow, cause permeability decrease, and slows the penetration of water from the oleophilic pore to the hydrophilic pore; (4) During well shut-ins, most of the water is imbibed from the oleophilic pores into the hydrophilic pores, and the gas can flow freely in the interconnected large lipophilic pores, increasing the permeability and productivity.

Fracturing fluids injected into underground may distributed near around the fracture face or far away into the matrix. For water wetting reservoirs, aqueous phase in fracturing fluids can entered far away into the matrix along the water film. These entered aqueous phase can increase formation pressure and simultaneously can be imbibed into smaller pores to displace oil. While for oil wetting reservoir, aqueous phase was blocked around the fracture surface. These fluids will flow back through the wellbore in the open flow period, and cause less harm to the formation. For high permeability sand reservoir, remained fluids in formation would form two phases flow and block oil flow out because of Jamin Effect. While for tight formation, because of the high capillary pressure caused by micro-scale pores and throats and because of their discontinuous distribution, two phases flow would not block all channels for fluids flow. What's more, they may become an advantage factors because of imbibition process and oil and gas increase.

5 CONCLUSION

In this paper, the core flooding system is used to study the reduction of permeability after water inrush and the improvement of permeability after shut-in. In addition, we also used the nuclear magnetic resonance method to study the water phase migration during the shut-in period to reveal its internal mechanism. Core flooding experiments show that leakage and water intrusion into the formation can cause a decrease in matrix permeability, and shut-in can restore permeability. The results of NMR showed that the water phase was mainly distributed in the macropores and mesopores after water intrusion. During the shut-in period, these water phases were redistributed and migrated from macropores to pores. The redistribution and migration of the water phase during shut-in can eliminate the water seal, reduce the discontinuity of the capillary, and increase the relative permeability of the oil phase.

REFERENCES

Al-Ameri, A., Gamadi, T., & Watson, M. 2018. The Impact of the Fracturing Fluid Spontaneous Imbibition on the Petrophysical Properties of Low Permeability Reservoirs. *In SPE International Conference and Exhibition on Formation Damage Control*. Society of Petroleum Engineers.

Birdsell, D. T., Rajaram, H., & Lackey, G. 2015. Imbibition of hydraulic fracturing fluids into partially saturated shale. *Water Resources Research*, 51(8),6787-6796.

Bertoncello A, Wallace J, Blyton C, et al. 2014. Imbibition and water blockage in unconventional reservoirs: well management implications during flowback and early production[C]//*SPE/EAGE European Unconventional Resources Conference and Exhibition*.

Chakraborty, N., Karpyn, Z. T., Liu, S., & Yoon, H. 2017. Permeability evolution of shale during spontaneous imbibition. *Journal of Natural Gas Science and Engineering*, 38, 590-596.

Cheng Y. 2012. Impact of water dynamics in fractures on the performance of hydraulically fractured wells in gas-shale reservoirs. *Journal of Canadian Petroleum Technology*, 51(02): 143-151.

Ehlig-Economides C A, Ahmed I A, Apiwathanasorn S, et al. 2012. Stimulated Shale Volume Characterization: Multiwell Case Study from the Horn River Shale: II. Flow Perspective[C]//*SPE Annual Technical Conference and Exhibition*. Society of Petroleum Engineers.

Fakcharoenphol P, Torcuk M A, Wallace J, et al. 2013. Managing shut-in time to enhance gas flow rate in hydraulic fractured shale reservoirs: a simulation study[C] *SPE Annual Technical Conference and Exhibition*. Society of Petroleum Engineers.

Gao, Z., & Hu, Q. 2016. Initial water saturation and imbibition fluid affect spontaneous imbibition into Barnett shale samples. *Journal of Natural Gas Science and Engineering*, 34, 541-551.

Haluszczak L O, Rose A W, Kump L R. 2013. Geochemical evaluation of flowback brine from Marcellus gas wells in Pennsylvania, USA. *Applied Geochemistry*, 28: 55-61.

Holditch S A. 1979. Factors affecting water blocking and gas flow from hydraulically fractured gas wells. *Journal of Petroleum Technology*, 31(12): 1,515-1,524.

Kathel, P., & Mohanty, K. K. 2018. Dynamic surfactant-aided imbibition in fractured oil-wet carbonates. *Journal of Petroleum Science and Engineering*, 170, 898-910.

Mirzaei, M., DiCarlo, D. A., & Pope, G. A. 2016. Visualization and analysis of surfactant imbibition into oil-wet fractured cores. *SPE Journal*, 21(01),101-111.

Morrow, Norman R., and Geoffrey Mason. 2001. Recovery of oil by spontaneous imbibition. Current Opinion in Colloid & Interface Science 6.4: 321-337.

Noe S, Crafton J W. 2013. Impact of Delays and Shut-Ins on Well Productivity, *SPE Eastern Regional Meeting*. Society of Petroleum Engineers.

Nooruddin, H. A., & Blunt, M. J. 2016. Analytical and numerical investigations of spontaneous imbibition in porous media. *Water Resources Research*, 52(9),7284-7310.

Sharma M, Agrawal S. 2013. Impact of liquid loading in hydraulic fractures on well productivity[C]*SPE hydraulic fracturing technology conference*. Society of Petroleum Engineers.

Schmid, K. S., Alyafei, N., Geiger, S., & Blunt, M. J. 2016. Analytical solutions for spontaneous imbibition: fractional-flow theory and experimental analysis. *SPE Journal*, 21(06),2-308.

Zaeri, M. R., Hashemi, R., Shahverdi, H., & Sadeghi, M. 2018. Enhanced oil recovery from carbonate reservoirs by spontaneous imbibition of low salinity water. *Petroleum Science*, 15(3),564-576.

Zhou, Z., Abass, H., Li, X., Bearinger, D., & Frank, W. 2016. Mechanisms of imbibition during hydraulic fracturing in shale formations. *Journal of Petroleum Science and Engineering*, 141, 125-132.

Youth Technical Sessions Proceedings – Litvinenko (ed)
© 2019 World Petroleum Council, London, UK, ISBN 978-0-367-34668-3

A downhole automatic water distribution technology with remote wireless communication system–a case study in Changqing Oilfield

Y. Liu, Z. Ji, L. Yang, G. Hu, Z. Wang & F. Bi
Oil & Gas Technology Research Institute, Petrochina Changqing Oilfield Company, Xi'an, China
National Engineering Laboratory of Tight Oil & Gas Field Exploration and Development, Xi'an, China

ABSTRACT: To solve the problem of discontinuous data obtained by concentric bridge layered injection technology due to its artificial controlled mechanical water distribution pattern, Changqing Oilfield has done enormous work on intelligent layered injection. At present, two technologies have been developed, which are intelligent layered injection technologies with cable and without cable. However, they are faced with two major difficulties of downhole communication and flow rate measurement. In this paper, based on throttling equation and remote wireless communication via pressure wave, that is widely used in Measurement while Drilling (MWD), a downhole automatic water distribution technology is proposed. By means of pressure wave signals transmitting in wellbore, the surface control system sends water regulation orders to intelligent regulator and receives injection data from it. Based on pressure difference before and after water nozzle, a flow rate measurement without flowmeter is developed. In last two years, this new technology has been applied in 138 water injectors in Chanqing Oilfield and proved to be effective. Field application demonstrated that the downhole remote wireless communication via pressure wave could realize real-time supervision of single layer pressure and flow rate, and the wireless communication distance reached to 2879m. According to analysis results of historical flow rate data, the qualified rate of single layer flow rate was maintained above 90% in long term. By successful application of this technology, waterflooding efficiency of the pilot area was improved remarkably. The average water absorbing thickness was increased from 19.3m to 20.4m, and the water drive reserves producing degree was increased from 65.2% to 67.7%. Meanwhile, the natural decline rate was decreased from 5.8% to 2.1%, and the water cut increasing rate maintained the same at 1.3%. This proposed technology provides an effective way of real-time monitor and downhole automatic control for layered waterflooding in low permeability reservoir. The real-time data obtained from it are meaningful in waterflooding performance evaluation and development optimization.

1 INTRODUCTION

Layered waterflooding is the most economic and effective way of maintaining formation pressure and enhancing oil recovery in low permeability reservoir. At present, the commonly used technology is concentric bridge waterflooding in Changqing Oilfield. However, it is faced with several serious problems. Firstly, the continuous growth in number of layered injectors leads to heavy workload of artificial regulation operations and high supporting costs, which increases the difficulty of injectors' management. Secondly, the water flow qualified rate reduces rapidly due to belated artificial regulation operation, and it constrains waterflooding performance. And thirdly, with a conventional regulation pattern of testing at fixed time, there is no continuous data can be used for waterflooding effect evaluation and development plan adjustment. Therefore, since 2012, numerous researches have been done on automatic water distribution technology in Chanqing Oilfield, and there has been two technologies developed, including layered water injection with cable and without cable. They have realized two major functions: automatic water regulation downhole and data monitoring. The layered water injection technology with cable uses cable installed outside the tube to power and transmit downhole data. Because it cannot meet the demand of operation under pressure, its application is restricted. The layered water injection technology without cable uses downhole battery pack to power and records data in downhole water distributor. It is need to run communication instrument into the wellbore to record data regularly. Moreover, the integrated electromagnetic flowmeter in the water distributor is costly, and has default of obvious zero drift which cannot be calibrated downhole. According to the advantages and disadvantages of these two technologies, and focusing on two technical difficulties including downhole communication and flow measurement, an automatic water distribution technology based on remote wireless communication by pressure wave and flow rate measurement by pressure difference before and after the water nozzle is proposed. It realizes multiple functions: two-way remote wireless communication between surface and downhole, layered automatic water regulation downhole, and real-time monitoring of layered data. With these functions, it provides an effective method to improve waterflooding effect.

2 RESEARCH OF KEY TECHNIQUES

A series research has been carried out aiming at the two technical difficulties of downhole communication and flow measurement. Based on the mature application of wireless communication by pressure wave in Measurement While Drilling (MWD), a two-way remote wireless communication technique between surface and downhole is developed. And inspired by throttling equation, a flow rate measurement without flowmeter is developed.

2.1 *Remote Wireless Communication by Pressure Wave*

2.1.1 *Information Telemetry Principle in MWD*

In the 1980s, information telemetry system by continuous pressure wave came into market, which drove the development of MWD. Taking pressure wave as signal carrier and drilling fluid as telemetry channel, this technology achieves the goal of real-time transmission of drilling data. In the drilling process, sensors, downhole controller and rotary valve or shear valve (used to generate pressure wave) are run into the wellbore as part of the lower drilling tool. All kinds of drilling data measured by the sensors are converted into electrical signal by the downhole controller. Then the controller controls rotary valve or shear valve to convert the signals into pressure wave, which can transmit in the drilling fluid to surface sensors. Finally, the surface sensor detects the pressure wave and decodes it into drilling data. In injectors, based on this principle, taking pressure wave as signal carrier and injected water as signal channel, a two-way wireless communication technique between surface and downhole is proposed.

2.1.2 *Two-way Remote Wireless Communication in Injectors*

Wireless Communication from Surface to Downhole. When communication from surface to downhole, the pressure wave needs to be generated at the wellhead and transmits to downhole, as shown in Figure 1. Changing pressure or flow rate in wellbore regularly by electronic control valve installed at the wellhead, the pressure wave can be created and transmits in injected water to downhole intelligent water distributor. By detecting and decoding the pressure wave, water distributor converts it into command signals, thus realizes command transmission.

Wireless Communication from Downhole to Surface. When communication from downhole to surface, the pressure wave needs to be generated form downhole and transmits to wellhead, as shown in Figure 1. Changing pressure or flow rate regularly by opening or closing water nozzle in the water distributor, the pressure wave can be obtained and transmits in injected water to wellhead. The surface control system can detect and decode it into injection data (layered flow rate, layered pressure, and temperature), thus realizes data remote transmission.

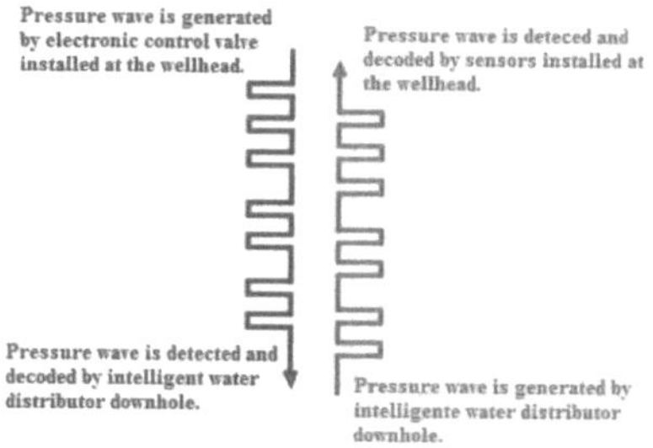

Figure 1. Schematic diagram of two-way remote wireless communication.

2.2 *Flow Measurement by Pressure Difference of Before and After Nozzle*

In the process of layered water injection, throttling effect will occur when injected water passes through the water nozzle with diameter of millimeter magnitude. According to the throttling equation, the relationship among flow rate, pressure and opening degree of water nozzle can be obtained. In order to solve the problem of incorrect flow measurement caused by zero drift of flow meter downhole, a flow measurement technique by pressure difference of before and after nozzle is developed.

2.2.1 *Flow Rate Calculation*

Two pressure sensors are installed in the intelligent water distributor to measure the pressure before and after water nozzle. With known pressure difference and open degree of water nozzle, the flow rate can be calculated by equation 1:

$$Q = C_d w x_v \sqrt{\frac{2}{\rho}(p_1 - p_2)} \tag{1}$$

Where, p1 is pressure before water nozzle, MPa; p2 is pressure after water nozzle, MPa; w is area gradient of water nozzle, m; Cd is velocity coefficient; xv is displacement of water nozzle, m.

2.2.2 *Flow Rate Calibration*

To improve the accuracy of flow calculation, the fuzzy mathematical theory is applied in self-learning of the relationship between flow rate, pressure difference and opening degree of water nozzle by surface controller, and the calculation model is constantly revised. The principle of flow calibration is to assume that the area of the nozzle does not change with the same opening degree, and the pressure difference error caused by the area change and the sensor can be eliminated by regular flow calibration. During the process of automatic water flow regulation downhole, the water nozzles are open separately, which means the surface flowmeter measures single layer flow rate. Surface controller will automatically record a large number of data during regulation, including flow rate, pressure difference and open degree of nozzle. Thus the calculation model can be revised constantly based on storage data, and the accuracy of flow calculation is improved.

3 A DOWNHOLE AUTOMATIC WATER DISTRIBUTION TECHNOLOGY WITH REMOTE WIRELESS COMMUNICATION SYSTEM

3.1 *System Composition*

The downhole automatic water distribution technology with wireless communication system consists of three major parts: intelligent water distributor, surface control system and remote monitoring system, as shown in Figure 2. Powered by built-in battery pack, the intelligent water distributor controls the motor to adjust the open degree of water nozzle to realize automatic water flow regulation. The surface control system is composed of a controller, a flow meter and an electric control valve. The main function of this part is to send control instructions to the intelligent water distributor downhole and receive layered injection data downhole. Supported by internal network of Changqing Oilfield, the remote monitoring system can transmit water injection data to office computers of relevant personnel, and help them to monitor injection data in real time.

3.2 *Laboratory Experiments of Functional Verification*

3.2.1 *Automatic Water Flow Regulation*

Before the experiment, the required flow rate was set at 25 m^3/d, and the automatic regulation time was set at 300s. Installing the water distributor in the flow test device and starting automatic regulation function. According to the preset flow rate and measured flow rate obtained from the water distributor, the opening degree of water nozzle was adjusted automatically to meet the preset permissible error of 20%. Figure 3 shows the automatic regulation curve of the intelligent water distributor, where the black curve represents the flow rate. It can be observed that there are two automatic regulation processes. During the first time, the measured flow rate was about 30 m^3/d, which exceeded 20% of the preset flow rate. Therefore, the automatic regulation function was activated. The opening degree of the water nozzle changed from 14 grade to 6 grade, making the flow rate changing to 19 m^3/d, which still couldn't meet the requirements of injection. During the second time, the water nozzle was opened up to 10 grade, and the flow rate was about 24 m^3/d, meeting the requirements of injection.

3.2.2 *Two-way Remote Wireless Communication Test*

Wireless communication test from surface to downhole. Sending commands of opening all water nozzles via the surface control system to verify whether all the two water nozzles can be opened. The experiment curve is shown in Figure 4. By sending two times pressure or flow wave hopping, two water nozzles were opened, which indicated that the control instruction was transmitted from surface to downhole successfully.

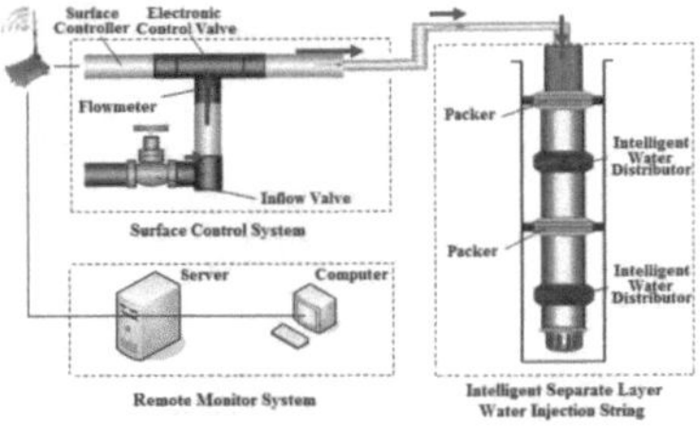

Figure 2. A downhole automatic water distribution technology with remote wireless communication system.

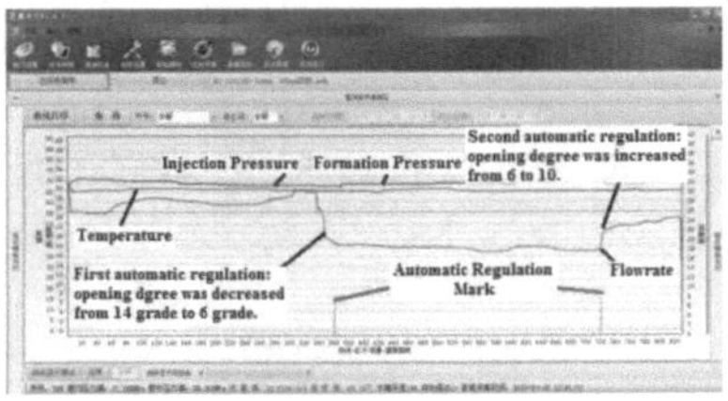

Figure 3. Automatic regulation curve of the intelligent water distributor.

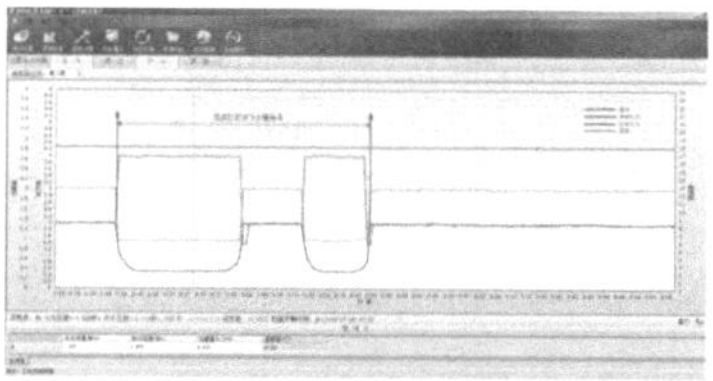

Figure 4. The pressure changing curve when sending commands of opening all the water nozzles.

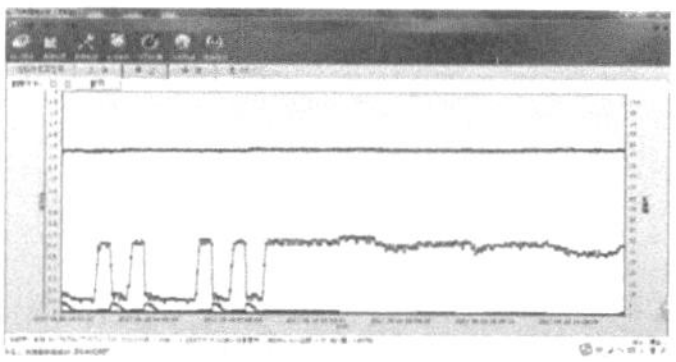

Figure 5. The pressure changing curve when sending commands of reading flow rate of the first layer.

Wireless communication test from downhole to surface. Sending commands of reading flow rate of first layer via the surface control system to verify whether the flow rate data can be uploaded by the water distributor. The experiment curve is shown in Figure 5. Sending the command of layer selection first (3 times of pressure or flow wave hopping), and then sending the command of flow reading (4 times of pressure or flow wave hopping). After a while, the surface control system received a flow rate value as same as the injection one of 23 m^3/d, which showed that the data transmitted from downhole to surface successfully.

4 FIELD APPLICATION

4.1 *Result Analysis*

By the end of 2018, the proposed technology has been successfully applied in 138 wells in Changqing Oilfield, with a wireless communication distance from surface to downhole reaching to 2879m. Through automatic regulation downhole, the water flow qualified rate of testing injectors was kept above 90% for a long time, which improves the efficiency of layered water injection. Supporting by internal network of Changqing Oilfield, it realized remote monitoring of testing well in real time. In the pilot test area of Bai 153, there were 69 testing injectors with 237 corresponding producers. The average water absorbing thickness was increased from 19.3m to 20.4m, and the water drive reserves producing degree was increased from 65.2% to 67.7%. Meanwhile, the natural decline rate was decreased from 5.8% to 2.1%, and the water cut increasing rate maintained the same at 1.3%. The effect of waterflooding development in the pilot test area was improved significantly.

4.2 *Cost Analysis*

According to the lifecycle of three years in injectors of Changqing Oilfield, the one-time costs and supporting costs of the proposed technology was compared with that of conventional bridge concentric injection technology and intelligent injection technology without cable, as shown in Table 1. Compared with the latter two, the application costs of the proposed technology can be saved at least 8000 RMB. Moreover, it avoids manual work and reduces operational risk.

Table 1. Comparison of capital costs and 3-years supporting costs of different injection technologies.

Injection technology		Bridge concentric injection	Intelligent injection without cable	The proposed technology
Cost of downhole tools/ RMB	Packer	10000	10000	10000
	Water Distributor	12000	200000	60000
Cost of Surface Tools/RMB		10000	30000	60000
Cost of artificial regulation	Number/year	4	2	/
	Cost/RMB	180000	90000	/
Total Cost/RMB		212000	330000	130000

5 CONCLUSIONS

The proposed downhole automatic water distribution technology realizes multiple functions, including two-way remote wireless communication between surface and downhole, automatic water flow regulation downhole, and real-time monitoring of injection data. With its pattern of regulation without manual work, this technology reduces the workload of artificial regulation and supporting test costs significantly.

This technology provides a large set of injection data for waterflooding effect evaluation and development plan adjustment, which meets the develop demands of big data and management requirements of digital oilfield. With a bright application prospect, it will drive oilfield waterflooding technology developing towards the direction of digitization, informatization and intelligence.

REFERENCES

Ai-Khodhori S. M., 2003. Smart well technologies implementation in PDO for production & reservoir management & control. *SPE 81486*.

Hou Aimin, 2015. Integrated application of reservoir performance analysis and layered water injection technology in Block Li 8. *Journal of Yangze University (Natrural Science Edition)* 12(14): 71-73.

Lei Chuang & Ma Yongzhong, 2016. Application of wireless intelligent separate layer injection technology in toothbrush-shape reservoir. *China Petroleum Machinery* 44(8): 54-57.

Li Donglei, 2016. Intelligent layered polymer injection technology with preset cable. *China Petroleum Machinery* 44(10): 93-96.

Liu Xiushan & Su Yinao, 2000. Investigation on the transmission behaviors o drilling fluid pulse signal. *Oil Drilling & Production Technology* 22(4): 8-10.

Pei Xiaohan & Yang Zhipeng, 2006. History and actuality of separate layer oil production technologies in Daqing Oilfield. *SPE 100859*.

Shen yue & Cui Shili, 2013. Delay differential detection and signal reconstruction of continuous pressure-wave signals of drilling fluid. *Acta Petrolei Sinica* 34(2): 353-358.

Su Yinao & Dou Xiurong, 2005. Measurement while drilling, logging while drilling and logging instrument. *Oil Drilling & Production Technology* 27(1): 74-78.

Yang Lingzhi & Ju Yafeng, 2014. Study and analysis of Flow Characteristics for digital separate-zone water flooding. *China Petroleum Machinery* 42(10): 52-55.

Yang Lingzhi, Yu Jiuzheng, 2017. An intelligent waterflood monitoring technology used for the ultra-low permeability reservoirs in Ordos. *Oil Drilling & Production Technology* 39(6): 756-759.

Youth Technical Sessions Proceedings – Litvinenko (ed)
© 2019 World Petroleum Council, London, UK, ISBN 978-0-367-34668-3

Mechanism investigation of nano technology for EOR of tight oil reservoir

X.F. Geng, B. Ding, C.M. Xiong & B.S. Guan
Research Institute of Petroleum Exploration & Development (RIPED), PetroChina, Beijing, China

J.H. Qin & J.Y. Xie
Xinjiang Oilfield Company, Petrochina, Karamay, China

ABSTRACT: Tight oil has become an important strategic replacement resource.Aiming at the problem of "water can not be injected, oil can not be extracted" for tight oil reservoir development, a in-situ nano permeability-increasing agent(in-situ NPIA, < 300 nm) with core-shell structure was prepared. The in-situ NPIA has a "small size effect", can significantly reduce the injection start-up pressure gradient, and it can release of the hydrophobic "core" when exposed to crude oil, inverting the "large oil" into "small oil", greatly improving the oil liquidity, and eventually increasing the EOR. It also showed excellent interfacial tension and lower injection pressure compared with water. The low-field nuclear magnetic resonance (NMR) flooding (0.02 mD core) experiments show that, compared with water flooding, in-situ NPIA can increase the oil displacement efficiency by 19.7 %. So, It may be used to help develop the unconventional resources such as tight oil/gas, shell gas and oil etc.

1 INTRODUCTION

Tight oil has become an important strategic replacement resource. According to the recent reports, oil production from tight formations including shale has occupied more than 50 % of total oil production in US (Yu 2016). Tight oil production has reversed a 24-year decline in U.S. oil production. The annual output has reached $3*10^7$ t in 2011, and is expected to reach $1.5*10^8$ t in 2020. North America has found significant tight oil resources in central Alberta and southern Texas, and other favorable zones include the Los Angeles area, the gulf of Mexico, the south and eastern Canada. It can be said that tight oil is one of the important pillars for the United States to maintain its long-term status as a major oil producer. In China, tight oil perspective resource of tight oil can reach about $70*10^9$ t, including technical recoverable resources about 20 ~ $2.5*10^9$ t, accounting for 9.3 % of total global tight oil resources. It mainly distributed in the Songliao, Ordos, Sichuan, Junggar et al. basins, and has a great potential for exploitation (Zou 2015).

Compared with conventional reservoirs, the reservoir formation mechanism of tight reservoir is more complex: pore throat is smaller, interstitial content is higher, and development is more difficult. For example, tight reservoirs in China are with a permeability less than 0.1 mD, porosity of 6 ~ 15 %, and 2 ~ 10 μm pores are the main storage space. Nowadays, the main development methods for tight oil are factory drilling and horizontal wells staged fracturing. In the early stage of horizontal well fracturing, general fracturing was adopted in the whole well section. In recent years, the fracturing technology of horizontal wells has developed rapidly, and the fracturing of open hole wells is the main method. The slurry water fracturing fluid is the most widely used fracturing fluid for tight oil and shale gas development in the US.. The formula includes polymer (the main agent), bactericide, surfactant, scale inhibitor and other auxiliary agents (Penny 2012). In addition, since Mathew S. (Mathew 1997) first reported polymer-free fracturing fluid and Chase B. (Chase 1997) reported viscoelastic surfactant fracturing fluid, the study and application of surfactant fracturing fluid in foreign countries have developed rapidly.

The existing fracturing liquid system in China mainly includes thickening agent (crosslinking agent), drag reducing agent, flowback agent, pH regulator, etc.. Chang Y.R. (Chang 2012) reported a fracturing fluid system composed of base fluid A and base fluid B. The base fluid A is composed of thickener, solubilizer, clay stabilizer, drainage aid, fungicide, demulsifier, low temperature activator and water, and the base solution B is composed of low-molecular weight polymer, clay stabilizer, drainage aid, bactericide, demulsifier, low-temperature activator and water. During fracturing, base solution A and base solution B are mixed in proportion. Liu Y.Q. (Liu 2011) reported a kind of rapidly recoverable slick-water in shale gas reservoir, which is composed of drag reducer, potassium chloride, drainage aid, anti-swelling agent and water.

These above mentioned liquid system has been used in the EOR of tight reservoir, but they all do not change the nature of crude oil, and can not improve the mobility of crude oil in tight reservoir. So, it is urgent to develop new materials and technologies to improve the EOR of tight oil.

2 METHODOLOGY

1.1 *Materials*

Diphenyl ether and its derivatives, alkane of $C_{10} \sim C_{12}$, salinity water (10900 mg/L), Heavy alkyl benzene sulfonate (HABS), betaine, commercial nano-SiO_2, foreign depressurization and injection augment agent, foreign nano micelles, kerosene, Xinjiang tight crude oil (the eight components information and viscosity is listed in Table 1), simulated crude oil (V (Xinjiang tight crude oil): V (kerosene) = 10: 4, the apparent viscosity is 63.4 mPa·s at 25 °C).

Artificial core (permeability is 0.02 mD, porosity is 10.03%), 2.5-dimension microfluidic model (Figure 1, self-made, made of glass, the size is 1.5 cm * 1.5 cm, the depth of pore and throat are 15 μm and 2 μm respectively, the pore volume is 10 μL), hydrophilic wetting and hydrophobic wetting capillary tubes (inner diameter is 300 nm, length is 60 cm).

1.2 *Experiment method*

The nonionic gemini surfactant was synthesized with diphenyl ether derivatives by acyl chloride reaction and esterification reaction.

The in-situ nano permeability-increasing agent (in-situ NPIA, < 300 nm) were prepared by simply stirring serveral additives in a beaker. It was composed of 10 ~ 75 wt% gemini surfactant, 5 ~ 25 wt% alkane of $C_{10} \sim C_{12}$ and 10 ~ 32 wt% solubilizer.

Interfacial tensions (IFT) between 0.1% in-situ NPIA, HABS and betaine with kerosene and simulated crude oil were measured with rotating drop interface tensiometer at 80°C.

Table 1. Eight components information and viscosity of Xinjiang tight crude oil

Item	Content, wt%
Saturated hydrocarbon	49.57
Light aromatic	2.28
Media aromatic	2.46
Heavy aromatic	7.47
Light colloid	5.45
Media colloid	3.15
Heavy colloid	5.22
Asphaltene	4.12
Viscosity under reservoir	58.14

Figure 1. Image of 2.5-dimension microfluidic model saturated with Xinjiang crude oil

Dynamic light scattering (DLS) experiments was conducted with a BI-DS1, Brookhaven Instruments Co. spectrophotometer by using a vertically polarized light of 532 nm from an argon laser as the incident beam and at a scattering angle of 90°. Before conducting the measurements, all the solutions were filtered through 0.45μm membrane filters (Geng 2013). The concentration of in-situ NPIA is 0.1 %.

Transmission electron microscopy (TEM) experiments was conducted with a H-600 IV transmission electron microscope, operated at an accelerating voltage of 75 kV. The concentration of in-situ NPIA is 0.1 %.

Injection ability and start-up pressure of variety chemicals were conducted with capillary action analysis system (Geng 2017), which was self designed. The inner diameter of capillary tube was 300 nm, length was 60 cm.The injection velocity was fixed at 0.1 mL/min and the concentration of in-situ NPIA was 0.1 %.

Core displacement experiments were conducted with MesoMR-HTHP low-field nuclear magnetic resonance (NMR) equipment at 25℃ (Ding 2018). The core was first saturated with kerosene, then the salinity water was injected, and finally the in-situ NPIA was injected after water flooding. The injection velocity was fixed at 0.01 ~ 0.02 mL/min and the concentration of in-situ NPIA was 0.1 %.

Function mechanism of in-situ NPIA was explained by microfluidic flooding experiments. The injection velocity was fixed at 0.1 μL/min and the concentration of in-situ NPIA was 0.1 %.

3 ANALYSIS

3.1 Structure and size of in-situ NPIA

The TEM results at the concentration of 1.0 % show that the structure of in-situ NPIA is expected "core-shell", as shown in Figure 2. The "core" is mainly composed of alkane of C_{10} ~ C_{12}, and the "shell" is mainly composed of gemini surfactant and solubilizer. The size shown in Figure 2 is about dozens of nanometer. Combining with the results of DLS at the

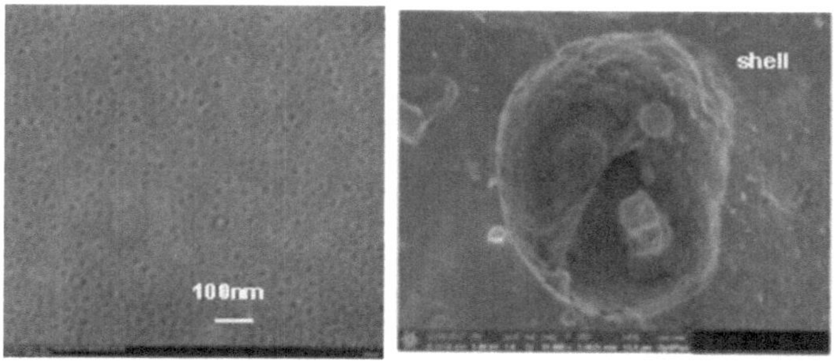

Figure 2. The TEM images of in-situ NPIA at the concentration of 1.0 %

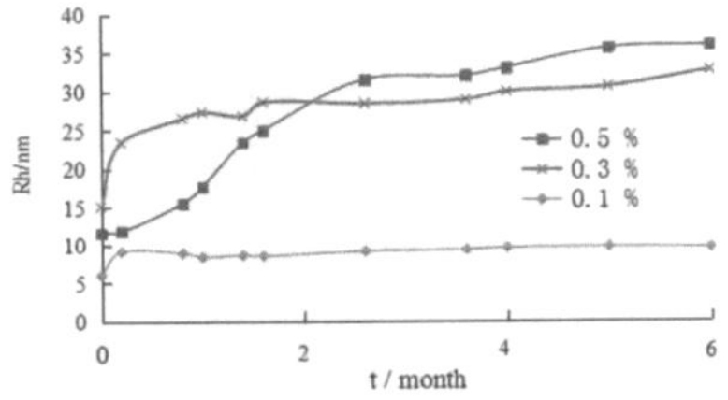

Figure 3. The size change as time of in-situ NPIA at the concentrations of 0.1 %, 0.3 % and 0.5 %

concentration of 0.1 % ~ 0.5 % shown in Figure 3, we can see that the in-situ NPIA is nano-scale. Furthermore, the size of in-situ NPIA is smaller than 40 nm when the concentration is lower than 0.5 % and cna keep stable for 6 months. It is worth mentioning that the size is smaller than 10 nm at the concentration of 0.1 %. Such small size is very in favor of injecting to the tight reservoirs.

3.2 *Interfacial activity*

Interfacial tensions (IFT) between 0.1% in-situ NPIA, HABS and betaine with kerosene and simulated crude oil were measured with rotating drop interface tensiometer at 80°C. The results are shown in Figure 4 and Figure 5, respectively. The equilibrium interfacial tension and the time need to reach equilibrium are listed in Table 2. Although the IFT between in-situ NPIA with kerosene and simulated crude oil are not lower, even higher than those between HABS and betaine with kerosene and simulated crude oil, the time need to reach the equilibrium IFT is the lowest, within 5 min. For example, it just need 1.5 min for the interfacial tension between in-situ NPIA with kereosene to reach equilibrium, indicating that the nano in-situ NPIA can operate more quickly with the crude oil in reservoirs.

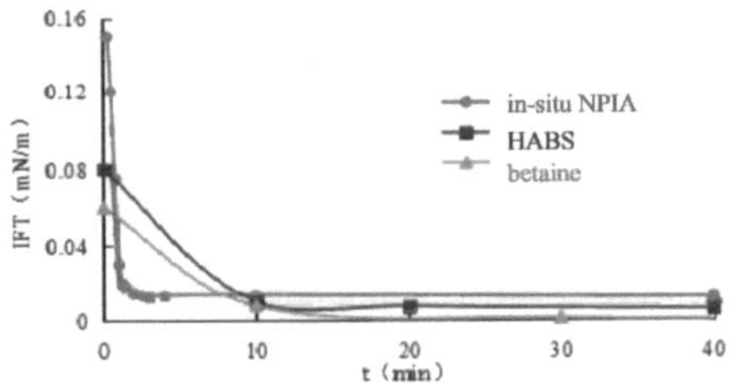

Figure 4. Interfacial tension diagram of 0.1% in-situ NPIA, HABS and betaine with kerosene at 80°C

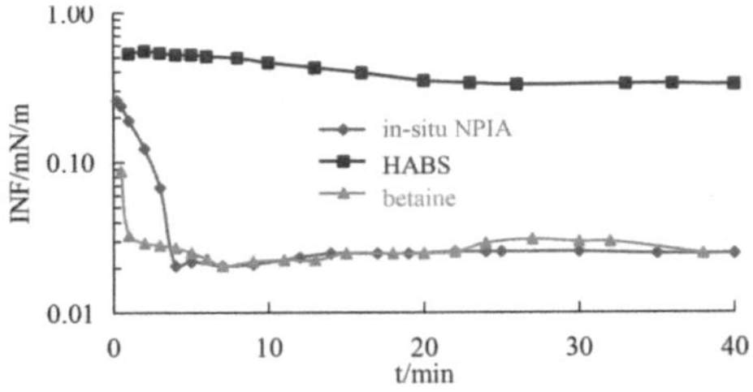

Figure 5. Interfacial tension diagram of 0.1% in-situ NPIA, HABS and betaine with simulated crude oil at 80°C

Table 2. The equilibrium interfacial tensions and the time need to reach equilibrium

	IFT, mN/m at 80 °C		Time to equilibrium	
Chemicals	Kerosene	Simulated crude oil	Kerosene	Simulated crude oil
HABS	0.067	0.33	40	26
Betaine	0.01	0.025	30	7
In-situ NPIA	0.09	0.025	1.5	4

3.3 *Depressurization and injection enhancement*

Capillary action exists in the whole process of low permeability and tight reservoirs development, and the ability of altering capillary action can directly influence the EOR. Based on the pore size distribution of tight reservoirs, we develop a capillary action analysis system cored with capillary tube/bundle, simulating the migration of flood fluids in the porous media as in the capillary tube/bundle, and evaluating the start-up pressure gradient of flood fluids, reflecting the injection ability and depressurization and injection enhancement of different flood fluids. Six oil displacing agents - water, betaine, nano-SiO_2, foreign depressurization and injection augment agent, foreign nano micelles and nano in-situ NIPA are selected as the objectives and evaluate their start-up pressure gradient in both hydrophilic and hydrophobic capillaries with inner diameter of 300 nm. The flooding rate is fixed at 0.1 mL/min. The results are shown in Figure 6.

No matter in hydrophilic and hydrophobic capillary tubes, the start-up pressure gradient of nano in-situ NPIA is the lowest compared with other five flood fluids. The start-up pressure gradient of nano in-situ NPIA in hydrophilic capillary tube is only 0.07 MPa/m, and in hydrophobic capillary tube is only 0.11 MPa/m. Compared with the values of water, the nano in-situ NPIA can decrease the waterflooding start-up pressure in 300 nm hydrophilic and hydrophobic capillary tubes by 98.3% and 98.9%, respectively, indicating that it can significantly improve the injection ability of low permeability and tight reservoirs.

3.4 *Displacement efficiency*

The displacement efficiency of water and nano in-situ NPIA are evaluated with low-field nuclear magnetic resonance (NMR) by measuring the oil signal of saturated oil, residual oil after water flooding and residual oil after in-situ NPIA. The result is shown in Figure 7.

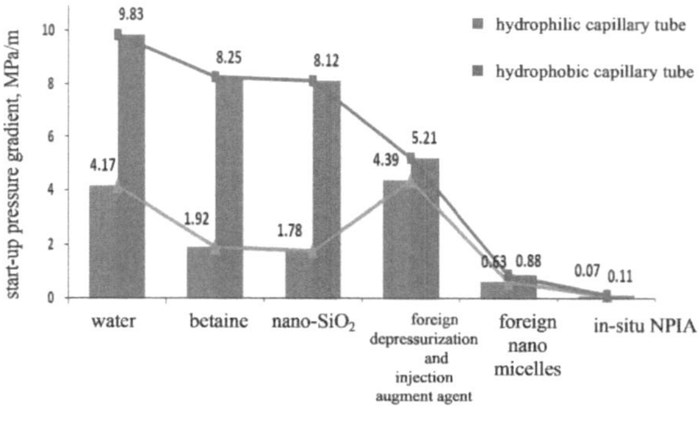

Figure 6. The start-up pressure gradient of six flood fluids in both hydrophilic and hydrophobic capillary tubes

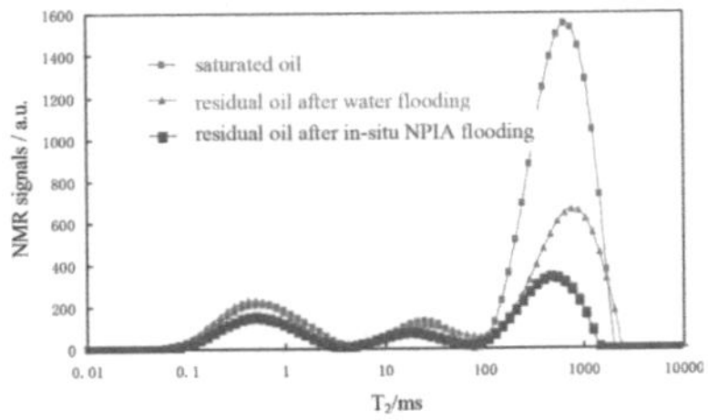

Figure 7. The low-field NMR flooding results of water flooding and nano in-situ NPIA flooding

We can see from Figure 7 that, the displacement efficiency of water flooding is 49.1 %, and in-situ NPIA can further increase the displacement efficiency by 19.7 %.

We convert the transverse relaxation time (T_2) in Figure 7 into pore throat radius (Ding 2018), and conclude the displacement efficiency of water flooding and nano in-situ NPIA flooding in different size pore throats in Figure 8. We can see that, in small pores (< 500 nm), the nano in-situ NPIA can significantly further increase the displacement efficiency after water flooding. The displacement efficiency of nano in-situ NPIA flooding is about 20 times of that of water flooding, indicating that the nano in-situ NPIA can enter the smaller pores easier, the same conclusion as part 3.3.

3.5 *Function mechanism*

Micro/nanofluidics technology is to online and visually evaluate the flow and migration features of fluids on the micro/nano dimension space. It has been widely used in EOR and to evaluate the oilfield agents (Godin 2008, He 2017). So, a 2.5 dimension throat model is used in this paper to online and visually characterize the function mechanism of water flooding and nano in-situ NPIA flooding for tight reservoirs.

The results are shown in Figure 9 (a), (b) and (c). We can see from Figure 9 (a) that the fingering phenomenon exsits in the process of water flooding and the displacement efficiency

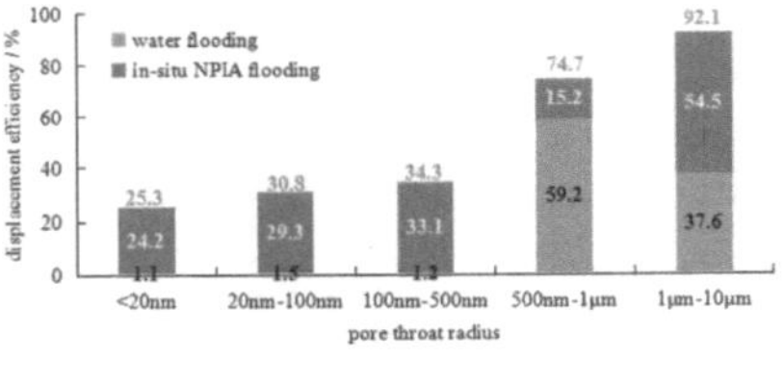

Figure 8. The displacement efficiency of water flooding and nano in-situ NPIA flooding in size different pore throats

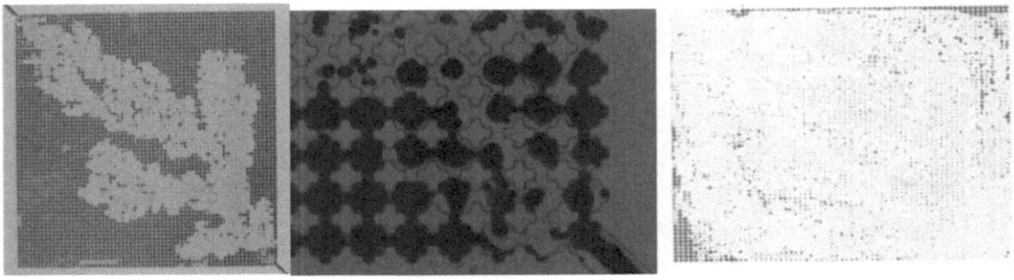

(a)   (b)   (c)

Figure 9. Micro/nanofluidics displacement effect of (a) after water flooding (b) in the process of nano in-situ NPIA flooding (c) after nano in-situ NPIA flooding

is only about 40 %, the same as the result in part 3.4. However, when nano in-situ NPIA is injected into the model after water flooding, the simulated crude oil is immediately divided into smaller oil drops, as shown in Figure 9 (b). The size of the oil drops is about dozens of and hundreds of nanoscale, thus significantly increasing the fluidity and migration ability of crude oil. The final oil displacement efficiency can be greater than 90%, as shown in Figure 9 (c).

Based on the above experiments results, we can conclude that the function mechanism of nano in-situ NPIA are: (1) the gemini surfactant composed of "shell" can help decrease the start-up pressure gradient and make it more easily to be injected into the tight reservoirs. (2) It can release of the hydrophobic "core" when exposed to crude oil in the formation, inverting the "large oil" into "small oil", greatly improving the oil liquidity, and eventually increasing the EOR.

4 CONCLUSION

A in-situ nano permeability-increasing agent (in-situ NPIA, < 300 nm) with "core-shell" structure was prepared in this paper. The minimum size of it can be lower than 10 nm at the concentration of 0.1 %. The "shell" is mainly composed of gemini surfactant and solubilizer, which makes the in-situ NPIA more easily be injected into the small pores. The "core" is mainly composed of alkane of $C_{10} \sim C_{12.}$ When exposed to crude oil in the formation, the "core" can invert the "large oil" into "small oil", greatly improving the oil liquidity, and eventually increasing the EOR.

The in-situ NPIA shows excellent interfacial tension between kerosene and simulated crude oil, shorting the time to reach equilibrium within 5 min. Also, it can decrease the waterflooding start-up pressure in 300 nm hydrophilic and hydrophobic capillary tubes by 98.3% and 98.9%, respectively, and increase the oil displacement efficiency by 19.7 % after water flooding, especially in small pores (< 500 nm), the displacement efficiency of nano in-situ NPIA flooding is about 20 times of that of water flooding, So, It may be widely used to help develop the unconventional resources such as tight oil/gas, shell gas and etc.. in the future.

REFERENCES

Chase, B., Chmilowski, W. & Marcinew, R. 1997. Clear fracturing fluid for increased well productivity. *Oilfield review* 9(3):20-33.

Chang, Y.R., Ran, X.Q. & Zhu, T.S. 2012. A fracturing fluid. Chinese Patent. CN201210347885.3.

Ding, B., Luo, J.H. & Geng, X.F. 2018. Visual evaluation method for fluids in cores based on low field nuclear magnetic resonance technology. *Oilfield chemicals* 35(1):170-174.

Geng, X.F., Hu, X.Q. & Xia, J.J. 2013. Synthesis and surface activities of a novel di-hydroxyl-sulfate-betaine-type zwitterionic gemini surfactants, *Applied Surface Science* 271(5): 284-290.

Geng, X.F., Luo, J.H. & Ding, B. 2017. Study on evaluation method of capillary action for low permeability reservoirs. *Oilfield chemicals* 34(4):717-721.

Godin, J., Chen, C.H. & Cho, S.H. 2008. Microfluidics and photonics for bio-system on a chip: a review of advancements in technology towards a microfluidics flow cytometry chip. *J Biophotonics* 1(5): 355-376.

He, K., Xu, L. & Yin, X.L. 2017. A rock-on-a-chip approach to study fluid invasion and flowback in liquidsrich shale formations. Oklahoma: the SPE Oklahoma Oil and Gas Symposium, 2017, SPE-185088-MS.

Liu, Q.Y., Chen, P.F. & Liu, S. 2011. A rapidly recoverable slick-water in shale gas reservoir. Chinese Patent. CN201110401452.7.

Mathew, S. & Schlumberger D. C. 1997. Polymer-free fluid for hydraulic fracturing. SPE 38622.

Penny, G.S., Crafton. J.W. & Champagne L. M. 2012. Proppant and fluid selection to optimize performance of horizontal shale frace; proceedings of the SPE Hydraulic Fracturing Technology Conference, Society of Petroleum Engineers.

Yu, Y., Li, L. & Sheng, J. 2016. Further discussion the roles of soaking time and pressure depletion rate in gas huffn puff process in fractured liquid-rich shale reservoirs. SPE-181471-MS paper presented at the SPE Annual Technical Conference and Exhibition held in Dubai, UAE, 26-28, September 2016.

Zou, C.N., Yang, Z. & Zhu, R.K. 2015. Progress in China's unconventional oil & gas exploration and development and theoretical technologies. *Acta gelogical sinica* 89 (6):919-1007.

Youth Technical Sessions Proceedings – Litvinenko (ed)
© 2019 World Petroleum Council, London, UK, ISBN 978-0-367-34668-3

Multiphase flow interfacial tension experiment research during CO_2 injected process

X. Wang, Z.M. Lun, R. Wang, C.Y. Lv, Y.Q. Tang & W. Hu
Sinopec Key Laboratory of Marine Oil & Gas Reservoirs Production, Beijing, China
Sinopec Exploration & Production Research Institute, Beijing, China

ABSTRACT: For reservoirs entering the middle and late stage of development, there is still a large amount of residual oil remaining in the reservoir after water flooding, and CO_2 flooding has been widely used as the main enhanced oil recovery technology at this time. To clear the interface characteristics and interaction relationship of oil-gas-water under high temperature and high pressure. The microscopic influencing factors of interfacial tension are analyzed and the mining effect of CO_2 flooding in the late stage of high water cut reservoir is revealed. Axisymmetric drop shape analysis (ADSA) method has been used to measure the interfacial features of oil-gas, oil-carbonated water and oil-water system under the simulated-formation condition. The influence of density of medium, intermolecular force and interfacial tension of various physical parameters are analyzed from the microscopic point of view. The effects of two-phase density difference, intermolecular force and interfacial tension of various physical quantities of oil-water are analyzed. In the oil-CO_2 system, the main factors affecting the interfacial tension are the density difference between two phases and the intermolecular force. In the oil-saturated carbonate system, the intermolecular force, the shape factor of suspended droplets and the square value of the maximum diameter of suspended droplets are the main factors to reduce the interfacial tension.

1 INTRODUCTION

In recent years, the CO_2 flooding technology is considered to be the most potential method to improve oil recovery, not just because it can effectively enhance oil recovery, but also because it can reduce greenhouse gas emissions (Ke 2011; Shen and Liao 2009; Yang and Gu 2005). Most of the oil field has entered the middle and later periods of the development in China, in these oil fields still have a large amount of residual oil after water flooding in reservoir, that water flooding unable to further improve oil recovery (Ao et al. 2014; Li et al. 2012; Qin et al. 2010, Bennion et al.2008). CO_2 flooding has been widely used to improve oil recovery technology after water flooding. In the United States, CO_2 flooding can increase oil recovery factor of 10% - 20% on the basis of water flooding (Zhang et al. 2009; Yu et al. 2011; Branko et al. 2002, Qin et al. 2015).

During the process of CO_2 flooding, CO_2 huff-puff and water- CO_2 alternating flooding, the interface of CO_2-oil-water is one of the most important factors that affect the effect of CO_2 flooding (Li et al. 2002; Wang et al. 2015; Ji et al. 2009). At present, there are many studies on the interfacial tension characteristics of the CO_2-oil system, but there are few studies on the interfacial interaction characteristics of the CO_2-formation water-oil three-phase system.

In this paper, the interface characteristics of CO_2-oil systems under different pressure conditions are studied by means of ASDA (Sun et al. 2006; Yang et al. 2004; Wang et al. 2011; Zheng et al. 2008). Under different reservoir pressure and different CO_2 solubility conditions, the interfacial tension variation of the oil-saturated CO_2 formation water system was accurately measured. The balance interfacial tension of crude oil-formation water system and crude oil-saturated CO_2 formation water system were compared, and the main factors affecting interfacial tension of three systems were analyzed. Moreover, the effects of different media density, intermolecular forces and physical parameters on the interfacial tension of the system are analyzed.

2 EXPERIMENTAL PRINCIPLES AND METHODS

2.1 *Experimental facilities*

Experiment using the high temperature and high pressure interface tensiometer produced by France Sanchez Technologies company. The highest working temperature is 200 °C, the highest working pressure is 70 MPa. The diameter of the needle used in the experiment is 0.81 mm, and the inversion needle diameter is 0.82 mm.

The density meter used in the experiment is produced Anton Paar company, including DMA 4500 atmospheric pressure module and high temperature and high pressure module. The scope of work stress: 0-700 bar, working temperature range is: - 10 °C to 200 °C.

2.2 *Experimental principles*

Experiments using axisymmetric drop shape analysis technology (ADSA), the technology is currently the most accurate method that measured the interfacial tension under high temperature and high pressure conditions.

The formula for hanging drop method is used for the determination of interfacial tension (Luo et al. 2008):

$$\gamma = \frac{\Delta\rho g d_e^2}{H}$$

where $\Delta\rho$= the density difference between the two phase fluid; d_e = the largest of hanging drop diameter; and $1/H$ = function of hanging drop shape factor S (S = d_s/ d_e).

2.3 *Experimental procedure*

In the first place, cleaning the whole experimental system, then do sealing inspection and vacuum. After that turn the oil, CO_2 and carbonation water sample into the equipment, then heating system and wait for system stability. In the second place, complete the experiment under the different experimental pressure, record the different moment images of the oil droplets, and calculate the interfacial tension.

In the process of calculation of interfacial tension, the density of CO_2 and crude oil under different pressure should be confirmed. Density measured by high temperature and high pressure module during the process of experiment.

3 EXPERIMENTAL PHENOMENA

3.1 *Characteristics of crude oil-formation water system*

Mass transfer between oil-formation water systems can be observed from Figure 1. It can be seen that under different pressure conditions, with the increasing of pressure, the interfacial interaction between crude oil and formation water system is not obvious, and there is no special interfacial phenomenon. Therefore, there is no medium transfer between crude oil and formation water.

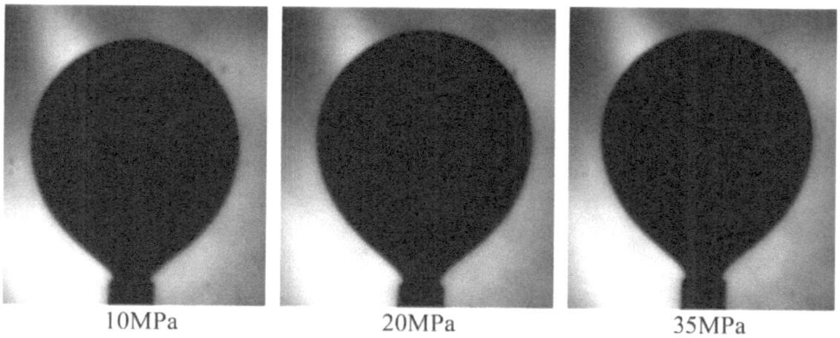

Figure 1. Crude oil-formation water interactions.

3.2 *Interface characteristics of crude oil-saturated CO_2 formation water system*

Mass transfer between oil-saturated CO_2 formation water system can be observed by Figure 2. As can be seen from the figure below, when the pressure is 5 MPa, there is no obvious light hydrocarbon layer at the interface between oil droplets and carbonate water. When the pressure rises to 14 MPa, a thin light hydrocarbon layer appears at the interface between oil droplets and carbonate water. In the gas-water system, CO_2 first contacts the gas-water interface, and finally dissolves and extracts oil droplets through the interface. When the pressure rises to 30 MPa, the phenomenon of light component extraction becomes more obvious. CO_2 extraction can be seen at the top of the oil drop to mention the light component column in the oil drop, and the light hydrocarbon layer around the oil drop becomes thicker and thicker. It can be seen that with the increase of pressure, the speed of CO_2 passing through the gas-water interface is accelerated, and the extraction effect with oil droplets is more obvious and rapid.

3.3 *Interface Characteristics of Crude Oil-CO_2 System*

As can be seen from Figure 3, the extraction of crude oil by CO_2 increases with the increase of pressure, and the interface between CO_2 and crude oil becomes unstable gradually. A small amount of light components were extracted at 16 MPa, and the extraction effect was obvious when the pressure reached 41 MPa.

3.4 *Difference Analysis of Interface Characteristics of Three Systems*

There is no mass transfer between crude oil and formation water system, which shows that crude oil and formation water are completely immiscible. For the crude oil-saturated CO_2 formation water system, CO_2 first slowly diffuses into the crude oil through the gas-water interface, and extracts the light components in the crude oil. There is medium exchange between oil and carbonate water, and there is partial miscibility. For oil- CO_2 system, CO_2 is directly

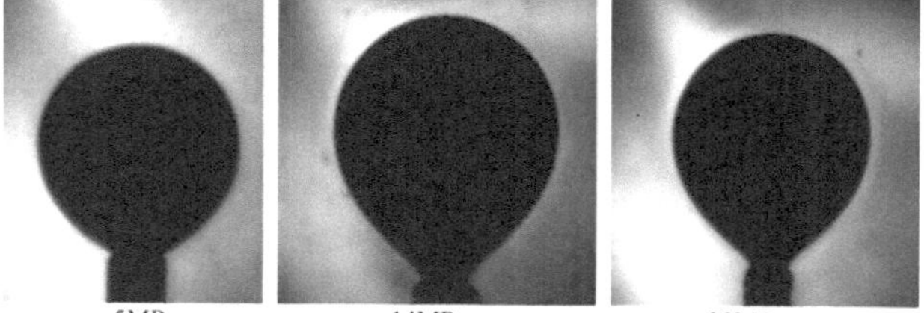

Figure 2. Interaction between crude oil and formation water saturated with CO_2

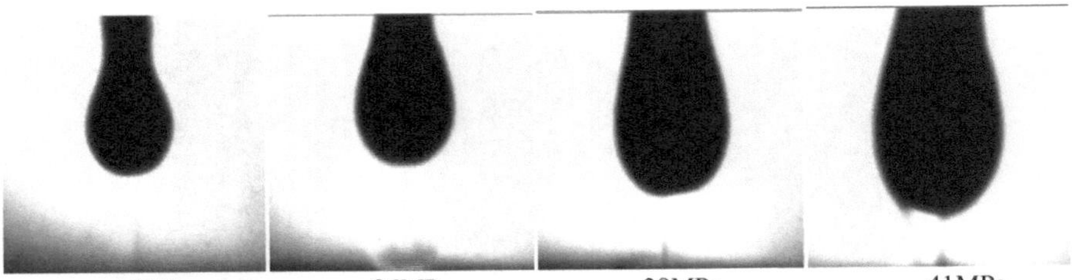

Figure 3. Images of the pendant oil drop in CO_2-crude oil system under different pressure

in contact with crude oil, and its diffusion and extraction effect is obvious. There exists obvious medium exchange phenomenon. With the increasing pressure, the phenomenon of light components in crude oil extracted by CO_2 becomes more and more obvious and rapid.

4 ANALYSIS OF INFLUENCING FACTORS OF INTERFACIAL TENSION

4.1 *Effect of density on interfacial tension*

In order to analyze the influence of density difference of different media on the interfacial tension of three systems, the densitometer of high temperature and high pressure was used to measure the density of crude oil, formation water, CO_2 and carbonate water at 74°C and different pressures. The results are shown in the figure below.

From Figure 4, it can be seen that the density of formation water, CO_2 and carbonate water increases in varying degrees with the increase of pressure in the experimental pressure range. Among them, the increase of CO_2 density is the largest, and the density difference between CO_2 and crude oil decreases sharply, followed by the increase of carbonate water density, and the increase of formation water density is the smallest. Combining the equilibrium interfacial tension analysis of three systems, the interfacial tension of crude oil- CO_2 system decreases sharply during the process of decreasing the density difference between crude oil and CO_2. It can be seen that the main factor affecting the interfacial tension of crude oil in oil and gas system is the difference of two-phase density. The difference of two-phase density between oil-saturated carbonate water system and oil-formation water system increases slightly, while the interfacial tension of oil-saturated carbonate water system is lower than that of oil-formation water system. Therefore, for oil-water system, the difference of two-phase density is not the main factor, but the interaction force between medium dissolved in formation water and crude oil and other physical quantities are the main factors affecting the interfacial tension.

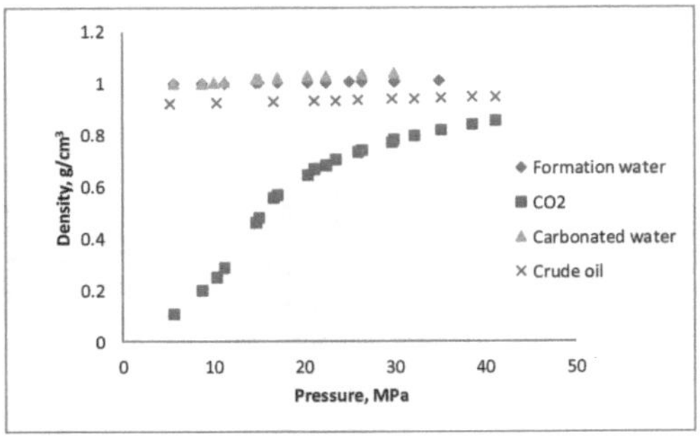

Figure 4. Images of the pendant oil drop in CO_2-crude oil system under different pressure

4.2 *Effect of Intermolecular Force on Interfacial Tension*

The fundamental cause of interfacial tension is the net attraction of molecules in the interfacial layer. At the same temperature, the main influencing factors are the polarity of the material and the pressure of the system.

For oil-formation water system without dissolved gas, the size of hydrocarbon molecules in crude oil and the amount of active substances have little effect on the interfacial tension of the system. This is because increasing pressure will change the intermolecular forces of oil and water at the same time, and the difference between them may remain unchanged, so the interfacial tension will remain unchanged.

For the interfacial tension of crude oil- CO_2 system, with the increase of pressure, the light hydrocarbon components in crude oil diffuse to CO_2, which makes the polar molecules in CO_2 increase continuously, the polarity of molecules between oil and gas approaches gradually, and the composition of oil and CO_2 system keeps close, which makes the mutual solubility of the two systems increase. At the same time, with the increase of pressure, the solubility of CO_2 in crude oil increases, the density of crude oil decreases gradually, and the intermolecular gravity decreases. Both of them reduce the net gravity of the molecules in the interfacial layer, increase the intermolecular force and decrease the interfacial tension.

For the formation water and crude oil system with dissolved CO_2, due to the slow diffusion of CO_2 into crude oil, the directional intermolecular force between carbonate solution and crude oil increases. And because CO_2 needs to pass through the gas-water interface to further diffuse into the crude oil, the diffusion rate is slow and the diffusion degree is low, which makes the interfacial tension of the system lower than that of the oil- CO_2 system and slightly higher than that of the oil-formation water system.

4.3 *The influence of physical parameters in the formula on interfacial tension*

According to the formula for determining interfacial tension by pendent drop method, there are three physical quantities affecting the interfacial tension value, namely, the density difference between two phases $\Delta\rho$, the shape factor of pendent drop1/H and the square value of the maximum diameter of pendent drop d_e^2.

In order to analyze the factors affecting the interfacial tension of oil- CO_2 system and oil-carbonate water system, the square values of the density difference between two phases, the shape factor of pendent droplets and the maximum diameter of pendent droplets were calculated and dimensionless. The results are shown in Figures 5 and 6.

For oil-CO_2 system, dimensionless droplet shape factor 1/H value is about 1, and the change range is small. The square value of dimensionless maximum droplet diameter decreases first and then increases, and the dimensionless density difference decreases by 87%. In oil-CO_2 system, the interfacial tension of oil-gas equilibrium decreases obviously with the increase of pressure. It can be seen that the main factor affecting the interfacial tension of the system is the density difference between two phases, while the shape of pendent droplets and the square

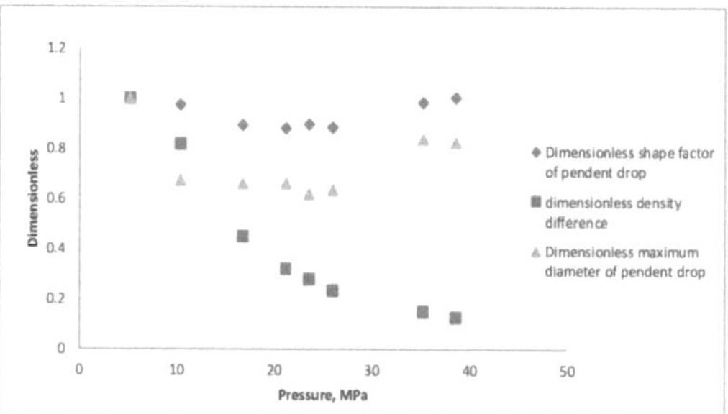

Figure 5. The dimensionless physical quantity of oil-CO_2 system changes with pressure

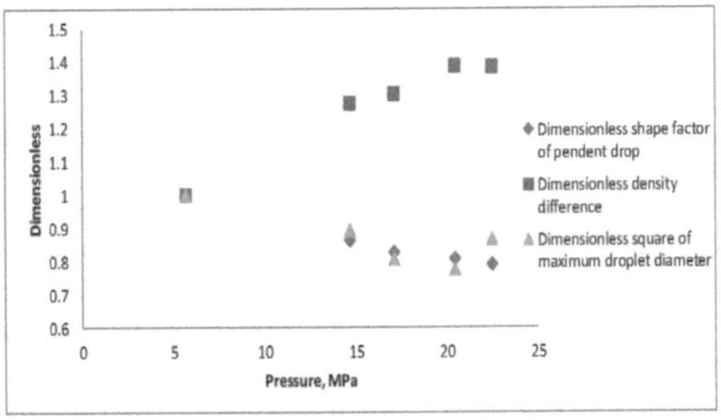

Figure 6. dimensionless physical quantity of oil-saturation carbonated water system changes with pressure

value of dimensionless maximum diameter of pendent droplets hardly affect the interfacial tension of the system.

For oil-saturated carbonate system, the equilibrium interfacial tension decreases with the increase of pressure by 13.94%, much less than that of oil-CO_2 system. The dimensionless density difference increases with the increase of pressure, while the dimensionless shape factor and the square value of dimensionless maximum diameter decrease with the increase of pressure. It can be seen that for this system, although the increase of density difference between two phases is not conducive to the decrease of interfacial tension, it has little effect on the results of interfacial tension, while the shape factor of pendent droplets and the square value of the maximum diameter of pendent droplets are the main factors to promote the decrease of interfacial tension.

5 CONCLUSIONS

1. There is no mass transfer between oil-water system. In oil-saturated CO_2 formation water system, with the increase of pressure, the extraction of crude oil by CO_2 becomes stronger. In the oil-CO_2 system, there is obvious medium exchange phenomenon. With the increasing pressure, the phenomenon of light components in crude oil extracted by CO_2 becomes more and more obvious and rapid.
2. The equilibrium interfacial tension of oil-CO_2 system decreases the most. After dissolving CO_2 in formation water, the interfacial tension of crude oil-saturated CO_2 formation water system is lower than that of undissolved CO_2, and the difference between the two systems increases with the increase of pressure.
3. The main factor affecting the interfacial tension of crude oil in oil-gas system is the difference of two-phase density. Moreover, the shape of pendent droplets and the square value of dimensionless maximum diameter of pendent droplets hardly affect the equilibrium interfacial tension of the system. For oil-water system, the difference of two-phase density is not the main factor affecting interfacial tension. For oil-saturated carbonate system, the increase of density difference between two phases is not conducive to the decrease of interfacial tension, and the shape factor of pendent droplets and the square value of maximum diameter of pendent droplets are the main factors to promote the decrease of interfacial tension.
4. In the oil-CO_2 system, the light components in crude oil are continuously extracted by CO_2, which makes the polarity of the two molecules approach gradually and the mutual solubility increase. With the increase of pressure, the interaction force between the directional molecules increases and the interfacial tension decreases gradually. In oil-saturated carbonate water system, CO_2 diffuses slowly to crude oil, which increases the intermolecular force between carbonate solution and crude oil, and decreases the interfacial tension. In oil-formation water system, the change of intermolecular force has little effect on interfacial tension.

REFERENCES

Ao, W.J.et al. 2014. Research and Evaluation of CO_2 Injection in Waterflooding Reservoirs. *Petrochemical Industry Application* 33(2): 13-17.

Bennion, et al.2008. Correlations for the Interfacial Tension Between Supercritical Phase CO_2 and Equilibrium Brines at In Situ Conditions. *114479-MS SPE Conference Paper-2008.*

Branko R, et al. 2002. Blunt. Effect of Composition on Waterblocking for Multicomponent Gasfloods. SPE 77697.

Ji, B.Y. et al. 2009. Re-understanding of Oil and Gas miscible phase characteristics during CO_2 Flooding. *Petroleum Geology and Oilfield Development in Daqing* 28 (3):103-109.

Ke, Z. 2011. Interfacial Characteristics and Application Research of CO_2-Formation Oil System. Institute of Porous Flow and Fluid Dynamics, *Chinese Academy of Sciences.*

Li, Z.C. et al. 2012. Laboratory Experiment of CO_2Injection in the Water-flooded and Abandoned Reservoirs. *Journal of Oil and Gas Technology* 34(4): 131-135.

Li, S.L. et al. 2002.Petroleum Geology and Recovery Efficiency. *Petroleum Geology and Recovery Efficiency* 9(2): 1-5.

Luo, M. et al. 2008.Method for Measuring Liquid Interfacial Tension - Method of Hanging-drop. *Journal of Materials Engineering* 10(2): 23-26.

Qin, J.S. et al. 2010. Mechanism of the CO_2 Flooding as Reservoirs Containing High Water. *Acta Petrolei Sinica* 31(5): 797-801.

Qin, J.S. et al. 2015. Application and Enlightenment of Carbon Dioxide Flooding in the United States of America. *Petroleum Exploration and Development* 42(2):209-216.

Shen, P.P.& Liao, X.W.2009. The Technology of Carbon Dioxide Stored in Geological Media and Enhanced Oil Recovery. *Petroleum Industry Press*, Beijing.

Sun, C.Y. et al. 2006. Interfacial Tension Experiment of Oil and Water, Oil and Gas for CO_2 Injected Reservoir Fluid System. *Journal of China University of Petroleum* 30(5): 109-112.

Wang, Xin. et al. 2015. Determination of CO_2/Crude Oil System Equilibrium Interfacial Tension and Dynamic Interfacial Tension by ADSA Method. *Bulgarian Chemical Communications* 47 (3): 945-948.

Wang, H.T. et al. 2011. Interfacial Tension of CO_2/Crude Oil and N2/Crude Oil at High Pressure and High Temperature. *Acta Petrolei Sinica* 32(1): 177-180.

Yang, D.Y. et al. 2004. Visualization of Interfacial Interactions of Crude Oil- CO_2 Systems under Reservoir Conditions. *89366-MS SPE Conference Paper.*

Yang,D.& Gu,Y.2005. A New Experimental Technique for Studying Gas Mass Transfer in the Crude Oil by Analysis of the Measured Dynamic and Equilibrium Interfacial Tensions. SPE 95844, Proceeding of SPE Annual Technical Conference and Exhibition, Dallas, TX.

Yu, H.W., et al. 2011. Experimental and Numerical Simulation Study on Single Layer Injectivity for CO_2 Flooding in Low Permeability Oil Reservoir. *SPE 144042.*

Zhang, S. et al. 2009. Near Miscible Flooding of CO_2 in Ultra-low Permeable Reservoir. *Petroleum Geology and Oilfield Development in Daqing* 28 (1):114-117.

Zheng, H. et al. 2008. Effect of injected gas and injected gas pressure on displacement characteristics in Particularly low permeability oil reservoir. *Petroleum Geology and Oilfield Development in Daqing* 27 (2): 110-112.

Youth Technical Sessions Proceedings – Litvinenko (ed)
© 2019 World Petroleum Council, London, UK, ISBN 978-0-367-34668-3

New surface tension meter by Rhebinder's method

V.S. Konopleva & A.P. Losev
Gubkin Russian State University of Oil and Gas, Moscow, Russian Federation

A.M. Kronin
LLC RTE Nedra-Test, Moscow, Russian Federation

ABSTRACT: Measurement of surface tension for oilfield surfactants is a routine task for petroleum engineers. The majority of enhanced oil recovery methods can't be implemented without the knowledge about the surface tension value. That's why accuracy and simplicity of such tests may cause significant effect for certain technologies. In this article a new instrument for precise surface tension measurement is described.

1 THE METHOD DISCUSSION

This paper is devoted to renovation of maximum bubble pressure method for surface tension measurement. The method was offered by Simon (Simon 1851) in 1851 and supplemented by Cantor (Cantor 1892) in 1892 and Jaeger (Jaeger 1917) in 1917.

The essence of the method is: air goes through a vertical capillary, which is mounted on a surface between two phases. A bubble, which has a form of sphere's segment, starts forming on a capillary tip. By the moment the bubble's radius is equal to the inner radius of the capillary, pressure in the bubble attains maximum value. Further grow makes the bubble unstable and it separates from the tip rapidly. Bubble generation cycle is repeated (Adam 1930). That's mean that pressure in the bubble has a maximum value per the cycle at the moment of separation. Using Laplace equation we can calculate surface tension:

$$\sigma = \frac{P_{\max} d}{4} \tag{1}$$

where σ = surface tension coefficient, N/m; $P_{\max}$= maximum bubble pressure per cycle, Pa; d = capillary diameter, m.

However this equation doesn't take into account all factors that affect on surface tension. For example, capillary wetting, liquid viscosity. Besides, pressure in the bubble is obviously always less than pressure that a meter shows. That occurs because of pressure losses: the pressure meter is set on the pipe entrance and the bubble is on the pipe tip. Based on these facts, capillary calibration, which allow to consider additional factors which impact on surface tension, is a necessary procedure for most accurate results (Fesan et al. 2014).

Surface tension of liquids significantly depends on temperature. Capillary material has a non-zero thermal-expansion coefficient. That is why it's important to maintain a temperature of all calibration liquids during experiments on a constant level. In this research a temperature of the liquids was maintained at 20.0±0.1°C level using a thermostat.

After the bubble separation from the capillary tip pressure oscillations arise. To reduce this negative effect we maximize time of the bubble's cycle. Lifetime of the bubble in all experiments exceeds 1 second so we can neglect the impact of liquid's viscosity on the results (Fainerman&Miller 1995).

Rhebinder offered an instrument, which is based on maximum bubble pressure method, for surface tension measurement (Kusakov 1936). Rhebinder invented the meter in the first half of the twentieth century so the accuracy of the results was not high. Pressure was measured with a U-tube liquid manometer, with accuracy of 1 millimeter of water column (approx. 10 Pa). There were a lot of fragile and unsteady parts in that instrument. The set up was fully manual. That is why in this study we attempted to improve the Rhebinder's method using modern technologies.

2 THE METER DESIGN

Pneumatic diagram for the designed instrument is presented on the Figure 1, electrical circuitry is on the Figure 2. Air from a compressor (2 on Figure 1, 2 on Figure 2) passing through a filter (1 on Figure 1) is injected into a capillary (4 on Figure 1). Air flow rate is controlled with a choke (3 on Figure 1). Pressure is measured with a precise pressure gauge (5 on Figure 1, 3 on Figure 2) Huba Control DTT5-420, which forms DC signal 4..20 mA. The gauge's full-scale valuation is 500 Pa, relative error is 0,1%. Direct current in the circuitry is maintained with programmable power supply (1 on Figure 2) AKTAKOM 0-18V/0-5A ATH-1441. The gauge's electrical resistance changes when pressure varies so current in the circuitry changes too according to the Ohm's low. Current is measured with a multimeter AKTAKOM ABM-4307 (4 on Figure 2), that works as a milliamperemeter. A signal obtained from the multimeter is transformed from analogue to digital form. The instrument is connected to a PC using COM port. Data about the timebased pressure in the bubble are written into a program, which was specially developed using an Object Pascal language. The data are transferred to Microsoft Excel where their processing takes place.

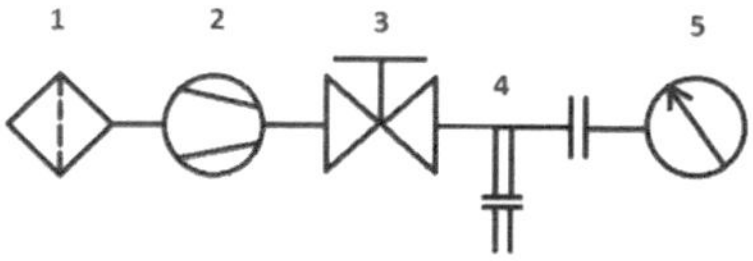

Figure 1. Pneumatic diagram.

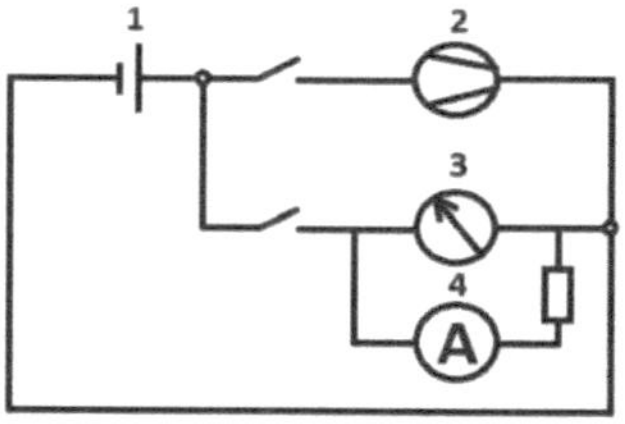

Figure 2. Electrical circuitry.

Screenshot of the running program is presented on the Figure 3. Timebased bubble pressure data are written into the table on the left. A graphic interpretation of timebased pressure is on the Cartesian plane on the right part of the screenshot. As follows from the graph on the Figure 3 bubble's separation from the capillary tip occurs at 470 Pa for this experiment with water. Time between nearby pressure peaks is approximately 15 seconds.

Pressure is maintained with a thermostat JULABO FP50-HL, which has range of operating temperatures (°C): -50 ...+200, stability of temperature maintaining is ±0.01°C. For thermostabilization a double-walled glass, through which ethanol circulates, is used.

The main part of the instrument is the capillary. That is why capillary's parameters selection is important part of the meter designing. The following parameters are recommended in the literature:

1. Capillary tip have to be cut square with axis. Tip melting is totally unacceptable because it breaks the cut sharpness. Capillary diameter have to be not less than 0.1 mm and not more than 0.5 mm (Kusakov 1936).

2. It's mentioned in the paper (Lylyk et al. 1998) that acceptable results (stable uniform bubble formation without any series or terminations at a given gas flow rate) were also obtained using capillaries with tips prepared by wet grinding. This was done using abrasive alumina powder with particle size less than 2×10^{-4}cm.
3. It is also written in the article (Lylyk et al. 1998) that the gas flow mode (aperiodic or oscillatory) is governed primarily with the geometric characteristics of the capillary, in particular the ration r^2/l. The application of capillaries with $r^2/l > 5 \times 10^{-5}$ cm doesn't lead to significant errors in the measured surface tension.

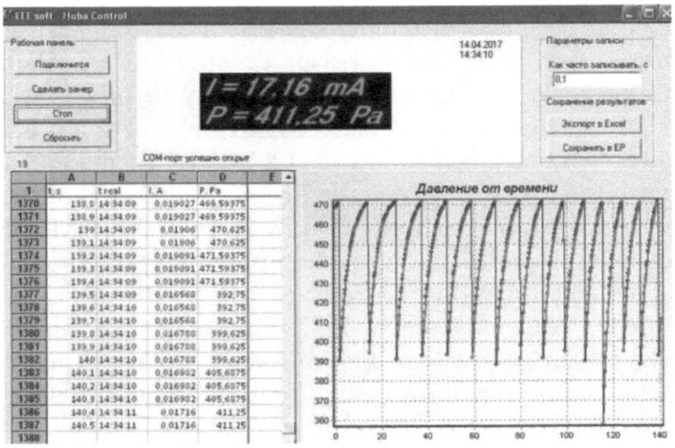

Figure 3. Running program.

However all capillary requirements can't be implemented in terms of technological capabilities. Capillaries used in this research were ordered from OOO Minimalno Invazivnie Technologii. Capillaries diameters were 536 μm and 634 μm. Smaller capillary is presented on Figure 4.

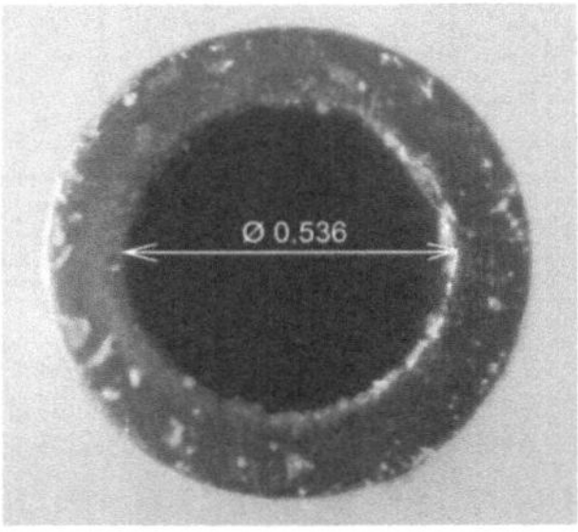

Figure 4. 0.536 mm capillary microphotograph.

The capillaries are made from stainless steel. The capillaries correspond to certain requirements:

1. Capillary tip is cut square with axis. The tip is not melted.
2. The tip is grinded.
3.

$$r_1^2/l = 2 \times 10^{-4}\ \text{cm}$$

$$r_2^2/l = 3 \times 10^{-4}\,\mathrm{cm}$$

where $r_1 = 0.536$ mm, $r_2 = 0.634$ mm

To sum up, the instrument improvement includes:

1. Measurement process is automated in the designed meter.
2. The accuracy of the results obtained using the meter is high because of modern equipment usage.
3. There are not fragile and unsteady compounds in the instrument.

As mentioned above the Laplace equation (1) doesn't take into account all factors which affect on surface tension. That's why application of this equation for surface tension calculation gives some errors compared with tabulated value. To diminish the error capillary calibration should be performed.

Liquids used for calibration (Table 1) were ordered from OOO Komponent-Reactiv:

Table 1. Calibration liquids.

Liquid's name	Purity qualification	TU/GOST	Production date
hexane	chemically pure	2-008-06	27.01.2017
n-decane	chemical reference subst.	2-512-12	26.12.2016
o-xylene	pro analysis	2-081-08	27.01.2017
benzyl alcohol	pro analysis	GOST 8751-72	09.01.2017
diethyleneglycol	pro analysis	2-289-11	02.11.2016
bidistilled water			

3 DATA PROCESSING

Data about pressure in the bubble obtained by the instrument are exported to Microsoft Excel. Pressure values which corresponds to the bubble separation are picked from the all values.

All maximum pressure values for water (smaller capillary) are presented on the Figure 5. As a result of a statistical data manipulation, average value of maximum bubble pressure and standard deviation were gained. Average value of maximum bubble pressure is 540.516 Pa and standard deviation is 6.39 Pa. All graphs were plotted in Easy Plot (MIT, USA). As can be seen from Figure 5, some maximum pressure values differ from the average one more than the others. It should be noted that these deviations have not any regularity. Due to the little amount of such deviations it can be concluded that they are connected with short-term changes of an environment. Pressure outliers are called such maximum bubble pressure values

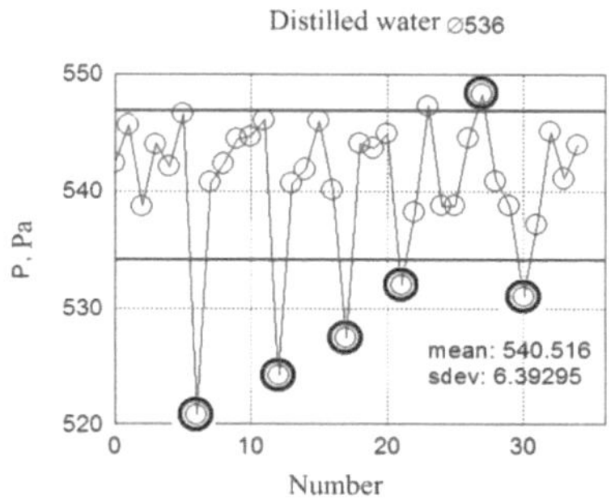

Figure 5. All maximum bubble pressure values for water (0.536 mm capillary).

which are differ in absolute value from the average pressure one more than the standard deviation. Pressure outliers were marked on the Figure 5. After all the outliers exclusion, new average maximum bubble pressure (543.059 Pa) and standard deviation (2.56 Pa) were received. The same operations were implemented for the others liquids.

For capillary calibration average values of maximum bubble pressure with cohered values of tabulated surface tension coefficients at 20 °C (Babichev et al. 1991; Ravdel & Ponomareva 1983; Nicolskiy et al. 1966) were plotted on the Cartesian plane. Due to the Laplace equation (1) the dependence of surface tension on maximum bubble pressure was approximate by linear function. Approximating function is presented on the graph (Figure 6 - for 536 μm capillary, Figure 7 - for 634 μm capillary).

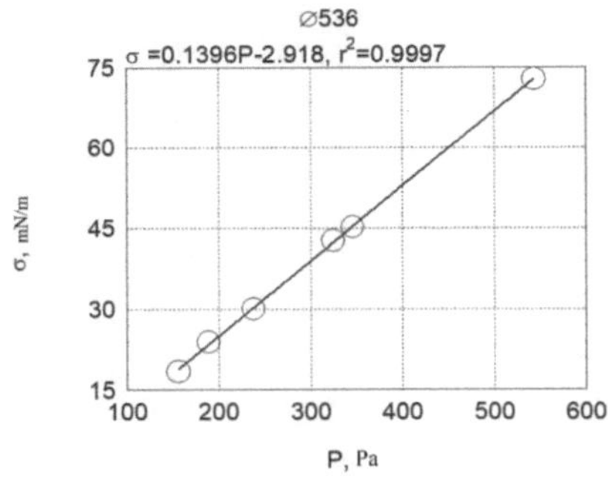

Figure 6. 0.536 mm capillary calibration curve.

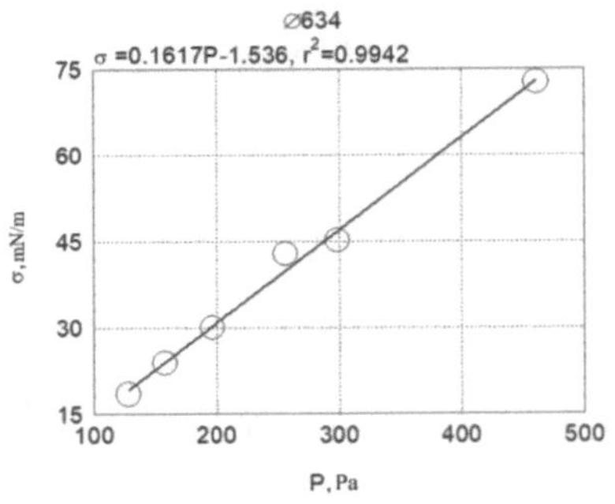

Figure 7. 0.634 mm capillary calibration curve.

4 CONCLUSIONS

A determination coefficient for 536 μm capillary is 0.9997 so a reliability of the approximation is very high. The approximation function for smaller capillary:

$$\sigma = 0.14\bar{P}_{\max} - 2.92 \tag{2}$$

Then

$$\bar{P}_{\max} = \sigma/0.14 + 2.94/0.14 \tag{3}$$

$$P = P_b + \Delta P \tag{4}$$

where P = pressure that gage fixes, P_b= pressure in a bubble, ΔP= pressure losses.

Formula (4) allows us to calculate average pipe pressure losses using (3):

$$\Delta\bar{P} = 2.94/0.14 = 21 \text{ Pa}$$

A determination coefficient for 634 μm capillary is 0.9943. An approximation function is

$$\sigma = 0.165\bar{P}_{\max} - 2.18 \tag{5}$$

Average pressure losses are $\Delta\bar{P} = 13.2$ Pa

As it was mentioned above bubble lifetimes were not the same for all the liquids due to the difficulties with a flow control. So it was necessary to realize how a flow velocity affects on pipe pressure losses. Darcy-Weisbach formula:

$$\Delta P = \lambda \frac{8Q^2 l}{g\pi^2 d^5} \tag{6}$$

where ΔP=pressure losses, λ=pipe friction factor, Q=volume flow, l=capillary length, g=gravity factor, d=capillary diameter.

An average volume flow (a bubble is considered as a hemisphere, one's lifetime is 3 seconds in average):

$$Q = \frac{V}{\tau} = \frac{2/3\pi r^3}{\tau} = 1.3 \times 10^{-11} m^3/s$$

Reynolds number:

$$\mathrm{Re} = \frac{4Q}{\pi d \nu} = 0.002$$

where ν= air cinematic viscosity.

Due to the received number it was considered that the flow was laminar. Then pipe friction factor is

$$\lambda = \frac{64}{\mathrm{Re}}$$

Therefore (6) can be transformed:

$$\Delta P = \frac{128 Q \nu l}{g \pi d^4} \tag{7}$$

That means pressure losses are in direct ratio to the volume flow. A constant volume flow corresponds to a constant bubble lifetimes. Then with constant bubble lifetimes for every liquid pressure losses have the same values and calibration curve accuracy is higher.

Mean approximation errors were calculated using the formula

$$\bar{A} = \frac{1}{n} \sum_{1}^{n} \frac{|\sigma_i - \sigma_x|}{\sigma_i} \cdot 100\% \tag{8}$$

For the 536 μm capillary error is 1.2%, for the 645 μm capillary one is 2.4%.

From the described experiments it can be concluded that the method which was offered by Rhebinder and improved by us allows to receive results of high accuracy. The meter can be used in further for solving some field problems which oil and gas industry faces, for example for surfactant selection.

REFERENCES

Adam N. K. The Physics and Chemistry of Surfaces. Oxford University Press, London: Geoffrey Cumberlege, 1930. 332p.

Babichev A.P., Babushkina N.A., Bratkovskiy A.M. et al. Physical values: critical tables [Fizichekiye velichini: spravochnic]. Under the editorship of Grigoreva I.S., Meylikhova I.Z. – M.; Energoatomizdat, 1991. – 1232 p.

Cantor // Z. Wied. Ann., 47, 399 (1892)

Fainerman V.B., Miller R. Dynamic surface tension measurements in the sub-millisecond range // J. Colloid and Interface Science. 1995. V. 175 (1). Pp 118–1121.

Jaeger // Z. Anorg. Ch., 101, 1 (1917)
Kusakov M.M. Methods of measurement petroleum products physical and chemical properties [Metodi opredeleniya fizico-himicheskih harakteristik neftyanih productov]; under the editorship of Rhebinder P.A. and with introduction from Rhebinder P.A., M.-L., 1936. 741p.
Lylyk S.V., Makievski A.V., Kovalchuk V.I., Schano K.-H., Fainerman V.B., Miller R. The effect of capillary characteristics on the results of dynamic surface tension measurements using the maximum bubble pressure method // J. Colloids and Surfaces A: Physicochemical and Engineering Aspects. 1998.V. 135.Iss. 1–3. Pp 27-40.
Nikolskiy B.P., Grogorov O.N., Pozin M.E. et al. Chemist's reference book [Spravochnik himika]. V.1. – 2 edition, M., 1966.
Ravdel A.A., Ponomareva A.M. Quick-reference book of Physical values [Kratkiy spravochnik fizico-himicheskih velichin]. Edition 8. M.:Himiya, 1983, p. 20-21.
Raymond I., Bendure. Dynamic surface tension determination with the maximum bubble pressure method // Journal of Colloid and Interface Science. 1971. V. 35. No. 2.
Simon M. // Z. Ann. Chim. Phys. 32 (1851) 5
Fesan A.A., Evdokimov I.N., Losev A.P., Kronin A.M., Lyubutina L.G., Panina T.N. Digital surface tension meter based on maximum bubble pressure method [Kompyuterizirovannaya laboratornaya ustanovka dlya opredeleniya poverhnostnogo natyageniya metodom maximalnogo davleniya v puzirke] // Collection of research papers of XIII international academic conference Modern Practical Physics. M.: Publishing house of Moscow physical society, 2014. P. 127-128.

Transportation Systems. Challenges and Solutions

Youth Technical Sessions Proceedings – Litvinenko (ed)
© 2019 World Petroleum Council, London, UK, ISBN 978-0-367-34668-3

Navigating in the future of electric vehicle battery

R. Samoylov
Vienna University of economics & business (WU), WU Executive Academy, Vienna, Austria

ABSTRACT: Energy transition to low carbon accelerates the move to electric vehicles (EV) in transportation sector. With 3 million EV in 2017 the expects to reach 100 million cars on the horizon of 2030. This could possibly generate 11 million spent EV battery by 2030. Electric battery is key component of EV and its sourcing (raw materials as lithium, cobalt, rare earths) faces challenges of environmental impact, ethics concern, uneven distribution of reserves and risk of supply. On the background of clean and "green" image of EV, battery creates concerns. The paper aims to find the solution of circular economy approach by closing the loop in battery lifetime by recycling, which would mean responsible and sustainable approach. Recycling industry misses spent batteries volumes and it is considered as not economically viable in current conditions. The paper defines potential uncertainties for recycling industry development by proposing probability tool to navigate better in electric vehicle recycling future.

1 INTRODUCTION

Nowadays energy industry experiences global transformation trend towards low carbon world. This trend has been accelerated after Paris agreement on climate change in 2015. According to this agreement majority of countries and companies commit to decrease environment impact through lowering greenhouse gases emissions (GHG). Along with industrial, residential, electricity and heat generation sectors, the transportation sector, being a blood veins of energy system, accounts for 14% of all GHG emissions, worldwide. In the USA and European Union the share of transportation in total energy consumption accounts for more than 30% (Eurostat, 2018) (EIA, 2018), while in China – for more than 20%. (Wang, 2014). Without transformation of transportation sector, the target of Paris agreement is unreachable. GHG emissions from transportation sector (air, road & marine) primarily come from fossil fuels origin, largely gasoline and diesel.

"Green" alternatives to conventional combustion vehicles can include biofuels, hydrogen, gas, hybrid electric vehicle (HEV) and electric vehicle (EV). The biggest attention & technology development is currently focused on HEV and EV. EV does not have tailpipe and does not emit neither carbon dioxide nor particles. This strong move towards EV can be explained not only by climate change engagement and the desire to escape from oil dependency, but also by the learning curve, which accelerates with EV cost decrease. EV is expected to play the key role in resolving pollution situation in cities. This becomes more & more relevant as the cities are forecasted to be responsible for more than 70% of CO_2 emissions in 2030 in business-as usual scenario. (IEA, 2016). The population of cities is expected to grow further as well. United nations expect 30% increase of urban population from 2015 to 2030. (United Nations, 2017)

The commercial sales of EV and HEV are breaking into the market with tremendous increase from almost none in 2010 to three million cars in 2017, with 50% of annual average increase since 2013. From current level of EV and HEV fleet, International Energy Agency (IEA) projections see 140 million vehicles in 2030 and 200 million in 2040 on roads, the latter constituting 30% of transportation sector worldwide. (IEA, 2017). Tesla being a pioneer in this industry created excitement about EV with its innovations, which attracted the public. Its new Model 3 electric car has projections of being "the biggest consumer product launch of all time". (Goliya & Sage, 2016). Other carmakers announce similar ambitious plans. No one can say if this will become the reality of our life or on the contrary EV fleet will stop to grow after certain moment, but nowadays this megatrend cannot be denied. The infrastructure starts to appear, customer fear disappears, technical characteristics improves, costs go down, drive range increases, governments provide incentives, regulation is under transformation. Even major Oil & Gas companies start to integrate EV business in their portfolio. The learning curve in all these aspects is thriving ahead every year. From one side, this looks very promising and inspiring. The world is preparing for potential massive penetration of EVs at least in certain regions – Western Europe, US, China. Transportation sector starts to play its role towards Paris agreement targets. From another side what are the challenges for EV penetration as a pathway to cleaner future? Will it be a high-speed highway or long and winding road?

2 AIM OF THE STUDY

2.1 *The context*

Rechargeable battery is crucial component of all digital and electric innovations in current world and there is no exception for EV. "Currently only 5% of batteries are recycled in Europe. Most of the current lithium is either dumped in landfill or incinerated," – Guardian magazine article says. (Gardiner, 2017). The striking thing is the image of EV as sustainable solution for transportation sector that goes apart from phrase from Guardian article. If EV fleet will grow to 100 million in 2030, then the amount of EV batteries should reach the same number. The production of EV batteries will require extraction of more & more of raw materials, specifically lithium and cobalt. It is common knowledge that after certain moment the battery loses its capacity and should be replaced. The typical life time is 8-10 years. Therefore, if most of EV batteries will be landfilled or incinerated, where is this sustainability principle of mobility sector? From general perspective I define "sustainability" term as not only the way how humanity deals with nature on long term basis, meaning no increase in extraction of natural resources, dangerous substance production and nature degradation caused by humanity and but also as a full satisfaction of humanity needs now and for next generations in the ethical and responsible way. If society continues to follow principle extract-consume-dispose in EV battery industry, then I would doubt that EV is a truly sustainable solution for transportation sector. Therefore, I consider the problematics and paradox of EV industry as sustainable solution for transportation sector is that the key component of it – EV battery - requires mining more and more raw materials, which after use will be landfilled or incinerated and not re-used or recycled.

2.2 *The aim*

The aim of the paper is to analyze the current state of EV battery market and future projections, provide overview of EV battery lifetime cycle from sourcing to end of lifetime treatment with focus on analyzing raw materials scarcity and EV recycling as a potential business towards circular economy. The ultimate target is to approach potential challenges and uncertainties in development of recycling business and create possible tool to navigate in it. The paper focuses specifically on the key component of EV – EV battery. The paper aims to create full rounded understanding of implications of the problem by addressing the following questions: What is the forecast for EV and EV battery growth? What is the lifetime of EV battery and its main implications in terms of sourcing raw materials and after lifetime treatment of spent EV battery? Is there any scarcity of raw materials resources for EV battery manufacturing? What is the current state of EV battery recycling industry and its likely future? What are alternatives to battery recycling? The ultimate objective is to define potential uncertainties which significantly impact the market of EV battery recycling and create a tool to navigate in it with better chance for success.

3 RESEACH RESULTS: MARKET REVIEW

3.1 *EV Market – current & future projections*

As has been said previously, the EVs including HEVs reached 3 million cars in 2017, starting from almost zero in 2010. EVs has two thirds of share in it. The market share of EV and HEV remains very low in total vehicle fleet, less than 2% with some exceptions in Norway and Netherlands, 30% and 5% respectively. 95% of EV and HEV sales happen in China, USA and several European countries (IEA, 2017). China fleet accounts for 40% market share and develops faster than others develop. The respectful agencies such as IEA, Bloomberg, BP forecast around 100 million cars in 2030 that corresponds to 33 times rise. Even with these figures, the fleet of EVs and HEVs will constitute only 30% of all vehicles worldwide. Car manufacturers are aligned with global expectations for EV quantities and have already started to transform the strategies to put millions of EVs to market.

EV is powered by rechargeable electric battery. It is considered that 70% of existing EV battery technologies will be based on lithium-ion technology at least until 2025-2030. There exists different EV battery formulations containing various amounts of raw materials. Typical EV contains not only such raw key materials as Lithium (5-7 kg), cobalt (5-15 kg), nickel (10-15 kg), but also graphite, rare earths (Desjardins, 2016). Majority of EV battery producers are located in China and South East Asia (over 65%), while US has about 20% of capacities. Capacities of production is expected to grow from 90 GWh in 2015 to 290 GWh in 2020 and 940 GWh in 2025 (Energy & Capital, 2017). Average EV needs 40-60 kW EV battery capacity. Technology of EV battery is being improved every year with cost decrease from 1000$ per kWh in 2010 to 273$ per kWh in 2016 (McKerracher, 2017).

3.2 *Sourcing raw materials for EV battery*

The situation in EV battery raw materials sourcing remains under question mark. 70% of lithium reserves, one of key raw material for EV battery are located in three countries of Latin America (Argentina, Chili, Bolivia). There is relative abundance of reserves and resources to satisfy lithium demand on long term. However, the existing extracting capacities should grow together with demand to mitigate price shocks. Distribution regions can potentially create geopolitical issue and uncertainty of supply. In 2017 about 40% of mined lithium was used for batteries (U.S.G.S, 2018). The lithium is expected to grow with at 15-25% rate annually. 60% of the production and reserves of cobalt, another key raw material, are located in Congo. 98% of cobalt is mined as by-product of copper and nickel mining (U.S.G.S, 2018). Cobalt reserves are not in abundance and rising demand could create scarcity of cobalt. Nickel availability should not pose a problem for EV as the reserves are relatively high and EV industry consumes only 3% of global consumption. However, the move to more nickel content-based batteries can provoke a boom in nickel demand, which might be difficult to satisfy with current capacities and other nickel-based applications. Other raw material with less content in EV battery as rare earth and graphite are also under question due to geographical distribution, which is concentrated in China with over 60%.

Pricing for raw material, specifically for lithium and cobalt shows potential image of unbalance of demand and supply which is probably due to slow reaction of miners to rapidly increase the rate of utilization of available capacities. Price for cobalt soared four times more during 2015-2017 while lithium prices experienced two times increase. Still price of raw material has limited effect on EV price, with most negative impact coming from cobalt price increase.

3.3 *Sustainability implications and solutions*

Environmental implication of EV battery is present on every step of lifetime chain. Extraction of raw material has serious consequences on water resources, pollution of soil and air in concerned areas. In-use period of battery lifecycle tells that EV remains as clean as the source of electricity that charges the battery. To claim that EV is clean & "green" solution for mobility sector, it is required to transform electricity generation source from fossil fuels to renewables energies. Sustainability implication of EV battery tells that by extracting raw materials, which are not in abundance and can be harmful to environment, we break this beautiful image of electric car with its sustainability claim. Recycling or reuse of EV battery should be the must if we want to commit to Paris agreement. In addition, ethic & social responsibility is at low level, taking the example of Congo, where 40 000 children are occupied in mining cobalt.

After use EV battery lifecycle implication tells that currently there is no recycling EV battery culture. Recycling of EV battery faces challenges of high costs (cheaper to extract more metals), economies of scales, absence of recycling in the business model of stakeholders, absence of dedicated regulation. Only 5% of lithium battery are recycled in European Union (Friends of the earth Europe , 2013), while 11 million of spent EV batteries are expected to exist by 2030. Closing the loop in EV battery lifetime with the mean of recycling should mean a responsible and sustainable approach for EV industry in terms of resources scarcity, social responsibility and environment. Legislation should guide stakeholders to integrate spent batteries after their lifetime into the whole industry value chain. Alternative use of energy storage will need more detailed approach and study in developing rules for battery reusing and applicability. Energy storage can have certain advantages over recycling in the sense of absence of additional emissions and waste generation which may happen during recycling process. Economics of EV battery recycling business remains unclear and very dependent on recycled metal prices that probably makes the final investment decision for serious R&D and installing recycling plant difficult to make. Current issue is that materials reclaimed value is only one third of recycling costs. Business start-up in EV battery recycling has better chance on the horizon of 2025 with larger volume of spent batteries available. However, Volkswagen, BMW, Chinese brands have recently announced projects for reusing/recycling spent EV batteries. Umicore put into operation semi-industrial recycling plant. Ideally recycled materials should serve to take market share from raw materials for EV battery production. Volkswagen believes in next 10 years, it will be able to recycle up to 97 % of all the raw materials used in the battery. Legislation should play an important role in pushing stakeholders to integrate recycling. Economics tells that cobalt price and volume of EV batteries will play a key role in attractiveness of the business. Alternative way to improve the sustainability of the EV battery use that is reuse for energy storage, which can be attractive is proper retrofit technology and legislation will be developed. Nissan is piloting the project of energy storage facility, capable to serve as back-up electricity source.

4 DISCUSSION AND SOLUTION PROPOSAL

4.1 *Uncertainty in the future of EV battery recycling*

One of key words with which I came up after the research work to characterize the future of EV battery industry and development of recycling industry is uncertainty. Navigating in uncertainty is widespread in business and it is applicable to EV industry as well. Taking notions of uncertainty from Mckinsey, EV industry falls into third level on uncertainty – "range of possible future outcomes". For EV and EV battery industry key area of uncertainties will include:

4.2 *Approach the uncertainty using probability tool*

The approach in navigating successfully in uncertainties of EV recycling business can be based on few tools. First, it is the top knowledge of the market and its mechanisms, understanding of main trends in supply and demand, innovations and legislation. This will let catch the market opportunities, secure competitive advantage and take better decisions, which would be in line with future industry trends. Second, permanent assumptions rechecking and context revisiting over time can help to navigate better in uncertainties and to adjust the move towards right direction. Thirdly, putting different scenarios/factors together with associated risks and

Table 1. Uncertainties & implications

Uncertainty	Implications
Uncertainty of EV fleet growth	Actual growth can be greater than expected on the wave of low carbon trend, technology innovations, cost decrease, competition between carmakers. From another side resources scarcity for EV battery can slow down the rapid growth of EV fleet because of slow down of the cost decrease for EV against conventional car cost.
Uncertainty of raw material availability	The evident question is that whether the reserves of lithium, cobalt, nickel, rare earths, graphite are sufficient to satisfy the demand of expected EV growth. In current EV battery formulations cobalt availability is thought to be the major uncertainty due to crucial dependence on supply from Congo and the country's capability to double the supply in 10 years. If not feasible, the cobalt price will continue to rise. This can drive R&D for alternative solutions, such as more nickel-based batteries or solid-state batteries. Lithium availabilities are considered to be in abundance, although its reserves are distributed mainly in Latin America. The availability of raw materials is characterized not only by reserves amount but also by geographical distribution. Congo, China, Argentina, Bolivia, Chile, Australia are those major raw materials suppliers. How reliable the supply on the long run would be?
Uncertainty of EV battery composition formulations	Uncertainty in battery formulation is highly dependent on raw materials scarcity. LNMC technology is expected to be 90% case technology. R&D targets to minimize cobalt content are already under development. If cobalt becomes a real barrier, then EV battery technology would probably search for cheaper alternatives. The question how fast will it happen? Such technologies like Solid-state technology and Li-air can be those game changers for resolving cobalt scarcity.
Uncertainty of Recycling, its legislation & economics	Increasing prices on raw materials can make recycling business profitable until alternative technologies becomes available. So, how fast alternative technologies will evolve, and will they leave the chance for recycling industry to take-off? The dedicated legislation to promote circular economy concept can be this trigger to oblige the companies to recycle and reuse spent EV batteries on early stage of EV industry development.
Uncertainty of alternative re-use of spent EV battery	Uncertainty for spent EV batteries destiny can become more complex with development of alternative destination of spent EV battery towards energy storage application. Will this technology be economically viable and if yes, how fast it will be available to market? Energy storage industry itself can provoke higher demand not only for recycled materials, but also raw ones. Energy storage application can endanger recycling feasibility.

attributing probabilities can create a tool to manage uncertainly better by creating chances of success. (Bradley et al, 2018) .

Here below I propose a tool to handle better the uncertainties by creating quantitative approach. The factors, which affect the recycling business, were assessed through "impact" and "probability" filters. Factors are the pieces of probable EV battery industry development scenario. The aim is to access the chances of success for recycling business by applying different factors and building the scenario.

"Impact" filter will have two directions – positive and negative. Each direction will have three grades (high, medium and low), which will have the following corresponding coefficients: 2.0/1.5/1.0 for positive direction and -2.0/-1.5/-1.0 for negative direction. The probabilities of different factors will have range from 0 to 1. For example, high cobalt price will have high positive impact on recycling industry (2.0) meaning the recycle could be become much more economically attractive to recycle and resell cobalt. Let us also assume that the probability that the price will be high (say more than 50 000 \$/ton) is 0.8 which can be due to scarcity of resource. This will give total grade of 1.60.

It is necessary to note that this tool has factors, which might be interconnected, when constituting the scenario of factors, hence these interconnections can affect the final average grade. For instance, if factor 6 would have high probability (table 2), then probability of the factors 5 and 7 is likely to have lower probability. On the contrary, if factor 6 has low probability, then the factor 5 or 7 is expected to have high probability. "Impact" filter aims to bring those probabilities in relation to recycling business development success. Positive coefficient will increase success, while negative decrease for developing recycling business.

Table 2. Probability tool factors – "Recycling industry future"

	Factors	Impact*	Probability**
1	Price of Cobalt grow 2x from current level on mid term	I1	P1
2	Price of Lithium grow 2x from current level on mid term	I2	P2
3	Breakthrough of game changer in lithium battery technology in mid-term	I3	P3
4	Breakthrough of game changer Lithium battery technology in long-term	I4	P4
5	EV fleet growth higher than expected	I5	P5
6	EV fleet growth as expected	I6	P6
7	EV fleet growth lower than expected	I7	P7
8	Battery technology improvement (lower cost and low content of Lithium and Cobalt)	I8	P8
9	Recycling legislation obliging carmakers to recycle	I9	P9
10	Energy storage application development for spent EV battery	I10	P10
11	Fast EV infrastructure development	I11	P11
12	New raw material reserves discoveries	I12	P12
13	Capacities of EV manufacturing do not catch up with EV battery demand	I13	P13
14	Raw material efficient replacement introduction	I14	P14

* Impact will be graded positively and negatively with corresponding coefficients. Positive: High (2.0)/ Medium (1.5)/Low (1.0) or Negative: High (-2.0)/Medium (-1.5.)/Low (-1.0)

** Probability will be graded from 0 to 1.

This list of factors will serve as list for constituting the possible scenario and accessing the success of recycling industry. Indeed, the list is not exhaustive and be completed in future improvements of the tool. The result of factors calculation grade will be cumulative average success grade of recycling industry development. For an example, let us have positive view on recycling industry and attribute "probabilities" and "impact" based on certain market situation:

Table 3. Probability tool factors – "Positive" scenario

Factor	Probability	Impact	Grade
1	0.8	P:2.0	1.60
9	0.9	P:1.5	1.35
7	0.6	N: -1.5	-0.90
12	0.3	N: -2.0	-0.60
Total Average			0.36

The total grade of 0.36 (calculated as sum of grades divided by number of factors) means the grade for success for battery recycling development based on attributed "Probabilities" and "Impact" to selected factors. This grade does not limit itself to grade 1.0. In this case, we considered that cobalt price remains high, recycling legislation will be put in place, EV fleet growth will be lower than expected due to cobalt prices and discoveries of raw materials will not likely to happen.

Now, let us take another example with negative view on recycling business development:

Table 4. Probability tool factors – "Negative" scenario

Factor	Probability	Impact	Grade
10	0.6	N: -2.0	-1.2
14	0.7	N: -1.5	-1.05
5	0.7	P: 1.5	1.05
12	0.7	N: -2.0	-1.2
9	0.9	P: 1.5	1.35
Total average			-0.21

The total grade -0.21 means the grade of success for battery recycling development based on attributed "Probabilities" and "Impact" for selected factors. The closer it is to positive value the better chances the recycling industry has for success. In this case, we considered that, recycling legislation will be put in place, discoveries of raw materials will likely happen, energy storage application will be developed, EV fleet will grow as expected, substitution material will be found for cobalt and lithium.

These two examples show that the tool reflects quantitatively the success grade of recycling industry based on market development factors. Indeed, the proposed tool can be very rough and questionable. Therefore, further work can be dedicated to access the applicability of this tool and possible improvements.

4.3 *Boundaries of the paper and further area of research*

While addressing topics of the paper, I set up several limits of the investigation for the sake of having a better precision as much as it is required for answering paper objectives. First, the paper is limited to analyze EV battery industry, but not EV industry itself. Where necessary, some EV industry data are included in the course of investigation. Focus is put on EV, not on HEV technology. Second, all future projections, including market growth, raw material reserves for recycling business calculation does not comprise new potential developments. New potential developments and factors could include, but not limited to break-through EV battery technologies with lower or none content of strategic raw materials (lithium, cobalt, rare earth), significant slowdown of EV fleet growth, update on raw materials reserves, expected growth of raw material production, rapid development of energy storage application, massive use of electric public transportation and so on.

Boundaries create potential new areas for research. While addressing the thesis problems I faced few topics, which were beyond the thesis framework, but which are worthwhile to investigate. First issue is technical assessment of such issues like standardization of recycling technology for various types of EV battery formulations, applicability of spent EV battery for energy storage and retrofit cost. These areas will let to have better understanding of technical complexity of the business. Second, profound look on recycling industry will be needed for revenue model precision calculations, notably operating expense and capital expenditures. As EV battery recycling industry is rather in its infancy, I believe new data will be available with time. Third, critical assessment of proposed tool for calculating success chances for recycling industry can create a valuable work for those who want to navigate better in EV battery recycling industry.

5 CONCLUSION

I consider the topics raised in this paper have great importance for comprehension of EV industry implications. EV is considered as main contributor to low carbon mobility and sustainability for transportation sector. EV battery, a key component of EV was centered in this paper. Cost for EV battery can become dependent on raw materials. Current availability of raw materials creates concerns in terms of supply risk, and hence threats the EV cost decrease and development of the whole industry. Extract-consume-dispose approach in EV battery life time cycle creates sustainability and ethical issues. Circular economy approach in form of recycling or reuse would be responsible response to these issues. Recycling industry for EV battery is yet to appear facing economies of scale issue and negative economics. Price shocks for raw materials create a field of uncertainties, comprising the development of alternative EV batteries, growth of EV fleet, future of recycling industry. To tackle the issue all industry chain entities (miners, governments, investors, policy makers, battery makers, car manufacturers, car users, refining companies) should be aligned on common target. Solution for recycling business in navigating in uncertainties can be based on perfect knowledge of market mechanisms, permanent update of taken assumptions and use of tool to access success probability.

ACKNOWLEDGEMENTS

This study is concise summary of research, which was carried in form of master thesis work with support and advice of prof. Dr. Jonas Puck, supervisor of MBA Energy management program master thesis at Vienna University of business and economics.

REFERENCES

Bradley, C./Hirt, M./Smith S. (2018): How to confront uncertainty in your strategy. https://www.mckinsey.com/business-functions/strategy-and-corporate-finance/our-insights/how-to-confront-uncertainty-in-your-strategy - retrieved on 1st June 2018

Desjardins, J. (2016): The Critical Ingredients Needed to Fuel the Battery Boom. http://www.visualcapitalist.com/critical-ingredients-fuel-battery-boom/- retrieved on 1st March 2018.

EIA. (2018):US Energy information Administration. US energy facts explained. https://www.eia.gov/energyexplained/?page=us_energy_home – retrieved on 1st March 2018.

Energy & Capital. (2017): Investing in Lithium Batteries. https://www.energyandcapital.com/resources/investing-in-lithium-batteries/15 - retrieved on 1st May 2018

European Environment Agency (2016): Electric vehicles in Europe. EEA Report No 20/2016

Eurostat (2017). Consumption of energy. http://ec.europa.eu/eurostat/statistics-explained/index.php/Consumption_of_energy - retrieved on 1st March 2018

Friends of the earth Europe (2013): Lithium. https://www.foeeurope.org/sites/default/files/publications/13_factsheet-lithium-gb.pdf - retrieved on 1st March 2018

Gardiner, J. (2017): The rise of electric cars could leave us with a big battery waste problem. https://www.theguardian.com/sustainable-business/2017/aug/10/electric-cars-big-battery-waste-problem-lithium-recycling - retrieved on 1st March 2018

Goliya, K./Sage, A. (2016): Tesla puts pedal to the metal, 500,000 cars planned in 2018. https://www.reuters.com/article/us-tesla-results-idUSKCN0XV2JL - retrieved on 1st May 2018

IEA. (2016): Cities are at the frontline of the energy transition https://www.iea.org/newsroom/news/2016/september/cities-are-at-the-frontline-of-the-energy-transition.htm – retrieved on 1st May 2018

IEA. (2017): Global EV Outlook 2017. 71 pages.

McKerracher, C. (2017): EV market trends and outlook. https://about.bnef.com/electric-vehicle-outlook/- retrieved on 1st March 2018

U.S.G.S, Mineral Commodity Summaries (2018): Lithium. Pages 98-99.

U.S.G., Mineral Commodity Summaries. (2018): Cobalt.Pages 50-51

United nations. (2017): World Population Prospects 2017. https://esa.un.org/unpd/wpp/ - retrieved on 1st May 2017

Wang, Y-F./Li, K-P./Xu, X. (2014): Transport energy consumption and saving in China. Renewable and Sustainable Energy Reviews. 29. 641-655.

Youth Technical Sessions Proceedings – Litvinenko (ed)
© 2019 World Petroleum Council, London, UK, ISBN 978-0-367-34668-3

Rheologic and thermodynamic aspects of oil structure formation during mid-stream

A. Boytsova & N. Kondrasheva
Saint Petersburg Mining University, Saint Petersburg, Russian Federation

ABSTRACT: The modern oil transportation system is characterized by mixing of different type of crudes such as heavy resin-asphaltenic and light paraffinic crude. The obtained blend is typically described as incompatible and leads to structure formation, blocking pipelines and equipment. The aim of research is a comparative assessment of structural properties and thermodynamic activation parameters for viscous flow in crudes of various composition in a broad range of temperatures and shear rates and determination of incompatibilities of crudes of different nature. The Rheotesy RN 4.1 was employed in all of the rheological examination tests. Obtained results can be used for predicting and preventing structure formation during production, transport and storage of crudes with various content of solid paraffins, resins and asphaltenes.

1 INTRODUCTION

Crude oil can be divided into two kinds according to their flowability. One is the light crudes with good flowability, and the other is that with high yield stress or high viscosity, including waxy crudes and heavy crudes. In Russia, more than 90% crude oil production is waxy or heavy oil. Paraffinic crude is characterized by high amount of wax (above 5%). At temperature below the wax appearance temperature (WAT), paraffin precipitates out from the crude gradually with decreasing temperature, and the crude exhibits non-Newtonian behaviors such as yield stress, thixotropy, etc [1]. It should be noted, that typically flow behaviors of paraffinic crudes are strongly depend on the shear rate and thermal prehistory. Heavy oil is characterized by its extra-high viscosity (might reach 10 Pa·s at 20 °C), low API gravity (less than 20), high content of asphaltenes, low amount of hydrocarbons with low molecular weight and complexity that make these oils difficult and expensive to produce, transport and refine [2-4]. Usually, the desired pipeline viscosity of oil mixture might not exceed 400 mPa·s [5-8]. Different studies were made to reduce the viscosity of heavy oil such as formation of oil-water emulsion [9,11], mixing with solvents or preheating the crude. Typically, to decrease viscosity of heavy oil its mixture with paraffinic crudes is used. It might lead to asphaltene-resin-paraffin deposits formation that can plug a pipeline during transportation of crudes mixture.

Rheological properties for petroleum are very important for processes where fluids are transferred from one location to another, for example oil migration inside the reservoir, oil wells, pipelines or refining units. Also it is one of the most useful ways to investigate heavy oil properties due to high content of resins and asphaltenes.

2 EXPERIMENTS

The original heavy crude oil was collected from Yarega deposit (Komi republic, Russia) with density 940 kg/m^3 at 20 °C. The light paraffinic oil from Kharyaga deposit (Komi republic, Russia) with density 824 kg/m^3 at 20 °C. The main properties of crudes are presented in Table 1.

Table 1. Properties of crude oils.

Parameters	Yarega	Haryaga
Density at 20 °C, kg/m^3	940	824
Viscosity at 40 °C, mm^2/s	562,2	5,4
Pour point, °C	-18	18
Solid paraffin, mass. %	0,5	23,0
Resins + asphaltenes, mass. %	49,0	25,0

The rheological behavior and thermodynamic calculations of crudes was studied using Rheotest 4.1. Measurements were carried out at temperature range from 10 °C to 70 °C and shear rate range from 0 s^{-1} to 300 s^{-1}. In accordance with guidance document of Transoil the used range covers the main values of oil transportation in Russia [12]. Energy of activation was calculated in accordance with logarithmic view of Arrhenius equation as a tangent of inclination of line of relationship between logarithm of viscosity and reciprocal temperature (1):

$$\mu = \mu_0 \cdot exp(E_a/RT) \quad (1)$$

μ – dynamic viscosity,
μ_0 – coefficient,
E_a – energy of activation,
R – gas constant;
T – temperature.

Thixotropic properties were tested by administering a shear rate on the crude from zero to 300 s^{-1} and then return at the same rate to zero.

3 RESULTS AND DISCUSSION

3.1 *Flow behavior*

The flow behavior of heavy and paraffinic crudes is investigated in terms of viscosity-shear rate. Figures 1 and 2 show a typical behavior for the measured results for heavy and was oil at different temperatures respectively. It is observed from Figures 1 and 2 that the crude oils have non-Newtonian shear behavior over the examined range of shear rate in which the apparent viscosity decreases considerably with temperature when it is heated from 10 °C to 70 °C for both types of crude. It is also demonstrated that the viscosity differences are larger at low shear rates than at high shear rates. As the temperature increases, the high molecular

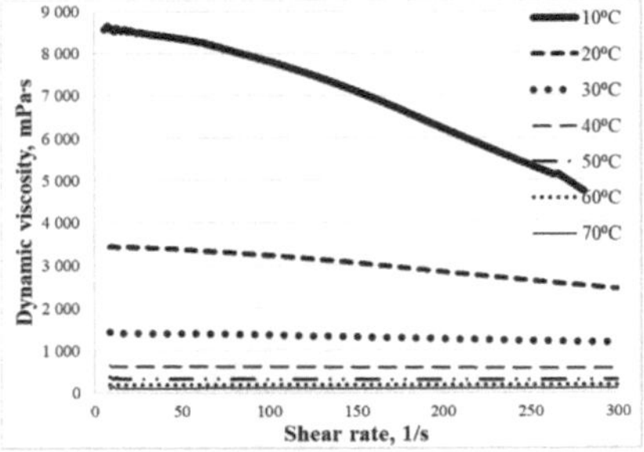

Figure 1. Viscosity profile of heavy oil at different temperature.

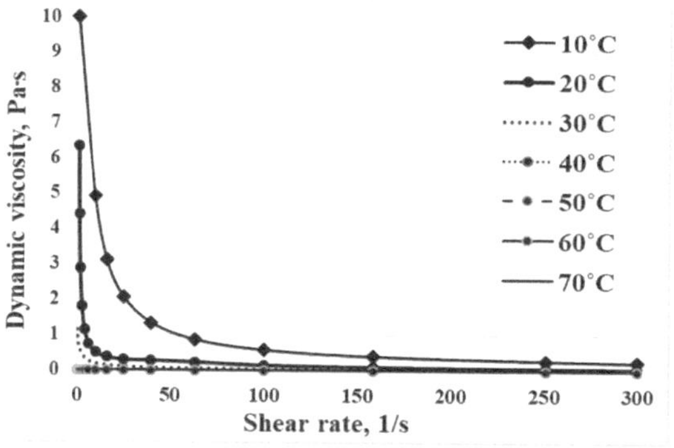

Figure 2. Viscosity profile of paraffinic oil at different temperature.

weight components of the heavy crude oil, such as asphaltenes and resins in heavy crude and waxes in paraffinic crude do not have the chance to agglomerate and form aggregates, and consequently affecting the bonds between the solid particles and eventually reducing the crude viscosity. From the viscosity investigation versus different temperatures, it is observed that there is a significant viscosity reduction over the tested temperatures. This can be attributed to the strong effect of temperature on the viscosity and chemical structure of heavy components of the crude oil such as wax and resins-asphaltenes. It should be noted that heavy oil has Newtonian behavior at temperature above 30 °C, in case of paraffinic crude it has Newtonian behavior at temperature above 40 °C and shear rate above 10 s^{-1}.

3.2 *Yield stress*

The mechanical characteristics of structured liquids modify significantly over a narrow low range of stress level. Below this stress level, the material deforms elastically with finite rigidity. Above this stress level, the applied stress leads to continuous deformation which causes the sample to start flow, i.e., viscous liquid-like behavior. The stress level at this transition stage is known as apparent yield stress. Existing method defines two type of yield stresses: static and dynamic. In this study static yield stress was investigated. It is observed that at 10 °C paraffinic crude has the much higher yield stress than heavy oil and it sharply decreases at temperature increasing. It could be explained by existing of strong structure in wax crude at low temperature.

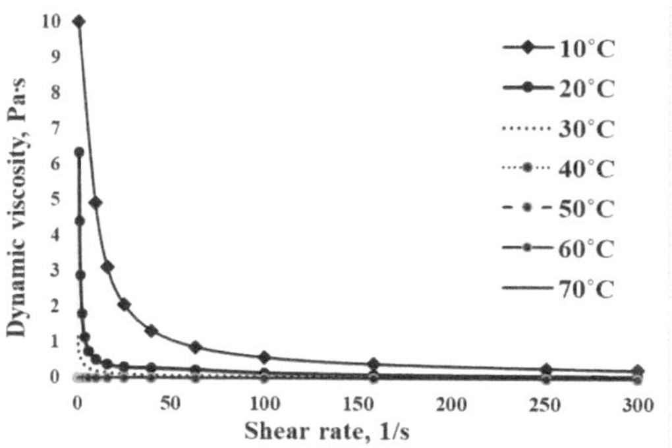

Figure 3. Yield point versus temperature.

3.3 *Thixotropy*

Thixotropy represents the property of fluid particles to form a loose association which is easily destroyed by shaking but re-establishes itself on standing. This parameter is important and pump performance in pipes is heavily dependent on the development of the velocity and pressure fields within that pipe. A thixotropic liquid which forms loose associations have the tendency of hampering the initiation of flow if it has been at rest for some time. This can easily lead to cavitation in the liquid within the pump. Difficulties in characterizing the flow can also ensure as a result of different flow profiles between the liquid layers under high shear on the walls of the pipe and fluid flowing in the middle of the pipe. It must be noted that in long pipes, issues arising from thixotropy are less of a concern as over time, the liquid is expected to reach equilibrium and the heterogeneous nature of the flow is less pronounced or non-existent. Figures 4 and 5 give the comparison of hysteresis loop of heavy and paraffinic crudes at 20 °C. It could not clearly observed the difference between thixotropies due to values of shear stress. Table 2 shows that the thixotropic characteristic of heavy oil is greater than those of paraffinic crude as a result of the different constituents including high-molecular components such as resins and asphaltenes.

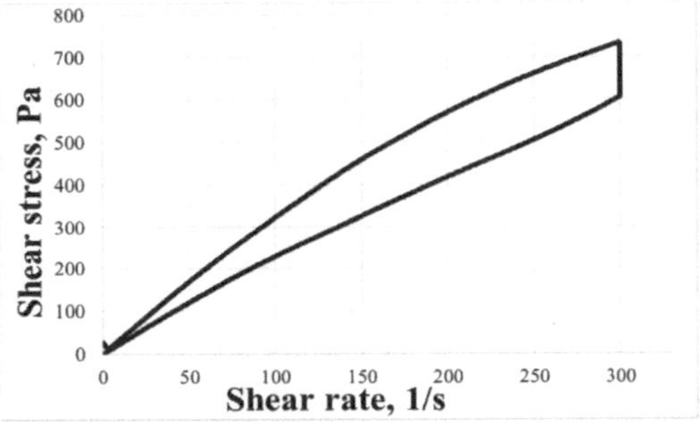

Figure 4. Hysteretic loop area of heavy oil.

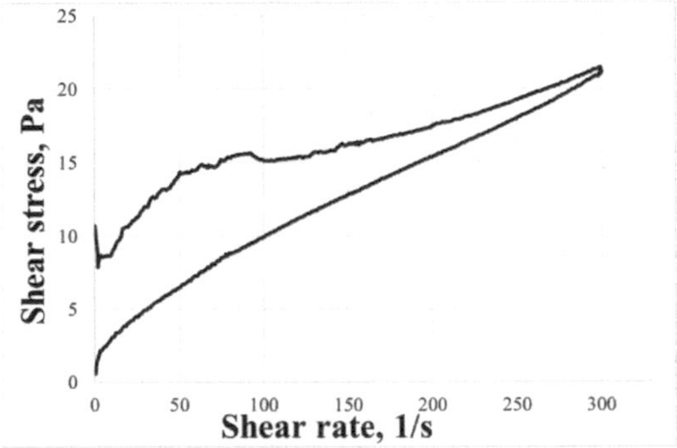

Figure 5. Hysteretic loop area of paraffinic oil.

Table 2. Thixotrophy energy at 20 °C.

Parameter	Yarega	Haryaga
Thixotropy energy at 20 °C, J/m^3·10^6	7130,2	0,3

3.4 *Arrhenius relationship*

In rheology experiments, the Arrhenius relationship (1) quantitatively shows the temperature dependence of the viscosity. The Arrhenius relationship, which addresses the kinetic properties of the material,naccount for phase behavior or molecular transformation within the material. Figure 6 shows a linear correlation between the logarithm of the viscosity and the inverse temperature quantified by the Arrhenius relationship. Squares of correlation greater than 98% were observed, but there was a noticeable angle trend on paraffinic crude's curve as wax appearance. It is considered that in this temperature liquid paraffin transforms to crystals that means temperature of phase transition (T_p). It is should be noted that heavy oil is characterized by absence of phase transition due to low content of wax. Paraffinic oil has phase transition at 40 °C. Thus the following calculations for wax oil were done in two ranges: from 10 °C to 40 °C (solid state) and from 40 °C to 70 °C (liquid state).

Thermodynamic parameters of viscous flow in Table 3 were determined in accordance with theory of activated complex. It has the great importance for understanding of structure formation during production, transportation and refining of crudes. Sharp decrease of activation energy, enthalpy and entrophy in response to the temperature increase is related to intensified thermal motion and, consequently, bond breaking in the wax oil, which increase values of thermodynamic parameters of initial state of inactive crude and approximate them to the values of thermodynamic parameters of the activated complex.

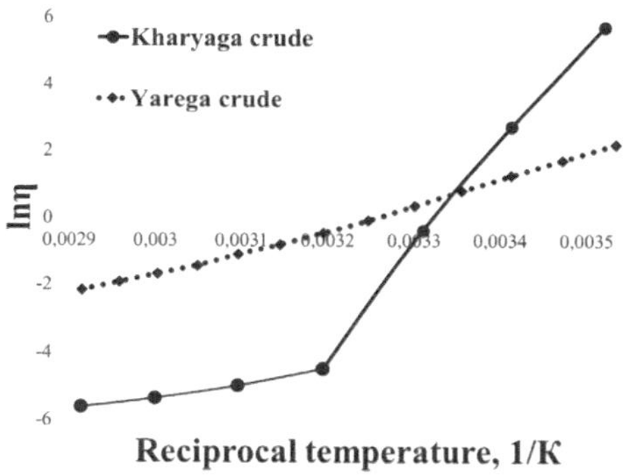

Figure 6. Arrhenius plot of heavy and paraffinic oils.

Table 3. Thermodynamic parameters of activation of viscous flow.

Parameter	Yarega	Haryaga	
		Below T_p	Above T_p
Activation energy, kJ/mole	57,9	247,2	33,7
Enthalpy, kJ/mole	60,5	249,7	36,3
Entropy, kJ/mole·K	79,1	715,7	39,1

4 CONCLUSIONS

The current study is carried out to investigate the comparison of rheological properties of the heavy and paraffinic crudes. The heavy and wax crude oils exhibit a non-Newtonian shear thinning behavior over the examined range of shear rates from 0 s^{-1} to 300 s^{-1}. The viscosity of the investigated oils decrease significantly with temperature over the range of 10–70 °C. The yield stress of the crudes decreases with temperature. The heavy crude oil shows a thixotropic energy of 7130.2 $\cdot 10^6$ J/m^3 and paraffinic crude – 0,3$\cdot 10^6$ J/m^3 at 25 °C.

Absence of liquidus temperature in heavy naphthene and aromatic crudes is conditioned by low content of solid n-alkanes. Activation energy of viscous flow of heavy crude is higher than activation energy of viscous flow of paraffin crude at temperatures above liquidus temperature of paraffin crude. Value of activation energy of paraffin crude is higher at lower temperatures due to the formation of strong cristallization structure.

REFERENCES

1. Zhang J., Liu X. 2008. Some advances in crude oil rheology and its application. *Journal of Central South University of Technology*. 15 (1). 288-292.
2. Castro, L.V., Vazquez, F., 2009. Fractionation and characterization of Mexican crude oils. *Energy and Fuel*. 23, 1603–1609.
3. Murillo-Hernandez J., Garcia-Cruz I., Lopez-Ramirez S., Duran-Valencia C., Dominguez J.M., Aburto J., 2009. Aggregation behavior of heavy crude oil-ionic liquid solutions by fluorescence spectroscopy. *Energy and Fuel* 23, 4584–4592.
4. Centeno G., Sanchez-Reyna G., Ancheyta J., Munoz J., Cardona N. 2011. Testing various mixing rules for calculation of viscosity of petroleum blends. *Fuel* 90, 3561–3570.
5. Fruman D.H., Briant J., 1983. Investigation of the Rheological Characteristics of Heavy Crude Oil-in-Water Emulsions. International Conference on the Physical Modeling of Multi-Phase Flow, Coventry, England.
6. Argillier J.F., Henaut I., Gateau P., 2006. Method of Transporting Heavy Crude Oils in Dispersion. US Patent Application 20060118467.
7. Schumacher, M.M., 1980. Enhanced Recovery of Residual and Heavy Oils. Noyes Press, Park Ridge, NJ.
8. Ghannam M.T., Hasan S.W., Esmail N. 2012. Rheological properties of heavy & light crude oil mixtures for improving flowability. *Journal of Petroleum Science and Engineering*. 81. 122–128
9. Dehkissia S., Larachi F., Rodrigue D., Chornet E. 2004. Characterization of Doba-Chad heavy crude oil in relation with the feasibility of pipeline transportation. *Fuel*. 83. 2157–2168.
10. Escudier M. P., and Presti F. 1996. Pipe flow of a thixotropic liquid. *Journal of Non-Newtonian Fluid Mechanics*. 62. 291–306.
11. Mortazavi-Manesh S., and Shaw J. M. 2014. Thixotropic rheological behavior of Maya crude oil. *Energy Fuels*. 28. 972–979.
12. Guidance Document 153-39.4-113-01 “Standard of technological design of oil pipelines” of Transoil

Youth Technical Sessions Proceedings – Litvinenko (ed)
© 2019 World Petroleum Council, London, UK, ISBN 978-0-367-34668-3

Modular mats west to reduce the cost of site preparation and roads access

F. Khazipov, L. Khazipova & A. Khazipov
PJSC TATNEFT, Almetyevsk, Russian Federation

ABSTRACT: Despite the moderately continental climate and the insignificant spread of swampy areas in the Republic of Tatarstan, the using of heavy equipment during the construction and repair of wells leads to the rapid destruction of isolated country roads and well sites, leading to their obstruction. This problem is particularly relevant in spring, autumn in the period of heavy rainfall and flood. The solution is to fill roads and well sites with crushed stone coating, which allows to organize efficient movement of special-purpose machinery. However, taking into account the average cost of construction of square meter of the site/road with crushed stone coating equal to 800 rubles, the total expenses of PJSC TATNEFT for the construction is hundreds of millions of rubles per year. Refusal of using crushed stone coating for roads construction in favor of the using the temporary road mats will provide annual savings up to 50 million rubles.

1 INTRODUCTION

In any country, roads always remain an important economy indicator. Bad roads lead to bad economy. The cost of an off-road problem for a state can be defined as the total amount of losses and lost profits associated with the insufficient development of the road network and its technical condition. Today the volume of loss is estimated at 450-500 billion rubles. This amount exceeds the annual amount of funds allocated for the needs of the entire road sector from the federal and the constituent entities budget of the Russian Federation (Yerygina 2006).

The problem is no less acute for oil and gas companies, incl. PJSC TATNEFT. The TATNEFT company's area of developed oil and gas fields of on the territory of the Republic of Tatarstan more than 7 thousand square kilometers, which are operated by more than 20 thousand production and injection wells, and this is without taking into account the objects for collecting, transporting and pumping well fluid. The geography of activity extends more than 200 km in the direction from the north-east to south-west and more than 100 km in the direction from the north-west to the south-east of the Republic.

Maintenance of these facilities requires road accesses, the total duration of which for the company TATNEFT exceeds thousands of kilometers. Since most of the access roads pass along country roads, this leads to inaccessibility of remote facilities during periods of rainfall in the autumn and spring periods.

Lack of the access to the remote facility makes it impossible to carry out repair works on the well with using of heavy machinery and the downtime of oil production facilities leads to losses of the company's profits. In case of off-road at the facility with working drilling rig the customer suffers losses in the form of equipment downtime of the contractor, which is usually full paid. In addition, the customer losses profit due to the delay in commissioning the well.

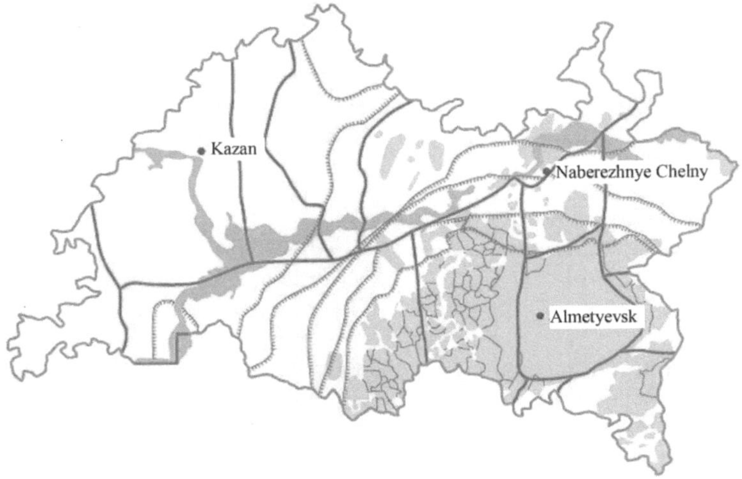

Figure 1. Oilfields in Republic of Tatarstan.

2 POSSIBLE SOLUTIONS

There are several solutions to this problem:

1. Planning activities in summer and winter periods. This solution is actively used in Western companies, including in Canada. However, in the conditions of PJSC TATNEFT this solution is limited because it will lead to the temporary disbanding of repair and drilling crews with losses of the profit for the service company.
2. The using of bulldozer equipment for towing special machinery. This solution does not eliminate the offroad problem and is not optimal due to the high cost of bulldozer equipment, low speed of movement, leading to downtime of drilling/repair crews, as well as due to high risks of damage to people health and special machinery.
3. The third solution is to fill roads and well sites with a crushed stone. This solution allows organizing the effective movement of special machinery, but has a higher cost. It is applicable only for roads with relatively high traffic.

Nowadays, in TATNEFT when the drilling wells are far from the roads, during the spring flood period the first solution used - works are stopping. For the facilities with a short distance, the roads are constructed with crushed stone. Taking into account the average cost of construction of a square meter of the site/road with crushed stone coating equal to 800 rubles, the total expenses in the TATNEFT conditions for the roads and site preparation are more than hundreds of millions of rubles per year.

Taking into account the annually growing plans for the drilling, the task of reducing the costs of construction roads and well sites is priority. It will result in reduction of the capital costs for construction and reconstruction of the wells. This will increase the profitability indexes thereby increasing the company's profit at the expense of additional oil.

To solve the above tasks is proposed to use temporary road mats to organize access to hard-to-reach areas when it is necessary. Today, there are hundreds of offers of temporary road mats around the world and in Russia, significantly differing in size, material, weight and price.

The following main categories of road mats can be distinguished:

- Reinforced concrete slabs. It used for the making of temporary roads (Road plates). The installation is fast enough, it needs to prepare the surface and put the slabs on top. The material is highly durable; therefore, it can withstand loads even from loaded trucks (GOST 21924.0-84 Reinforced concrete slabs for urban roads. Technical conditions). The disadvantage of the solution is requirements for preliminary surface preparation, a limited cycles of reuse.

- Rig mats (floorings) are intended for use as bedrock supports (cushions) under the base of the drilling rig and rig equipment, as well as for temporary road construction and well sites on soils with a loading capacity of at least 1.2 kg/cm². Due to the harsh operating conditions they have large size and heavy weight (Production of drilling mats).
- High-molecular polymer road mats. The lightest type of temporary pavement. Polymer mobile road mats, manufactured by pressing. They have an internal honeycomb structure, as well as technological openings along the perimeter for connecting the products together. Polymer road mats have a wide range of sizes, methods of attachment. However, have special requirements on the quality of base preparation due to the fragility (2017 Xtreme Matting Engineering, Mobile pavement).
- Wooden mats. They are made entirely of wooden planks, fastened together with special glue or nails. The advantage of this type of mats is comparative lightness, low price, availability of material, environmental friendliness (Southern Hardwood Company). Meanwhile, there are a number of drawbacks, the main one is being low durability of use in humid environments.
- Rubber road mats. These type of heavy pavements based on rubber granules. Allows creating driveways and platforms as quickly as possible for the operation of heavy equipment in hard-to-reach areas including swamps. Rubber can be used at air temperatures from -60 C to + 60 C. Advantages of the rubber mats are following: resistant to moisture, heat, acids and alkalis, petroleum products, UV radiation; almost no requirements for the preparation of a base. Rubber mats are made from recycled and environmentally friendly materials (Nanotechnology center of composites). Meanwhile, there are a number of drawbacks, the main one is high cost and heavy weight in comparison with polymer plates.
- Pre-fabricated road from aluminum profiles. Pre-fabricated road has an original design, which consists of aluminum profiles, combined with a polymer composition and special locks. The profile has a large number of lamellae and stiffeners, which additionally enhance the bearing capacity of the road and improve the adhesion properties of the coating. This type of coating has the following advantages of use: compact design during delivering into the place; high speed of installation and deinstallation (Faun trackway). However, the disadvantages are also obvious: the high cost of manufacture, requirements of unique special machinery for installation; requirements for the preparation of the base surface; limited cycles of reuse.

Each type of coatings on the market has its own advantages and disadvantages. The main characteristics of different types of mats are shown on Table 1.

Table 1. The number of officially reported plague cases in the world.

Parameters/application	Concrete	Rig Mats	Polymer	Wood	Rubber
Cost	Yes	No	No	Yes	Yes
Low weight	No	No	Yes	No	No
Easy installation	No	No	Yes	No	No
Caterpillar equipment	No	Yes	No	Yes	Yes
Heavy trucks	Yes	Yes	No	Yes	Yes
Winter time	Yes	Yes	No	Yes	Yes
Environmentally friendly	Yes	Yes	No	No	Yes

It is two types of modular mats is needed for TATNEFT:

1. High-strength Rig mats installed under the drilling equipment.
2. Road and site mats to ensure access roads and work safety.

Figure 2. Drilling mats under the base of drilling rig equipment.

Mats for rig equipment have already been successfully used in the company (Figure 2). Due to the fact that the need for this equipment is relatively insignificant, it was decided to purchase drilling mats of the required characteristics through standard tender procedures.

Such solution is effectively applied in Western countries (USA and Canada), as well as in several facilities of Siberia in Russia. The technological potential of Tatarstan and TATNEFT allow to start the production of this type of mats at the facilities of PJSC Nizhnekamskshina - the leader of the domestic tire industry (PJSC Nizhnekamskshina) and with technical and technological potential of LLC "Tatneft-Presscomposite". The proposed solution, in addition to reducing the costs of the enterprises of the oil and gas sector, it will increase the plant's workload and create new jobs in the Republic. The expected EMV of the project will be 26.3 million rubles.

To achieve this goal it is necessary to solve the following tasks:

1. Develop within the framework of R & D composite material that provides the required characteristics of the modular mats. The source material for the manufacture of mats will be rubber from used tires.
2. Implement experimental program of manufacturing small batch of products during the bench tests and tests in real conditions.
3. To make search and arrangement of the place for installation of the equipment and warehouse for finished goods. Nowadays the required facilities and building are available and this task is solved.
4. Search and train personnel responsible for: technological process, loading raw materials and finished products, the logistics of the delivery raw materials and the shipment of finished products.
5. Make a replacement of the investment program for drilling wells: replace the costs for the construction sites and roads to the costs of purchasing modular mats.

3 MARKET ANALYSIS

The market geography is represented almost by the entire territory of the Russian Federation. A consumer of modular road and site mats can be any company whose activity is related to off-road operations. The main consumers in the Tatarstan Republic will be primarily the oil and gas sector, and secondly, construction companies. The competitive prices and efficient logistics will increase demand for products as an alternative to the applicable solutions as crushed stone coating. The availability of a delivery system of finished products, as well as rental services with offering of turnkey services, will allow to take advantage of the products produced for small companies that do not have sufficient resources to purchase and use coatings on an ongoing basis.

Such large companies as represent competitors in Russian Federation:

- RusTEK;
- RusComposite;
- Tech polymer
- NCC, etc.

The competitive advantages of starting proposed business in TATNEFT are:

1. The size of TATNEFT Company and the geography of its activity make it possible to produce mats only for its own needs. As a result it allows the use of products cheaper due to the close to zero internal rate of return.
2. Technological and production base of LLC "Tatneft-Presscomposite". and PJSC Nizhnekamskshina. allows designing material with the desired properties and quality during R & D in-house, thus ensuring the safety of intellectual property.
3. The production base of LLC "Tatneft-Presscomposite". and PJSC Nizhnekamskshina. allows organizing industrial production in the shortest time.
4. The experience of selling own products, our own and affiliate sales network allows to enter the foreign market as soon as possible.
5. Further expansion of the product range based on the developed technologies will further improve the profitability of production and the stability of PJSC TATNEFT by diversifying its business.
6. Last but not least important advantage is the fact that there are a large number of raw materials in the Republic. Plant for recycling of used tires to a small crumb rubber is scheduled to open. This will make it possible to obtain raw materials without a margin on profitability, thereby increasing the competitive advantages of the proposed business.

4 RISK ANALYSIS

The project implementation involves the following risks:

1. The high cost of components (raw materials) for the manufacture of modular mats. Nowadays the problem is not relevant, the cost of raw materials is low (production waste is used). The raw material is also planned to be produced at the production bases of the TATNEFT.
2. Competition from other manufacturers. The advantage of the implementation of this business within the TATNEFT is the reduction of profitability rates for domestic needs, which will allow to lower the cost and have a high competitiveness during the tender procedures. The internal need of the TATNEFT for this type of mats allows having full load of equipment without the need to enter the foreign market.
3. A significant reduction in the cost of existing crushed stone coatings is not expected; therefore, the relevance of the technology will continue in the future.
4. A sensitivity analysis of changing price parameters for the project was made. The analysis showed a high degree of business reliability under adverse market conditions (increasing the cost of raw materials, a sales dip, decreasing the prices for finished products, etc.).
5. High cost for mat making machine. This risk have relatively high affect to NPV. Risk management can be made by thorough market analysis and concluding a SMART contract.

5 ECONOMIC EFFICIENCY CALCULATION

Evaluation of the economic results is one of the most difficult task, however, it must be necessarily made due to the significant costs of the restructuring project. To analyze and compare the effectiveness of projects, we propose to use the indicators outlined in the "Methodological guidelines for evaluating the effectiveness of investment projects" approved by the Ministry of Economy and the Ministry of Finance of the Russian Federation (Kossov 2000).

The feasibility study was calculated taking into account all possible costs. Lifetime of the project was determined to be 60 months. Expected monetary value over the life of the project is expected to be 26.3 million rubles at payback period equals 3 years and index discounted yield equal to 1.3.

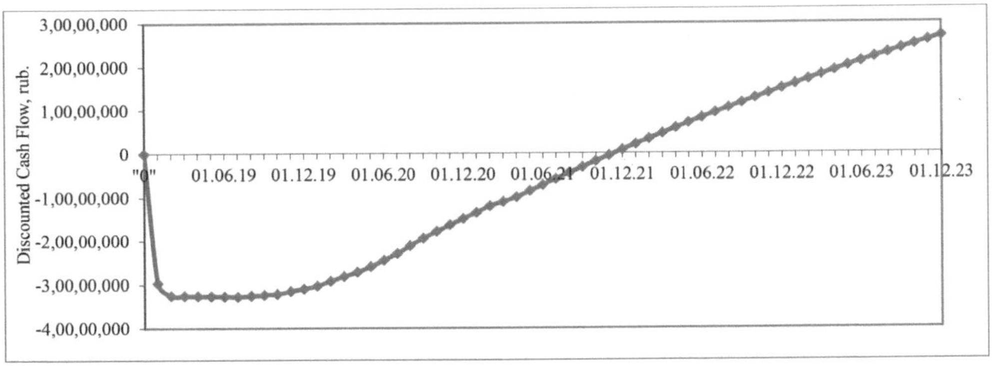

Figure 3. Discounted cash flow.

The sensitivity analysis showed a high degree of business reliability under adverse market conditions (an increase in the cost of raw materials, sales dip, decreasing in prices of finished products, etc.).

6 CONCLUSION

The economic feasibility of organizing a permanent road access (most often crushed stone coating) should be assessed for each well cluster individually according to the following simplified formula:

$$E = C/(P\ L + S) \tag{1}$$

Where:

E (Efficiency) - the effectiveness of laying a permanent road with the use of macadam coating.

C (Cost) – the cost of providing access roads, taking into account mainly the remoteness of the facility.

P (Profit) – the acquisition of the benefits from the using of the road, including taking into account the risk of loss of the profit.

L (Load) – the ratio of the road use.

S (Safety) – coefficient, taking into account the usefulness of the object, including safety needs, environmental protection.

The scatter of the values of these variables may differ in 10 or more times depending on the geological conditions, the type of work, and the geographical location of the facility. In this regard, the calculation of the expected efficiency must be done individually. This formula for deciding of necessity of using permanent pavement can also be used to determine the feasibility of installing a temporary pavement. The only difference in the formula will be due to the absence of load factor.

It has been proven that the implementation of the project in TATNEFT will make it possible to increase the profitability of the construction and reconstruction of wells, which will expand the perimeter of the candidates and ensure the growth of oil production due to the involvement of previously non-drained reserves.

Finally, it should be noted that availability of access to oil production facilities is dictated not only by the task of costs saving. We have not to forget that oil companies work on high-risk facilities and any delay in an ambulance or a fire truck can cost human life.

REFERENCES

2017 Xtreme Matting Engineering, viewed 10.09.2019, http://www.xtremematting.com/brochures/2017%20Xtreme%20Matting%20Engineering.pdf

GOST 21924.0-84 Reinforced concrete slabs for urban roads. Technical conditions [GOST 21924.0-84 Plity zhelezobetonnyye dlya pokrytiy gorodskikh dorog. Tekhnicheskiye usloviya], viewed 23.10.2018, http://www.vashdom.ru/gost/21924_0-84/

Faun trackway, viewed 05.10.2018, https://fauntrackway.co.uk/

Kossov, V., Livshits, V. & Shahnazarov, A. 2000. Guidelines for evaluating the effectiveness of investment projects. Ministry of Finance of Russian Federation. *Economy*: 11.

Mobile pavement, [Mobil'nyye dorozhnyye pokrytiya], viewed 28.09.2018, http://www.texpolimer.ru/production/mobilnye-dorozhnye-plity/

Nanotechnology center of composites, viewed 05.10.2018, http://www.nccrussia.com/en/products/construction/267-road-mats.html

PJSC Nizhnekamskshina, viewed 09.11.2018, http://shinakama.tatneft.ru/?lang=en

Production of drilling mats, [Proizvodstvo burovykh matov], viewed 12.12.2018, http://nordtekh.ru/catalog/burovye-maty/

Road plates [Dorozhnyye plity], viewed 23.10.2018, http://www.pliti.ru/pliti.html

Southern Hardwood Company, viewed 05.10.2018, http://www.southernhardwood.net/oilfield-mats.html

Yerygina, M. 2006. *The role of highways in the development of the economy* [Rol' avtomobil'nykh dorog v razvitii ekonomiki]. viewed 10.02.2019, http://www.cs-alternativa.ru/text/1631

Well construction, Drilling and Completion. Era of New Technologies

Youth Technical Sessions Proceedings – Litvinenko (ed)
© 2019 World Petroleum Council, London, UK, ISBN 978-0-367-34668-3

Improved waterflood performance by employing permanent down-dole control devices: Iran case study

Sh. Mohammadpourmarzbali, Y. Rafiei & J. Fahimpour
Amirkabir University of Technology, Tehran, Iran

ABSTRACT: Horizontal and multilateral wells are becoming a principal well architecture in the current field developments. (Al-Khelaiwi and Davies, 2007).However, the increase in wellbore length and exposure to different reservoir facies come at a cost. Excessive frictional pressure drop due to high flow rates in horizontal sections, compared to vertical wells, causes higher pressure drawdown in the heel section of the completion, resulting in an unbalanced fluid inflow along the horizontal well length. As a result, coning of water and gas toward the heel of the well is an inevitable problem. (Al-Khelaiwi and Davies, 2007).Over the last decades, use of Advanced Well Completions employing Downhole Flow Control technology such as Inflow Control Devices (ICDs) and Interval Control Valves (ICVs) have equipped the operators with effective techniques to tackle the aforementioned challenges (Al-Khelaiwi, 2013). This paper aims to investigate the technical advantages of using ICVs to control and improve the production from a horizontal well, which is under waterflood, in one of Iranian oil fields. For this purpose, the reservoir was first divided into different hydraulic flow units and then ICVs were installed in each identified unit. Following this, Integrated Production Modelling was employed to simulate and study the improvement in the oil recovery by using those downhole valves to manage the production from different zones. The results from this work demonstrate the effective application of Down-hole Control Devices for improving waterflood performance in terms of more oil recovery and less water production.

1 INTRODUCTION

The case under this study is an Iranian oil field located in the south-west part of Iran. The field is spread out over an area of 23 km by 6 km and includes two Asmari and Bangestan oil reservoirs.

To compensate for the considerable decline in reservoir pressure observed over the last few years, waterflooding has been initiated in the field. In addition, horizontal wells have been drilled in order to enhance the well productivity and reduce the risk of water coning. On the other hand, premature breakthrough of water from highly permeable rock types and channels is a persisting issue in producing from those horizontal wells. Furthermore, another downside of horizontal well application is the uneven distribution of the injected fluids along the horizontal wellbore section (Al-Khelaiwi, 2013).

In recent years Intelligent Well Completion has been employed as a potential solution to some of these production challenges (Rafiei, 2014). Intelligent Well Technology can provide the engineers and operators with a zonal flow control management system based on flow rate and phase cut measurements. It flexibly controls production and/or injection scenarios, reducing well counts and workover operation costs (Elmsallati, 2005).

In this technology, the reservoir inflow into the well is controlled by restricting the fluid's flux from annulus into the tubing. (Birchenko, 2010). The incorporation of these advanced downhole tools into conventional well completions can successfully be used for optimizing the production/injection strategies during implementation of the IOR and EOR projects (Asl et al., 2017, Rafiei et al., 2017, Cheperli and Rafiei, 2018, Rafiei et al., 2018, Rahimbakhsh and Rafiei, 2018).

An Inflow Control Valve (or Interval Control Valve) is a remotely controlled downhole device for choking the flow, in a partial or complete manner, at different production/injection intervals. (Al-Khelaiwi et al., 2008).It should be noted that the number and location of the ICV controlled zones dictate the added value associated with an Intelligent Well. While using too many valves lead to unnecessary and excessive cost and potential for reduced reliability, the use of limited number of valves will also not provide adequate flexibility for efficient control (Davies and Ebadi, 2006).

This study initially aims to investigate the application of Hydraulic Flow Unit concept in determining the number and location of installing ICVs in the reservoir.

Hydraulic flow units is defined as a mappable portion of the reservoir within which the geological and petrophysical properties that affect the flow of fluid are consistent and predictably different from the properties of other reservoir rock volume (Tiab and Donaldson, 2016). The flow unit concept can provide a means of uniquely subdividing a reservoir into intervals that approximate the architecture of a reservoir at a scale consistent with reservoir simulations (Rafiei, 2007).

HFU is defined based on the following parameters and equations (Tiab and Donaldson, 2016):

$$RQI = (\phi_n \times FZI) \ in\ which:\ \phi_n = \left(\frac{\phi_e}{1-\phi_e}\right),\ FZI = \frac{1}{\sqrt{F_s\tau S_{gv}}},\ RQI = 0.0314\sqrt{\frac{k}{\phi_e}} \quad (1)$$

In which RQI is reservoir quality index, φ_n is normalized porosity and FZI is flow zone indicator. Also, k is permeability in μm^2 and φ_e is effective porosity in fraction. The term $Fs\tau^2$ is known as the Kozeny constant, which is usually between 5 and 100 in most reservoir rock. The term $Fs\tau^2S^2_{gv}$ is a function of geological characteristics of porous media and varies with changes in pore geometry (Tiab and Donaldson, 2016).

Taking the logarithm of Equation 1 on both sides yields:

$$\log RQI = \log \varphi_n + \log FZI \quad (2)$$

Equation 2 yields a straight line on a log-log plot of RQI versus φ_n with a unit slope. The intercept of this straight line at $\varphi_n = 1$ is the flow zone indicator (FZI). Samples with different FZI values will lie on other parallel lines. Samples that lie on the same straight line have similar pore throat characteristics and, therefore, constitute a flow unit.

2 CASE STUDY

This study has been conducted in three major steps as follow:

1. The reservoir was divided into different flow units based on the HFU concept.
2. The Integrated Production Modelling (IPM) approach was employed to design the production system and also to model the ICVs in the production network.
3. The performance of ICVs were evaluated by comparing two production scenarios; one with conventional completion and the other one with ICVs for controlling production from different flow units using the integrated production model.

4.1 *HFU identification*

Iterative muliti-linear clustering regression technique is a widely used approach for partitioning the data into appropriate number of clusters, according to which the optimum number of flow units is determined (Rafiei, 2007, Stolz and Graves, 2003).

During a HFU classification process, using some core data (core permeability and porosity), the following steps should be followed (Rafiei, 2007, Stolz and Graves, 2003):

1. Compute the reservoir quality index (RQI) and normalized porosity φ_n from routine core data.
2. Plot RQI versus φ_n in logarithmic scale.
3. Use a reasonable initial guess of the intercept for each straight-line equation, the mean FZI value of each HFU.
4. Assign core sample data to the nearest straight line.
5. Recalculate the intercept of each HFU using Levenberg- Marquart regression equations.
6. Compare the new and old values of the intercept for every straight line. If the difference is within the acceptable tolerance, update intercept values and go to Step 4.
7. The procedure above should be repeated until the optimal location of each straight line is found in which the error sum of squares is a minimum for the desired number of hydraulic flow units Figure 1.

To find the optimum number of flow units, a sensitivity analysis was carried on the number of flow units. Adding more HFUs should cause a reduction in the Sum of Square Error (SSE). However, as we continue to add HFUs, we will reach a point where reduction in error gets smaller. The SSE in this case can be used as a criterion for selecting the optimum number of HFUs. A plot of SSE versus the number of HFUs is shown in Figure 2. According to this, an optimum number of four HFUs was determined for the reservoir under study. The measured values of FZI for different HFUs are shown in Table 1.

4.2 *Designing of integrated production model*

After determining the optimum number of the flow units, the next step was to model and simulate the application of interval control valves in the identified flow units. For This purpose, the integrated production modelling approach was employed. The Integrated model allows to monitor how a change in one part of the system will affect the behavior of other components in the system, which are in pressure communication with each other. In order to do this IPM software package developed by Petroleum Expert company was used.

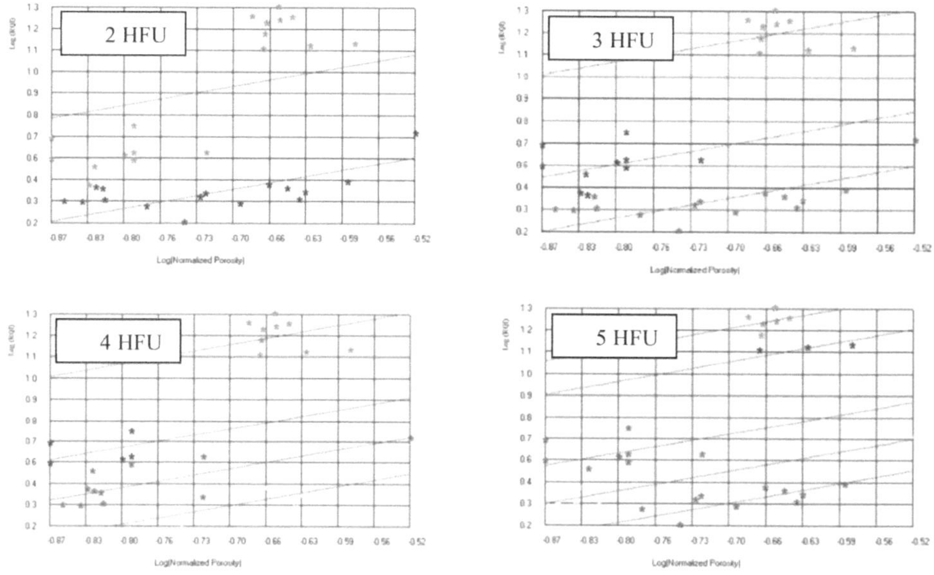

Figure 1. Plot of reservoir quality index versus normalized porosity for different numbers of flow units.

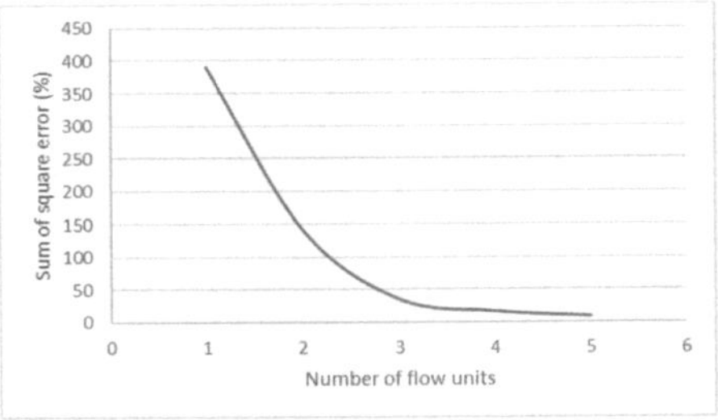

Figure 2. Plot of SSE versus number of HFUs.

Table 1. Measured values of FZI for each flow units.

	FZI	Length (ft)
HFU 1	76.1	1304
HFU 2	32.2	983
HFU 3	64.5	1054
HFU 4	28.4	1103

The reservoir is simulated with Mbal software, which is an analytical reservoir modelling toolkit. Schematic of the reservoir model with its associated flow units are shown in Figure 3. Tables 2 illustrate the PVT properties of the corresponding reservoir fluid.

Prosper software was used to model and evaluate the performance of the well. Then, Gap software was employed to integrate the reservoir and well model and also to simulate ICVs within the production network. Figure 3 shows different elements of the overall production network designed in Gap. As it is shown, the production system is composed of reservoir with four flow units each equipped with an ICV. The inside diameter of casing, tubing and ICV is given in Table 3.

After designing the integrated production model, two production schemes were studied.

4.3 *Production scenarios*

In the first scenario, the production well was controlled only by surface choke and all the downhole chokes kept fully open. In other words, there was no control over fluid influx from

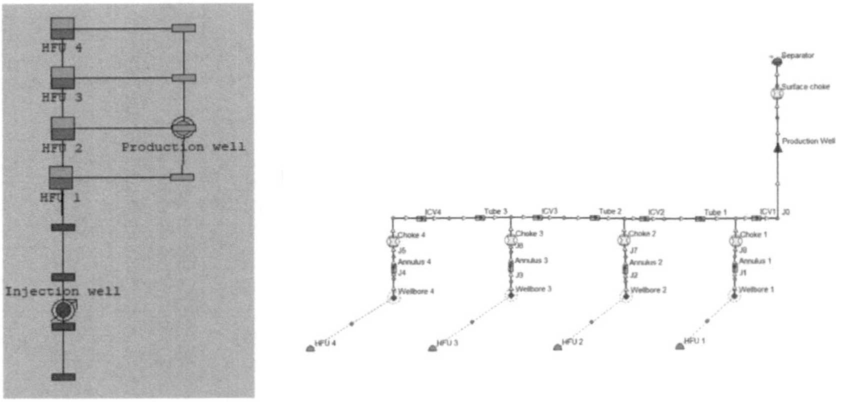

Figure 3. Schematic of the reservoir model with four flow units in mbal (left) and the schematic of the production network including reservoir, ICVs and well in Gap (right).

Table 2. Reservoir fluid properties.

GOR (SCF/STB)	446
Oil API Gravity	24
Gas Density (g/lit)	1.55
Water Salinity (ppm)	75000
Mole Percent of CO_2	2.25
Mole Percent of H_2S	0.26
Mole Percent of N_2	7.53
Saturation Pressure (psi)	~1370
Temperature (F)	260-275

Table 3. Casing, tubing and ICV inside diameter.

	Inside diameter (in)
Casing	6.09
Tubing	3.96
ICV	2.99

each reservoir flow unit in to the well. In the second scenario, downhole chokes were set to be controlled by Gap internal optimiser. The objective was to reduce the water influx as much as possible and to maximize oil offtake from the reservoir. Therefore, the production from different flow units was managed in a manner that increase the oil production and reduce the water influx from the reservoir.

Both scenarios were carried on for 20 years with a wellhead pressure of 200 psi. The results will be discussed in the next section.

4.4 *Results and discussions*

Figure 4 shows the cumulative oil and water production from the reservoir for both production scenarios. The results demonstrate significant improvement in cumulative oil production

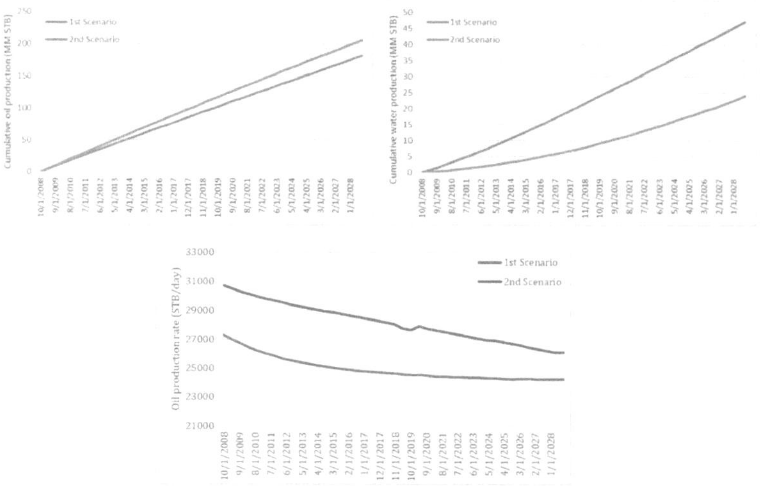

Figure 4. Plot of cumulative oil (left), cumulative water production (right) and oil production rate (bottom) for both production scenarios.

in the second scenario by 11% after 20 years by applying ICV technology. Moreover, it should be noted that, beside an increase in oil recovery, cumulative water production reduced by 46% as a result of using downhole controlled valves. This shows that in this case ICV were more effective in controlling water production rather than increasing oil recovery.

Looking at the results oil production rate and water cut from different layers in the first scenario (Figures 5) show that most of the water production is coming from HFUs 1 and 3 while oil is mainly produced from zones 2 and 4.

Employing downhole controlling devices in different layers reduced liquid production from layers 1 and 3 in the second scenario by reducing the choke size of their associated ICVs (Figure 6). In contrast, the liquid production increased considerably from zones 2 and 4. This resulted in less water production from first and third flow units, whereas oil production from layers with less detrimental water cut during first scenario (layers 2 and 4) increased significantly.

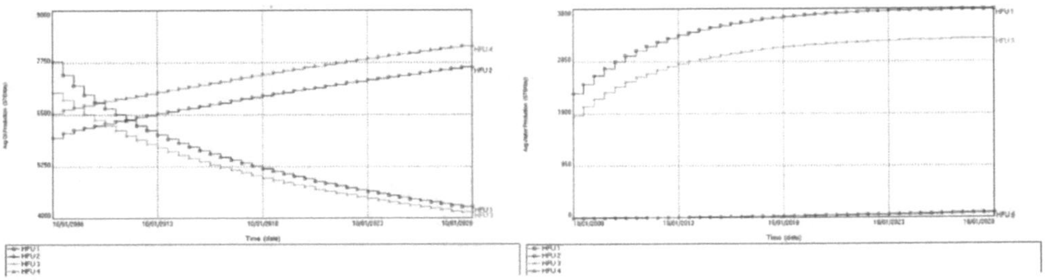

Figure 5. Produced oil rate (left) and water cut (right) from different HFUs in the first production scenario.

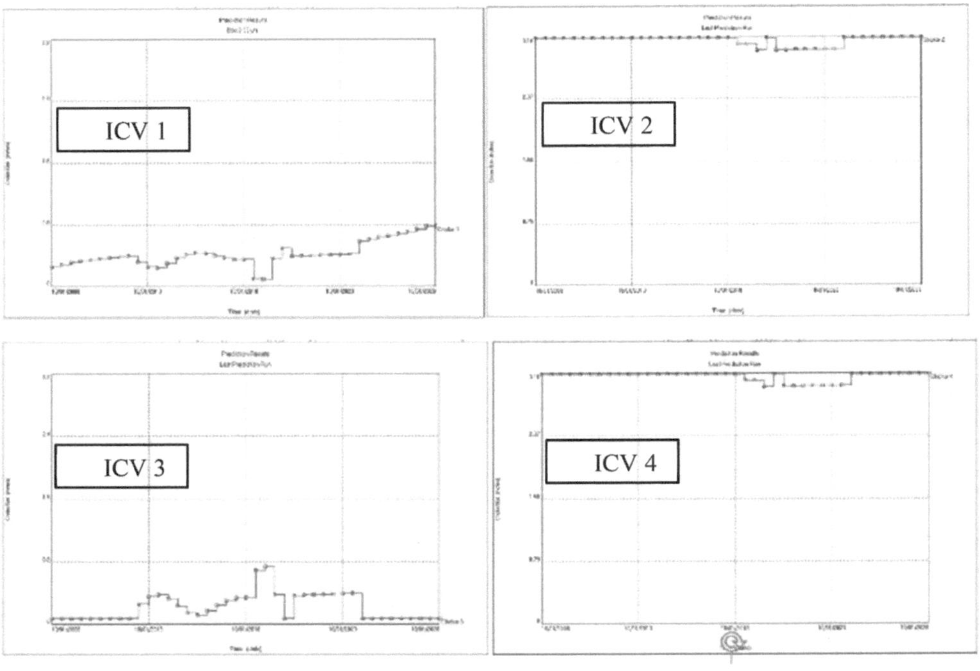

Figure 6. Change in the choke size of each ICV in the second scenario.

3 CONCLUSION

Hydraulic flow unit classification can be used as an effective tool to determine the number and the location of installing downhole control valves.

Application of interval control valves helps to improve waterflood performance not only by increasing the oil recovery, but also by reducing the water production.

It will not be necessary to complete all the wells in a given reservoir or all the reservoir flow units with downhole control valves. It is recommended to complete only the problematic wells or layers with downhole devices for reducing the cost and the risk of mechanical failure.

IPM suite is a powerful package to monitor the interaction of downhole and surface parameters for better well and reservoir management and also to model intelligent well completion.

REFERENCES

Al-Khelaiwi, F. T., Birchenko, V. M., Konopczynski, M. R. & Davies, D. R. 2008. Advanced Wells: A Comprehensive Approach to the Selection between Passive and Active Inflow Control Completions. *International Petroleum Technology Conference*. Kuala Lumpur, Malaysia: International Petroleum Technology Conference.

Al-Khelaiwi, F. T. & Davies, D. R. 2007. Inflow Control Devices: Application and Value Quantification of a Developing Technology. *International Oil Conference and Exhibition in Mexico*. Veracruz, Mexico: Society of Petroleum Engineers.

Al-Khelaiwi, F. T. M. 2013. *A Comprehensive Approach to the Design of Advanced Well Completions* Doctor of Philosophy Heriot-Watt University.

Asl, F. N., Rafiei, Y., Hasanzadeh, N. M. & Abbasi, S. Controlling Water Production by Application of Smart Well Technology. Rotterdam, Netherland, Conference: The 9th Interpore International Conference on Poros Media & Annual Meeting, 2017.

Birchenko, V. 2010. *Analytical Modelling of Wells with Inflow Control Devices*. Doctor of Philosophy, Heriot-Watt University.

Cheperli, A. & Rafiei, Y. 2018. New Analytical Method for Improving Water Allocation Management in Water Flood Projects. *Petroleum Research*, 28, 9.

Davies, D. R. & Ebadi, F. 2006. Techniques for Optimum Placement of Interval Control Valve(s) in an Intelligent Well. *SPE Europec/EAGE Annual Conference and Exhibition*. Vienna, Austria: Society of Petroleum Engineers.

Elmsallati, S. M. 2005. *A Study of the impact of Intelligent Well Technology on Reservoir Development* Doctor of Philosophy Heriot-Watt University.

Rafiei, Y. 2007. *An Extensive Study on Effect of Different Hydraulic Flow Unit Classification on Normalizing of Capillary Pressure Curves And Permeability Estimation of one of Iranian Heavy Oil Reservoir.* Master of Science, Petroleum University of Technology, IFP School.

Rafiei, Y. 2014. *Improved oil production and waterflood performance by water allocation management.* Doctor of Philosophy, Heriot-Watt University.

Rafiei, Y., Nazari, F., Davies, D. & Vahabzadeh, E. A New Analytical Approach for Water Allocation Management by Determining Inter-well Connectivity: North Sea Case Study. 80th EAGE Conference and Exhibition 2018, 2018.

Rafiei, Y., Nazarisafarzadeh, F. & Cheperli, A. Improved Oil Recovery by Real-time Water Allocation Management. The First Interpore International Conference on Improved/Enhanced Oil and Gas Recovery, 2017. International Society for Porous Media.

Rahimbakhsh, A. & Rafiei, Y. Investigating the Effect of Employing Inflow Control Devices on the Injection Well on the SAGD Process Efficiency Using an Integrated Production Modeling. SPE EOR Conference at Oil and Gas West Asia, 2018. Society of Petroleum Engineers.

Stolz, A.-K. & Graves, R. M. 2003. Sensitivity Study of Flow Unit Definition by Use of Reservoir Simulation. *SPE Annual Technical Conference and Exhibition*. Denver, Colorado: Society of Petroleum Engineers.

Tiab, D. & Donaldson, E. C. 2016. *Petrophysics: theory and practice of measuring reservoir rock and fluid transport properties.*

Youth Technical Sessions Proceedings – Litvinenko (ed)
© 2019 World Petroleum Council, London, UK, ISBN 978-0-367-34668-3

Investigation of cement stone self-healing process in oil and gas wells

F.A. Agzamov, A.S. Maskenov & N.N. Shantasov
Ufa State Petroleum Technological University, Ufa, Russian Federation

ABSTRACT: The article discusses the concept of using modern materials for self-healing of cement stone, which will improve the quality of construction of oil and gas wells, as well as provide a safer work from an environmental point of view.

To date, self-healing of cement stone is represented by three conceptual approaches: self-healing system based on capsules (granules), in the shell of the vessel (fiber) and without shells.

This paper presents the results of experimental studies on the effect of adding rubber particles on the self-healing ability of cement stone. The paper discusses the self-healing of cement stone after creating artificial cracks. The effect of various environments on the kinetics of swelling, as well as the effect of temperature on the self-healing process has been studied. The research results prove the fundamental possibility of the existence of self-healing cements and the possibility of the restoration of cement sheath.

1 INTRODUCTION

Despite the rapid pace of development of technologies in the oil and gas industry, the problem of cementing of oil and gas wells remains unresolved. Various complications, such as fluid flows between the layers, leakages in the annulus lead to flooding of the production .

As is known, the most vulnerable element of the well is the cement sheath, which is exposed to various technological loads. These loads lead to the formation of a network of cracks and, as a result, to the loss of tightness. Completion of the reservoir is characterized by perforation of cement stone, which, in turn, reduces its strength. The nature of the decrease in strength is determined by various factors and directly depends on the type of perforation. Another factor affecting the cementing process and the quality of the cement produced is the increase in temperature and pressure with the depth of the well. Since the construction of wells is characterized by the lack of admission of a person into it, inspection and examination of the state of the sheath during the operation of a well is an impossible task. Thus, to solve this problem, it becomes necessary to apply new technologies. The most promising and unexplored is the activity aimed at the development of cement material capable of self-repair. This will not only improve the quality of well construction, but also reduce the costs associated with their repair and restoration.

2 LITERATURE REVIEW

Self-healing cements are being developed in some foreign and domestic companies. The principle of "self-healing" is to restore the integrity of the cement stone using special complex and modifying additives.

The concept of self-healing polymeric materials has been known since the 1980s. In 1981 Donald Jude reflected the possibility of healing invisible microcracks with the help of polymer components in order to prolong the life of the cement stone.

The principle of autonomous "healing" of cracks in cement stone without human access for cementing oil and gas wells was successfully demonstrated by Schlumberger. The "Futur" technology blocks gas migration channels, cement is capable of multiple activation of components, which allows operating the well without the need for repairs during the entire working period.

Nearly nine years of research have been conducted at Deft University (the Netherlands) under the guidance of microbiologist Henk Jonkers. A group of scientists has developed self-healing concrete, the successful application of which was announced on May 14, 2015. The concept of the development was to introduce a special type of bacteria bacilli into building concrete.

Self-healing materials can be divided into three large groups: encapsulated (granules), in the vessel shell (fiber) and without shells. Each group has a different mechanism used to activate "self-healing" agents.

Obviously, the recovery effect cannot appear from nowhere - this is a complex phenomenon that is activated by some external force (for example, water). Activation proceeds in several stages. At the first stage, when the load on the stone rises, a channel (crack) appears in it, at the second stage, water appears, which in our case plays the role of a trigger in the process of "self-healing"; the third stage is the blockage of the crack itself. The absence of water showings is the result of the effective work of the agent in the third stage.

Although the occurrence of cracks in the cement stone is taken into account during the design of the well, it does not guarantee the complete absence of additional repair costs. Some of them can seriously affect the durability and stability of the sheath. Thus, the use of self-healing cement will avoid the direct and indirect costs caused by repair work. In turn, the observations of many scientists prove the fundamental possibility of healing cracks in a cement stone in a natural way, without any specific modifiers, this phenomenon is caused by "autogenous" healing.

3 METHODOLOGY

We have chosen the principle of "self-healing" of damaged cement stone by blocking the channels with swellable particles added to the cement slurry. Modifying additives after the interaction with water are activated and for a certain period of time they block the cracks. At the same time, the most important requirements for the modifying additive are a significant amount of swelling for sealing large cracks and the absence of swelling in the cement mortar in the period from mixing to hardening. In addition, the additive should have the properties of reusable swelling, should not affect the technological properties of the cement slurry and the cement stone. Water swelling rubber was considered as a modifying additive. A schematic representation of the structure of this additive is presented in Figure 1.

It is a highly porous rubber frame, in the channels of which there is a water-swelling polymer. Since water is an activator of this polymer,after penetrating into the channels it causes its

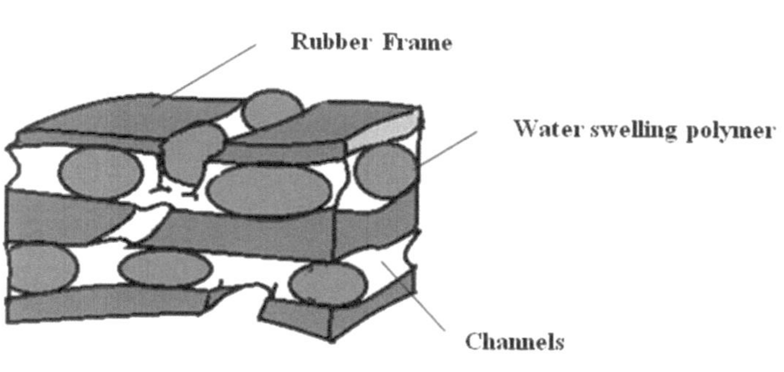

Figure 1. The structure of water-swelling rubber

hydration. Increasing in volume, the polymer also expands the rubber frame and blocks cracks in the cement stone, preventing further inflow of formation water.

4 EXPERIMENTS

At the first stage of the study, the kinetics of swelling was carried out using the Zhigach-Yarov method, which showed that the additives began to swell immediately when dissolved in water, in saline water and in a saturated solution of hydrated lime. Thus, in its neat form, it cannot be used as a modifying additive. To prevent premature swelling (in the process of preparing and pumping cement slurry), it becomes necessary to create a protective waterproof shell that should temporarily suspend the activation of the polymer. As a protective component, several types of water repellents were considered, among which powdered - calcium stearate and liquid water repellent *GF-12* (Figure 2).

The most effective retarder in the process of swelling of rubber crumb is the water-repellent agent *GF-12*. Most likely, this is due to the fact that the liquid water-repellent agent penetrates better into the channels of crumb rubber and forms a stable water-repellent film on its surface. Calcium stearate powder, locally adsorbed on the surface of the additive, only partially blocks the access of water to the water-swellable polymer.

At the next stage of the study, the self-healing ability of the cement stone containing rubber crumb was evaluated. The experiment used cement type G, the experiments were carried out on a specially designed installation. At the same time, artificial channels with a diameter of 0.3 mm were created in the cement stone. During the experiment, samples of cement stone with different concentrations of crumb rubber, treated with water-repellent agent, and a control sample of cement stone with normal density and water-cement ratio equal to 0.5 were tested. The time of passage of water through channels was recorded. The results of the experiments are presented in Figure 3.

According to the obtained results, we can conclude that the artificially created channels in the cement stone with the addition of water-swelling rubber crumb seals over time. The increase in the duration of the passage of water through artificial channels is the result of the effective operation of the reagent.

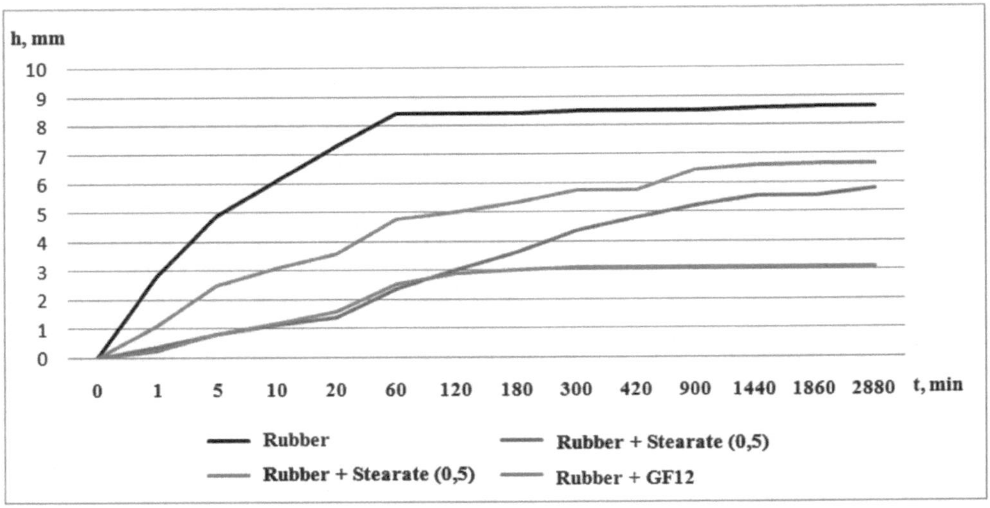

Figure 2. The dependence of the amount of swelling of rubber treated with various types of water-repellent from time

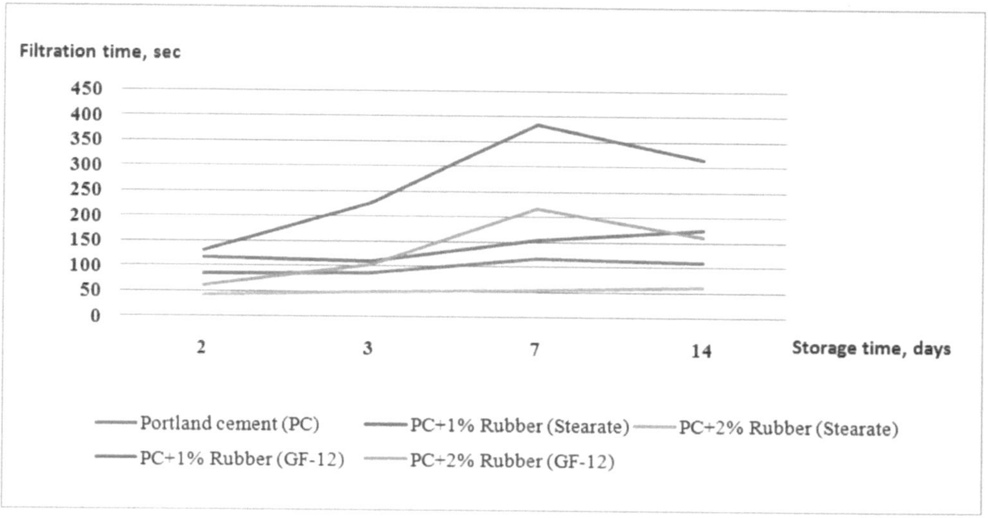

Figure 3. The effect of rubber on self-healing of cement stone

Since under reservoir conditions, the temperature of the environment is higher than the surface temperature, the effect of temperature on the self-healing process was investigated. During the experiments, the samples were divided into two groups: some were kept at normal temperature, others - at a constant elevated temperature of 60 °C (the average temperature in the wells in the fields of Western Siberia). In the samples, artificial channels with a diameter of 0.6 mm in the amount of 6 pieces were created using metal wire.

A Pure Portland cement and cement with untreated rubber were investigated. Each of the above compositions was prepared in duplicate: some were hardened at 20 °C, others at 60 °C. At the initial stage, after the setting period, all 6 channels were included in the work.

The obtained results show that for samples numbered 1 and 2 there is a significant increase in filtering time in the period between 2 and 7 days, which indicates the narrowing of the channels. This is due to the ongoing process of hydration of cement stone and, as a consequence, an increase in the volume of hardening products.

We got a different picture for samples with rubber treated with water-repellent agent: a significant change in the filtration time of the liquid is not observed. The results can be explained by the fact that the water repellent blocks the channels in the rubber and blocks the access of water to the water-swellable polymer, however, after 14 days, the water still begins to penetrate the channels, activating the polymer.

In the course of the experiment, the most unfavorable situation was simulated, when a crack passes not through an additive, but tangentially to it, and, accordingly, the full potential of the modifying additive was not taken into account.

After studying the self-healing ability of cement stone with the addition of crumb rubber and obtaining positive results, a series of experiments were conducted to study the effect of the above additive on the rheological properties of cement stone and the structural and mechanical properties of cement stone.

5 CONCLUSIONS

1. The fundamental possibility of the existence of self-healing cements and the ability to restore lining wells was proved.
2. The choice of additives for the production of self-healing cements is substantiated, formulations of single-layer shell are developed, their advantages and disadvantages are analyzed.

3. Conducted experiments showed that the effect of modifying additives on the rheological properties of the cement slurry is insignificant.
4. When modifying additives are added, a significant decrease in the permeability of cement stone is achieved with an increase in the shelf life of samples.
5. It has been experimentally shown that the influence of modifying additives on the strength characteristics of the cement slurry is insignificant.
6. Despite the positive results of the experiment, there is a need to conduct additional series of experiments, it is also necessary to study in more detail the effect of the autogenous healing of cracks.

REFERENCES

Agzamov, F.A., Izmukhambetov, B.S., Tokunova, E.F. 2011. *Khimiya tamponazhnykh i promyvochnykh rastvorov* [Chemistry of Cement and Drilling slurries]. St.Petersburg: Nedra.

Agzamov, F.A., Izmukhambetov, B.S. 2005. *Prochnost tamponazhnogo kamnya v agressivnykh sredakh* [Durability grouting stone in corrosive environments]. St.Petersburg: Nedra

Bellabarba, M., Bulte-Loye, E., Frelich, B., Le Rya-Delage, S.,Van Keik, R., Ziru S. 2008. Ensuring the effective separation of the seams after the operation of wells. *Oil and gas review*, 20(122): 37-42.

Breugel, K. 2009. Self-healing concepts in civil engineering for sustainable solutions: potential and constraints. Proceedings of the *Second International Conference on Self-Healing Materials*, Berlin, 8-10 October.

Budnikov, V.F., Bulatov, A.I., Makarenko, P.P. 1996. *Problemy mekhaniki bureniya i zakanchivaniya skvazhin* [Problems of drilling and completion]. Moscow: Nedra.

Danyushevsky, B.C. 1978. *Proyektirovaniye optimalnykh sostavov tamponazhnykh tsementov* [Designing the optimal composition of cements]. Moscow: Nedra.

De Belie, N., Schlangen, E. 2013. *Self-healing Phenomena in Cement-based Materials*, London: Springer.

Homma, D., Mihashi, H., Nishiwaki, T. 2009. Self-healing capability of fiber reinforced cementitious composite. *Journal of Advanced Concrete Technology* 7(2): 217-228.

Qian, S. 2009 Self-healing behavior of strain hardening cementitious composites incorporating local waste materials. Cement and Concrete Composites 31(9): 613-621.

Turner, L. 1937. The autogenous healing of cement and concrete: its relation to vibrated concrete and cracked concrete. Proceedings of *International Association for Testing Materials*, London Congress, London, pp. 122-127.

Yang, Y. 2009. Autogenous healing of engineered cementitious composites under wet-dry cycles. *Cement and Concrete Research* 39(5): 382-390.

Youth Technical Sessions Proceedings – Litvinenko (ed)
© 2019 World Petroleum Council, London, UK, ISBN 978-0-367-34668-3

The technology of completion reservoir by drilling a network of branched channels under controlled trajectory

A.A. Liagova
Saint-Petersburg Mining University, Saint-Petersburg, Russian Federation

I.A. Liagov
LLC «Perfobur», Moscow, Russian Federation

ABSTRACT: At present, the share of hard-to-recover reserves in Russia and in the world is growing. Production on the many wells are stopped, due to the lack of flow of formation fluid, or due to the high water content of the production. The unprofitability of oil production at wells with hard-to-recover reserves leads to their conservation or liquidation. In this article, we offer an innovation Perfobur system. It is modern technology of non-damaging perforation for increase of the flow rate of producers, enhancement of the injection capacity of injectors and elimination of cones of water-gas-oil contacts due to expansion of the filtration area. In the article the questions of advantages of drilling of small-diameter channels, capabilities and technical efficient of Perfobur systems are studied.

1 INTRODUCTION

The quantity of hard to recover reserves in Russia and the world is currently growing. By now, well production has been suspended in many wells because of absence of the fluid inflow, or because of a high water-cut of the product produced. Unprofitability of well production in wells with hard to recover reserves results in their suspension and abandonment; the number of such wells is approximately 25-30 thousand. The main problem with old, marginal wells is the absence of proper and reliable connectivity in the "well-formation" system, which is caused by external impact on the bottom-hole area during well construction and operation: clogging of the reservoir by filtrate of drilling mud, slurry or killing fluid.

2 LITERATURE REVIEW

Let us make a review of current widely used methods of well stimulation.

Perforation (cumulative, slotted, drilling) can make channels with the length of up to 2 meters but does not always go through wall packing areas. The actual recovery of a formation, which was drilled-in this way, is much less than the target estimated on the basis of simulation results and this is a typical situation (Basarugin 2000; Shamov & Liagov 2012).

Chemical methods of enhanced oil recovery have highly effective, but only in homogenous permeability formations. If there is even one layer with higher permeability, the chemical compound starts moving along it as the lowest resistance path, which significantly reduces the general contact area between the acid and the formation (Rogachev & Myxametshin 2018).

Jet radial drilling-in is a modern technology, developed and implemented by non-Russian companies. The method enables to construct long radial channels. This technique has some crucial limitations: the channel cross-section is of irregular shape and re-entry into the channel is impossible; the channel trajectory cannot be measured and controlled; the method works only in carbonate reservoirs; this technique requires perfect quality of cementing in the kick-off point (*Technology of jetting radial drilling* n.d.).

Drilling with Fishbones titanium needles is innovation of Norwegian experts. Prior to running in hole, it is necessary to remove a multi-meter section of casing, which takes a lot of time and money. Works only in carbonate reservoirs. Requires a perfect alignment in the well. Re-entry into the drilled channel is impossible (*Technology of Fishbones titanium needles drilling* n.d.)

Frac is characterized by high costs and serious limitations in application (cannot be used in thin layers and in case of close occurrence of water-bearing horizons), requires high quality of cement bond and complete absence of behind-casing leaks. Crack propagation trajectory is estimated approximately. That is why, even though this method is very effective, later a high growth of the well water-cut is observed and remedial cementing is necessary.

Side-tracking is nowadays a widely used method of recovery of old wells. Applied only in thick sections of the pay zone. Side-tracking is possible in intervals of close occurrence of water-bearing strata but this results in a significant increase of the cost. The long horizontal hole does not guarantee a high growth of the production rate (Liagov & Chistov 1995).

3 ADVANTAGES OF DRILLING OF SMALL-DIAMETER CHANNELS

Pay zone stimulation is one of the most important and complicated stages of well completion, whose crucial task is to establish perfect connectivity between well and its pay zone without negative effect on reservoir properties of the pay zone and destruction of the cement plug.

Nowadays cumulative perforation, which does not meet the above-described requirements, is widely used. There are non-damaging methods such as slot and abrasive jet perforators that do not cause catastrophic destruction of the cement plug but at the same time are not able to make deep channels for establishment of the perfect connectivity between the formation and the well. One of the modern ways of non-damaging well stimulation is drilling of channels with a small diameter and a radius of curvature along the projected trajectory. It has become possible to construct such highly deviated channels due to development of the Perfobur system.

Application of modular-type Perfobur enables to drill in without preparatory operations, such as removal of a casing part, placing a cement plug and drilling out an auxiliary channel in it, which also contributes to enhancement of the economic efficiency from usage of this method, and the system of removable whipstocks reduces time for multilateral drilling (Lyagov et al. 2014)

Drilling of small-diameter channels has a number of geological, technical and economic advantages. Geological advantages are thoroughly justified in the studies of N.F. Kagarmanov and M.U. Muratov (1959, 1965). Dependence of economic and technical indicators of the well construction of various diameters was for the first time revealed in the 1950s by V.S. Fedorov (1947) who drew attention to acceleration and cost-cutting of well drilling due to reduction of applied power, which, according to V.S. Fedorov, was proportional to the bit diameter. V.I. Tarasevich, A.F. Sukhanov, A.I. Medvedko, F.M. Guseinov and other experts agree with rate of penetration increase with reduction of the bit diameter. In addition, better conditions for bottom cleaning from drilled cuttings are created because of increase of the upward flow rate in reduced clearances between the hole walls and the casing with a lower mud pump rate.

Perfobur is designed in such a way as to eliminate the necessity of preparatory operations (removal of a casing section and placing of a cement plug). Usage of two different assemblies – for commencement of channel drilling and for drilling itself – will enable to expand the effective filtration area, perform a directional frac and apply the technology as an alternative to side-tracking. Combination of positive effects from Perfobur utilization will put many shut-in wells again into operation. The proposed method allows multiple re-entry into the drilled channel for the trajectory control, channel repair and stimulation of the formation fluid flow.

However, consumers are now not inclined to wide application of mechanical methods because of their high cost in comparison with cumulative perforation. But one does not take into account that mechanical well stimulation does not require additional expenses for permeability build-up in the bottom-hole area of a pay zone by chemical/physical methods or frac as part of well completion, which may be much more expensive than the cost of the perforation process itself (*Technology drilling network of extensive channels by system Perfobur* n.d.).

4 SYSTEM CAPABILITIES

The Perfobur technical system is intended for quality improvement of the connectivity between the pay zone and the drilled well at the final stage of well construction or workover by deep perforation of the producing zone of oil and gas wells. The system enables to increase the flow rate of producers and the injection capacity of injectors. We systematise key parameters of Perfobur technical system in Table 1.

Table 1. System Parameters.

Parameter	Meaning	Unit
Diameter of perforation channels*	58-68	mm
Channel deviation from the main hole	5-10	m
Length of perforation channel	up to 50	m
Number of channels on one layer	up to 4	ea.
Curvature radius of channel**	3.5-13	m
Circulation rate	0.5-3.5	l/sec.
Motor shaft torque	200-400	Nm
Pressure differential in Perfobur BHA	4-8	MPa
Axial load on bit	2-6	kN

* Channel length being 14 meters
** Build-up rate at a radius of 10.5m makes 5.5 deg./m

Formation depth – up to 3000 meters, casing diameter – 146mm, wall thickness (steel of any grade) - 7-9mm.

The target is wells with a potential ability of flow rate (injection rate) increase by making perforation channels with a sufficient deviation from the main hole, including:

- wells with low-permeability reservoirs for drainage improvement;
- wells with a bottom-hole area with a low skin-factor as a result of wall packing;
- wells, in which applied stimulation methods (for example, acid treatment) do not achieve the planned efficiency;
- wells with pay zones of a small thickness (if the formation section is very much wall-packed);
- wells with close occurrence of water-bearing and producing horizons.

Perfobur enables to (Figure 1):

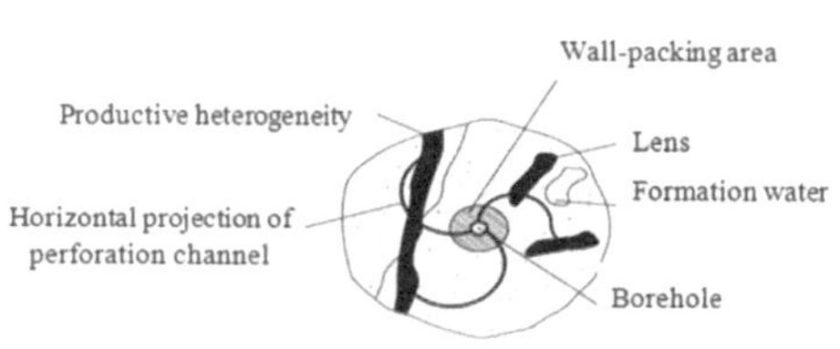

Figure 1. The layout of channels in well.

- penetrate into thin formations with weak boundary lines;
- go through deposits of paraffin films and asphaltens in the bottom-hole area that reduce the pressure differential in the "hole-formation" system.

5 TECHNICAL EFFICIENCY OF THE PERFOBUR SYSTEM

Deep perforation of wells helps to increase the oil recovery ratio by several times, which is confirmed by calculations and supported by the reservoir simulation model.

5.1 *Method of efficiency calculation*

Simulation using the below method has been carried out in order to assess the effect from application of the Perfobur radial drilling system and compare it with other ways of oil recovery enhancement.

To study the method efficiency, three wells with typical characteristics common for Urals-Volga region were chosen. The following assumptions were admitted for the calculation:

1. The well flow rate was calculated as a sum of flow rates from the wellbore and one drilled side channel.
2. The wellbore is vertical, the drilled channel is horizontal – at an angle of 90 degrees to the borehole. For calculation of the flow rate of a vertical wellbore - Dupuis formula was used, for that of horizontal borehole – Joshi-Economides formula.
3. The following incoming parameters were taken: rock permeability – 35mD, porosity – 16%, oil weight – 0.846 g/cm^3, radial channel diameter – 5.8 cm, oil viscosity – 0.6 cPs, well flow rate without drilled channels – from 7 to 58 m^3/day, drainage radius – 250m.

To determine existence of Darcy's linear filtration law during inflow of formation fluid to drilled perforation channels of a small diameter, the following Schelkachev's Reynolds number (Equation 1) was used:

$$Re = \frac{10W\sqrt{k}\rho}{m^{2.3}\mu} \qquad (1)$$

where Re – Reynolds number; W – linear filtration velocity of fluid inflow to the borehole; m/sec; k – average rock permeability, m^2; ρ – fluid weight, kg/m^3; μ – fluid viscosity, cPs.

Critical value of Schelkachev's Reynolds number is within the range from 0.032 to 14. The calculated value of Reynolds' number with account of provided incoming data makes 0.00015, which means conservation of Darcy's filtration law for drilled 58-mm channels. The diagram (Figure 2) shows dependence of the increase factor of flow rate from the channel

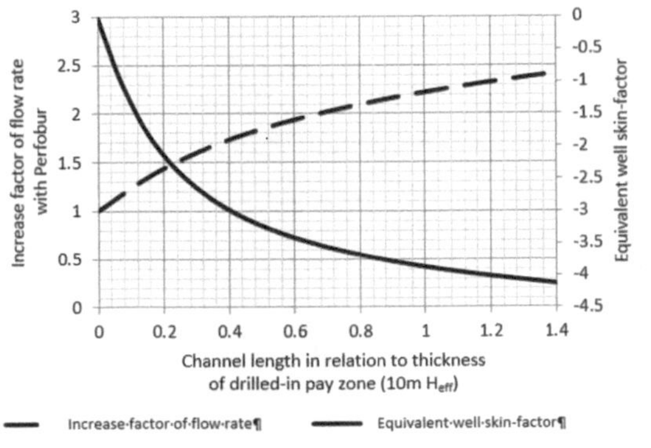

Figure 2. Dependence of the increase factor of flow rate from the channel length and the corresponding curve of equivalent skin-factor.

length and the corresponding curve of equivalent skin-factor. One can see that the well delivery growth rate goes down with the channel length increase. These calculation results should be considered as an assessment of the maximum growth of well delivery after drilling of one channel (Liagov 2014; Liagov et al. 2015).

5.2 *Comparative evaluation of the filtration area of drilled channels*

Below one can see a comparative Table 2 of the most popular perforation methods.

Channels drilled with Perfobur in the effective filtration area are by 1.6 times bigger than those resulting from radial drilling, by 48 times bigger than those drilled by slot perforators, by 97 times bigger than those made by cumulative perforators. Though the Perfobur parameters are close to those of jet radial drilling (RDS), the Perfobur system has an undeniable advantage – controlled drilling trajectory, which will help to cover those particular formation zones that contain the maximum quantity of hydrocarbons and avoid dry layers and impermeable rock.

5.3 *Assessment of inflow to drilled channels as to additional wellbores of smaller diameter*

Initial parameters: curvature radius of well – 12m; cylindrical shape of channels; length of drilled channel – 12-14m; diameter of channel cross-section – 0.06-0.058m; number of channels in one section - 4. All calculations do not take into account contamination of the pay zone and resistance inside channels. The borehole is open.

Fluid inflow to drilled channels is calculated independently of each other and of the wellbore. Horizontal orientation of the formed solution channels is not considered. Fluid inflow to the wellbore is calculated using Dupuis formula. The total well flow rate is calculated as a sum of flow rates from the wellbore and each of the channels. Inflow to horizontal side channels is calculated as per Joshi formula.

According to Joshi-Economides formula, unitless productivity index of a horizontal well is calculated as follows:

$$J_{hor} = \left(\ln \frac{a + \sqrt{a^2 - (L/2)^2}}{L/2} + \frac{\beta h}{L} \left[\ln \frac{\beta h}{r_w(\beta + 1)} + S \right] \right)^{-1} \tag{2}$$

$$a = \frac{L}{2} \left(0,5 + \sqrt{0,25 + \left(\frac{2r_e}{L} \right)^4} \right)^{0,5}, \beta = \sqrt{\frac{k_h}{k_v}} \tag{3}$$

Table 2. The most popular perforation methods.

Parameter	Unit	Perforation methods Perfobur	Radial drilling	Slotted perforation	Cumulative perforation
Pay zone thickness	m	4	4	4	4
Number of channels	ea	4	4	4	20
Length of one channel	m	50	50	2	2
Channel diameter	m	0.058	0.045	0.031	0.015
Filtration area of one channel	m^2	9.11	5.7	0.19	0.094
Ability of trajectory control		Yes	No	Yes (partially)	Yes (partially)
Ability of trajectory check		Yes	No	Yes (indirectly)	Yes (indirectly)
Cement plug integrity		Yes	Yes	Yes	No
Possibility of channel re-entry		Yes	No	No	No
Ability of operation in terrigenous reservoirs		Yes	No	Yes	Yes

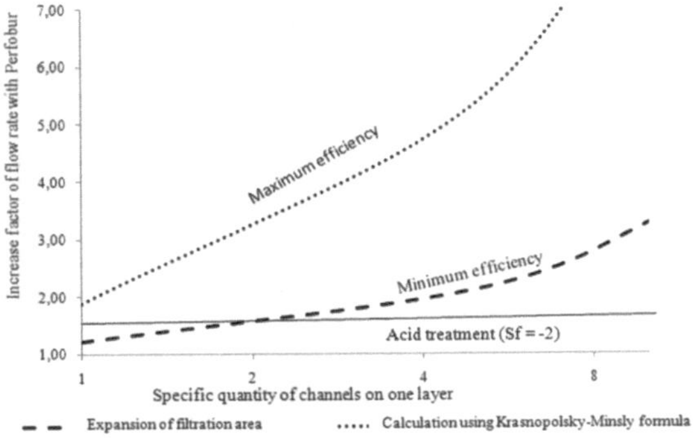

Figure 3. Dependence of the increase factor of well delivery versus the specific quantity of channels.

where L – length of horizontal wellbore of a radial channel drilled with Perfobur, m; r_w - is radial distance, m; k_h - horizontal permeability; k_v – vertical permeability; r_e – is the equivalent radius in a presumed circular shape of a given drainage area.

This calculation should be considered as the upper estimate of the technology efficiency for an ideal well.

Results of the calculations by two methods showed dependence of the increase factor of well delivery versus the specific quantity of channels as follows (Figure 3).

According to the study results, more than double increase of well delivery is observed in case of drilling of two channels with 10-meter deviation from the well axis. The efficiency of Perfobur application under the same conditions is higher than that of deep cumulative perforation with subsequent acid treatment. Drilling of four radial channels with 10-meter deviation gives the increase factor of well delivery equal to approximately 3.

6 DYNAMICS OF WELL OPERATION INDICATORS BASED ON SIMULATION RESULTS

Three wells were chosen on a field with carbonate reservoirs in Urals-Volga region for using Perfobur. Drilling of 4 holes with a deviation of 14-meter length was planned for all wells. The following incoming parameters were taken: rock permeability – 35mD, porosity – 8%, oil weight – 0.843 g/cm^3, diameter of radial channel – 58mm, oil viscosity – 0.6 cPs, well flow rate without drilled channels – from 7 to 58 m^3/day, drainage radius – 250m.

For calculation of a flow rate of the vertical wellbore - Dupuis formula was used, for that of horizontal borehole – Joshi-Economides formula (2).

Increase of the productivity index (PI) is observed in all wells after the operations. The planned well delivery has significantly changed after Perfo-drilling. This means that the bottom-hole area was subjected to lost circulation. High efficiency of the technology is related

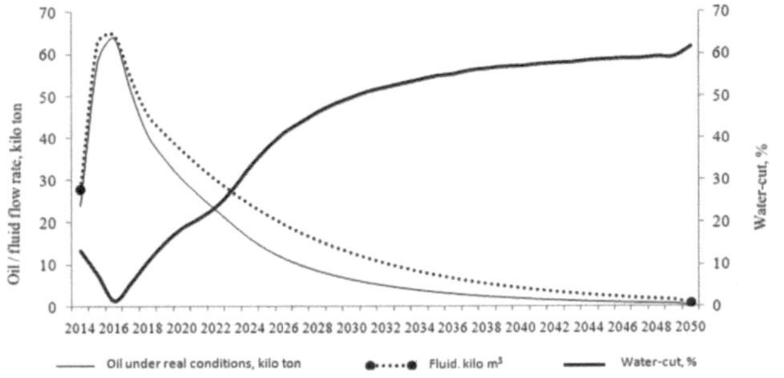

Figure 4. Predicted rates of flow rates respect to time.

to sufficient geological exploration (Figure 4). A considerable increase of water-cut is observed in all wells, which is also caused by changes in the pay zone (Liagov 2014; Liagov et al. 2015).

7 CONCLUSION

Economic efficiency of the Perfobur system application is achieved due to the following:

- increase of the production rate of producers and of the injection capacity of injectors thanks to expansion of the effective filtration zone and penetration into high-permeability layers of pay zones that are characterized by high heterogeneity in layers and zones;
- decrease of the oil water-cut due to reduction of pressure drawdown during production and, consequently, approximation of salt water into the near-wellbore area of the pay zone;
- reduction of the fluid injection pressure in the formation pressure maintenance system of injectors due to expansion of the filtration area;
- prolongation of the mean time between well servicing and workover due to reduction of rate and time of deterioration of reservoir properties of the pay zone, where a big filtration zone will be formed by application of the proposed technology.

The next step in the product development will be transition from the self-contained directional survey system to the online data transition system (using an electromagnetic communication channel). Ability of application of online logging (gamma-resistivity) is planned, which will enable the system operation without additional geophysical activities in well.

REFERENCES

Basarugin, Y.M. 2000. *Drilling Completion.* Moscow: Nedra.

Fedorov, V.S. 1947. *Technical and economic advantages of small diameter drilling. [Technical and economic advantage of drilling small diameter].* Groznyy: GrozNII.

Kagarmanov, N.F. & Muratov, M.Y. 1959. *Drilling of small diameter wells in Bashkiria [Burenie skvazhin malogo diametra vBashkirii].* Ufa: UPI.

Liagov, V.A. & Chistov, D.I. & Shaidakov, V.V. & Ishemgugin, I.E. 1995. Engineering workover of the well by drilling the sidetracks from the production string. [Inzhiniring kapital'nogo remonta skvazhin bureniyem bokovykh stvolov iz ekspluatatsionnoy kolonny]. *Oilfield* 4: 16-17.

Liagov, I.A. & Liagov, A.V. & Suleimanov, I.N. & Kachemaeva, M.A. 2015. Manufacturing of the technical system «Perfobur» and research of it's work in channel with small radius of curvature *Oil and Gas Business* 5: 45-105 viewed 09.03.2019 http://ogbus.ru/files/ogbus/issues/5_2015/ogbus_5_2015_p45-105_LyagovIA_ru.pdf

Liagov, I. A. 2014. *Substantiation and development of the technology of the secondary opening of productive layers by ultra-small-well-branched wells. [Obosnovanie i razrabotka tekhnologii vtorichnogo vskrytiia produktivnykh plastov raz-vetvlennymi skvazhinami sverkhmalogo diametra].* Saint-Petersburg: Mining University.

Lyagov, I.A. &Vasilev, N.I. & Reich, M. & Mezzetti, M. 2014. Analytical research and experimental tests on the technology fov drilling small diameter channels with small radius of curvature. *Oil Gas European Magazine* 40(3): 124-129.

Rogachev, M.K. & Myxametshin, V.V. 2018. Control and regulation of the hydrochloric acid treatment of the bottomhole zone based on field-geological. *Journal of Mining University* 231: 275-280, viewed 09.03.2019, http://pmi.spmi.ru/index.php/pmi/article/view/7797/8427.

Shamov, N.A. & Liagov, A.V. & Panteleev, D.V. & Vasilev, A.V. & Liagova, M.A. & Liagov, I.A. & Nazarov, S.V. & Aseev, E.G. 2012. Equipment and technology creation of ultra-deep perforations. *Oil and Gas Business* 2: 131-174.

Technology drilling network of extensive channels by system Perfobur viewed 09.03.2019:http://www.perfobur.com.

Technology of jetting "Radial drilling" viewed 09.03.2019, https://www.radialdrilling.com/.

Technology of Fishbones titanium needles drilling viewed 09.03.2019 http://fishbones.as/.

Zamytina, A.F. & Kagarmanov, N.F. 1965. Theoretical background of small-diameter drilling [Teoreticheskie predposylki bureniya skvazhin malogo diametra]. *Collection of studies UfaNII*, Ufa: Bashkirskoe knizhnoe izdatel'stvo.

Youth Technical Sessions Proceedings – Litvinenko (ed)
© 2019 World Petroleum Council, London, UK, ISBN 978-0-367-34668-3

Increasing the durability and corrosion protection of cement stone in aggressive media

R.R. Sabirzianov, F.A. Agzamov & E.F. Tokunova
Ufa state petroleum technological university, Ufa, Russian Federation

ABSTRACT: The cement slurries applied when fastening wells must provide leakproofness of a timbering and protect an upsetting column from aggressive formation fluids. At the same time for cementation, slurry based on cement and water are used, at water-cement ratio 0.45-0.9 that predetermines their high porosity and low corrosion resistance. The technologies of increase in a durability of concrete applied in structural practice when fastening wells aren‘t acceptable. One of the most aggressive components of formation fluids is a hydrogen sulfide. We are offered for decrease in porosity of a cement stone use of liquid polysulfide sulfur (further PSS) which can be besieged in a stone time when concreting a cement slurry, colmatate the pore space, reduce the diffusion rates of aggressive ions in a cement stone. Besides, PSS being adsorbed in the pores on cement concreting products, will provide pore space inhibition.

1 INTRODUCTION

Aggressive influence on the cement stone is having an impact by all acids acid and numerous salts. Formation waters of many fields contain various dissolved salts are extremely aggressive with respect to the cement stone. The salinity of formation water depends on many factors and varies widely.

Analysis of various types of corrosion shows that the kinetics of many of them is determined by the diffusion of aggressive ions into the stone. It belongs to corrosion under the action of acidic environments, sulphate, magnesia corrosion. During leaching corrosion, the kinetics of diffusion of calcium ions from the cement stone also has diffuse control. From the practice of industrial and civil construction, it is known to use dense concretes with a low water-cement ratio (W/C) to increase the durability of structures or impregnate products with special compositions (Volgushev 1976).

In the construction of wells, these technologies are unacceptable, since cement and water-based solutions are used for cementing, with W/C=0.45-0.9, sometimes containing modifying additives. The decrease in the water-cement ratio of cement slurries inevitably leads to an increase in the density of the cement slurry, which can often lead to drilling problems or an accident during cementing. In addition, a decrease in the W/C of the slurry leads to a loss fluidity of cement paste, and the use of plasticizers to maintain fluidity may cause sedimentation segregation of the slurry (Agzamov 2011).

The use of impregnating materials to improve the durability of cement stone in downhole conditions is impossible, because the cement-water slurry used in cementing wells is pumped for 1-4 hours inside the casing and in the annulus with small diameters of pipes and small (15-20 mm) annular space at depths a few kilometers. For wells cementing, consisting of a casing and a top cement of small thickness, the weakest link is cement stone, which works in constant contact with aggressive formation fluids and is vulnerable to their effects.

2 LITERATURE REVIEW

One of the most aggressive components of reservoir fluids is hydrogen sulfide, the amount of which in reservoir gas can reach up to 25% (Izmuhambetov 2005). In oil or water reservoirs, depending on thermodynamic conditions, hydrogen sulfide can be either dissolved or gaseous. Depending on the state of aggregation of hydrogen sulfide, the mechanism of cement stone damage may also change.

When exposed to gaseous hydrogen sulfide, the destruction process, as a rule, is accompanied by a large increase in the volume of the cement stone, caused by the development of internal stresses due to the accumulation of slightly soluble products. The necessary conditions for the destruction of the stone in the gas hydrogen sulfide aggression is a high porosity of the stone, ensuring the penetration of the aggressor deep into the stone, and a high pH of the products of hardening, penetrating the formation of slightly soluble calcium sulfide. The accumulation of the latter leads to the development of tensile stresses under restraint in the cement stone. A secondary factor in the appearance of tensile stresses may be the formation of ettringite during the transition of sulphide sulfur to sulphate sulfur. However, for the implementation of this transition requires oxygen, which in the well conditions is missing. The presence of iron oxide in the composition of cement and hardening products enhance the processes of stone destruction in gaseous hydrogen sulfide, as it reacts with iron in any form. The most vulnerable to gaseous hydrogen sulfide is a stone from Portland cement, whereas, stone samples with pH≤11, obtained from drossy binders at high temperatures, showed high resistance. Conducted field tests confirmed these ideas (Danjushevskij 1977).

Under the effect of dissolved hydrogen sulfide on the cement stone, corrosion of the first type is observed (according to V.M. Moskvin), and stone samples from all types of binders, with the exception of acid-resistant cement, are destroyed, and it is only about the speed of the process (Mavljutov 1987). At the same time, the process of damage is occur layer by layer inherent in acid corrosion.

The kinetics of the corrosion process is determined by the concentration of the corrosive medium at the boundary with the stone, the cement composition and density of the layer of corrosion products, the reaction capacity of the cement stone, and can be described by various kinetic curves. Rakhimbaev Sh.M. when describing the processes of corrosion of concrete identifies 4 types of kinetic curves (Rakhimbaev 1996).

For the cement stone this particular corrosion process is specific, and therefore the most effective way to decrease corrosion is to control the structure of the pore space, which has become the main task of research.

It is known that the porosity of cement stone is determined by the water-cement ratio, the surface specific of the cement, the degree of its hydration (Nevill 1972). Unfortunately, these techniques, widely used in construction, are almost impossible to use when cementing casing strings in deep wells, although laboratory work has been done in this direction. In particular, it is possible to note the work (Kuncevich 1992), which describes the experience of using the method of chemical condensation of microdisperse particles in stone pores, achieved by adding salts of polyvalent metals together with organic reagents, stabilizers and growth inhibitors of the sizes of the formed particles.

A similar effect is also observed with the use of nanosilica additives (Urkhanova 2014). A positive result of the synthesis in the slurry, and then in the cement stone, of the organomineral microdispersed phase in the form of a gel filling the pore space is shown in using the example of an organic plasticizer of the «RDN-U» type, an alloy of salts with polymeric stabilizers (Dorovskih 2009).

A decrease in the permeability of cement stone can also be achieved by a micro-silica additive that forms a microgel in the pores of the cement stone due to the exchange reaction of silica with calcium hydroxide (Babkov 2002).

3 METHODOLOGY

To reduce the porosity of the cement stone, we have proposed the use of liquid polysulfide sulfur (further referred to as PSS), the solutions of which are successfully used in construction as impregnation and acting as a protective membrane. Polysulfide sulfur (liquid sulfur) in the processing of concrete products reduces permeability, increases the strength of concrete and freezing resistance (Massalimov 2015).

The idea of using PSS in cement slurries was that PSS can be deposited in the pores of the stone during the hardening of the cement slurry. This process will result in clogging of the pores, reducing their size, decreasing the rate of diffusion of aggressive ions into the cement stone, which should significantly reduce the corrosion rate.

Another aspect of the application of the PSS was that, by adsorbing in the pores on the hardening products of the cement, it will prevent their contact with aggressive ions, ensuring inhibition of the pore space.

In this case, three ways of input liquid PSS into the cement slurry were tested: into the grouting fluid (water), directly into the prepared slurry and the dry method, through cement flour, which was subsequently subjected to disintegrating treatment. The results of the experiments showed that the last method turned out to be the most effective, although more laborious.

4 EXPERIMENTS

4.1 *The effect of PSS on the rheological properties of cement slurry and properties of cement stone*

At this stage of the research, the PSS was introduced directly into the cement slurry prepared from the grouting Portland cement «PCT-I-50», with a water-cement ratio of 0.45. The concentration of PSS varied from 0 to 5% by weight of cement. To determine the rheological properties of cement slurries (plastic viscosity η and dynamic shear stress τ_o), an atmospheric «OFITEModel 900» viscometer was used (Konesev 2017). The cement spreadability of the cement slurry (2R), measured by the «AzNII cone» (see Table 1).

The results of the experiments showed that the PSS does not affect the rheological properties of cement slurries.

The next stage of the experiments was to assess the effect of the PSS on the consistance (pumpability) of the cement slurry, determined on the «OFITE Model 60» atmospheric consistometer and measured in Berden units (Vs) (Tokunova 2011). The experiment ended when the consistency of slurry reached 30 Vs. At the same time, the effectiveness of various methods of introducing PSS into cement slurry was checked (see Table 2).

Increasing the concentration of PSS in the cement slurry increases the time of pumpability. When the additive input into the grouting fluid or in the cement slurry, the permissible time for pumping the solution is almost the same, but when introducing the reagent through the disintegrator, this value is already 1-1.5 hours less. We believe that the processes occurring with cement during its disintegrating treatment play an important role in accelerating the structure formation of cement stone (Hafizov 2017).

The effect of PSS on the strength characteristics of cement stone are given in table 3.

Table 1. The effect of PSS on the rheological properties of cement slurry

Addittive PSS, %	η, mPa·s	τ_0,dPa	CHC_{10sec}, dPa	CHC_{10min}, dPa	2R, mm
0	68.5	10.0	8	10↓	220
1-	63.5-	10.8	9-	12	220
3	63.1	10.5	9	14 ↓	240
5	62.6	10.6	8	15↑	240

Table 2. The effect of liquid sulfur on the pumpability of cement slurry

Additive PSS, %	Pumpabillity of cement slurry, hours		
	in grouting fluid (water)	in cement slurry	through disintegrating treatment
0	4	4	3
1	5	5	4
3	5.9	6	4.5
5	6.4	6.5	5

Table 3. The effect of PSS on the strength of cement stone

Additive PSS, %	Flexural strength, MPa	Compressive strength, MPa
0	1.16	20.9
1-	1.16	21.2
3	1.29	24.5
5	2.32	25.6

The test results showed that the introduction of the PSS cement slurry into the composition increases the strength of the obtained stone. Moreover, with increasing concentration of the PSS in the slurry, the compressive and flexural strength becomes significantly higher. For example, in a sample with 5% PSS, the flexural strength increased by 2 times compared with the original sample, and the compressive strength increased by 1.23 times. The reason for increasing the strength of cement stone, perhaps, is the clogging of its pores.

Proofs of this idea were the results of the evaluation of the permeability of cement stone. For the experiment, 4 samples of cement stone were obtained, obtained from Portland cement «PCT-I-50» at W/C=0.4. The input method of PSS is through the disintegrator, the age of the samples is three days. As a result of the experiments, it was found that input of the PSS into the cement slurry, the permeability of the cement stone formed is sharply reduced (more than 2 times for the sample with 5% PSS) as compared with the control sample. This means that the additive changes the structure of the pore space of the cement stone.

Since most types of corrosion of cement stone under the action of aggressive fluids are controlled by diffusion of the aggressor deep into the stone, the change in the structure of the last one should significantly affect the rate of corrosion processes.

4.2 *Effect of PSS on acid corrosion of cement stone*

In the experiments, the corrosive medium was represented by a one-molar solution of hydrochloric acid (HCl) with a ratio of the volumes of the acid solution to the volume of cement stone samples 10:1. The water-cement ratio of all samples is 0.4, the cement grade of cement is «PCT-I-50», the age of the samples prior to installation in an aggressive environment is three days. Replacement of the corrosive medium was carried out in three days with the duration of the experiment 35 days. In addition, every 7 days, samples were taken from the corrosive medium, which were subjected to research.

It was also assumed that as the temperature of sample formation increases, the PSS can melt and, spreading, will cover a larger surface of the pores, increasing the degree of their inhibition. To this end, several samples with the PSS concentration of 3% and 5% were boiled in water for two hours. This experiment partially modeled the working conditions of cement stone in wells at bottomhole temperatures close to 100 °C, corresponding to depths of more than 4000 m.

After the end of the experiment, samples were removed from the corrosive medium and analyzed. The following results were obtained.

1. Samples of additive-free cement during the test completely collapsed.
2. Samples of cement with the addition of 1% PSS during the test were almost completely destroyed.
3. Samples of cement with the addition of 3% PSS during the experiment were partially destroyed. On one sample in the corner a "corroded" layer broke off.
4. Samples of cement with the addition of 5% PSS during the test did not practically corrode, with the exception of the breakaway angle.
5. Samples of cement with the addition of 3% PSS, subjected to two-hour boiling, after 35 days of testing did not collapse and practically did not undergo corrosion. Cement cracking and segregation of cement stone was not observed.
6. Samples of cement with the addition of 5% PSS, subjected to two-hour boiling, after 35 days of testing did not collapse and did not undergo corrosion.

Prior to placing the samples in an aggressive environment, the permeability of the cement stone was measured, and during the experiment the corrosion depth of the test samples was determined (Table 5).

The experiments carried out confirmed the positive effect of PSS on the acid corrosion kinetics of cement stone. The hypothesis of inhibition of the surface of the pores with molten sulfur was also confirmed.

4.3 *Effect of PSS on hydrogen sulfide corrosion of cement stone*

The final stage of the experiments was the study of corrosion of cement stone in the conditions of hydrogen sulfide aggression. Hydrogen sulfide dissolved in water was obtained by dissolving sodium sulfide in a special sealed container, into which the tested samples of cement stone were dipped. In the experiment, seven cement slurries were used: without additives; with PSS at a concentration of 1%, 3%, 5% input method - in the cement slurry; with PSS in the same concentrations, but the input method is through a disintegrator. Portland cement grade «PCT-I-50», water-cement ratio 0.4, sample size 4x4x4 cm.

After one and a half months, the samples were removed from the corrosive environment, sawed, and the depth of corrosion was evaluated.

In table 6 summarizes the results of the study the hydrogen sulfide corrosion of cement stone samples.

Analysis of the data obtained during the experiment, confirmed that the PSS has anti-corrosion properties to aggressive media. With an increase in the concentration of liquid sulfur in the composition, the penetration depth of aggressive components decreases, therefore, the corrosion process significantly slows down.

Another proof that the PSS, being distributed in the pore space, blocks their surface, limiting the contact of corrosive ions with hardening products. Micrographs clearly show that the

Table 5. Summary of Acid Corrosion Research

Additive PSS, %	Permeabillity, mDa	Corrosion depth, mm			
		7	14	21	35
0	0.34	5	8	30	-
1	0.20	5	10	18	30
3	0.18	3	5	6	10
5	0.14	5	8	30	-
3+boiling	0.14	5	8	30	-
5+boiling	0.12	5	8	30	-

Table 6. Generalized data on the study of hydrogen sulfide corrosion

Additive PSS, %	Input method of PSS	Corrosion depth, mm	Volume of corroded cement stone, mm^3
0	-	6	324
1	in cement slurry	5	270
3	in cement slurry	4	216
5	in cement slurry	4.5	243
1	through disintegrator	3	162
3	through disintegrator	2	108
5	through disintegrator	0,3	16,2

PSS fills the pore space of the cement stone during its hardening and does not allow the components of the aggressive medium to penetrate into the stone and destroy it.

5 CONCLUSION

1. The use of polysulfide sulfur in cement slurries is an effective way to increase the durability of cement stone when exposed to aggressive fluids, due to clogging of the pores and inhibiting their surface.
2. The most effective way is the technology of inputting the polysulfide sulfur into a dry grouting material during its disintegrator treatment.
3. Polysulfide sulfur does not impair the rheological properties of cement slurries and their pumpability, allowing for normal casing cementing processes.
4. The addition of polysulfide sulfur to cement slurry increases the strength and reduces the permeability of the cement stone.

REFERENCES

Agzamov F.A., Izmuhambetov B.S., Tokunova E.F. *Chemistry of grouting and flushing solutions* [Himija tamponazhnyh i promyvochnyh rastvorov] – S-Pb.: Nedra, 2011. – 268 p.

Agzamov F.A., Izmuhambetov B.S. *Durability of a grouting stone in the corrosive environments* [Dolgovechnost' tamponazhnogo kamnja v korrozionno-aktivnyh sredah]. – S-Pb.: Nedra, 2005. – 318 p.

Agzamov F.A., Konesev G.V., Hafizov A.R. *Application of disintigratory technology for the modification of materials used in the construction of wells.* Part II. Nanotechnologies in Construction. 2017, Vol. 9, no. 3, p. 96–108. viewed 15.02.2019, dx.doi.org/10.15828/2075-8545-2017-9-3-96-108.

Babkov V.V., Mohov V.N., Kapitonov S.M., Komohov P.G. Structurization and destruction of cement concretes [Strukturoobrazovanie i razrushenie cementnyh betonov]. – Ufa: GUP 2002. – 376 p.

Danjushevskij V.S., Tarnavskij A.P. *Gas hydrosulphuric corrosion of grouting cements* [Gazovaja serovodorodnaja korrozija tamponazhnyh cementov]//Gas industry. – 1977. – no. 6. – p. 46–48.

Dorovskih I.V., Zhivaeva V.V. *Perfecting of processing behavior of corrosion resistant grouting solutions for the purpose of increase in an overhaul life of work of oil and gas wells* [Sovershenstvovanie tehnologicheskih svojstv korrozionnostojkih tamponazhnyh rastvorov s cel'ju uvelichenija mezhremontnogo perioda raboty neftjanyh i gazovyh skvazhin]//[Drilling and oil]. – 2009. – no. 11. – p. 44–46.

Ignat'ev A.V. *Increase in durability and water tightness of solutions and concrete impregnation by their liquid silica glass* [Povyshenie prochnosti i vodonepronicaemosti rastvorov i betonov propitkoj ih zhidkim steklom] – Kiev. – 1963. – p. 283-296.

Kind V.V. *Corrosion of cements and concrete in hydraulic engineering constructions* [Korrozija cementov i betona v gidrotehnicheskih sooruzhenijah]. – Moscow: Gosjenergoizdat, 1955. – 230 p.

Kravcov V.M., Kuznecov Ju.S., Mavljutov M.R., Agzamov F.A. *Cementing of high-temperature wells in the corrosive environments* [Kreplenie vysokotemperaturnyh skvazhin v korrozionno-aktivnyh sredah]. – Moscow: Nedra, 1987. – 190 p.

Kuncevich O.V., Mahinin B.V., Shangina N.N. *Structure of a cement stone with additives of supersoftener and microsilicon dioxide* [Struktura cementnogo kamnja s dobavkami superplastifikatora i mikrokremnezema]//Cement. – 1992. – no. 6. – p. 30–36.

Mamadzhanov U.D. *Corrosion of grouting cements in the carbon dioxide environment* [Korrozija tamponazhnyh cementov v uglekislotnoj srede]//Trudy AN UzSSR, tehn. science Works of academy of Sciences of UZSSR, office of technical science. – 1976. – p. 69-73.

Marchukajtis G.V. *Concrete impregnation influence by monomers on its stress-strain properties* [Vlijanie propitki betona monomerami na ego deformativnye svojstva]//Concrete and reinforced concrete. – 1977. – no. 6. – p. 30-31.

Massalimov I.A., Yanakhmetov M.R., Chuykin A.E. *Strength and durability of Concrete modified by sulfur-based impregnating compounds.* Nanotechnologies in Construction. 2015, Vol. 7, no. 3, pp. 61–75. viewed 15.02.2019, dx.doi.org/10.15828/2075-8545-2015-7-3-61-75.

Moskvin V.M., Ivanov F.M., Alekseev S.N., Guzeev E.A. *Corrosion of concrete and reinforced concrete, methods of their protection* [Korrozija betona i zhelezobetona, metody ih zashhity]. – Moscow: Strojizdat, 1980. – 536 p.

Nevill' A. M. *Properties of concrete* [Svojstva betona]. – Moscow, 1972. – 344 p.

Ovchinnikov V.P. *Technology of drilling of oil and gas wells* [Tehnologija burenija neftjanyh i gazovyh skvazhin: uchebnik dlja studentov vuzov]. – Tyumen: 2014. – 418 p.

Polak A.F. *Mathematical model operation of process of corrosion of concrete in liquid environments* [Matematicheskoe modelirovanie processa korrozii betona v zhidkih sredah]//Concrete and reinforced concrete. –1988. – no. 3. – p. 30–34.

Rahimbaev Sh.M., Avershina N.M. *Model operation of processes of chemical corrosion of structural materials* [Modelirovanie processov himicheskoj korrozii stroitel'nyh materialov]//» Reports of the international conference "New Technology Solutions and Economic Problems in Production of Concrete, Other Structural Materials and Products. – Belgorod, 1996. – p. 167–172.

Rahimbaev Sh.M., Karpacheva E.N., Tolypina N.M. *About the choice of type of cement on the basis of the theory of a colmatation at the complex structure of a severe atmosphere* [O vybore tipa cementa na osnove teorii kol'matacii pri slozhnom sostave agressivnoj sredy]//Concrete and reinforced concrete. – 2012. – no. 5. – p. 25–26.

Rahimbaev Sh.M., Tolypina N.M. *Justification of the choice of type of the organic origin knitting for severe atmospheres on the basis of the theory of heterogeneous physical and chemical processes* [Obosnovanie vybora tipa vjazhushhego dlja agressivnyh sred organicheskogo proishozhdenija na osnove teorii geterogennyh fiziko-himicheskih processov]//Vestnik Belgorodskogo gosudarstvennogo tehnologicheskogo universiteta im. V.G.Shuhova [Bulletin of the Belgorod state technological university of V.G. Shukhov. – 2016. – no. 9. – p. 159–163.

Rahimbaev, Sh.M., Tolypina N.M. *Increase in rust resistance of concrete by the rational choice knitting and fillers* [Povyshenie korrozionnoj stojkosti betonov putem racional'nogo vybora vjazhushhego i zapolnitelej]. – Belgorod: Izd-vo BGTU, 2015. – 321 p.

Shmelev P.S. Drilling of deep wells in the conditions of the abnormal influence of corrozive environments [Burenie glubokih skvazhin v uslovijah anomal'nogo vozdejstvija korrozionnoaktivnyh sred]. – Moscow: Science, 1998. – 351 p.

Urkhanova L.A., Lkhasaranov S.A., Rozina V.Ye., Buyantuev S.L., Bardakhanov S.P. *Increased corrosion resistance of basalt reinforced cement compositions with nanosilica.* [Nanotechnologies in Construction]. 2014, Vol. 6, no. 4, p. 15–29. viewed 15.02.2019, http://nanobuild.ru/en_EN/

Volgushev .A.N., Paturoev V.V., Putljaev I.E. Krasil'nikova O.M. *Use of sulfur for impregnation of steam structure of structural materials* [Primenenie sery dlja propitki porovoj struktury stroitel'nyh materialov]//Concrete and reinforced concrete. – 1976. – no. II. – p. 38-39.

Yanakhmetov M.R., Chuykin A.E., Massalimov I.A. *Pore structure modification of cement concretes by impregnation with sulfur-containing compounds.* Nanotechnologies in Construction. 2015, Vol. 7, no. 1, pp. 63–72. viewed 15.02.2019, dx.doi.org/10.15828/2075-8545-2015-7-1-63-72.

Youth Technical Sessions Proceedings – Litvinenko (ed)
© 2019 World Petroleum Council, London, UK, ISBN 978-0-367-34668-3

Challenges during oil processing caused by back-production of polymers

A. Brikov
JSC "Neftegazholding", Moscow, Russian Federation

A. Markin
Nizhnevartovsk branch of Tumen Industrial University, Nizhnevartovsk, Russian Federation

ABSTRACT: During oil processing operation companies encounter deterioration of oil and water separation, generation of separate layer at oil/water interface and degradation of produced water quality due to increase of suspended solids and hydrocarbons concentration. This significantly affects operation of oil processing units, might lead to frequent clogging of filters and reduces repair interval of equipment. Studies of causes of produced water quality deterioration of "Piltun-Astokh-A" offshore oil platform revealed that oil/water interface in process equipment contains layer of floating solid particles. These particles significantly affect separation of oil and water. Analysis of solid particles revealed that main component of organic part of the sample contains polymer with composition similar to Novon which is biodegradable polymer with altering polysaccharides and polyolefins chains. It was found out that on the platform number of chemicals contain polysaccharides molecules. These chemicals are used as lost circulation materials during drilling, workover and completion operations, for modification of rheological properties of drilling mud and during fracture treatment. Probable mechanisms of polymer back-production and its impact on oil processing for offshore oil and gas facilities are proposed. Recommendations on mitigation of this impact are provided

1 INTRODUCTION

Piltun-Astokh oil field was discovered in 1986. The field is considered large with regards to oil resources. Length of the field is 35 km and width 5-10 km. The field is located at northeast shelf of Sakhalin island, at the latitude of south end part of Astokh bay, 10-14 km away from the shore line between Odoptu-sea field from the north and Arkutun-Dagi from the south. Exploration and production of the field is executed in frames of Sakhalin-2 project [1].

For the effective development of the field it was divided it two parts: Astokh and Piltun. Offshore platform Piltun-Astokh-A (PAA) was installed at Astokh part of field, Piltun-Astokh-B (PAB) platform was installed at Piltun part [2].

PAA platform is modified drilling platform, used before at arctic waters near Canada shoreline. There are 14 production wells, 4 well are used for pressure maintenance and 1 for drilling cuttings reinjection.

Oil processing equipment at PAA shall maintain oil export quality parameters on water content – 0.25 % vol. Principal oil processing scheme is provided at figure 1. Produced oil enters two-phase high pressure separator (1), designed for removal of major part of associated gas. Then fluid is directed to three-phase middle pressure separator (2), form which separated water goes to degasser (6) – two-phase separator for removal of residual gas and oil before water enters injection system. Gas form separator (2) is directed to gas compression system, oil - to low pressure separator (3) for the final oil degassing. Then oil goes to two parallel electrical dehydrators (ED) (4 and 5). Separated in ED water is also directed to degasser [3]. After

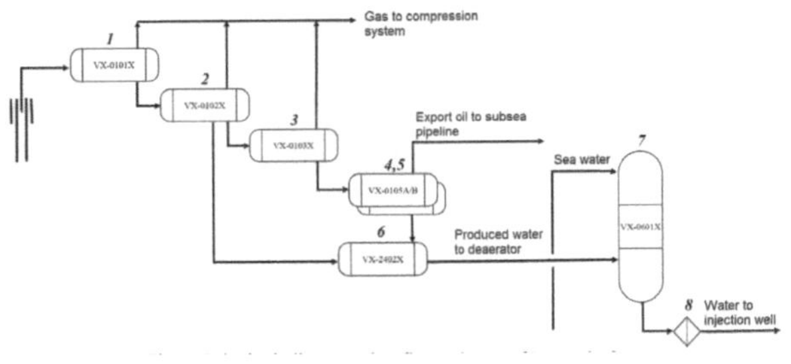

Figure 1. Principal oil processing flow scheme of PAA platform

degasser water flows to deaeration column (7) where it mixes with deaerated sea water and passing through filtration system (8) enters pressure maintenance system.

1.1 *Aim of paper*

Main purpose of the paper is to understand the impact of polymers applied during well drilling, completion and operations to surface equipment, oil and produced water quality, separation efficacy.

1.2 *Problem description*

During 2013-2014 significant degradation of produced water quality was noticed at PAA platform from ED and degasser water outlets. Concentration of suspended solids increased to 1.8 g/l, hydrocarbons to 3.5 g/l.

High concentration of suspended solids led to frequent (1 time per month) clogging of filter (8) shown on figure 1, installed on suction line of high pressure pumps. This caused pressure maintenance system trips. Operator noticed that for complete chocking of filter only 3-5 kg of solid deposits is required.

Suspended solids have been accumulating on the water/oil interfaces in separators, degasser and ED, which stabilized water-oil emulsion and led to emulsion layer formation. This layer significantly reduced separation efficacy [4].

2 MATERIALS AND METHODS

During studies of causes of oil separation efficacy reduction and produced water quality degradation it was found that at the water/oil interface in ED and degasser there was a layer, consisted of solid particles, preventing phase separation. Solids did not dissolve in toluene and hydrochloric acid at room temperature, melted at 130-140 °C.

For identification of composition of solids from ED water/oil interface a sample was taken. Sample consisted of two phases: black coloured liquid (oil) and insoluble black deposits.

For identification of nature of organic matter of sample taken pyrolytic chromatography-mass spectrometry was used [5–7] – gas chromatography-mass spectrometer Shimadzu GCMS QP-2010 with pyrolizer Double-Shot Pyrolyzer PY-2020iD. Pyrolizer oven temperature 600 ° C, interface PY/GC temperature – 320 °C. Product separation was conducted with Ultra ALLOY-5 chromatographic column (30 m, internal diameter 0.25 mm, phase 0.25 mcm), carrier gas – helium, rate 1 cm^3/min, sample splitter 1:100. Column temperature programming from 40 °C (stabilization 3 min), with following increase to 320 °C with 20 °C/min rate. Mass range (m/z) 35-800, mass-spectrometric detector interface temperature 280 °C, ions source temperature 250 °C, electrical charge of detector 1 kV. Comparison of pyrograms was conducted with FSearch «All-In-One» Ver. 3.10 software.

3 RESULTS

FSearch «All-In-One» Ver. 3.10 software search showed that sample taken from interface of ED at PAA platform contains polymer with structure similar to Novon (matching 93%) and also hydrocarbons, identified as polyethylene and its derivatives (matching 91%).

Novon polymer is manufactured by Novon International Inc. (USA). This is biodegradable polymer. It consists of alternating chains of polysaccharides and polyolefins. Polymer is used for manufacturing of materials, decomposing under sunlight, microorganisms in soil or water [8–10].

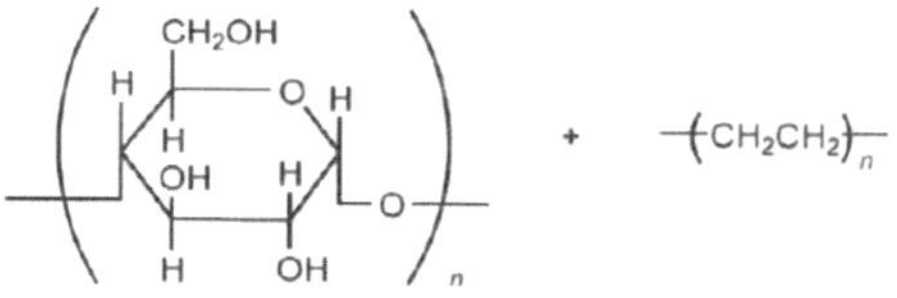

Figure 2. Molecular structure of Novon from FSearch «All-In-One» Ver. 3.10 database

4 ANALYSIS

Data on polymer composition were compared with composition of chemicals and reagents used in well drilling, completion and workover. It was confirmed that polyethylene and its derivatives are not used in mentioned operations. However as a lost circulation materials of drilling mud or kill brine in well drilling, completion and workover, for drilling mud viscosity adjustment and for hydraulic fracturing operations at PAA platform different chemicals with polysaccharide molecules (for instance Flowzan, guar gum, xanthan gum, modified starch etc.) were often applied. That was the reason analysis of organic part of sample taken form ED with pyrolytic chromatography-mass spectrometry was identified Novon.

Thus it was confirmed that organic part of sample taken from water/oil interface of ED at PAA platform is a mixture of oil and polysaccharides. It needs to be mentioned that polysaccharides (hydroxyethilcellulose) was found in sample of deposits of heat exchanger at PAB platform [2].

Quantitative analysis of sample taken from water/oil interface of ED at PAA platform with inorganic part included is provided in table 1.

Table 1. Chemical composition of sample taken from water/oil interface of ED at PAA platform

Component	Concentration, % mass.
$CaCl_2$	1,26
NaCl	0,69
FeS	0,51
SiO_2	0,07
$BaSO_4$	0,06
Total inorganics	2,70
Water	77,40
Oil	19,10
Novon	0,80

5 DISCUSSION

After identification of solids composition in sample taken from water/oil interface of ED it was necessary to determine the mechanisms of polymers back-production from wells to surface equipment. Most probable source of polymers back-production are activities related to wells drilling, hydraulic fracture and losses prevention.

5.1 *Wells drilling*

Well drilling is conducted with drilling fluids circulation. The main purposes of fluid circulation is to remove cuttings from well bottomhole and prevent well walls collapse. To change rheological properties of drilling mud (viscosity increase) different additives, containing polysaccharides, are applied. Concentration of that additives could reach 2-5 % vol. Taking into account that for effective functioning of drilling mud its hydrostatic pressure is maintained above reservoir pressure, there are always drilling mud losses into reservoir. As a result with drilling mud losses polysaccharide based polymers penetrate into reservoir. This is called contamination of formation pay zone.

Most probably in that case polysaccharide based polymers are located not more that 0.5-1.0 m depth in reservoir. Thus main portion of polymer will be back-produced within short period of time. This can cause problems for oil processing at the topside facilities. Taking into account high concentration of polymers into drilling mud and its possible significant losses some portion of polymer can strongly adhere to reservoir rock. Polysaccharides have low solubility in hydrocarbons and high solubility in water. That is why part of polymer could start to produce back to surface only after water production commences.

5.2 *Hydraulic fracture*

Standard procedure of hydraulic fracture consists of following steps:

- Hydrochloric acid treatment;
- Minifrac test;
- Main frac.

Polysaccharide based polymers are used at every phase of hydraulic fracture operations as a diverter (during acid treatment) or are the part of gelling fluid. Main part of polymeric gel is used during third phase of hydraulic fracture. For instance, volume of xanthan gum used for main frac at one of the wells of PAA platform was 7.5 m^3. At this stage frac fluid (polymeric gel without proppant) is used for fracture generation. Following pumping of proppant-carrier fluid (polymeric gel mixed with proppant) at the high pressure (24-26 MPa) is required for packing of created fracture with proppant.

Together with polymeric gel mixing with proppant gel breaker is added in the mixture at surface. This chemical is added in special capsules, which destroys after certain period of time under reservoir temperature. Thus viscosity of main part of gel reduces and low viscosity fluid ("broken gel") back-produces during well clean-up operations.

In process of hydraulic fracture losses of fracture fluid and proppant-carrier fluid occurs. This is called filtration of fluid. It leads to filtrate formation on wells of the fracture after hydraulic fracture operations. Main part of that filtrate consists of polysaccharide based polymer, which is a component of polymeric gel. Obviously, polymer breaker will not be able to react with gel filtrate because breaker activates after 24 hours when hydraulic fracture job is completed, well is not under fracturing pressure and no fluids filtration in reservoir occurs. It means that after well clean-up fracture contains some polymer, which was not destroyed by breaker.

Authors suppose that low solubility of polysaccharide based polymers in hydrocarbons is the reason of significant time gap between hydraulic fracture and polymer back-production. For several wells of Piltun-Astokh field back-production of polymer was observed after 3-5 years after hydraulic fracture. Operator noticed that polymer back-production starts after produced water breakthrough.

5.3 *Fluid losses prevention*

After well perforation loss of kill brine is possible. At the PAA platform loss circulation material (polysaccharide based polymer) was applied on single well only. Volume of loss circulation material was 5 m^3, concentration of polymer 80 g/l. As in case of drilling polysaccharide based polymer in that case locates into formation for 0.5-1.0 m depth. However due to low concentration of polymer impact of this mechanism in overall polymer back-production scheme shall not be significant.

6 CONCLUSION

1. It is shown that problems with oil processing at PAA production platform were caused by solids accumulated at water/oil interface into separation equipment.
2. It was determined that solids samples from water/oil interface of ED consist of polysaccharide based polymers.
3. Source of polysaccharide based polymers are components of drilling mud, loss circulation materials and hydrofracturing fluid.
4. Most probable mechanisms of polymers back-production from reservoir to surface are proposed.
5. It is shown that due to low solubility of polysaccharide based polymers in hydrocarbons its back-production starts only after produced water breakthrough.
6. It is recommended that during clean-up operation of well where polysaccharide based polymers were applied not to allow water phase to enter mail process before complete polymer back-production.
7. It is recommended to assess effectiveness of use of viscoelastic frac fluids and proppant-carrier fluids for hydraulic fracture operations. These liquids are based on surface active substances and does not contain polymers.

REFERENCES

1. Kosyak D.V., Markin A.N. Experience of wax control in subsea pipelines of Sakhalin-2 project Territoria neftegaz. 2011. No.6. P 12-18.
2. Tarabarina K.Yu., Sukhoverkhv S.V., Markin A.N. Formation of solid deposits in heat exchanger of oil production platform Piltun-Astokh-B (Sakhalin island) and its removal Oilfield engineering. 2013. No. 8. P 51-55.
3. Kosyak D.V., Lutsev D.S., Markin A.N. Savin K.I. Experience of demulsifier application at offshore planform of Sakhalin-2 project Territoria neftegaz. 2010. No.8. P 48-51.
4. Tsyganov D.G., Bashkirtseva N. Yu., Sladkovskaya O. Yu. Formation of stable oil emulsions because of use of chemicals for enhanced oil recovery of Kamennoye and Em-Eganskoye field Oilfield engineering. 2015. No. 5. P 38-43.
5. Khmelnitsky R. A., Lukashenko I.M., Brodsky E.S. Pyrolytic chromatomass spectrometry of high molecular substances. – Moscow.: Khimiya, 1980. – 279 p.
6. Zayikin V.G., Mass-spectrometry of synthetic polymers. – Moscw.: VMSO, 2009. -332 p.
7. F-Search "All-In-One". Mass Spectral Libraries for Polymers and Additives, and Search Software. – Frontier Laboratories Ltd., 2008. – 61 p.
8. Biodegradable polymers and plastics. Edited by Emo Chiellini and Roberto Solaro. 2003 Springler Science+Business Media New York. Originally published by Kluwer Academic / Plenum Publishers, New York in 2003. P. 195.
9. Biopolymers: Processing and Products. Michael Niaounakis. 2015 Elsevier Inc. P 534.
10. M. Flieger, M. Kantorova, A. Prell, T. Rezanka, J. Votruba. Biodegradable Plastics from Renewable Sources Folia Microbiol. 48 (1), 27044 (2003).

Youth Technical Sessions Proceedings – Litvinenko (ed)
© 2019 World Petroleum Council, London, UK, ISBN 978-0-367-34668-3

Reservoir oil-bearing evaluation based on machine learning

Y.L. Ren, Z. Feng & L. Luo
Petrochina Research Institute of Petroleum Exploration and Development, Beijing, China

ABSTRACT: For reservoir oil-bearing evaluation, the traditional method relies on rich expert experience, which is subjective and inefficient. With the steady accumulation of logging data, one of the most important problems for oil exploration and development is that how to fully mining and analysis geological data to reduce human workload. Machine learning provides an effective way. In this paper, we use machine learning method to analyze logging curves, and then get the evaluation results of reservoir oil-bearing property. We establish serval prediction models and give the corresponding parameter optimization mechanism. The experiment results show that the machine learning algorithm and optimization mechanism proposed in this paper can well evaluate reservoir oil-bearing property based on logging data. This work provides a reference for improving the automation and efficiency of logging interpretation.

1 INTRODUCTION

The basic idea of reservoir evaluation is taking full advantage of various materials including well logging data to evaluate reservoir properties, such as oil bearing evaluation, layer classification, reservoir parameters prediction, capacity estimation etc. With the steady accumulation of logging data, one of the most important problems for oil exploration and development is that how to fully mining and analysis geological data. Machine learning provides a new effective method. As to well logging interpretation, one of advantages of machine learning is that professional knowledge, such as rock physics mechanism, is not needed in advance. It only needs a small amount of background knowledge and constructs model directly from geological data.

In recent years, machine learning has been more and more widely used in well logging interpretation. It is successfully used in reservoir real-time dynamic monitoring (Alimonti & Falcone 2002), best practice recognition of petroleum engineering work (Mohaghegh 2005), reservoir description (Nikravesh 2004) etc. It has been found that these nonlinear methods perform better than linear statistical analysis technologies which are widely used in oil exploration and development. Classification is the most widely used method related to reservoir evaluation. It is widely used for lithology identification (Al-Anazi & Gates 2010; Konate & Pan & Fang 2015; Silva & Neto & Missagia 2015), sedimentary facies division(Chai &Li &Xiao 2009; Coleou & Poupon & Azbel 2003), reservoir parameters prediction(Le & Yuan 2005) and oil, gas or water layer identification(Shi 2013; Shi 2018) etc.

In recent years, China National Petroleum Corporation (CNPC) pays a lot of attention to artificial intelligence, especially in oil exploration and development area. As the inner IT department of CNPC, we devote ourselves to the application research of artificial intelligence in oil exploration and development area.

In this paper, we put forward a framework for reservoir oil-bearing evaluation based on machine learning. We recommend some suitable algorithms and give the corresponding model optimization mechanism. The experimental results show that the proposed machine learning framework works well and can quickly and accurately predict the results of reservoir evaluation, thus improving the automation and intelligence level of logging interpretation.

2 A FRAMEWORK FOR RESERVOIR OIL-BEARING EVALUATION BASED ON MACHINE LEARNING

The main task of reservoir evaluation is using original data collected by well logging to evaluate its properties, such as porosity, oil saturability, lithological characters etc., and finally give a comprehensive evaluation.

Original data collected by well logging mainly includes well logging curves, such as potential curve (SP), gamma ray curve (GR), resistivity curve (RT, RI, RXO), density log (DEN), acoustic logging (AC), compensated neutron log (CNL) etc. Besides original well logging data, we also collect experimental data of critical wells. For critical wells, we not only conduct a comprehensive logging to collect original data, but also construct experiments on rock samples collected from cored wells. We obtain lithology, oil bearing properties by analyzing the experimental data, such as porosity, permeability etc. For non-critical well, we only have well logging data, such as SP, GR, RT, RI, RXO, AC, since it is quite expensive to core from cored wells. We can't get experimental data for non-critical well since there are no rock samples.

Traditional procedure for logging reservoir evaluation is shown below.

- Simplify the complexity of the strata and presume that it is homogeneous and isotropic physical model.
- Construct mathematical model F_{org} with undetermined parameters according to physical theory of fluid and solid.
- Calculate the undetermined parameters of F_{org} using data from critical wells which both have original data and experimental data. Here, mathematical model F_{org} with determined parameters becomes well logging interpretation model $F_{Decision}$.
- Set the threshold value and grade division standard D_{org} for porosity, oil saturability and lithological characters.
- Construct data analysis on non-critical well data by using well logging interpretation model $F_{Decision}$ and use D_{org} to do comprehensive assessment.

Traditional method relies too much on expert experience and is inefficient, which is not conducive to its popularization and application.

2.1 *Reservoir Parameters Prediction*

As the foregoing, traditional method for measuring porosity, saturability and permeability is constructing a mathematical model F_{org} with predetermined independent variables. For example, the independent variables of porosity are usually AC, CNL and DEN. Model parameters of F_{org} are calculated by training coring data. This is quite dependent on expert experience since the final comprehensive assessment will be given by professional experts.

As already mentioned, regression algorithm can be used to predict reservoir parameters, such as multiple linear regression, Least Square Support Vector Machine. The advantage of this method is that it doesn't need to predetermine the independent variables, but find the relationship between reservoir parameters and well logging curves by learning training data collected form the cored wells. It is a data driven method.

The procedure of porosity, permeability and saturability interpretation are shown in Figure 1.

2.2 *A Framework of Reservoir Logging Evaluation Based on Machine Learning*

After computing reservoir parameters, oil exploration and development experts give the final evaluation conclusion through taking into account both computing results and professional knowledge. This is quite dependent on expert experience.

Data mining provides an automatic and data-driven method for reservoir logging evaluation. The framework is shown as follows.

In this framework, train data is comprised of logging curves and experimental analyzing data collected from cored wells. The biggest difference between this framework and traditional

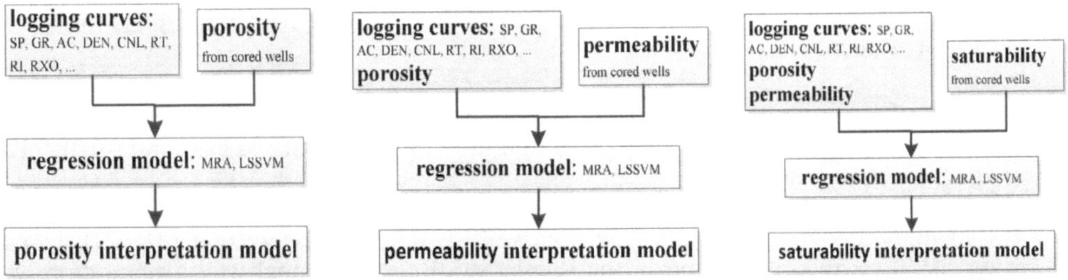

(a) porosity prediction (b) permeability prediction (c) saturability prediction

Figure 1. Reservoir parameters prediction procedure

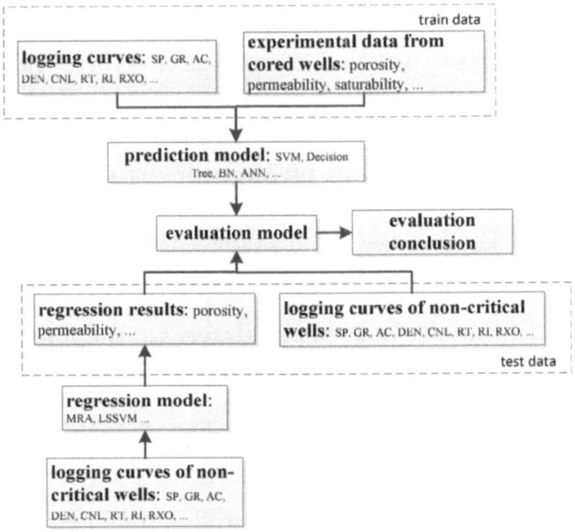

Figure 2. A framework of machine learning for logging reservoir evaluation

method is that it is automatic and not so dependent on expertise. It can be used as expert recommendation system of oil exploration and development.

3 OPTIMIZATION MECHANISM FOR OIL-BEARING EVALUATION PREDICTION

There are some key points for the framework put forward in part 2.

3.1 *Feature Selection*

Since the dimension of input data is much higher than traditional method, it may lead to overfitting. It is necessary to do feature selection while modelling. Common feature selection methods for well logging interpretation are cross plot technique, correlation analysis or expert experiences.

In the process of machine learning, there are 3 kinds of working patterns between feature selection and classification, embedded approach, filter approach and wrapper approach. Feature selection appears as a part of the machine learning algorithm for embedded approach. Machine learning algorithm is considered as a black box for wrapper approach. Filter approach is the method which is independent of classifier, such as Correlation-based Feature Selection (CFS) and Las Vegas Filter (LVF). The evaluation standards for CFS are that the correlation between feature and class labels is high while the redundancy between different features is low. LVF evaluate features by computing the consistency of feature sunset.

3.2 *Model Optimization*

The performance of different algorithms depends largely on the setting of model parameters. So it is necessary to do model optimization. In most cases, it can only rely on experience or contrast experiment to set parameters. There are also few researches on model parameters optimization. An effective method is the combination of k-Cross Validation and Grid Search.

k-Cross Validation
Training dataset will be split into k separate files of equal size. k-1 is selected as training dataset and the rest as validation set. Modeling process will be repeated k times and choose the average value of MSE after k iterations to estimate expected generalization error.

Grid Search
Grid search is a practical method of searching for data. It is quite suitable for searching the multidimensional data from different growing directions. Take an example to illustrate the principle, when we choose RBF kernel function for SVM, we should confirm two parameters, penalty parameter C and kernel function parameter σ. Grid Search choose change step Cs for C∈[C1,C2] and σsfor σ∈[σ1,σ2]. We use each pair of parameters (c′,σ′) for modeling and choose the best one which performs best as the final parameters.

3.3 *Model Evaluation*

Since it is quite expensive to core by drilling, there are few data which can be used for train data. Cross validation should be used for model evaluation. In addition to the evaluation of classifier, statistical test, such as X^2 test and F test, is necessary, because statistical model is used for predicting reservoir parameters.

3.4 *Solutions for Class Overlap and Imbalance Problem*

Due to the heterogeneity of formation, there are serious sample imbalances and overlapping phenomena in logging curve data collected. In order to solve the above difficulties, we add a class robust regularization term on the basis of classical *softmax* classification loss function.

$$L = L_S + \gamma C = -\log(e^{w_y^T x_i + b_y} / \sum_k^m e^{w_k^T x_i + b_k}) + \gamma(||c_y||_2^2 - \frac{1}{m}\sum_{k=1}^m ||c_k||_2^2)^2,$$

In this formula, L_S is the classification loss function of *softmax*, C is a class robust regular term, γ is a balanced parameter, c_y is the center of a class. The class robust regular term in the above objective function punishes the difference of each class center, so that each class center can be evenly distributed on the surface of a hypersphere, so that the sample points of a class can shrink as much as possible and the sample points of overlapping classes can be separated.

4 EXPERIMENT AND ANALYSIS

In this part, we verify the effect of machine learning in reservoir oil-bearing evaluation from three aspects. Here are the illustration of the three strategies.

STRATEGY I:

In this strategy, we only use original logging curves to predict oil-bearing evaluation results. That means the attributes of our data include natural potential, natural gamma ray, density, neutron, acoustic time difference, resistivity and other original logging curves.

STRATEGY II:

In this strategy, we use the logging interpretation data to predict oil-bearing evaluation results, they are porosity, permeability and saturation.

STRATEGY III:

For the third strategy, we predict oil-bearing evaluation results by combining original logging curves and logging interpretation conclusions. That means the attributes of our data not only include potential, natural gamma ray, density, neutron, acoustic time difference, resistivity etc., but also include porosity, permeability and saturation.

The prediction models are SVM, Decision Tree, Nave Bayes and KNN. We use F-measure, ROC and other indicators to evaluate the prediction effect.

4.1 *Description of Our Data*

We collected logging data and logging interpretation conclusion data from two oilfield blocks, and aggregated them into two data sets, named data1 and data2. There are more than 5000 pieces of data in the two datasets. Examples of data sets for the three validation methods are as follows.

Table 1. Data set for strategy I

No.	original logging curves natural potential	resistivity	......	density	oil test
1	64	206		48.6	poor oil layer
2	140	208		41.5	dry layer
3	49	207		67.5	poor oil layer
4	80	207		50.9	dry layer
......					
3536	109	199		17.8	dry layer

Table 2. Data set for strategy II

No.	logging interpretation conclusions Porosity	Permeability	saturation	oil test
1	10.1	0.8652	74.1	poor oil layer
2	6.17	0.4782	50.3	dry layer
3	7.03	0.3526	66	poor oil layer
4	5.57	0.3312	51.4	dry layer
......				
2985	8.31	0.3762	37	dry layer

Table 3. Data set for strategy III

	original logging curves				logging interpretation conclusions			
No.	natural potential	resistivity	......	density	porosity	permeability	saturation	oil test
1	64	206		48.6	10.1	0.8652	74.1	poor oil layer
2	140	208		41.5	6.17	0.4782	50.3	dry layer
3	49	207		67.5	7.03	0.3526	66	poor oil layer
4	80	207		50.9	5.57	0.3312	51.4	dry layer
......								
3536	109	199		17.8	8.31	0.3762	37	dry layer

In order to see the data structure of our dataset, we use t-SNE method to do data visualization. T-SNE is a non-linear dimension reduction algorithm, which is very suitable for high-dimensional data to be reduced to two or three dimensions for visualization. The results are shown in Figure 3.

(a) logging data with interpretation conclusion

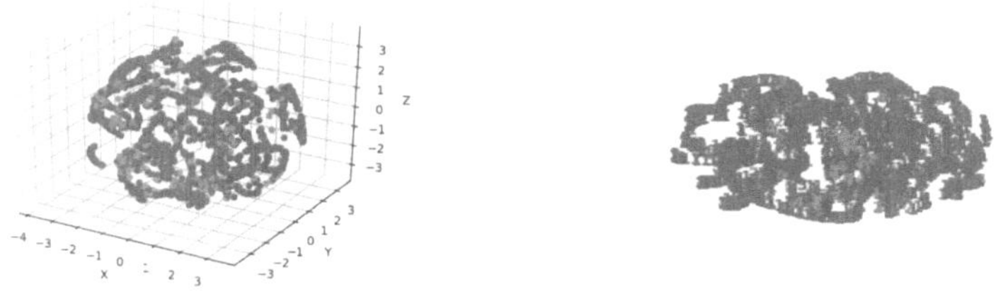

(b) original logging data

Figure 3. Distribution of our dataset

From the results of data visualization, it can be seen that the class overlap problem of original logging curve data is quite serious. When logging interpretation conclusions such as porosity and permeability are included, the phenomenon of class overlap is reduced a lot. We can also see that the problem of sample imbalance is common in both datasets.

4.2 *Experiment Results and Analysis*

The prediction models are KNN, SVM, Decision Tree, Nave Bayes and Logistic Regression. At the same time, grid search and k-cross validation are used to optimize the model parameters. The experimental results are shown in Figure 4.

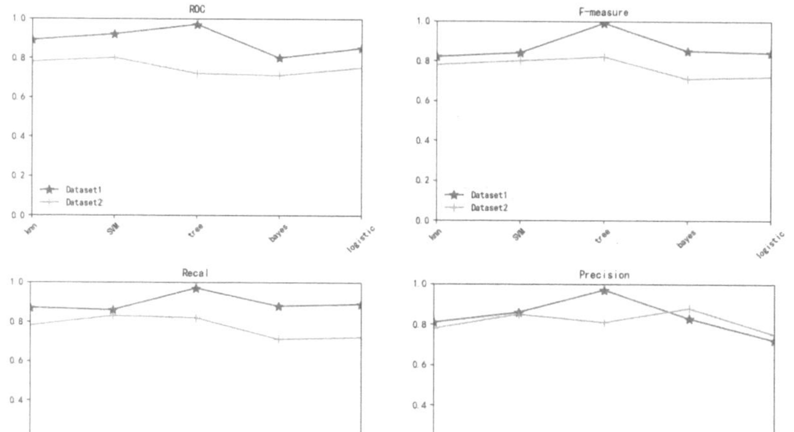

Figure 4. Experiment results for the two dataset. Dataset2 only contains original logging cures while dataset1 contains original logging cures and interpretation conclusions

From the experimental results, we can see that the evaluation results based on machine learning prediction are very accurate for mines with calculated porosity, permeability and saturation. The experimental results show that when we use both original logging curve and log interpretation conclusion to predict, or only use log interpretation conclusion to predict, the four indexes can reach more than 85%. It is slightly worse to use only the original logging curve to predict the oil-bearing evaluation results, but the ROC index can still reach 80%. In a word, the machine learning algorithm and optimization mechanism proposed in this paper can well evaluate reservoir oil-bearing property based on logging data.

5 CONCLUSION

In this paper, a framework for reservoir oil-bearing evaluation based on machine learning is put forward. At the same time, a set of machine learning algorithms and corresponding optimization mechanism are proposed. We collect logging data and interpretation conclusions from two oil fields and verify the effect of machine learning in reservoir oil-bearing evaluation from three aspects: one is to predict oil-bearing evaluation results directly by using original logging curves; the other is to predict oil-bearing evaluation results by using porosity, permeability and saturation; and the third is to predict oil-bearing evaluation results by combining original logging curves and logging interpretation conclusions. The experiment results show that when the data set contains logging interpretation conclusions such as porosity and permeability, the prediction effect of the model is good and can meet the production demand. But when the data only contain the original logging curve, the prediction effect of the model is slightly worse than that with logging interpretation conclusions. This work provides a reference for improving the automation and efficiency of logging interpretation. In the future, we will try to improve the prediction effect based on original logging curves by constructing deep CNN network.

REFERENCES

Alimonti, C. & Falcone, G. 2002. Knowledge discovery in databases and multiphase flow metering: the integration of statistics, data mining, neural networks, fuzzy logic, and Adhoc flow measurements towards well monitoring and diagnosis, *SPE Annual Technical Conference and Exhibition*, Texas

Mohaghegh, S. D. 2005. A new methodology for the identification of best practices in the oil and gas industry, using intelligent systems, *Journal of Petroleum Science and Engineering*, vol. 49, pp. 239-260.

Nikravesh, M. 2004. Soft computing-based computational intelligent for reservoir characterization, *Expert Systems With Applications*, vol. 26, pp. 19-38.

Al-Anazi, A. & Gates I. D. 2010. On the capability of support vector machines to classify lithology from well logs. *Natural Resources Research*, 19(2): 125-139.

Konate, A. A. & Pan, H. & Fang, S. 2015. Capability of self-organizing map neural network in geophysical log data classification: case study from the CCSD-MH[J]. *Journal of Applied Geophysics*, 118:37-46.

Silva, A. A. & Neto, I A L. & Missagia, R M. 2015. Artificial neural networks to support petrographic classification of carbonate-siliciclastic rocks using well logs and textural information[J]. *Journal of Applied Geophysics*,117:118-125.

Chai, H. Li, N. Xiao, C. et al. 2009. Automatic discrimination of sedimentary facies and lithologies in reef-bank reservoirs using borehole image logs. Applied Geophysics, 6:17-29

Coleou, T. & Poupon, M. & Azbel, K. 2003. Unsupervised seismic facies classification: a review and comparison of techniques and implementation. *The leading Edge*, 22:942-953.

Le, Y. & Yuan Q. 2005. The application of SVM in reservoir parameters prediction. *Geophysical Prospecting for Petroleum*, 5:388-392.

Shi, G. 2013. Data Mining and Knowledge Discovery for Geoscientists, *Petroleum Industry Press*

Shi, G. 2018. Superiorities of support vector machine in fracture prediction and gassiness evaluation, *Journal of petroleum exploration and development*, 35(5):588-594.

Youth Technical Sessions Proceedings – Litvinenko (ed)
© 2019 World Petroleum Council, London, UK, ISBN 978-0-367-34668-3

Design of domestic compositions of drilling fluids for drilling wells in shales

I.V. Chudinova, N.I. Nikolaev & A.A. Petrov
Saint Petersburg Mining University, Saint Petersburg, Russia

ABSTRACT: For retention of the leading positions of Russia in the oil and gas sector it is necessary to support production growth. It can be reached due to use of new technologies in the field of drilling wells in difficult mining-and-geological conditions. There can be various problems connected with instability of rock in borehole during of construction of wells.This problem often arises during the drilling process in shales, one of the widespread rock of the most part of fields of the world. Ensuring stability of shales in a borehole is the difficult task demanding a good technological equipment and wide experience of employees. As a result of research, drilling mud which provides stability of borehole during drilling process in shales was developed. The positive effect of the developed drilling mud is reached due to a combination of the inhibiting reagents which allow to lower extent of interaction of the dispersive environment of drilling mud and the breeds composing well walls.

1 INTRODUCTION

Today Russia is one of the largest players on oil and gas production in the world. The oil and gas extraction industry is bringing about 50% of income in the budget of the country and has huge value for social and economic development of Russia.

However, on September 12, 2014 due to geopolitical situation, USA and European Community put the sanctions for the largest Russian oil and gas companies and restrict delivery of technologies for production and investigation on the shelf of the Arctic seas and the deep-water shelf. In this connection the structure of the Russian oilfield service market exposed to significant changes. Russian oil service companies replaced foreign service companies in separate segments (Kamyshnikov & Kolpakov 2018, Blinov & Dvoynikov 2018).

For preservation of the leading positions it is necessary to support high level of oil and gas production. It can be reached by means of introduction of new technologies, searching and development of new fields, tutoring of personnel. Effective work with already found fields and searching of new fields requires use of the modern inventory and advanced technologies in the field of services (Blinov 2019). Level of production of oil and gas directly depends on quality of services, such as drilling and repair of wells, conducting geological technical actions. Import substitution becomes one of the most important priorities of the Russian oil and gas service companies and producers of an equipment. Representatives of the profile ministries assume that this task mostly must be solved within several years. In response to sanctions against Russia in 2014 the Ministry of Energy of the Russian Federation prepared the plan of import substitution in the oil and gas sector (Malikova & Davydenko 2011).

Drilling of oil and gas wells and also their further maintenance are the most capital-intensive component in the oil and gas sector. Under the modern conditions – an unstable political situation, limited access to foreign technologies, reduction of production according to the agreement of OPEC, rigid ecological conditions – optimization of expenses becomes a priority (Blinov et al. 2017, Melekhin et al. 2016).

Losses of time at construction of the well can be reduced by 15-25% at the expense of perfectings of technologies (6-10%), rational use of drilling equipment (for 4-6%), improvements of coordination of works (for 4-8%), decrease of breakdown rate (for 1%) (Dvoynikov et al. 2017). The analysis of balances of time of drilling and fastening of operational and exploratory wells showed that loss of unproductive time amount to 19-20% in development and exploring drilling.

2 MATERIALS AND METHODS

Due to the increase in number of wells with the difficult spatial profile, especially in the worsened mining-and-geological conditions, places high demands on the equipment, drilling muds, technology of casing and completion of the well. The instability of a borehole of the well during the opening of rocks is caused by redistribution of tension in a near wellbore zone and transition of breeds to a new equilibrium state. The new equilibrium state of rocks will be defined by a number of factors: specific gravity of drilling mud, pore pressure, cycle variations of temperature, etc (Dvoynikov & Blinov 2016, Nutskova & Kupavyh 2016).

Among sedimentary rocks shales are subject to the greatest influence of a water phase of drilling mud. Because shales are 80% of all sedimentary rocks, selection of component composition of drilling mud for opening of clay breeds has the defining value in preservation of a steady condition of a borehole of the well (Ulyasheva et al. 2016, Leusheva & Morenov 2017).

Borehole instability in shale formations is one of the main reasons of complications in the process of drilling. Borehole stability is influenced by such factors as orientation and trajectory of the well, high values of stresses on the well walls, type of drilling fluid, hydrostatic and pore pressures, mechanical properties of the shale rocks and also the interaction between the dispersion phase of the drilling mud and the shale. Effects of these factors can be shown differently: soft clays are squeezed out in a borehole, while brittle shales collapse under the influence of stresses and fall off to the bottom hole. All this leads to increase in the borehole radius, pack-off the shale debris inside the well and dogleg of the drilling tools (Chudinova & Nikolaev 2017).

The efficiency of maintenance of a wall of the well in a steady state decreases in process of penetration of a filtrate of drilling mud into breed. If drilling in shales happens to use of incompatible drilling fluid, its components getting into rock cause swelling and a softening of shales (Morenov & Leusheva 2017, Morenov & Leusheva 2016).

As a result of this work the compounding of the inhibiting drilling mud was developed for drilling wells in unstable shales.

Stability of a borehole is provided due to use of polyacrylates of sodium or potassium at their combination to glycerin and with sodium formate as a part of drilling mud. The preliminary depolymerization of liquid glass by dissolution of glycerin in it promotes increase in the inhibiting ability to shales. Polyvinyl pyrrolidone (PVP) shows tendency to complex formation and make low-molecular connections. Introduction to composition of water-based drilling mud of a polyvinyl pyrrolidone (PVP) promotes decrease of the activity of the continuous phase of drilling mud, control of growth of pore pressure in the breeds of borehole and by that increases stability of a borehole, during drilling in shales(Chudinova et al. 2018, Morenov et al. 2018).

3 RESULTS

Table 1 shows results of researches by influence of the designed drilling mud on stability of clays in comparison with the known solutions. The assessment of the inhibiting property of solution by amount of the absorbed water in samples of shales was carried out: if the amount

Table 1. Results theinhibitive properties test

№	Composition of drilling mud, mass %	Behavior of examples of shales kept in solution in 5 days	Variation mass of sample, g		
			initial	in 5 days	Δm
Comparison drilling mud					
1	83.2% water + 10% bentoniteclay + 3% polyglycol + 0.5% stabilizer + 3% liquid silica + 0,3% chloride of potassium	overswellingand dispersion	12.0	32.6	20.6
2	66% water + 10% bentonite clay + 5% polyglycol + 3% stabilizer + 15% liquid silica + 1% chloride of potassium	swellingand dispersion		26.4	14.4
Designed drilling mud					
3	83.9% water + 5% bentoniteclay + 5% PVP + 3% sodium formate + 3% liquid silica + 5% glycerine + 0.1% foam suppressant	swellingand dispersion	12.0	23.3	11.3
4	80.3% water + 5% bentoniteclay + 6.3% PVP + 5,3% sodium formate + 3% liquid silica + 5% glycerine + 0.1 % foam suppressant	low-level swellingand dispersion		19.8	7.8
5	76.9% water + 5% bentoniteclay + 7.5% PVP + 7,5% sodium formate + 3% liquid silica + 5% glycerine + 0.1% foam suppressant	low-level swellingand deficiencyof dispersion		18.2	6.2
6	73.3% water + 5% bentoniteclay + 8.8% PVP + 8.5% sodium formate + 3% liquid silica + 5% glycerine + 0.1% foam suppressant	low-level swellingand deficiencyof dispersion		17.6	5.6
7	69.9% water + 5% bentoniteclay + 10% PVP + 9.2% sodium formate + 3% liquid silica +5 % glycerine + 0.1% foam suppressant	deficiencyof swelling, low-level of dispersion		17.3	5.3
8	69.9% water + 5% bentoniteclay + 10% PVP + 10.6% sodium formate + 3% liquid silica +5 % glycerine + 0.1% foam suppressant	deficiencyof swelling, dispersion		17.5	5.5

Notice: positive result is Δm<8 г

of the absorbed water in a sample is less, then the inhibiting ability of solution is higher. For carrying out researches we used the samples of dry shale made of a bentonite by pressing under loading of 16 MPas for receiving tablets with an initial weight 12.0 g.

At contents of the sodium formate up to 3% and PVP up to 5% samples of shale swell and disperse. At contents of the sodium formate of 5.3-9.2% and PVP of 6.3-10% samples of shale keep stability and don't collapse (table 1, item 4, 5, 6, 7). Increase in contents of the sodium formate until 9.2% leads to destruction of a sample of shale (table 1, item 8) (Chudinova et al. 2018).

Results of researches on influence of content in solution of the sodium formate and PVP on a fluid loss indicator are given in Table 2. At contents of the sodium formate of 5.3-9.2% and PVP of 6.3-10% the minimum value of a fluid loss indicator is provided (table 2, item 4, 5, 6, 7) (Chudinova et al. 2018).

Table 2. Results the experimental tests of the drilling fluid rheology

№	Composition of drilling mud, mass %	Indicators of drilling mud		
		F, cm^3	$\eta_{пл}$, mPa·s	τ_0, Pa
Comparison drilling mud				
1	83.2% water + 10% bentoniteclay + 3% polyglycol + 0.5% stabilizer + 3% liquid silica + 0.3% chloride of potassium	9	25	6.4
2	66% water + 10% bentonite clay + 5% polyglycol + 3% stabilizer + 15% liquid silica + 1% chloride of potassium	6	38	7.9
Designed drilling mud				
3	83.9% water + 5% bentoniteclay + 5% PVP + 3% sodium formate + 3% liquid silica + 5% glycerine + 0.1% foam suppressant	6	23	6.2
4	80.3% water + 5% bentoniteclay + 6.3% PVP + 5,3% sodium formate + 3% liquid silica + 5% glycerine + 0.1 % foam suppressant	4	28	6.4
5	76.9% water + 5% bentoniteclay + 7.5% PVP + 7,5% sodium formate + 3% liquid silica + 5% glycerine + 0.1% foam suppressant	4	34	6.7
6	73.3% water + 5% bentoniteclay + 8.8% PVP + 8.5% sodium formate + 3% liquid silica + 5% glycerine + 0.1% foam suppressant	3	42	7.2
7	69.9% water + 5% bentoniteclay + 10% PVP + 9.2% sodium formate + 3% liquid silica +5 % glycerine + 0.1% foam suppressant	3	48	7.5
8	69.9% water + 5% bentoniteclay + 10% PVP + 10.6% sodium formate + 3% liquid silica +5 % glycerine + 0.1% foamsuppressant	3	46	7.3

Notice: F – fluidlossindicator; $\eta_{пл}$ – plasticviscosity; τ_0–yield point.

4 CONCLUSION

Thus, the analysis of results of the experiments allowed establishing an optimum compound of the drilling mud having the increased inhibiting ability to shales and the lowered of a fluid loss filtration. From this it can be concluded that use of liquids with low ion-molecular properties promotes decrease of the activity of the continuous phase of drilling mud, control of growth of pore pressure in the shales near borehole walls and by that increases the stability of a borehole during drilling of wells in shales.

REFERENCES

Chudinova I.V. & Nikolaev N.I. 2017. Justification of the choice of the inhibiting reagents for increase in stability of clay breeds. Oil Engineer 2: 10-12.

Chudinova I.V. et al. 2018. Patent of the Russian Federation No. 2675650. Polimer clay drilling mud.

Blinov P.A. Determining the stability of the borehole walls at drilling intervals of loosely coupled rocks considering zenith angle // Journal of Mining Institute. 2019. V. 236. P. 24-29. DOI: 10.31897/PMI.2019.2.24

Blinov, P.A., Dvoynikov, M.V. Rheological and filtration parameters of the polymer salt drilling fluids based on xanthan gum // Journal of Engineering and Applied Sciences 2018: 13(14), pp. 5661-5664

Blinov, P.A., Dvoynikov, M.V., Sergeevich, K.M., Rustamovna, A.E. Influence of mud filtrate on the stress distribution in the row zone of the well // International Journal of Applied Engineering Research 2017: 12(15), pp. 5214-5217

Dvoynikov, M., Syzrantsev, V., Syzrantseva, K. Designing a high resistant, high-torque downhole drilling motor // International Journal of Engineering, Transactions A: Basics 2017: 30(10), pp. 1615-1621
Dvoynikov, M.V., Blinov, P.A. Analysis of incident causes while directional and horizontal wells drilling // International Journal of Applied Engineering Research 2016: 11(20), pp. 10039-10042
Kamyshnikov G. & Kolpakov A. 2018. Deloitte. Review of the oilfield service market of Russia
Leusheva, E., Morenov, V. Research of clayless drilling fluid influence on the rocks destruction efficiency (2017) International Journal of Applied Engineering Research 12 (6), p. 945-949
Malikova, O. & Davydenko A. 2011. Prospects of development of the Russian oilfield service market Public service 5: 35-39.
Melekhin, A.A., Chernyshov, S.E., Blinov, P.A., Nutskova, M.V. Study of lubricant additives to the drilling fluid for reducing the friction coefficient during well construction with rotary steerable system // *Neftyanoe Khozyaystvo - Oil Industry* 2016: (10), pp. 52-55
Morenov, V., Leusheva, E., Martel, A. Investigation of the fractional composition effect of the carbonate weighting agents on the rheology of the clayless drilling mud (2018) International Journal of Engineering, Transactions A: Basics 31 (7),p. 1152-1158 DOI: 10.5829/ije.2018.31.07a.21
Morenov, V., Leusheva, E. Development of drilling mud solution for drilling in hard rocks (2017) International Journal of Engineering, Transactions A: Basics 30(4), p. 620-626 DOI: 10.5829/idosi.ije.2017.30.04a.22
Morenov, V., Leusheva, E. Energy delivery at oil and gas wells construction in regions with harsh climate (2016) International Journal of Engineering, Transactions B: Applications 29(2), p. 274-279 DOI: 10.5829/idosi.ije.2016.29.02b.17
Nutskova, M.V., Kupavyh, K.S. Improving the quality of well completion in deposits with abnormally low formation pressure // International Journal of Applied Engineering Research 2016: 11(11), pp. 7298-7300
Ulyasheva N.M. et al. 2016. Technology solutions for construction of wells in hardpan clay breeds. Oil economy 4: 82-86.

Youth Technical Sessions Proceedings – Litvinenko (ed)
© 2019 World Petroleum Council, London, UK, ISBN 978-0-367-34668-3

Topology and dynamic characteristics advancements of liner casing attachments for horizontal wells completion

A.A. Kunshin, M.V. Dvoynikov & P.A. Blinov
Saint Petersburg Mining University, Saint Petersburg, Russia

ABSTRACT: Construction of extended-reach wells is accompanied by significant capital expenditure and risks at all stages of construction. Lots of drilling problems and accidents arise during drilling and completing of such wells. When a filter is running into a well, the most common problems are slacking off and its inability to reach the projected depth. These problems may be related to the uncontrollable geomechanical processes in the hole, which are caused by the deformation of rocks (Dvoynikov 2017). Liner design includes a filter that is installed directly in the productive interval zone (Hossain & Abdulaziz 2015). To reduce the risk of such accidents, the liner can be connected to rigid power sub or to a swivel. The liner swivel rotates together with the stationary part including the filter. The operational characteristics and the total life of the well are decreased when the liner's structural integrity is destroyed.

1 MISSION

Designing the filter part of the liner with optimal parameters in stress-deformation state will enhance its reliability and durability during well production.

2 RESEARCH

Spiral surfaces withstand the impact of strong rocks without deforming plastically and they are able to provide a torque to the filter, preventing it from sticking (Morenov & Leusheva 2016, Nikolaev & Leusheva 2016).

The concept of the developed filter design is based on the following provisions: preservation of the filter integrity; filter is equipped with right screw surface; the helix angle must rotate filter during running.

The new filter design development. It is necessary to analyze existing models and select the criteria for evaluation according to its strength factors.

The digital design was carried out by using a computer-aided design (CAD) system, which is used for creating digital industrial prototypes of products. Subsequent mathematical calculations for strength and momentum were produced in Microsoft Excel.

Designing filter OTTM-102 with a multicast arrangement of holes and with a density of 20 holes per 1 meter is considered.

The model is divided into finite elements (finite element analysis – FEA), which together form a grid (in our model: 29,562 nodes; 40,834 elements) to calculate the stress distribution on the filter body during static, parametric and modal analysis. The FEA in CAD is based on Delaunay Triangulation algorithms and that makes it possible to create absolutely automatic finite element CAD. Delaunay Triangulation helps to reduce miscalculations because in FEA error is inversely proportional to the sine of the minimum/maximum grid angle.

In course of the work, a digital industrial prototype of the filter acting as a digital twin was modeled to reduce costs at subsequent stages of actual product operation. The developed product can reduce the likelihood of accidents (like a sticking, jamming and etc.), to improve technological, technical and economic indicators of well completion and well operation Tabatabaee Moradi & Nikolaev 2018). The final stage of product creation consists in the application of additive metal 3D printing technologies.

Additive technologies – layer-by-layer building up and synthesis of an object using 3D computer technologies, 3D printing technology.

Presented a technical solution aimed at one of the new and rapidly developing areas of Industry 4.0 – printing products from steel and alloys, which has analogues in the world, but has no analogues in the technology of manufacturing products and cheapness of production.

2.1 *Main results of the research*

The results of the calculations shown that the offset is 0-3 mm, margin of safety is 0.67-15 units, and the fluctuation of body filter is 0-556.1 mm.

Additionally, the following parameters are calculated:

polar moment of inertia, м^4; polar modulus (Chernilevskiy 1982), m^3; moment of friction (Dvoynikov & Ovchinnikov 2014), Nm; the sum of torques, Nm; rotation power used, W; maximum shear stress, MPa.

Calculation of stresses, fluid rheology and modeling of sticking in different sections of borehole makes it possible to identify zones, where the filter has a high-risk of plastic deformation (Lyagov et al. 2014).

The bottom hole assembly of completion with filter prototype presented on Figure 1 and Figure 2.

The developed mathematical tool enables to analyze the filters with arbitrary geometric complexity, select the criteria for swivel installation between sections of the filter and modify the liner running parameters (Lebedev at al. 2015). Graphs showing the distribution of summed torsional moments shown on Figure 3 and the maximum shearing strength shown on Figure 4

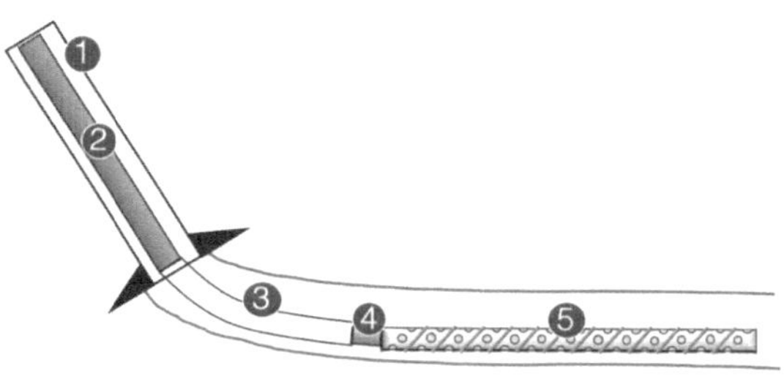

Figure 1. Completion assembly (simplified view). 1 – casing; 2 – drill string; 3 – liner; 4 – swivel; 5 – filter.

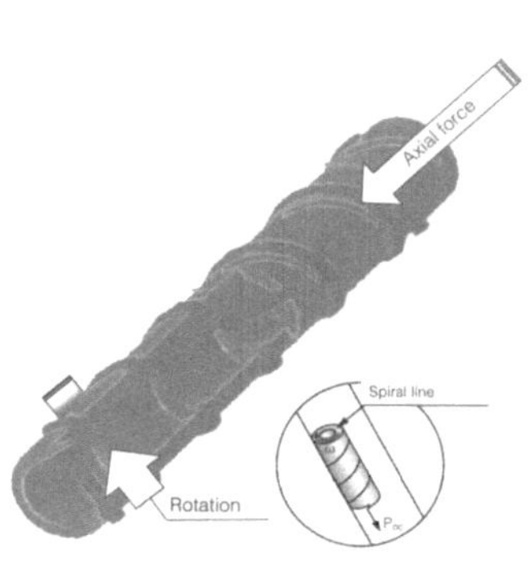

Figure 2. Design of the filter. Poc – load; ω – rotational speed.

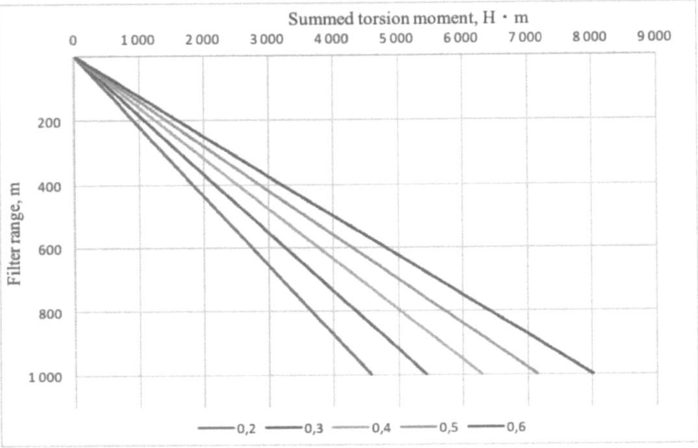

Figure 3. Summed torsional moments-filter range at resistance factor graph. K_c ranging from 0.2 to 0.6.

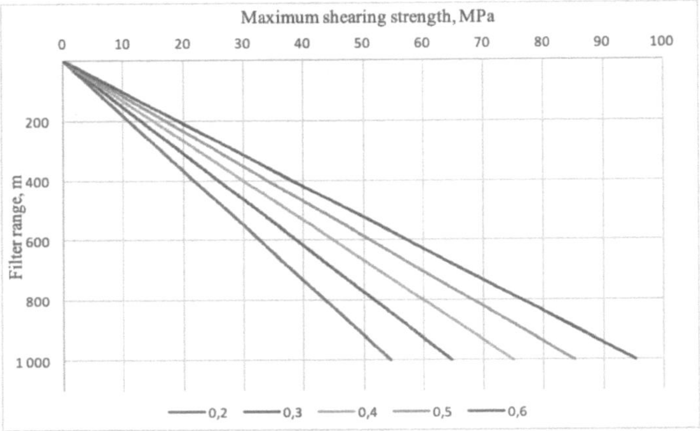

Figure 4. Maximum shearing strength distribution-filter range at resistance factor graph. K_c ranging from 0.2 to 0.6.

were drawn using results of calculations to make adjustments in the assembly of the filter. These results make possible to identification of zones with increased risk of deformations.

2.2 *Guiding turbulator development*

The guiding turbulator with adjustable angle of extrusions spiral blades to simplify production and reduce production cost is proposed as an alternative of filter developed (Bratvold & Begg, 2010, Kozhevnikov et al. 2015). Parametric analysis using to simulate stress state and select materials, a digital and grid model of guiding tabulator were created and shown on Figure 5.

Splines which has been placed on the inner surface of models are designed for completion mud motion and operation fluid motion stabilization (Pospehov et al. 2017).

The guiding tabulator installation on the filter is carried out by using of pins, which screwed and provide for its reliable fixation.

Product was printed by PLA-plastic, based on the grid model and shown on Figure 6.

PLA-plastic (polyactide) is an ideal material for prototypes and products of 3D-printing, as it is biodegradable, clear for usage and utilization (Pospehov et al. 2018).

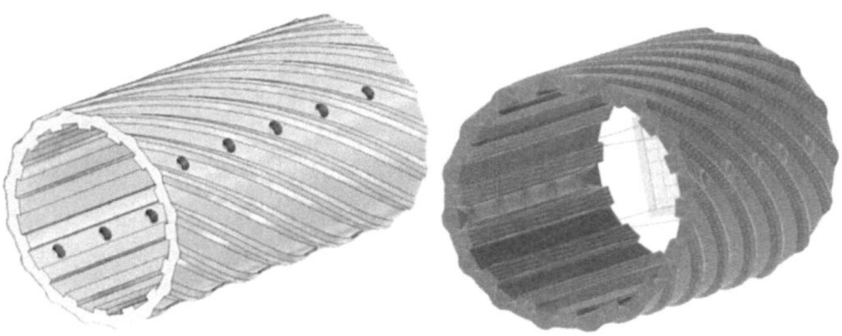

Figure 5. Digital and grid models of guiding tabulator.

Figure 6. Guiding turbulator prototype.

Advantages in comparison with the classical metal products manufacture are: complex work, consisting of technical and technological operations is absent (turning process, drilling, carbonization, hardening, tempering, determining accuracy grades, etc.), working process simplification and 8-10 times decrease in manufacturing cost (given the average estimated value), flexible property distribution along the product is possible (Leusheva & Morenov 2015).

The developed product is planned to be printed in steel on an assembled 3D printer capable of printing with metal. 3D model of printer working head presented on Figure 7. A different manufacturing technologies were applied, such as replacing the existing one – the technology of direct metal sintering with a laser.

Technology of noncontact heating of the conductor by high frequency and high values currents. The electromagnetic field induces eddy currents in the workpiece, which heat the conductor under the action of Joule's heat (Leusheva & Morenov 2017a, b).

Before submitting the model to print, it is processed by artificial intelligence (AI) algorithms, which are based on the deep processing of the product by parametric analysis (Grinfeld, 2013).

The process of 3D metals printing consists of:

1. Creating a model, its correction and refinement after CAD.
2. Model optimization for additive production – in terms of ratio of the mass, extent, strength, constructability. Here, the model can take new properties – elasticity (including anisotropic), porosity, thermal conductivity, hygroscopy.

   Model optimization process stages: Monolithic model -> system of shells and lattices -> calculation of loads. Optimization process based on the results of calculation.

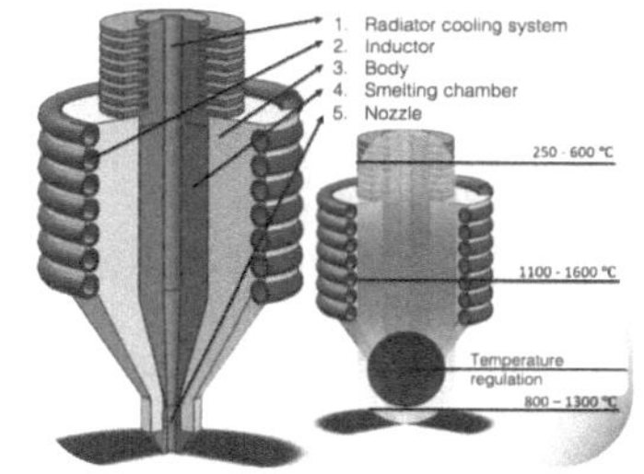

Figure 7. 3D printer working head for metals printing development (simplified form).

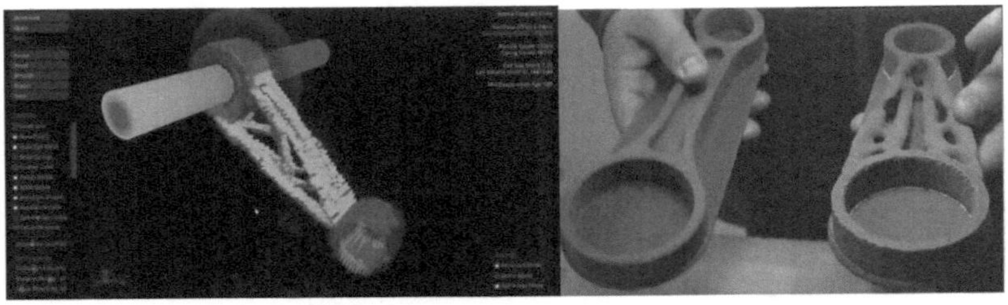

Figure 8. Preliminary analysis of the model and physical implementation of the production.

3. Preparing the model for printing – creating a support system for stabilizing the model in workspace, packing the parts into the printer‘s working area for simultaneous printing of several products at the same time, etc. Developing process with AI and CAD algorithms presents on Figure 8.
4. Model analysis and simulation of printing processes can predict and eliminate possible errors.
5. The stage of final product processing operation.

2.3 *Determine of guiding turbulators distance*

The distance between guiding turbulators is determined to avoid rock contact because of filter zone dead load deflection.

Input parameters are: filter diameter – 102 mm; bit diameter – 139.7 mm; external diameter of turbulator – 124 mm. Bending deflection should not exceed value more than 11 mm.

The calculations shown that in the given diametrical ratios between the bit and the filter with the maximum deflection of the middle part, the distance between guiding turbulators must not exceed more than 45.75 m.

It should be also noted that when 100 m long liner running in the horizontal section of the borehole with all guiding turbulator installed, for every 45.75 m, the moment of friction drops from 171.58 Nm to 14.2 Nm.

The proposed design and technology of liner running into extended reach well will improve the efficiency of well completion, and also increase the service life of the filter elements in the pay horizon during subsequent field operation.

Attention should not be limited only efficiency of the design description and we can to analyze the working process in a longer period of time.

Considering the filter as a body that does not touch the walls of the well, implies that the efficiency of the filter increases due to an increase in its effective surface area. The term effective surface area means that the filter completely allows the formation fluid through itself and does not lie on the wall of the well but is suspended by the guiding turbulators. The guiding turbulators are diametrically larger than the filter. This conclusion confirms the fact that the proposed filter design will increase the efficiency not only the completion operation at the stage of well construction, but also entire life of the well.

ECONOMIC RESULTS

The filter integrity is preserved. The time spent on the completion operation is reduced. The risk of accidents during filter running is minimized, due to its independent rotation through the swivel in the narrow areas of the wellbore. The surface of the filter is completely drains the formation fluid. This is achieved through spacing the guiding turbulators at 45.75 m from each other. A significant amount of cash is saved: $ 20.6 million per well and $ 260 million for 100 wells.

3 CONCLUSIONS

Progress is being made in designing complicated technologies. However, oil and gas projects are costly, and therefore optimizing cash costs and increasing work efficiency are very essential. The project fulfills these two main points: it saves money and increases the efficiency of well construction and its operation throughout the entire development of the field.

Products for analyzing and simulating metals printing processes will drastically reduce the cost and speed up the production process, eliminating a lot of mistakes at the stage when significant financial and material resources are not spent on them.

Automating creation of support structure tools, automatic packing of many parts into the printer workspace and fine-tuning of control programs will further accelerate and reduce the cost of production.

REFERENCES

Bratvold, R.B. & Begg, S.H. 2010. *Making Good Decisions*. USA: SPE.

Chernilevskiy, D.V. 1982. *Technical mechanics*. Moscow: Nauka.

Dvoynikov, M.V. 2017. *Research on technical and technological parameters of inclined drilling*. Journal of Mining Institute. Vol. 223. Saint-Petersburg: SPMU.

Dvoynikov, M.V. & Ovchinnikov, V.P., et al. 2014. *Technology of drilling oil and gas wells: textbook for-University students*. Vol. 1. Tyumen: TYUIU.

Grinfeld, P. 2013. *Introduction to Tensor Analysis and the Calculus of Moving Surfaces*. USA: Springer.

Hossain, M.E. & Abdulaziz Abdullah Al-Majed. 2015. *Fundamental of Sustainable Drilling Engineering*. Wiley: Scrivener Publishing LLC.

Kozhevnikov, E.V., Nikolaev, N.I., Melekhin, A.A., Turbakov, M.S. Studying the properties of cement slurries for cementing oil wells with long horizontal section drilled with rotary steerable system Neftyanoe Khozyaystvo - Oil Industry 2015: (9), pp. 58-60

Lebedev, V.A., Leusheva, E.L., Morenov, V.A. Complex power supply at well drilling in complicated climate conditions Journal of Mining Institute. 2015. V. 213. P. 47-53

Leusheva, E.L., Morenov, V.A. Combined oilfield power supplying system with petroleum gas utilization as an energy carrier // Neftyanoe khozyaystvo - Oil Industry 2015: (4), pp. 96-100

Leusheva, E., Morenov, V. Research of clayless drilling fluid influence on the rocks destruction efficiency (2017) International Journal of Applied Engineering Research 12(6), p. 945-949

Leusheva, E.L., Morenov, V.A. Development of combined heat and power system with binary cycle for oil and gas enterprises power supply // Neftyanoe Khozyaystvo - Oil Industry 2017: 7, pp. 104-106

Lyagov, I., Vasilev, N., Reich, M., Mezzetti, M. Analytical research and experimental tests on the technology fov drilling small diameter channels with small radius of cuvature // Oil Gas European Magazine 2014: 40(3), pp. 124-129

Morenov, V., Leusheva, E. Energy delivery at oil and gas wells construction in regions with harsh climate // International Journal of Engineering, Transactions B: Applications 2016: 29(2), pp. 274-279

Nikolaev, N.I., Leusheva, E.L. Increasing of hard rocks drilling efficiency // Neftyanoe Khozyaystvo - Oil Industry 2016: (3), pp. 68-71

Pospehov, G., Pankratova, K., Straupnik, I. Geoengineering researches for the restoration of the lands disturbed by mining operations 2018 Engineering and Mining Geophysics 2018-14th Conference and Exhibition

Pospehov, G.B., Straupnik, I.A., Pankratova, X.V., Kamenchukova, E.V. Geoengineering specifics of the tailings characteristic for their storaging 2017 Engineering Geophysics 2017

Tabatabaee Moradi, S.S., Nikolaev, N.I. Assessment of the cement failure probability using statistical characterization of the strength data Journal of Petroleum Science and Engineering 2018: 164, pp. 182-188

Youth Technical Sessions Proceedings – Litvinenko (ed)
© 2019 World Petroleum Council, London, UK, ISBN 978-0-367-34668-3

Combined cooling heat and power supplying scheme for oil and gas fields development

E.L. Leusheva, V.A. Morenov & A.S. Martel
Saint Petersburg Mining University, Saint Petersburg, Russia

ABSTRACT: Problems of combined power supply at oil extraction process in complicated climate conditions are considered in the article. Option of associated petroleum gas utilization in gas turbine power units for energy supply of oil recovery operations is offered. Schemes of electricity, heat and cold generation for oil field power management are presented. Combined cooling heat and power system allows performing of energy distribution corresponding to workloads and consumer needs to create efficient enterprise power structure. Implementation of simultaneous power supply of oilfield objects can be advised on sites with considerable amounts of extracted associated petroleum gas.

1 INTRODUCTION

Power supply at oil recovery operations in complicated climate is characterized by increased required amounts of electric and heat energy. Furthermore, power generation and consumption is greatly affected by changes of electric and heat loads and ambient air temperature drops (Oshibkov et al. 2015a, b). Despite such harsh conditions there should be constant supply of high-quality power to consumers. At the same time centralized power grid can't always guarantee stable and efficient power supply: cases of emergency powerdowns have frequented, power lines construction costs have risen, besides quality of electricity supplied by centralized system may vary greatly (Dvoynikov et al. 2016). In addition, if oilfield is constructed at remote locations it is not feasible to build new power lines due to high expenditures. So often, the only possible way of supplying oilfield objects with energy is to use self-sufficient power sources. Diesel power units are often used for these purposes, but long transportation distances may increase cost of diesel fuel by more than 200% making their maintenance quite expensive (Dvoynikov et al. 2017).

Oil recovery is usually accompanied by vast amounts of associated petroleum gas (APG), with extracted volume up to several thousand cubic meters. APG has value as an energy source as well as a resource for petrochemistry, but conducted research revealed that more than 50 billion m^3 of gas is burnt annually in Russia (Pospehov et al. 2017a, b).

Increasing fees for excessive APG flaring are making oil and gas operators to be more rational about its utilization. While gas transfer to gas refinery plants remains to be unprofitable be-cause of the undeveloped transportation network, effective usage of APG as a fuel for gas energy generating plants allows providing consumers with cheap and quality electrical power. Russian government act #1148 of 08.11.2012 "About fee peculiarities for air pollution emissions of burnt associated petroleum gas at flaring installations and dissipation" defines obligatory level of APG utilization at oil and gas fields development as 95%. Some of the Russian operators were able to satisfy this demand on several oil fields, but still the majority of exploited objects, especially beyond the Polar circle, hasn't reached the required minimum. There extracted APG is mostly flared away and energy supplying is done by means of diesel stations (Kupavikh & Nutskova 2017).

Power generation using APG may be considered one of the most promising solution for power problem on remote oil production objects. Trend in transition from unified power grid to local self-sufficient power stations has become actual during last decade. This is due not only to system cost and quality parameters decreasing, such as cases of sudden drops in energy supplying, aging of the equipment, energy taxes rising, increased cost of power lines construction, de-creased energy quality, but also due to considerable distances of new oil and gas fields from centralized power lines (Podoliak & Blinov 2016, Dmitriev et al. 2016).

APG can be utilized as an energy carrier in gas generating units, such as gas turbines (GT) and gas reciprocating engines. Implementation of GT is seen as most effective, because of such advantages like flexibility of operation, work rate, environmental compatibility. The most ad-vanced type of GT is microturbines (MT), operating on gaseous or liquid fuel. Coefficient of performance (COP) of these devices is up to 35%, which can be compared with diesel or gas reciprocating stations (Kotlyakov et al. 2013, Litvinenko et al. 2014).

2 MATERIALS AND METHODS

GT have the highest coefficient of heat production on output power, thus devices used on oil and gas extraction objects are mostly efficiently utilized in combined cooling heat and power (CCHP) mode (or trigeneration mode). The process is shown in figure 1. This mode allows efficient use of primary energy source and is better suited for regions characterized by frequent and formidable temperature drops (Merzliakov & Podoliak 2017).

As COP of thermodynamical system is defined as (Boyce 2012):

$$\eta = \frac{W_{out}}{Q_{in}} \tag{1}$$

where W_{out}– work done by all system components, Q_{in}– energy introduced into the system
Then COP of CCHP structure based on MT:

$$\eta = \frac{N_{gt}^{e} + N_{r}^{h} + N_{ch}^{c}}{Q_{g}^{c}} \tag{2}$$

where N_{gt}^{e}, N_{r}^{h}, N_{ch}^{c} - electric, heat and cooling power of gas turbine, recuperator and chiller respectively, Q_{g}^{c} - heat of combustion of gas burnt in MT combustion chamber

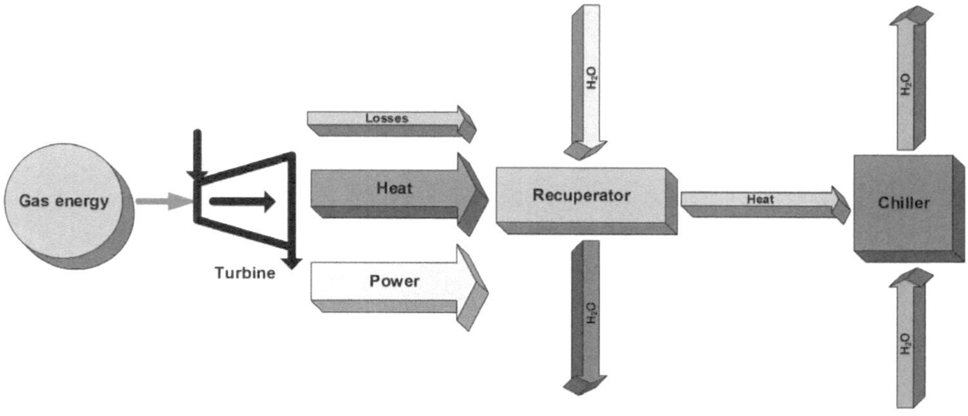

Figure 1. CCHP process.

3 DISCUSSION

Depending on power needs, charts of heat and power loads and ambient air temperature consumer may require different amounts of output power. With this feature in mind, the idea of the current project is to develop a power structure based on MT using APG in CCHP mode with ability of flexible regulation of output parameters. Conducted analysis showed the most effective exploitation method of power structure is combined use of turbines in Brayton-Renkine cycle and absorption chiller. In accordance with revealed functions, trigeneration power supplying scheme was developed utilizing APG as an energy source (figure 2).

In Brayton-Renkine cycle GT and steam turbine (ST) are operating in parallel, while steam is generated by utilizing high-temperature exhaust of GT.

In this case efficiency of GT part of the installation will be defined by the work of the turbine itself and the compressor (Boyce 2012).

Work of the compressor:

$$\dot{W}_c = \dot{m}_a(h_2 - h_1) \tag{3}$$

where $\dot{m}_a$-mass air flow, h_n-specific enthalpy of the fluid at n-point of T, S - diagram (figure 3).

Work of the turbine:

$$\dot{W}_t = (\dot{m}_a + \dot{m}_f)(h_3 - h_4) \tag{4}$$

where $\dot{m}_f$ - mass fuel flow

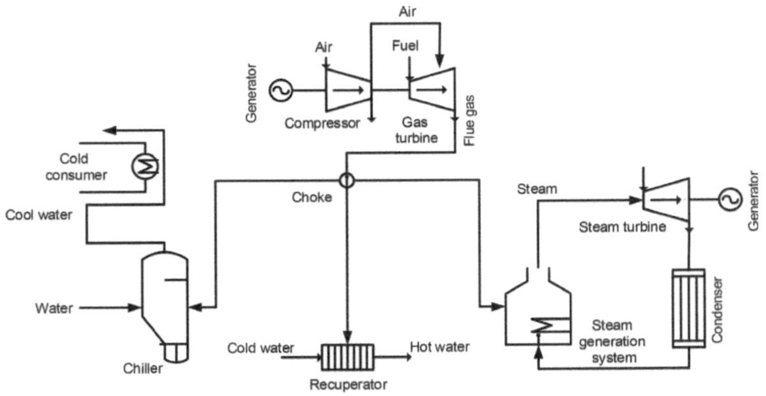

Figure 2. Basic power supplying scheme consisting of gas and steam turbine in combination with absorption chiller.

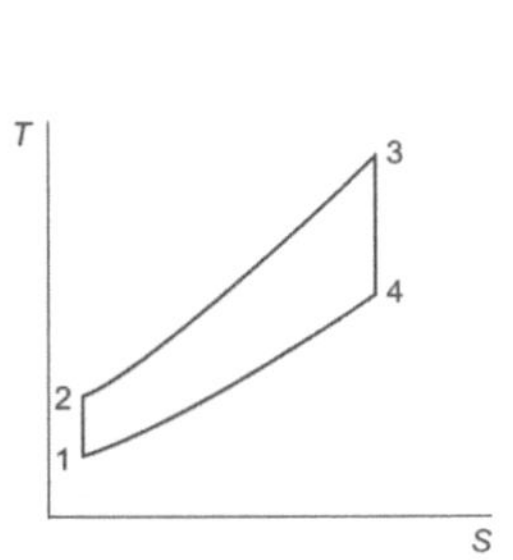

Figure 3. T, S – diagram of standard Brayton cycle.

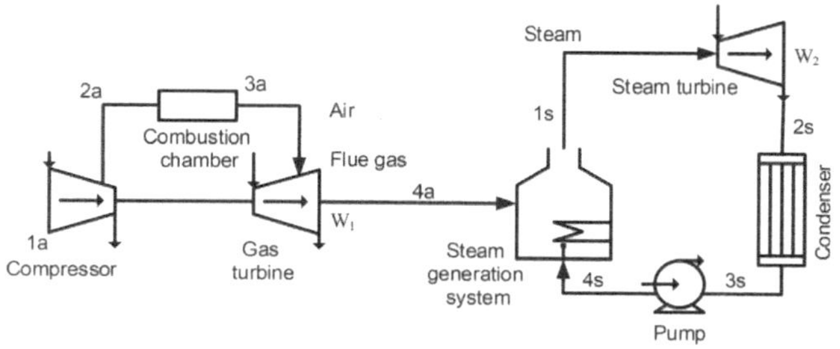

Figure 4. Brayton-Renkine cycle using MT and ST.

Efficiency of ST part of the installation will be defined by the work of the steam turbine and the pump, delivering condensed fluid from condenser to steam generation unit (figure 4).

Work of ST:

$$\dot{W}_{tS} = \dot{m}_s(h_{1S} - h_{2S}) \tag{5}$$

where $\dot{m}_s$- mass steam flow, h_{nS}- specific enthalpy of the fluid in n-area of the Brayton-Renkine cycle

Work of the pump:

$$\dot{W}_p = \frac{\dot{m}_s(h_{4S} - h_{3S})}{h_p} \tag{6}$$

Summarized work of the whole turbine installation:

$$W_{sum} = W_t + W_{tS} - W_c - W_p \tag{7}$$

If there is a source of cheap heat, which is high-temperature exhaust of GT, it is reasonable to use absorption chillers for cooling, which generate cold water from two products, water and lithium bromide salt. Heat is used to drive the device. Electrical energy is only used to power pump and electronics.

Chiller efficiency is defined by thermal COP:

$$\zeta = \frac{Q_0}{Q_h} \tag{8}$$

where Q_0- heat quantity taken from low temperature source, Q_h- heat quantity used to drive the chiller.

4 RESULTS

Acquired cold energy may be used for both industrial and living needs, such as air conditioning at hot periods. As an industrial example, GT efficiency improvement may be mentioned. Rated power of gas turbine is calculated at ISO conditions - ambient air temperature is 15 °C, specific humidity is 60 %, atmospheric pressure is 1,01 Bar. During hot periods ambient air temperature may exceed that value. With temperature growth efficiency of GT drops, but vice versa is also true.

In order to prevent the drop of power output during hot season scheme was developed which utilizes generated cold for GT inlet air cooling (figure 5).

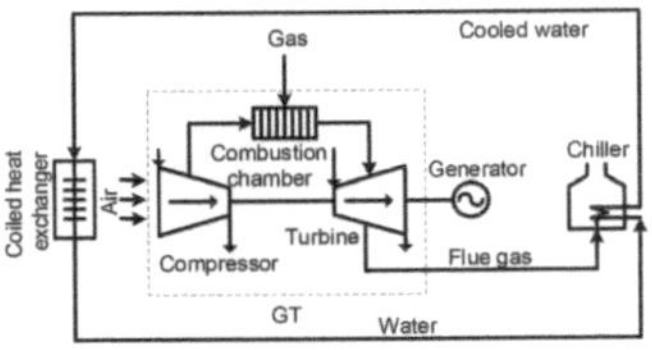

Figure 5. GT inlet air cooling scheme.

5 CONCLUSION

Concluding, it may be outlined that power supplying schemes under consideration may satisfy energy needs of oilfields while providing parameters corresponding with consumer load charts. Exploitation of turbine structure in Brayton – Renkine cycle allows generation of additional amount of electricity in case of abundance of heat. Adding chiller to power structure enables application of generated cold in GT inlet air cooling system that prevents power loss at high temperatures of ambient air during hot seasons. At the same time, APG is effectively utilized for energy efficiency increase of oil production and preparation. Efficiency of CCHP may exceed 90%. Power structure functioning without downfalls of produced energy during the year guaranties most full and efficient use of energy carrier. Applying such structure is relevant for remote areas with vast reserves of associated petroleum gas, where centralized energy grid is not available, such as regions located beyond the Polar circle.

REFERENCES

Boyce M.H. (2012). Gas turbine engineering handbook. Elsevier inc., p. 993

Dmitriev, A.N., Bolshunov, A.V., Podoliak, A.V. Assessment of ice borehole temperature conditions at interface with subglacial lake Vostok (Antarctica) // International Journal of Applied Engineering Research 2016: 11(11), pp. 7230-723

Dvoynikov, M., Syzrantsev, V., Syzrantseva, K. Designing a high resistant, high-torque downhole drilling motor // International Journal of Engineering, Transactions A: Basics 2017: 30(10), pp. 1615-1621

Dvoynikov, M.V., Blinov, P.A. Survey results of series-produced downhole drilling motors and technical solutions in motor design improvement // International Journal of Applied Engineering Research 2016: 11(10), pp. 7034-7039

Kotlyakov, V.M., Lipenkov, V.Y., Vasil'ev, N.I. Deep drilling in central Antarctica and penetration into subglacial Lake Vostok // Herald of the Russian Academy of Sciences 2013: 83(4), pp. 311-323

Kupavikh, K.S., Nutskova, M.V. Ecological features of oil well repair at low-permeability reservoir International Journal of Applied Engineering Research 2016: 11(11), pp. 7505-7508

Litvinenko, V.S., Vasiliev, N.I., Lipenkov, V. Ya., Dmitriev, A.N., Podoliak, A.V. Special aspects of ice drilling and results of 5G hole drilling at Vostok station, Antarctica // Annals of Glaciology 2014: 55 (68), pp.173-178

Merzliakov, M.I., Podoliak, A.V. Terms of melting the permafrost when cementing of boreholes plugged with gas-liquid mixtures with hollow microspheres // International Journal of Applied Engineering Research 2017: 12(9), pp. 1874-1878

Oshibkov A.V., Vodorezov D.D., Syzrantseva K.V., Frolov S.A., Dvoynikov M.V., Bakirov, D.L. Analysis of sidetracking characteristics influence on multilateral well construction reliability. *Neftyanoe Khozyaystvo - Oil Industry*. 2015. Issue 11. pp. 121-123

Oshibkov A.V., Vodorezov D.D., Syzrantseva K.V., Frolov S.A., Dvoynikov M.V., Svetashov V.N. The stress-strain condition estimation of construction of multilateral well junction. 2015. Neftyanoe khozyaystvo - Oil Industry. Issue 6. pp. 83-85

Podoliak, A.V., Blinov, P.A. The technology of directional drilling in ice via drills on carrying cable // International Journal of Applied Engineering Research 2016: 11(9), pp. 6411-6417

Pospehov, G.B., Pankratova, K.V., Straupnik, I.A., Ustiugov, D.L. Problems of land reclamation during liquidation of coalmining enterprises, 2017 IOP Conference Series: Earth and Environmental Science 87 (4),042015

Pospehov, G.B., Straupnik, I.A., Pankratova, X.V., Kamenchukova, E.V. Geoengineering specifics of the tailings characteristic for their storaging 2017 Engineering Geophysics 2017

Youth Technical Sessions Proceedings – Litvinenko (ed)
© 2019 World Petroleum Council, London, UK, ISBN 978-0-367-34668-3

Development of cement slurries for oil and gas wells lining in aggressive environment

M.Y. Merzlyakov, Jennifer R.R. Hernandez & Ch.A. Zhapkhandayev
Saint Petersburg Mining University, Saint Petersburg, Russia

ABSTRACT: The paper presents the results of laboratory studies of the influence of aggressive fluids containing $MgSO_4$, Na_2SO_4, $MgCl_2$ and $CaCl_2$ as well as porosity of hydrated cement have on its strength. Cement samples had been put into brine and after a fixed period of time were tested to measure ultimate compression strength and ultimate flexural strength.

This work represents data on physical and mechanical properties of polymer–cement slurries and cement stone formed on that basis, suitable for cementing wells in conditions of brine aggression. In order to increase corrosion resistance of produced concrete, plug-back mixtures were made with lowered water to cement ratio.

INTRODUCTION

In today's world, with the constant growth of the oil industry, there is a shortage of energy resources associated with the depletion of available and relatively cheap for the development of hydrocarbon reserves. It is necessary to search for new oil and gas fields located at great depths, where concentrated accumulations of aggressive substances such as hydrogen sulfide (H_2S), carbon dioxide (CO_2), sulfates (SO_4^{-2}) and chlorides (Cl^-) are much more common. Well construction in these conditions requires complex solutions (Blinov & Dvoynikov 2016, Blinov & Dvoynikov 2018).

The presence of hydrochloric acid components in formation fluids leads to corrosion destruction of the cement ring. Solution to this problem is particularly relevant to the following fields: Astrakhan, Kashagan, Zhetybai, Verkhnekamsk, East-Kamyshanskoe, Kalinin, Kurgan, Sostinskoe, Costanero in Bolivar (Venezuela) and a number of other deposits (Blinov et al. 2017, Blinov & Podoliak 2016).

The efficiency of well construction is inextricably linked with the quality of their attachment. Currently, grouting materials used for fixing wells in fields with aggressive reservoir fluids often do not provide the formation of a reliable cement ring and its necessary strength in a hostile environment (Dvoynikov et al. 2017, Dvoynikov & Blinov 2016). Under these conditions, there is a corrosive destruction of cement stone, which leads to a violation of the tightness of the annular space with the appearance of annular flows. It is obvious that the creation of grouting compounds resistant to aggressive environments is an urgent task. The creation of a sealed support is a very difficult task due to the significant degree of aggressive action of reservoir fluids (Leusheva & Morenov 2017, Morenov & Leusheva 2017). Penetrating through the capillaries and pores of the cement stone, they cause corrosion, leading to loss of stability of the casing and the destruction of the cement stone. Leakage of the lining leads to the formation of annular flow of fluids, which is unacceptable from the point of view of industrial and environmental safety. Unfortunately, existing methods, such as the use of protective coatings on the surface of the casing, are ineffective. The real way to protect metal structures is to develop ways to modify the cement stone in contact with the surface of the casing, in order to give it properties that reduce the intensity of corrosion processes (Agzamov 2005).

One the main purposes of cementing is to create a barrier, which would shield casing from corrosion by formation fluids and as a consequence from losing load capacity. The most aggressive components within formation fluids are: hydrogen sulfide, salts of magnesium, carbon dioxide, alkali metals sulphides, sulphates, chlorides and etc.

One of the main goals of cementing is to create a barrier that protects the casing from corrosion by formation fluids and, as a consequence, increase the bearing capacity.

Analysis of the studies on cement powder, carried out by Russian and foreign researchers, has showed that despite the rather broad range of corrosion-resistant powders, in most cases they do not provide an adequate cementing quality within aggressive rock formations and fluids (Mukhametshin 2017, Melekhin et al. 2016).

It is well known that corrosion-resistance of hydrated cement is directly dependent on its permeability, which can be reduced by lowering water to cement ratio (Nikolaev 2004).

In practice of casing cementing water to cement ratio is generally taken between 0,5 and 1,0 and even more, in order to provide effective pumping of cement slurries in harsh geological and thermo-baric conditions deep down in the well. However in so doing, equally important characteristics of cement sheath such as strength and tightness inevitably worsen (Nutskova & Kupavyh 2016, Nutskova et al. 2017).

In studies in conditions of aggressiveness of various salts it is proposed to use widespread cement grade portland cement 50 (PCT-I-50) and as a reagent-plasticizer it is proposed to use a high-molecular polymer - polyvinylpyrolidone (PVP) as a plasticity agent, due to its accessibility, non toxicity, organic solvent and aqueous solubility, good adhesion properties, stability and so on.

RESEARCH METHODS

In well conditions cement stone is affected by evolving temperature and pressure, that cause tensile, compressive and bending stresses which occur both simultaneously and separately. Furthermore because of low plasticity, cement stone undergoes unstable failure during perforation process and pressure testing, what leads to cross flows and premature water flooding. In this regard researchers has paid particular attention to strength properties of set cement.

In order to more objectively assess the corrosion resistance of cement stone, three different compositions were prepared. The compositions of grouting mixtures are presented in table 1. Salt solutions in different concentrations were prepared as aggressive media.

Laboratory studies were aimed at studying the physical and mechanical properties of cement stone and grouting mixtures with different concentrations of reagents. The main parameters included consistency, setting time, plastic strength, uniaxial compression and bending strength.

To assess the strength of cement was experimental study of cement grade PCT-I-50 with additives, PVP octanol and PLABET in different concentrations. To combat the increased foaming, octanol was introduced into the composition of the grouting mixtures in concentrations of 0.025 and 0.03 to the mass of dry cement, and as an increase in the strength and corrosion resistance of the cement stone, the plasticizer PLACBET was used, which also contributes to an increase in water resistance.

Table 1. Setting time of cement mixtures

№	The composition of the cement slurry	W/C	Beginning of setting, h	The end of the stiffening, h	Spreadability, cm
1.	PCT-I-50+PVP0,25%+1- octanol 0,025%+ PLACBET 0,99%	0,4	14	16,5	20,75
2.	PCT-I-50+PVP0,25%+1-octanol0,025 +PLACBET0,9%	0,45	14	18,5	20,75
3.	PCT-I-50+ quartz 30%+PVP0,25%+1-octanol 0,03%	0,5	6,5	9,3	20,25

The optimal percentages of reagents with the volume of mixing water and with each other were selected empirically using the theory of experiment planning, taking into account their compatibility of the main parameters of the resulting grouting mixture.

RESEARCH RESULTS

Based on the work on the study of physical and mechanical characteristics of cement stone and grouting solution, we have the following results.

From figures 1 and 2 it follows that the grouting solution of the first composition has the best performance. The curves show a more rapid development of the structure formation process for the third composition compared to the first and second, which can be explained by the content of the third composition of the addition of dust-like quartz, significantly increasing its structural viscosity. The second composition over time and has not gained sufficient plastic strength. It can also be noted that the greatest increase in plastic strength is observed in the period coinciding with the beginning of the setting of the cement paste, determined by the Vica device.

Studies of the spreadability of cement mortar show that when the concentration of PVP reagent in the polymer-cement composition is equal to 0.25%, its spreadability can be sufficient for values of water-cement ratio less than 0.5 (table 1). The introduction of PVP also leads to a positive change in the structural and rheological parameters of grouting compositions, such as the setting time (table 1) and consistency (figure 2).

Data on the strength characteristics of cement stone for uniaxial compression and bending show (figure 3, figure 4, figure 5, figure 6) that the first composition at a salt concentration of 3% and 5% has a greater advantage over other compositions. This behaviour can be explained by the fact that the polyvinylpyrrolidone has inhibitory properties and, together with the reagent, PLACBET at a concentration of 0.99% helps to reduce the filtering cement slurry, and permeability of cement stone, thereby increasing corrosion resistance.

This statement is confirmed by researches of V. S. Danyushevsky and A. I. Bulatov (Danyushevsky 1987, Bulatov 1987). Their work shows that when filling the pore space with a liquid or solid hydrophobic substance, the permeability of the cement stone decreases, which positively affects the corrosion resistance of the cement stone to all mineral aggressive media.

As the strength of the cement stone of the first composition increases, aggressive media can be arranged in the following sequence: $CaCl_2 \leq MgCl_2 \leq Na_2SO_4 \leq MgSO_4$. The strength value of the first composition in these media is much higher than other compositions.

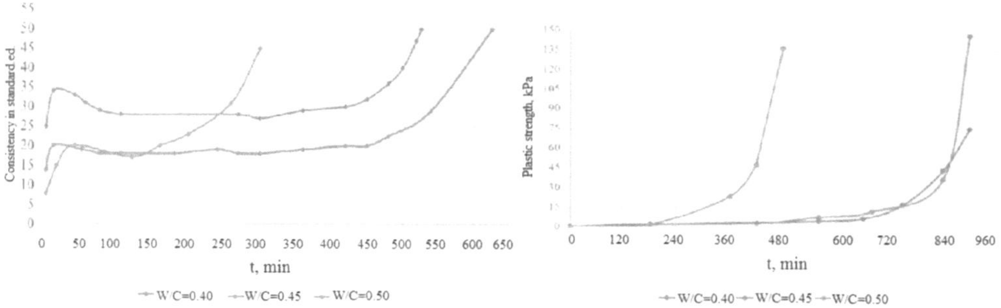

Figure 1. Graph of changes in plastic strength.

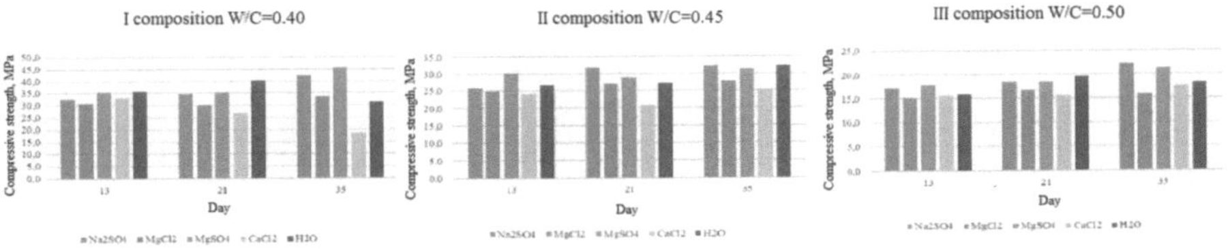

Figure 2. Graph of changes in the consistency of cement slurries

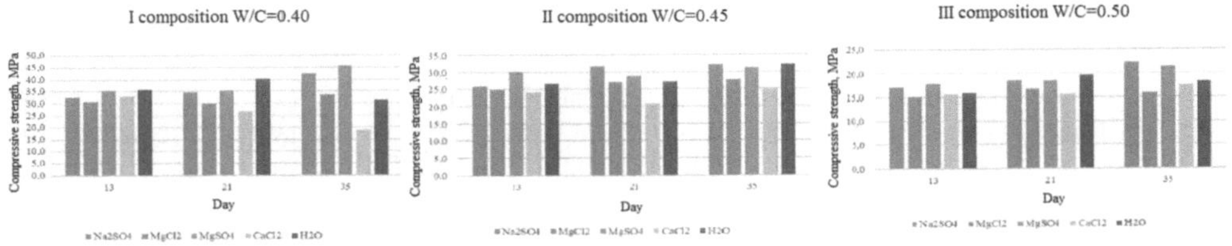

Figure 3. Results of compressive strength tests at salt concentration 3%

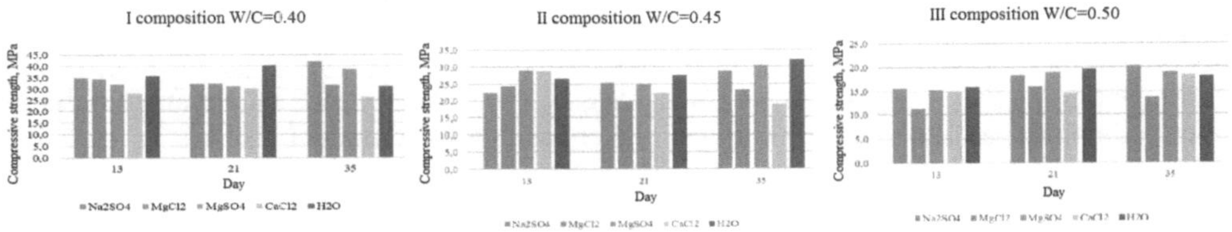

Figure 4. Results of compressive strength tests at salt concentration 5%

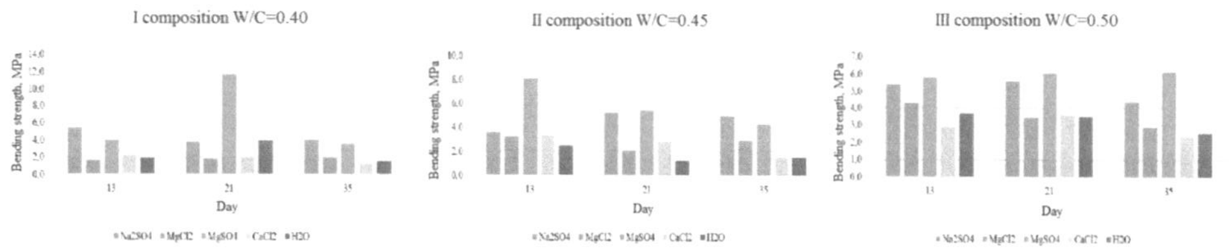

Figure 5. Results of bending strength tests at salt concentration 3%

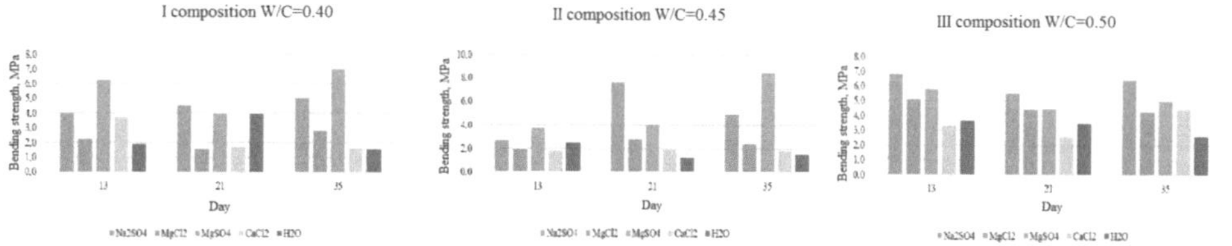

Figure 6. Results of bending strength tests at salt concentration 5%

Conclusion

Development of cement slurries resistant to aggressive formation fluids exposure, is an important complex task whose solution would increase maintenance-free lives of wells, and not least provide a greater environmental security.

Improved corrosive resistance of cement sheath can be achieved by reducing permeability (porosity) of cement stone by using cement-slurry mixtures with lowered content of dispersion medium .

Application in the composition of cement mixtures of high-molecular reagent-plasticizer PVP helps to improve the pumpability of cement mortar at/C less than 0.5. It was established experimentally that the addition of PVP and PLACBET increase corrosion resistance and less chemically active compared to salts.

The developed composition improves the quality of well fixing due to the absence of destructive processes under the influence of salts, as well as the absence of through channels in the volume of cement stone, due to sedimentation resistance.

It should be noted that the improvement of formulations of cementing compositions can be pursued through the formulation of the problem of complex processing of backfill slurry during the solidification of Portland cement, the traditional brands, and improve corrosion resistance.

REFERENCES

Agzamov F.A., Izmukhambetov B.S. Dolgovechnost' tamponazhnogo kamnya v korrozionnoaktivnykh sredakh./Agzamov F.A., Izmukhambetov B.S.//- SPb.: OOO «Nedra», 2005.-318.

Blinov, P.A., Dvoynikov, M.V. Rheological and filtration parameters of the polymer salt drilling fluids based on xanthan gum // Journal of Engineering and Applied Sciences 2018: 13(14), pp. 5661-5664

Blinov, P.A., Dvoynikov, M.V. The process of hardening loose rock by Mud Filtrat // International Journal of Applied Engineering Research 2016: 11(9), pp. 6630-6632

Blinov, P.A., Dvoynikov, M.V., Sergeevich, K.M., Rustamovna, A.E. Influence of mud filtrate on the stress distribution in the row zone of the well // International Journal of Applied Engineering Research 2017: 12(15), pp. 5214-5217

Blinov, P.A., Podoliak, A.V. The method of determining the effects of drilling fluid on the stability of loose rocks // International Journal of Applied Engineering Research 2016: 11(9), pp. 6627-6629

Bulatov A.I., Danyushevskiy V.S. Tamponazhnye materialy: ucheb. posobie dlya vuzov./Bulatov A.I., Danyushevskiy V.S. // M.: Nedra, 1987.– 280.

Danyushevskiy V.S., Aliev R.M., Tolstykh I.F. Spravochnoe rukovodstvo po tamponazhnym materialam./Danyushevskiy V.S., Aliev R.M., Tolstykh I.F.//– M.: Nedra, 1987.– 373.

Dorovskikh I.V. Obosnovanie i poluchenie korrozionnostoykikh tamponazhnykh materialov so smeshannoy kondensirovannoy fazoy dlya stroitel'stva skvazhin: Avtoref. dis. . . .kand. tekhn. nauk. – SPb.: SPGGU, 2011. – 20.

Dvoynikov, M., Syzrantsev, V., Syzrantseva, K. Designing a high resistant, high-torque downhole drilling motor // International Journal of Engineering, Transactions A: Basics 2017: 30(10), pp. 1615-1621

Dvoynikov, M.V., Blinov, P.A. Analysis of incident causes while directional and horizontal wells drilling // International Journal of Applied Engineering Research 2016: 11(20), pp. 10039-10042

Kravtsov V.M., Kuznetsov Yu.S., Mavlyutov M.R., Agzamov F.A. Krepleniye vysokotemperaturnykh skvazhin v korrozionnoaktivnykh sredakh/Kravtsov V.M. [i dr.], Kuznetsov YU.S., Mavlyutov M.R., Agzamov F.A./.- M.: Nedra, 1987-190.

Leusheva, E., Morenov, V. Research of clayless drilling fluid influence on the rocks destruction efficiency // International Journal of Applied Engineering Research 2017: 12(6), pp. 945-949

Melekhin, A.A., Chernyshov, S.E., Blinov, P.A., Nutskova, M.V. Study of lubricant additives to the drilling fluid for reducing the friction coefficient during well construction with rotary steerable system // Neftyanoe Khozyaystvo - Oil Industry 2016: (10), pp. 52-55

Morenov, V., Leusheva, E. Development of drilling mud solution for drilling in hard rocks (2017) International Journal of Engineering, Transactions A: Basics 30(4), p. 620-626

Mukhametshin V.V., Kadyrov R.R., Zhirkeyev A.S., Zeygman Yu.V. Novyye tamponiruyushchiye materialy na osnove modifitsirovannogo tsementa i polimerov./Mukhametshin V.V., Kadyrov R.R., Zhirkeyev A.S., Zeygman YU.V. // Neftepromyslovoye delo 1/2017. 41-46.

Nikolayev N.I./Nauchnaya osnova dlya beztrubnogo perekrytiya razvedochnykh skvazhin./Nikolayev N.I. [i dr.]//SPb: Nauka v Gornom universitete. 1997.
Nikolaev, N.I., Nifontov, Y.A., Derrnov, D.A., Toyb, R.R. Generatsiya otechestvennykh polimerov dlya bureniya skvazhin./Nikolaev, N.I., Nifontov, Y.A., Derrnov, D.A., Toyb, R.R. 2004 // Sovremennaya industriya (1): 2004-39–44.
Nutskova, M.V., Dvoynikov, M.V., Kuchin, V.N. Improving the quality of well completion in order to limit water inflows // Journal of Engineering and Applied Sciences 2017: 12(22), pp. 5985-5989
Nutskova, M.V., Kupavyh, K.S. Improving the quality of well completion in deposits with abnormally low formation pressure // International Journal of Applied Engineering Research 2016: 11(11), pp. 7298-7300
Osadchaya I.L. Povyshenie korrozionnoy stoykosti tamponazhnykh materialov v usloviyakh serovodorodnoy agressii: Avtoref. dis.... kand. tekhn. nauk. – Stavropol': OAO «SevKav. NIPIgaz», 2013. – 25.
Tabatabaee moradis S.H., Nikolaev N.I. Optimizatsiya modeli reologii tsementnykh prokladok s ispol'zovaniyem geneticheskogo algoritma./Tabatabaee moradis S.H., Nikolaev N.I. // Mezhdunarodnyy inzhenernyy zhurnal (1): 43-51.

Youth Technical Sessions Proceedings – Litvinenko (ed)
© 2019 World Petroleum Council, London, UK, ISBN 978-0-367-34668-3

Investigating of compositions for lost circulation control

M.V. Nutskova, E.Y. Rudiaeva, V.N. Kuchin & A.A. Yakovlev
Saint Petersburg Mining University, Saint Petersburg, Russia

ABSTRACT: In this work, compositions for isolation of the thief zones are investigated. To define the conditions for polymer application, the kinetics of it's swelling in different mediums are assessed. Laboratory investigations showed that the swelling property of the reagent depends on the medium pH, which can be used to adjust the dynamics of structure formation in different technological schemes of lost circulation control.

1 INTRODUCTION

Drilling fluid lost circulation is considered as one of the most common types of difficulties during well drilling. After defining the lost circulation category, it's control action will be selected, while the criteria for selection are economical effectiveness, implementation simplicity and work experience in similar geological conditions (Leusheva 2017, Turitsyna 2010, Vasilev 2016).

Researchers suggest different classifications of lost circulation and different methods to select the proper action of their prevention and control. Criteria are the lost circulation intensity (or the coefficient of lost circulation intensity), coefficient of fluid intake capacity (or the specific injectivity index) and the fracture opening. Based on fluid intake capacity the technology is selected and based on dimensions of the intake channels the filler agent is selected. The existed classifications are usually valid for specific regions and therefore for other geological and techno-technological conditions have informative character for selecting methods, which roughly can be used in every specific case. There are several methods of lost circulation control: decreasing the hydrodynamic and hydrostatic pressure on the wellbore wallby stopping circulation, decreasing the flow rate or density of the drilling fluid, mud fill of the channels in the thief formations by adding fillers into drilling fluid, tamping by viscoelastic compositions, pastes, cements mixtures (non-thickening, fast-setting etc.) (Blinov 2018, Kozhevnikov 2015, Kupavikh 2016, Melekhin 2012, Merzliakov 2017, Moradi 2017, Moradi 2018a, Moradi 2018b, Nikolaev 2009a, Nikolaev 2009b, Nikolaev 2016a, Poliakov 2009, Tabatabaee Moradi 2016, Tabatabaee Moradi 2018, Turitsyna 2009, Turitsyna 2010), applying main sealer or running intermediate casing (including the ones without access to surface) (Dvoynikov 2017).

In order to decrease the time of lost circulation control, it isreasonable to use the temporary isolation of permeable layers. To do this, the flush fillers, pumping viscoelastic compositions, non-thickening cement and fast-setting compositions are usually used (Dvoynikov 2017, Morenov 2017, Morenov 2018, Nikolaev 2009a, Nikolaev 2016b Nutskova 2017, Poliakov 2009, Turitsyna 2009, Turitsyna 2010). One of the important directions is to develop the water-swellable polymeric compositions for efficient tamping (Nikolaev 2009b).

2 RESEARCH METHODOLOGY

To develop the compositions for temporary isolation of lost circulation layers the following method of investigations are suggested:

Step I. Investigating the effect ofdrilling fluid medium on the performance of the water-swellable polymer:

1. Defining the performance character of the water-swellable polymer in the water medium with different pH values (in static and dynamic conditions);
2. Investigating the performance of the polymer by changing the medium property (mixing the fluid with polymer and formation water model, abandonment in rest);
3. Defining the performance of water-swellable polymer in the composition of the drilling fluid (in static and dynamic conditions);

Step II. Investigating the performance of the water-swellable polymer in simulated lost circulation medium:
1. Developing bench model of lost circulation layer;
2. Saturating the model by the fluid of water-swellable polymer with different properties;
3. Defining the maximum pressure, above which the polymer exits from the model of lost circulation layer.

The consistency or mobility of the fluid, which is a time dependent value, is considered as an important characteristic during pumping the tamping composition into the lost circulation interval. The other important parameter for compositions that are used to control lost circulation, is the rheological properties, i.e. the plastic viscosity and dynamic shear stress (yield point), which should have minimum value until the fluid reaches the tamping zone, and maximum value, when the fluid is filling the channels of the permeable layer. Therefore, changes of plastic viscosity and dynamic shear stress with respect to time are studied to evaluate the performance of the water-swellable polymer in the water medium and in the composition of the drilling fluid. Following time intervals are considered in the investigations:

1. 0-36 minutes –from the preparation of the fluid until it's pumping into the lost circulation interval;
2. 36 minutes after preparation – time of fluid penetration into the lost circulation layer and it's interaction with the formation fluid;
3. 15 minutes in the rest and 24 hours after preparation –simulating the fluid staying in the layer.

The rheological parameters are measured on a6-speed rotational viscometer Fann 35SA with constant displacement in the first time interval (simulating the fluid movement in the drill string). During the investigations of the polymer water solutions, the rheological measurements are not conducted after 24 hours. The super absorbent reagent «Petrosorb»is used in the investigations. The reagent currently is not used as the lost circulation control agent, but has a perspective in this field (Nikolaev 2009a, Nikolaev 2009b). The 2% concentration of «Petrosorb» is selected for the investigations, as it has been shown in the past conducted researches that concentration for sealing the lost circulation layers is 1-3% during application of the mud fluids. However, to select the appropriate concentration of the reagents in the drilling fluid to control the lost circulation, investigation on the compatibility of the fluid with the reagents and kinetics of the fluid properties change with the time, should be conducted.

To analyze the effect of medium acidity on the performance of the «Petrosorb», water solutions of the reagent with pH values of 4, 7 and 10 are investigated.

The past studies of «Petrosorb» showed its effectiveness in the mud fluids, which is reasonable during drilling underconductor, surface casing and in some cases intermediate casing. However, such composition is not suitable in the drilling intervals under the production casing. Therefore, it is reasonable to conduct further investigations on the drilling fluids that are applicable for the well completion (polymer and biopolymer fluids, calcareous, carbonate, fluids on the hydrocarbon basis etc.)

3 EXPERIMENTAL

This section shows studies of biopolymer drilling mud, weighted up by the barite (to provide the possibility of investigating compositions with different pH). Investigation is conducted on the plastic viscosity and dynamic shear stress changes in the moment of fluid penetration into the lost circulation layer and mixing with the formation water:

1. In the moment of adding «Petrosorb» into receiver tank,
2. Moment of achieving the bottom of the well, before penetration into the lost circulation layer,
3. Moment of interaction with formation fluid (for two pH values),
4. After resting in the layer for 15 minutes,
5. After resting in the layer for 24-hours period.

Similar to the investigations in the water medium, until penetration into the lost circulation layer the rheological indexes are more stable in the acidic drilling fluid medium (pH = 4) in comparison to basic medium (pH = 12). The relation is presented in figures 1 and 2.

Neutral water formations are rarely available at depths of more than 1000 meters, therefore 2 cases are analyzed:

1. Acidic drilling fluid penetrates into the lost circulation layer with basic water formation;
2. Basic drilling fluid penetrates into the lost circulation layer with acidic water formation.

The analysis is carried out similar to the investigations in the water solution of polymers, however, one more time interval is added: to abandon the fluids in rest (simulation of fluids being in the layer).

Results are presented in the figure 3.

Before pumping the drilling fluid with polymer into the lost circulation layer, plastic viscosity and dynamic shear stress have high enough values. Decrease of these indexes in the time of penetration into the layer is caused as result of water content increase, while the «Petrosorb» particle swelling is not stopped. By passing time, the polymer, locating in the lost circulation layer, continues to grow in sizes to control the lost circulation by mud filling the voids.

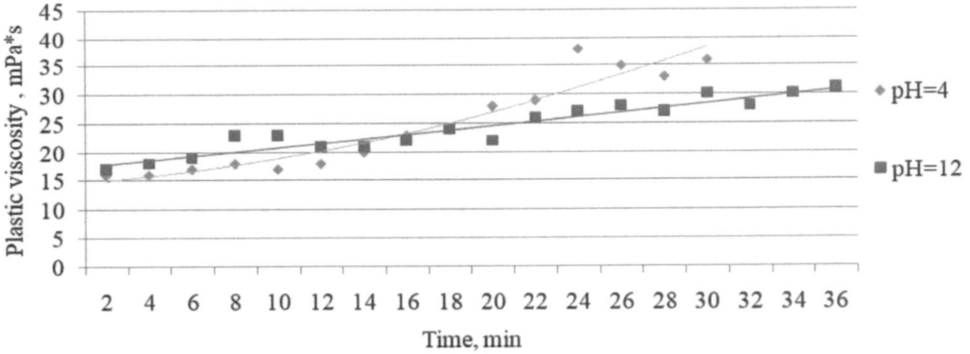

Figure 1. Relation between dynamic shear stress and time for «Petrosorb» water solutions with different pH values.

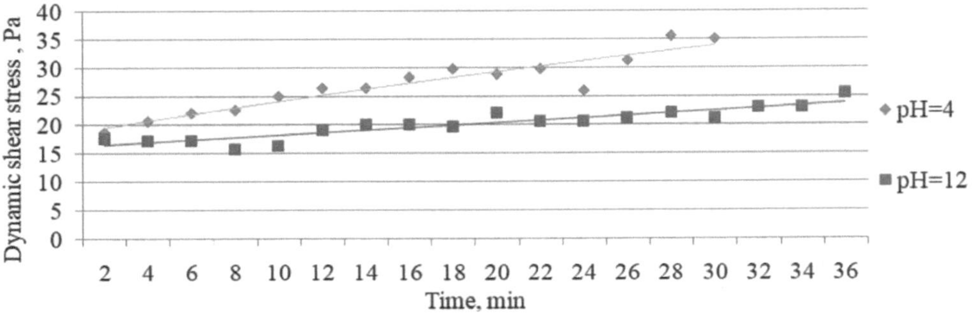

Figure 2. Relation between dynamic shear stress and time for «Petrosorb» water solutions with different pH values.

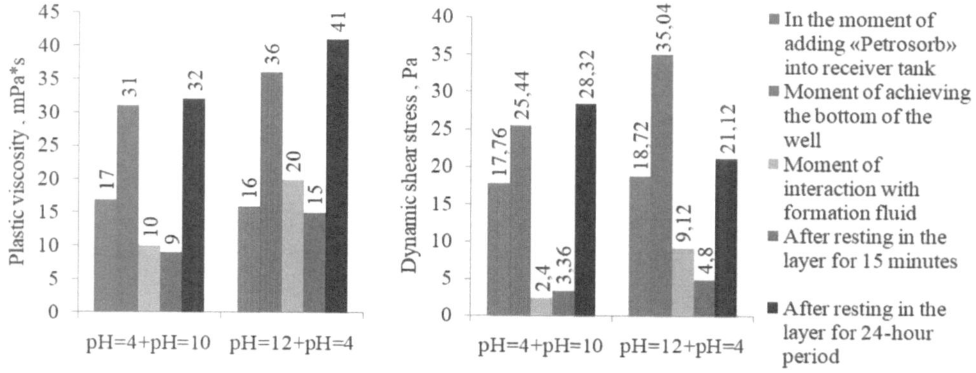

Figure 3. Relation between dynamic shear stress and time for «Petrosorb» water solutions with different pH values.

4 RESULTS AND DISCUSSION

For pumping into deep layers (when the carbonate and acid-soluble formations are absent in the section), the drilling fluid medium should be acidic, as in such mediums the intensity of «Petrosorb» particle swellingis less, which makes it possible to drive the fluid down to bottom hole in mobile condition for 25-35 minutes.

During application of basic fluids (from the moment of adding «Petrosorb») an efficient pumping takes 10-15 minutes.

Due to the addition of «Petrosorb» into the composition of the drilling fluids, the lost circulations can be controlled in short times without stopping the drilling process, at different depths of lost circulation layer occurrence and with application of different technological schemes of pumping.

However, this priority in the pH indexes is only effective during application of specific technology of drilling fluid pumping (if the water-swellable polymer is added in the receiver tanks). Adding initial fluid up to acidic medium can negatively affect on the properties of the drilling fluid. As a result, it is effective to use the drilling fluid without changing the initial pH. But in this case, delivering the polymer solution with the required consistency to the deep layers is almost impossible. Therefore, the technology of polymer delivery to the lost circulation layers should be changed by using containers-packets with special compositions, which make it possible to provide the mixing of the drilling fluid and polymer dry mixture at bottom hole and pumping the solution into the lost circulation layer.

Further bench tests of the lost circulation control process are suggested, which can be used to define the additional characteristics of the polymer solutions (for example the maximum pressure that holds the layer sealed with the use of polymer compositions). Also it is recommended to carry out the analysis of economical effectiveness during the application of the suggested compositions and technology.

REFERENCES

Blinov, P.A. & Dvoynikov, M.V. 2018 Rheological and filtration parameters of the polymer salt drilling fluids based on xanthan gum, *Journal of Engineering and Applied Sciences* 13(14): 5661-5664.

Dvoynikov, M.V. & Nutskova, M.V. & Kuchin, V.N. 2017. Analysis and justification of selection of fluids to be used for water shut-off treatment during well completion, *Perm Journal of Petroleum and Mining Engineering* 16(1): 33-39.

Kozhevnikov, E.V. & Nikolaev, N.I. &Melekhin, A.A. & Turbakov, M.S. 2015. Studying the properties of cement slurries for cementing oil wells with long horizontal section drilled with rotary steerable system, *Neftyanoe Khozyaystvo - Oil Industry* 9: 58-60.

Kupavikh, K.S. & Nutskova, M.V. 2016 Ecological features of oil well repair at low-permeability reservoir, *International Journal of Applied Engineering Research* 11(11): 7505-7508.

Leusheva, E. & Morenov, V. 2017. Research of clayless drilling fluid influence on the rocks destruction efficiency, *International Journal of Applied Engineering Research* 12(6): 945-949.

Melekhin, A.A. & Chernyshov, S.E. & Turbakov M.S. 2012. Rasshiriaiushchiesia tamponazhnye sostavy dlia likvidatsii pogloshchenii pri kreplenii obsadnykh kolonn dobyvaiushchikh skvazhin [Expanding compositions of cement slurries for lost circulation control under cementing of casing of producing wells], *Neftyanoe Khozyaystvo - Oil industry* 3: 50-52.

Merzliakov, M.I. & Podoliak, A.V. 2017. Terms of melting the permafrost when cementing of boreholes plugged with gas-liquid mixtures with hollow microspheres, *International Journal of Applied Engineering Research* 12(9): 1874-1878.

Moradi, S.S.T. & Nikolaev, N.I. & Chudinova, I.V. & Martel, A.S. 2018a. Geomechanical study of well stability in high-pressure, high-temperature conditions, *Geomechanics and Engineering* 16(3): 331-339.

Moradi, S.S.T. & Nikolaev, N.I. & Leusheva, E.L. 2018b. Improvement of cement properties using a single multi-functional polymer, *International Journal of Engineering, Transactions A: Basics* 31(1): 181-187.

Moradi, S.S.T. & Nikolaev, N.I. 2017 Free fluid control of oil well cements using factorial design *Journal of Engineering Research* 5(1): 220-229.

Morenov, V. & Leusheva, E. & Martel, A. 2018. Investigation of the fractional composition effect of the carbonate weighting agents on the rheology of the clayless drilling mud, *International Journal of Engineering, Transactions A: Basics* 31(7): 1152-1158.

Morenov, V. & Leusheva, E. 2017. Development of drilling mud solution for drilling in hard rocks, *International Journal of Engineering, Transactions A: Basics* 30(4): 620-626.

Nikolaev, N.I. & Ivanov, A.I. 2009a. Povyshenie effektivnosti bureniia neftianykh i gazovykh skvazhin v oslozhnennykh usloviiakh [Higher efficiency in drilling of oil and gas wells under complicated conditions], *Zapiski Gornogo instituta* 183: 308-310.

Nikolaev, N.I. & Ivanov, A.I. 2009b. Rezultaty analiticheskikh i ekspluatatsionnykh issledovanii zakuporivaiushchei sposobnosti polimerglinistykh tamponazhnykh sostavov pri burenii neftianykh i gazovykh skvazhin [Results of analytical and operational studies of blocking ability of polymer-based cements compounds in the drilling of oil and gas wells], *Stroitelstvo neftianykh I gazovykh skvazhin na sushe i na more*, 5: 8-11.

Nikolaev, N.I. & Leusheva, E.L. 2016a. Increasing of hard rocks drilling efficiency, *Neftyanoe Khozyaystvo - Oil Industry* (3): 68-71.

Nikolaev, N.I. & Tianle, L. 2016b. Sovremennye tehnologii burenija i kreplenija skvazhin pri razvedke gazovyh gidratov [The modern technology of drilling and casing of well during the exploration of gas hydrates], *Zapiski Gornogo instituta*, 218: 206-214.

Nutskova, M.V. & Dvoynikov, M.V. & Kuchin, V.N. 2017. Improving the quality of well completion in order to limit water inflows, *Journal of Engineering and Applied Sciences* 12(22): 5985-5989.

Poliakov, V.N. & Mnatsakanov, V.A. 2009. Prichiny nizkoi effektivnosti metodov borby s pogloshcheniiami v burenii [Reasons for low efficiency of methods for shut-off the circulation loss in drilling], *Stroitelstvo neftianykh i gazovykh skvazhin na sushe i na more*, 3: 14-17.

Tabatabaee Moradi, S. & Nikolaev, N.I. 2016. Optimization of cement spacer rheology model using genetic algorithm, *International Journal of Engineering, Transactions A: Basics* 29(1): 127-131.

Tabatabaee Moradi, S.S. & Nikolaev, N.I. 2018. Assessment of the cement failure probability using statistical characterization of the strength data, *Journal of Petroleum Science and Engineering* 164: 182-188.

Turitsyna, M.V. & Dolgikh, L.N. & Chernyshev S.E. 2009. Investigation of effective and operational removal areas insulation facilities during wells drilling in Trifonovskoye field, *Perm Journal of Petroleum and Mining Engineering*, 8(4): 45-52.

Turitsyna, M.V. 2010. Analysis methods of absorption zones isolation during well drilling on Trifonovskoye oilfield, *Perm Journal of Petroleum and Mining Engineering*, 9(5): 37-44.

Vasilev, N.I. & Dmitriev, A.N. & Podoliak, A.V. & Lukin, V.V. & Turkeev, A.V. 2016. Maintaining differential pressure in boreholes drilled in ice and the effect of ice hydrofracturing, *International Journal of Applied Engineering Research*, 11(19): 9740-9747.

Youth Technical Sessions Proceedings – Litvinenko (ed)
© 2019 World Petroleum Council, London, UK, ISBN 978-0-367-34668-3

Development of cement composition with enhanced properties with the addition of microsilica

D.A. Zimina
Saint Petersburg Mining University, Saint Petersburg, Russia

R.Y. Kuznetsov
Gazprom International, Saint Petersburg, Russia

ABSTRACT: This research investigates the problem of well cementing in areas with the presence of permafrost. The analysis of the main complications arising from attaching wells under permafrost conditions is carried out, as well as the results, showing an indicator of the adhesion of cement stone with a casing string. The importance of the correct choice of cement material with regard to its thermophysical and technological properties in order to reduce operating problems is emphasized. The results of studies of the influence of micro silica on the structure and properties of cement stone are presented. The expediency of using silicon production wastes to increase stone strength indicators is shown.

1 INTRODUCTION

The main fuel and energy base of Russia is the number of north deposits of Western Siberia, which are characterized by the presence of permafrost in the geological section. Design technologies for casing cementing under permafrost conditions have significant drawbacks, which can lead to the emergence of behind-the-casing flows and gas plumes, inter-layer flow of fluids, melting of frozen rocks, slow strength generation of cement rock, absorption and circulation loss when there is no leakage of behind-the-casing and annular space (Kozhevnikov et al. 2014, Kozhevnikov et al. 2015).

Casing cementing is a necessary and important operation in the process of well construction. The success and quality of cementing determine the duration, failure-free operation and well operation efficiency (Morenov at al. 2018). Traditional technologies and materials for lining the hole in such conditions do not always ensure the creation of reliable and durable lining in the annular space. First of all, it refers to the sealing of areas, conductors and technical columns, which are located in permafrost areas.

For instance, in the Urengoy and Bovanenkovo oil and gas condensate deposits, when cementing conductors and directions, the presence of permafrost in the range of 0–400 m creates great complications. Due to the low frac gradient, the combined cementing technology with overlapping of the lower part with a cementing slurry of normal density and the upper - with lower density, facilitated by microspheres is most often used in these fields (Tabatabaee Moradi & Nikolaev 2016, Tabatabaee Moradi & Nikolaev 2018).

2 STATEMENT OF THE PROBLEM

The analysis of the results of casing cementing while drilling production directional wells indicated a number of problems arising from the adhesion of the cement stone to the rock. The small amount of adhesion, in addition to the presence of frozen rocks, is also explained by the deformation of shrinkage during cement hardening. (Blinov & Dvoynikov 2018) The same

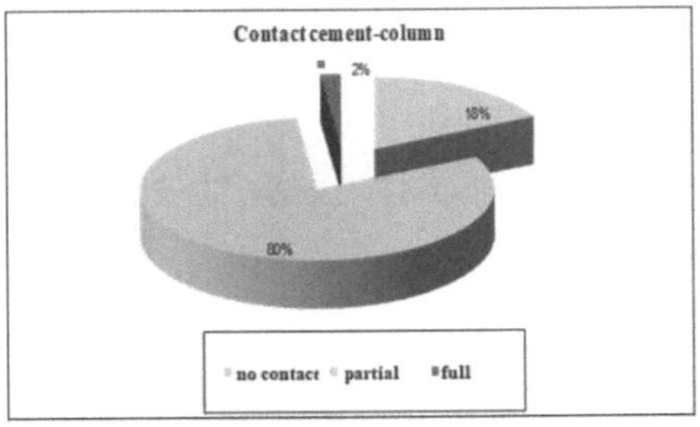

Figure 1. Cementing quality results for the Urengoyskoye and Bovanenkovskoye deposits.

situation is present in fields with similar natural mining and geological conditions. Gas wells of the Medvezhye and Yamsoveyskoye deposits, which are located in the permafrost area of the Nadym-Purovsky interfluve, were drilled based on existing standards that did not fully consider the specific conditions of the cryolithozone, which subsequently led to deformation of the wellbore space and, as a result, to the loss of stability of the wellhead.

The results of well cementing at the deposits in the Far North, where complex environment contribute to an increase in water separation in the cementing slurry, show that the use of traditional materials does not ensure the proper quality of well casing. (Samsonenko et al. 2014) Figure 1 shows the results of geophysical studies of the quality of adhesion of the described cementing technology with the casing string. Portland cement TsTRS-50 (AWP) was used for cementing - cement-based expanding stabilized cement with reinforcing additives and lightweight cement oil CTTR cement (Leusheva & Morenov 2017, Morenov & Leusheva 2017). The diagram in Figure 1 shows the continuous contact of cement with a column is only 2 %.

Violation of the thermal regime of frozen rocks by transferring heat from the flushing fluid and cement slurry to the bore hole walls during the time of drilling a well causes collapses of the melted rocks and sticking of drilling tools (Nikolaev & Leusheva 2016). In the permafrost areas, a cement stone that is located in the annular space is formed, at the same time being exposed to negative (from the rocks) and positive (from the casing) temperatures (Merzliakov & Podoliak 2017, Moradi et al. 2018). At the temperature gradient of 10–15C, between the inner and outer layer of cement stone, free water, which is in an unfrozen state, begins to move to the region of negative temperatures, then it turns into ice, and the solid phase moves to the column and sets. As a result, the pore walls are destroyed and the permeability of the stone increases several times.

3 DISCUSSION

Contraction of the cement stone, low water retention properties, sedimentation stability of the cement slurry, poor adhesion of cement stone with adjacent agents, typical for most traditional cement materials, are the main causes of poor-quality lining.

Cementing slurry used for cementing wells in frozen rocks should set at low temperatures, have a low freezing point of the mixing fluid (-4 ..- 5C), a pumping period of at least 2 hours with minimal setting time, low water separation, no contraction during hardening cement stone, to have a high crack resistance, water resistance and frost resistance, high heat release rate at the beginning of mixing (during hydration) with a minimum amount of heat generated in low thermal conductivity, which would reduce heat transfer from the wellbore to the rocks.

Due to a number of required properties, several experiments were conducted to find out the properties of the cement composition with the addition of silicon production waste. (Munkhtuvshin & Balabanov 2017) The aim of this research was to examine the effect of microsilica on the strength and structure formation of cement stone, as well as to obtain high-strength cement with enhanced frost-resistant properties.

At the moment, it is known that microsilica is being used in manufacturing and in the production of structural concrete for obtaining strong durable stone. Practical application has shown that 1 kg of microsilica provides similar strength as when using 4-5 kg of ordinary portland cement. High properties of microsilica improves concrete characteristics such as compressive and flexural strength, wear hardness, frost resistance and significantly reduce permeability.

Microsilica is a by-product of silicon production, ultrafine material, in other words, it is silicon dioxide, formed as a result of gas cleaning of metallurgical ore-thermal furnaces. The production of silicon is currently inextricably linked with the formation of a large amount of dust waste (for example, the output of waste microsilica varies from 30 to 1000 kg per 1 ton of silicon produced). (Holland 2005) Thus, each year, after the production of metallurgical silicon, about 30000 tons of fine dust (micro and nanosilic) are formed and then transferred to the mud basin for long-term storage, which indicates a lack of efficiency in the utilization of microsilica and exhaust gas. (Gamaliy et al. 2009) Table 1 presents the chemical composition of the sample microsilica.

According to European and American microsilica usage standards (EN 13263 and ASTM C 1240), the silica content (SiO_2) in microsilica should be at least 85%. The Table 1 shows that the largest mass fraction is SiO_2 (at least 98%). (ASTM International 2017)

According to the examination of the silica fume class II (SiO_2 ≥85%, dark gray powder) on a laser analyzer of particle size "Microsizer 201", we can conclude that the grain size distribution of the presented microsilica sample which is presented in Table 2.

The particle size of silicon dust in percentage terms suggests that silicon dust is a conglomerate of solid particles, ranging in size from submicron to several tens of microns. Investigated silicon dust refers to inorganic, mixed type, with complex chemical and phase composition. An image (Figure 2) from a microscope gives us an idea of the diversity of the grain size distribution of silicon dust.

Microsilica is a highly active pozzolan additive with a fine particle size, and when mixed with a cement mortar, conditions are created for the conversion of fragile calcium hydroxide (formed by mixing cement mixture with water and hydrating clinker material) into calcium silicate hydrate, which creates conditions for increasing the strength of cement stone, and microspheres fill the space vacated by water. (Butakova et al. 2017) The increasing density of

Table 1. The chemical composition of the nanostructure concentrate based on silicon dioxide.

Substance identification	Mass fraction	Measuring technique
	%	-
SiO_2	98	EN 196-2
CaO	0.4	EN 451-1
K_2O	0.3	EN 196-2
Na_2O	0.2	EN 196-2
Fe_2O_3	0.1	GOST 2642.5-97
AL_2O_3	0.3	GOST 2642.4-86
MgO	0.3	GOST 2642.8-97

Table 2. The particle size of microsilica.

Particle size	Size	The content of particles in the sample
	microns, μm	%
Separate nanoparticles	less then 1	less than 1
Agglomerates small	more then 1	1-10
Agglomerates middle	more then 10	10-45
Agglomerates large	more then 45	more than 45

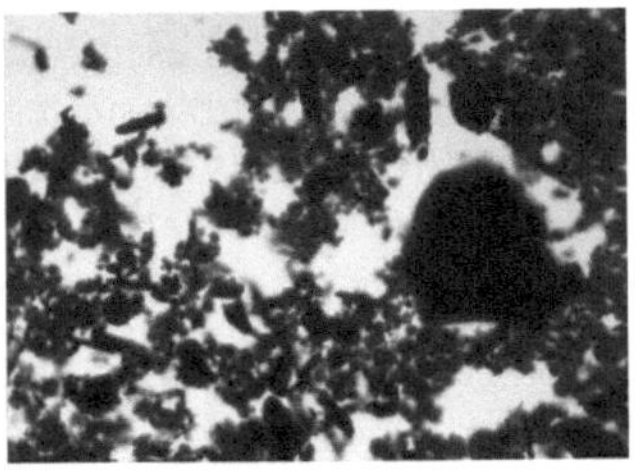

Figure 2. Silica Nanostructure Concentrate.

the composition increases both the strength and water resistance of the stone, and subsequently its durability.

4 CONCLUSIONS

Cement mortar with a water-cement ratio of 0.55 was used as the basic formula, including cement powder, microsilica and an accelerator of setting time (4% solution $CaCl_2$).

Low and negative temperatures slow down the hardening of cement, which lengthens the setting time in the winter period. In this case, it is advisable to use additives that accelerate the setting. Under standard downhole conditions at normal temperatures, the $CaCl_2$ electrolyte serves as a set accelerator. $CaCl_2$ is also often used in drilling at low temperatures; however, if this accelerator is used excessively in negative temperatures, the thawing of frozen rocks during its exothermic reaction may be caused. When using 4% $CaCl_2$ solution, the solution has a high heat release rate at the beginning of mixing (during hydration) with a minimum amount of heat generated in general. Tables 3-4 presents the results of studies of the strength of cement stone after 2 and 7 days of hardening at normal and low temperatures.

According to the tables 3-4, we can conclude that the addition of silicon production waste to the cementing slurry, with the use of calcium chloride, as mentioned above, has a positive

Table 3. Cement strength test results at normal temperatures.

	Maximum compressive strength		Maximum bending strength	
Indicator	MPa (2 days)	MPa (7 days)	MPa (2 days)	MPa (7 days)
Standard composition	10.2	22.8	3.8	9.4
Composition with microsilica	15.4	27.4	5.3	12.3
Composition using microsilica and calcium chloride	19.1	30.3	4.7	12.6

Table 4. Cement strength test results in low temperature conditions.

	Maximum compressive strength		Maximum bending strength	
Indicator	MPa (2 days)	MPa (7 days)	MPa (2 days)	MPa (7 days)
Standard composition	8.0	15.3	1.1	3.5
Composition with microsilica	12.3	18.9	1.9	4.1
Composition using microsilica and calcium chloride	14.3	21.2	3.5	5.5

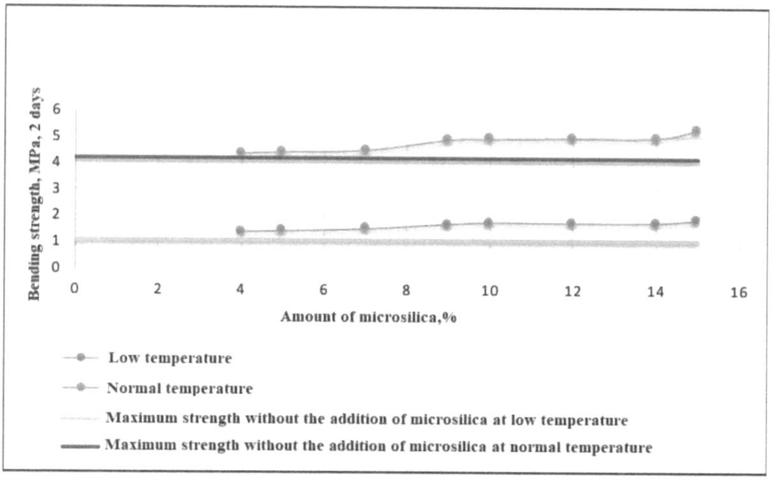

Figure 3. The dependence of the bending strength of cement stone on the amount of microsilica injected.

effect on the properties of the cement stone and improves indicators such as strength, frost resistance, permeability, provided that the cement content is low.

With this aforecited water-cement ratio, the density of the solution was 1.87 g/cm3, the spreadability was 185 mm. Microsilica additives was introduced into the dry mixture in the amount of 4, 5, 7, 9, 10, 12, 14 and 15% by weight of cement powder. The study of the strength of the cement stone for bending and compression was carried out on samples of size 40x40x160 mm and 40x40x40 mm, respectively, maintained at temperatures of 24°C and 6-8°C in the refrigerating chamber. The strength characteristics of the stone were determined from the test results as an arithmetic average of 3 results. (Dessouki & Mitri 2011) Tests were carried out after 2 and 7 days of setting. Below is a graph (Figure 3) of the data obtained for comparison of the results of the research.

Based on the data obtained, we can draw the following conclusions: the material cracks when more than 15% of microsilica is added to the composition, so the further studies are not expedient, adding 10–12% of microsilica to the solution increases the strength of the cement stone, by using as a tempering fluid, a 4% $CaCl_2$ solution strength is increased at low temperatures. On the basis of the researches performed, it is possible to recommend the addition of calcium chloride to the composition of the cement slurry together with microsilica in order to increase the strength characteristics.

REFERENCES

ASTM International. Available at: https://www.astm.org/cis/ru/index.html (accessed 29.01.2017).

Blinov, P.A. & Dvoynikov, M.V. 2018. Rheological and filtration parameters of the polymer salt drilling fluids based on xanthan gum. Journal of Engineering and Applied Sciences, 13(14): 5661-5664.

Butakova, M.D., Mikhailov, A.V., Saribekyan, S.S. 2017. Influence of Silicon-Containing Additives on the Property of the Watertightness of Concrete Samples. Bulletin of the South Ural State University. Ser. Construction Engineering and Architecture. Vol. 17. № 2: 34–41.

Dessouki, A. E. & Mitri, H. 2011. Rock Breakage Using Expansive Cement. Montreal, Canada: 167-173.

Gamaliy, E.A., Trifimov, B.Y., Kramar, L.Y. 2009. Structure and properties of cement stone with admixtures of silica fume and polycarboxylate plasticizing agent. Vestnik YUUrGU. №16:29-35.

Holland, T.C. 2005. Silica Fume User's Manual. Technical Report. Silica Fume Association. USA.

Kozhevnikov, E.V., Nikolaev, N.I., Melekhin, A.A., Turbakov, M.S. Studying the properties of cement slurries for cementing oil wells with long horizontal section drilled with rotary steerable system//Neftyanoe Khozyaystvo - Oil Industry 2015: (9), pp. 58-60

Kozhevnikov, E.V., Nikolaev, N.I., Ozhgibesov, O.A., Dvoretskas, R.V. Studying of sedimentation influencing on cement stone properties//Neftyanoe khozyaystvo - Oil Industry 2014: (6), pp. 23-25

Leusheva, E., Morenov, V. Research of clayless drilling fluid influence on the rocks destruction efficiency (2017) International Journal of Applied Engineering Research 12 (6), p. 945-949

Merzliakov, M.Y., Podoliak, A.V. Improving the efficiency of well cementing in permafrost regions by using gas-liquid cement mixtures, International Journal of Applied Engineering Research, Volume 12, Issue 9, 2017, pp. 1879-1882

Moradi, S.S.T., Nikolaev, N.I., Leusheva, E.L. Improvement of cement properties using a single multi-functional polymer//International Journal of Engineering, Transactions A: Basics 2018: 31 (1), pp. 181-187

Morenov, V., Leusheva, E., Martel, A. Investigation of the fractional composition effect of the carbonate weighting agents on the rheology of the clayless drilling mud (2018) International Journal of Engineering, Transactions A: Basics 31 (7),p. 1152-1158 DOI: 10.5829/ije.2018.31.07a.21

Morenov, V., Leusheva, E. Development of drilling mud solution for drilling in hard rocks (2017) International Journal of Engineering, Transactions A: Basics 30 (4),p. 620-626 DOI: 10.5829/idosi.ije.2017.30.04a.22

Munkhtuvshin, D. & V.B. Balabanov, K.N. 2017. Experience of use of adds of micro- and nanosilica from the wastes of silicone production in concrete technologies. Putsenko Proceedings of Universities. Investments. Construction. Real estate Vol. 7. N. 3:107-115

Nikolaev, N.I., Leusheva, E.L. Increasing of hard rocks drilling efficiency//Neftyanoe Khozyaystvo - Oil Industry 2016: (3), pp. 68-71

Samsonenko, A.V., Simonyants, S.L., Samsonenko, N.V. 2014. Novye materialy i innovatsionnaya tekhnologiya krepleniya skvazhin v usloviyakh Kraynego Severa. Stroitel'stvo neftyanykh i gazovykh skvazhin na sushe i na more. VNIIOENG. № 11: 40–47.

Tabatabaee Moradi, S., Nikolaev, N.I. Optimization of cement spacer rheology model using genetic algorithm//International Journal of Engineering, Transactions A: Basics 2016: 29(1), pp. 127-131

Tabatabaee Moradi, S.S., Nikolaev, N.I. Assessment of the cement failure probability using statistical characterization of the strength data//Journal of Petroleum Science and Engineering 2018: 164, pp. 182-188

Author Index